INNOVATIVE PRODUCTION AND CONSTRUCTION

Transforming Construction Through Emerging Technologies

Other Related Titles from World Scientific

Design for Maintainability: Benchmarks for Quality Buildings
by Michael Yit-Lin Chew, Ashan Senel Asmone and Sheila Conejos
ISBN: 978-981-3230-59-0
ISBN: 978-981-3232-95-2 (pbk)

Facilities Planning and Design: An Introduction for Facility Planners,
Facility Project Managers and Facility Managers
by Jonathan Lian
ISBN: 978-981-3278-81-3
ISBN: 978-981-3278-96-7 (pbk)

Maintainability of Facilities: Green FM for Building Professionals
(Second Edition)
by Michael Yit-Lin Chew
ISBN: 978-981-4725-64-4
ISBN: 978-981-4725-75-0 (pbk)

Naturally Animated Architecture: Using the Movements of the Sun, Wind,
and Rain to Bring Indoor Spaces and Sustainable Practices to Life
by Kevin Nute
ISBN: 978-1-78634-438-0

3D Printing and Additive Manufacturing: Principles and Applications
(Fifth Edition)
by Chee Kai Chua and Kah Fai Leong
ISBN: 978-981-3146-75-4
ISBN: 978-981-3146-76-1 (pbk)

INNOVATIVE PRODUCTION AND CONSTRUCTION

Transforming Construction Through Emerging Technologies

Editors

Peng Wu
Curtin University, Australia

Xiangyu Wang
Curtin University, Australia

NEW JERSEY · LONDON · SINGAPORE · BEIJING · SHANGHAI · HONG KONG · TAIPEI · CHENNAI · TOKYO

Published by

World Scientific Publishing Co. Pte. Ltd.

5 Toh Tuck Link, Singapore 596224

USA office: 27 Warren Street, Suite 401-402, Hackensack, NJ 07601

UK office: 57 Shelton Street, Covent Garden, London WC2H 9HE

Library of Congress Cataloging-in-Publication Data
Names: Wu, Peng, editor. | Wang, Xiangyu, editor.
Title: Innovative production and construction : transforming construction through
    emerging technologies / edited by Peng Wu (Curtin University, Australia) and
    Xiangyu Wang (Curtin University, Australia).
Description: New Jersey : World Scientific, [2018] | Includes bibliographical references.
Identifiers: LCCN 2018032350 | ISBN 9789813272484 (HC : alk. paper)
Subjects: LCSH: Building--Technological innovations.
Classification: LCC TH155 .I56 2018 | DDC 624.068--dc23
LC record available at https://lccn.loc.gov/2018032350

British Library Cataloguing-in-Publication Data
A catalogue record for this book is available from the British Library.

For any available supplementary material, please visit
https://www.worldscientific.com/worldscibooks/10.1142/11048#t=suppl

Desk Editors: Herbert Moses/Jennifer Brough/Koe Shi Ying

Typeset by Stallion Press
Email: enquiries@stallionpress.com

About the Editors

 **Peng Wu** received his B.S. degree in Project Management from Tsinghua University, China in 2006. He received his M.S. degree in Construction Management from Loughborough University, UK in 2007 and his Ph.D. degree in Project Management from National University of Singapore, Singapore in 2012. He is currently an Associate Professor with the Department of Construction Management and an Associate Director with the Australasian Joint Research Centre for Building Information Modeling, Curtin University. His research areas include sustainable construction, lean production and construction, production and operations management, and life cycle assessment. In 2016, he received the Discovery Early Career Research Award from the Australian Research Council, which is a prestigious award to support excellent basic and applied research by early career researchers.

Xiangyu Wang is a Chair Professor at the School of Design and the Built Environment at Curtin University. Since March 2012, he has taken the role of Director at the Australasian Joint Research Centre for BIM. Prof. Wang's research interests focus mainly on BIM, visualization and information technology in construction engineering and management. His academic achievements are highlighted by Google Scholar Citations, 3,000+ with H-index as 30, and over 100 top-quality refereed journal articles. In the last six years, Prof. Wang has secured more than $10 million in funding from the industry and the commonwealth government, demonstrating his research impact. He is the Founder and Editor-in-Chief for the journal *Visualization in Engineering*, and also serves on the board for over 10 journals and has reviewed 40 journals across various engineering and technology disciplines.

Contents

Chapter 1

Implication of a Construction Labour Tracking System for Measuring Labour Productivity

Sara Shirowzhan, Samad M. E. Sepasgozar*,
Fahim Ullah and Pavneet Singh Minhas

*Faculty of Built Environment, The University of New South Wales,
Sydney, NSW 2052, Australia*
**samad.sepasgozar@gmail.com*

Abstract

Tracking systems are known as valuable approaches for monitoring vehicles and heavy equipment in infrastructure construction and mining sites. Tracking job-site equipment and monitoring the workforce during construction are critical for improving productivity. The current locating practices are based on real-time positioning sensors by alerting staff when entering specific areas within construction sites. However, this approach has not been fully used for measuring labour productivity, and there is much room for improvement. In addition, not much attention has been given to individuals' two-way communication (e.g. among workers, site engineers and supervisors) in conjunction with tracking them in

---

*Corresponding author.

an interactive Geographic Information System (GIS) environment for productivity purposes and the daily performance evaluation. The aim of this chapter is to implement an online tracking system for indoor positioning using radio communication systems, called the Construction Labour Tracking System (CLTS), and integrate it with GIS. The CLTS is used for monitoring labour in the working area and processing the data for identifying heat maps of the crowded areas of a construction site at various time intervals. The CLTS process includes determination of the functionalities of the system components and the interrelationships between them and GIS. A series of field experimentations were carried out at a selected building. The results of this ongoing study show that CLTS has the capability to monitor the approximate position of workers in detail. The developed CLTS tool is highly valuable to construction project managers because it increases the quality of the collected data and enables them to measure labour productivity on construction sites.

Keywords: Construction Labour Tracking System; Geographic Information Systems; Online Tracking Systems.

1. Introduction

The construction industry lagged behind other industries in technology implementation (CII, 2008; Sepasgozar *et al.*, 2016). However, recently the construction industry has experienced many internal and external forces for improving productivity, safety and increasing sustainability. The recent publications show that there is a growing interest to utilize new technologies for different purposes in the construction industry (Sepasgozar *et al.*, 2016). Hooper and Haris (2010) describe a wide range of technologies which pioneers in the construction industry have already started to use. These technologies include building information modelling (Sepasgozar *et al.*, 2015), virtual and augmented reality, mobile and wearable technologies, real-time locating systems (LSs) including Radio-frequency Identification (RFID)- and Global Positioning System (GPS)-guided plant and machinery, LiDAR technologies (Shirowzhan and Trinder, 2017) and sensors embedded into buildings and infrastructure and robotics. In particular, locating and tracking systems have been utilized in construction and mining sites for monitoring and controlling purposes.

These technologies are generally known as highly effective tools for tracking and monitoring the location of individuals and vehicles in different businesses, such as healthcare, hospitality, mining and construction. For example, the Veterans Affairs Department of the US Government intends to spend $550 million for implementing advanced LSs to improve efficiency and monitor their employees in hospitals. LSs emerge as one of the latest key technological advancements in modern times. It is an effective way to identify and track the location of an object in both indoor and outdoor environments (Li *et al.*, 2016). To date, various LSs and its adaptive models have been developed, tested and made commercially available for general and public use. The LSs responsible for reshaping the digital world have not been explored much in the field of construction (Vähä *et al.*, 2013). The exploration, till date, is investigative or tentative in nature and detailed real-time and real-life case project-based investigations are missing.

LSs can be used for tracking and reporting the current position of an individual, equipment or pieces of materials to facilitate the data tracking and aid the subsequent management (Ding *et al.*, 2013). It is considered as a groundbreaking innovation in the construction industry that has changed and modified its traditional hardcore practices over the last decade or two (Guo *et al.*, 2017). LS can be defined as an amalgamation of hardware and software systems that automatically and in real time determines the coordinates of an object within a target area (Li *et al.*, 2016). The LS-collected data can be used for both real-time purposes as well as for further analysis after a dataset is collected. Generally, an LS consists of receiver sensors and remote tags that communicate with the sensors for signal reception. The location of an object is determined by the received signal strength and its arrival time (Jiang *et al.*, 2015). Some modern types of LS are based on visual positioning and do not require the tags. It can be used in both indoor and outdoor environments. The indoor applications are more suited towards construction and hence targeted in this chapter.

LS has multiple applications in construction. It is often constrained by the expense of the equipment and associated expertise requirements, otherwise the applications can be far more than the currently existing ones. Previously, it has been used for tracking the location of materials. According to Grau *et al.* (2009), LS-based materials tracking can improve traditional tracking time from 36.8 to 4.56 min. This resulted in a saving of $121,507 over the project life. Jang and Skibniewski (2009) argue that materials tracking based on LS save up to 64% of labour costs in a two-year construction project.

Despite LS receiving good attention recently, its practical applications and deployments are limited mainly due to the sophisticated and costly equipment and lack of operational understanding. The implementation side is targeted in this chapter, and a model is presented to show the potential implications and usage of CLTS tool.

2. Review of the Implementation of Tracking Systems

LS usage is on an increase globally and it is being adopted and utilized in various industrial sectors. The construction sector, although one of the leading global GDP contributors, is at the lowest when it comes to utilization of LS technologies. The use of construction LS is mainly exploratory and experimental in nature, and its practical applications are limited so far.

2.1. *Definitions and technology types*

LS is used in various shapes and technologies in construction. Modern-age LS comes in several types starting from the traditional GPS systems to the more advanced and latest 360 cameras and laser scanners. Different types of LS are mentioned and described along with their applications in construction in Table 1. The table lists LS under three key headings: starting from the more traditional geospatial technologies to modern imaging and the latest interactive technologies.

2.1.1. *Geospatial technologies*

Geospatial technologies are an amalgam of traditional and modern LS technologies. It offers a completely different way of mapping to manage our communities, towns, suburbs and industries. In the last decade and a half, these technologies have evolved into a network of national research, scientific study, security, cyber control and commercially operated satellites that are usually complemented by powerful desktop GIS tools

Table 1. Real-time locating technologies and their usage in construction.

Technology	Sub-technology	Description	Application in Construction
Geospatial	Barcoding	A series of parallel and adjacent bars are scanned with a stationary or hand-held barcode scanner or reader that is used to capture and read the information contained in a barcode about the item in question. It allows collection of real-time data	Materials and machinery tracking, inventory, project progress and labour tracking, and project document control
	Radio-frequency Identification (RFID)	An automatic identification technology that uses radio frequencies for capturing and transmitting field data. The data can be stored and retrieved from the test items by using read-and-write-enabled small tags	Progress measurement, machinery and labour tracking, inventory and shipping, tracking earth moving operations
	Ultra-wide band (UWB)	It is a network of tags and receivers that communicate over a bandwidth of >500 MHz with each other. An embedded tag transmits UWB radio signals enabling the system to find its 3D position and coordinates regardless of multipath effects	Material- and activity-based progress tracking, safety, real-time location sensing and resource-tracking
	Geographic information system (GIS) & Global positioning system (GPS)	GIS helps in visualizing, questioning, analysing and interpretation of data for understanding the relationships, patterns and trends across it GPS is a satellite-based and widely used application that considers the position and navigation of construction activities. It locates the position of a specific object attached to a tag based upon satellite signals	Visualize and review a construction project schedule, visualize the position and orientation of building components in erection, track steel structural materials throughout construction, data collection and positioning for a tunnel boring machine

(*Continued*)

Table 1.   (*Continued*)

Technology	Sub-technology	Description	Application in Construction
Imaging	Photogrammetry	It is an accurate method of generating 3D or cloud point models of a construction site using its digital photos. Thus, it compares an as-built 3D model to 3D CAD models and automatically calculates the percent complete of each component that is used to measure the progress of construction projects	Automated collection of data, record as-built information, temporal estimation of quantities of work performed between two scans, pothole detection in pavements
	Laser scanning (Shirowzhan and Trinder, 2017)	In this method, a laser emits a pulse of laser light to a target and calculates the distance to it by timing the round-trip time of the light pulse (Sepasgozar *et al.*, 2014a, 2014b)	Capture status of construction projects, develop complete as-built BIM, progress tracking
	Videogrammetry	It extracts features from video recordings using the sequential frames of videos and progressively reconstructs the pixels in each frame based on the previous frame	Damage detection and safety evaluation, produce 3D point clouds
	Automated drones	In this technique, high-quality cameras are mounted on automated drones for capturing both real-time videos and high-quality images	Monitoring, progress tracking, safety

Interactive technologies	Range images	They are acquired with range sensors (range cameras) and offer an inexpensive and accurate means for digitizing the shape of 3D objects. 3D range cameras are useful for tracking moving objects and construction equipment and materials	Obstacle avoidance, safety, avoiding hazardous gestures and postures, spatial modelling and sensing applications
	Augmented reality (AR)	It is the superimposition of a computer-generated image on a user's view of the real world to provide a composite view for better understanding and apprehension	Field construction monitoring, compare laser-scanned surface data during excavation against 3D plan designs for conformance in the field, underground utilities management, infrastructure cracks and failure models
	Virtual reality (VR)	It is a computer technology-based application using virtual reality headsets and gadgets in combination with physical spaces for generating realistic sensations, images and sounds simulating a user's physical presence in an imaginary or near reality environment	Walkthroughs, BIM models, construction failures models and simulations
	360 Cameras	It uses a gadget or phone mounted rotatory camera that captures both 360 videos and images enabling the users to look and cover all site angles simultaneously	Progress tracking and monitoring, schedule updating

(Jiang *et al.*, 2015). Additionally, a recent innovation, aerial remote sensing platforms, including drones and other unmanned aerial vehicles are used for commercial use. These are of higher use offering more competitive advantages in remote areas or coverage of larger sites, especially road networks. With the increased use, high-quality reliable hardware and data are now available to new audiences including corporations, universities, construction solution providers and other non-governmental organizations (Omar and Nehdi, 2016). These technologies are invading multiple fields and sectors at a rapid pace and are initiating quality debates over topics, such as water body estimation and monitoring, industrial engineering, flood water monitoring, agricultural monitoring, biodiversity conservation, urbanization check and deforestation, forest fire suppression, green plots monitoring, humanitarian relief and others. There are now a variety of types of geospatial technologies potentially applicable in construction, including the GIS, RFID, UWB, GPS and other sub-types. In construction, these have been successfully used and tested in inventory and machinery tracking, project progress and labour tracking, earth moving operations, and project schedule control and monitoring (Grau *et al.*, 2009; Jiang *et al.*, 2015; Guo *et al.*, 2017).

2.1.2. *Imaging*

Imaging technology is mainly the application of methods and materials to create, duplicate or preserve images. The images, whether mobile or stationary, are used to generate 3D models and replicate the otherwise unreachable or occupied sites. It is not only restricted to plain images, but modern imaging incorporates both mobile devices and videos as well. Ranging from the traditional 2D plain images to the latest laser-scanning and drone-mounted cameras capable of detecting heat and thermal radiations, modern imaging has transformed construction operations (Omar and Nehdi, 2016). The advanced 3D cameras are even capable of recording moving objects that are later converted into point clouds ready to act and serve as BIM models. It has many types, but the most common ones that are focused in this chapter are photogrammetry, laser scanning, videogrammetry, automated drones and range images (Ding *et al.*, 2013; Li *et al.*, 2016).

2.1.3. *Interactive technologies*

Interactive technologies are one of the latest innovations making its way into construction. It facilitates digital interaction between people, allowing for user content creation and manipulation based upon their instincts. Software systems designed for interactive technologies support users task performance in a collaborative manner that is more pleasing, involving and playful in nature (Guo *et al.*, 2017). It allows for a two-way flow of information through an interface between the user and the technology; the user usually communicates a request for data or action to the technology with the technology returning the requested data or result of the action back to the user (Grau *et al.*, 2009). The aim of these technologies is to replicate the near-real or virtual environment of the construction site to detect and observe the potential problems in advance and plan for its non-occurrence on site. These technologies exist in various types, such as VR, AR and 360 cameras (Grau *et al.*, 2009, 2017; Sepasgozar *et al.*, 2015).

3. Applications in Construction

This section discusses the application of LSs in construction and develops a framework for implication based on the previous literature. Ding *et al.* (2013) presented a real-time safety early warning system to prevent accidents and improve safety management in underground construction using the "internet of things" (IoT) and RFID-based labour tracking system for the metro tunnel project in China. The results indicated a significant improvement in underground construction safety management efficiency in the form of real-time detection, monitoring and early warning of safety risks leading to accident prevention. Zhou and Ding (2017) proposed an IoT-based safety barrier warning system to achieve a safer underground construction site. A hazard energy monitoring system was established that uses IoT to generate early warnings and alarms, thus acting as a dynamic safety barrier for avoiding underground hazards. The IoT technologies, including meter level of RFID-based location and tracking technology, centimetre level of ultrasonic detection technology and infrared access technology, were used and developed in three-tier network architecture to help workers change their risky behaviours and avoid accidents in the changing construction site.

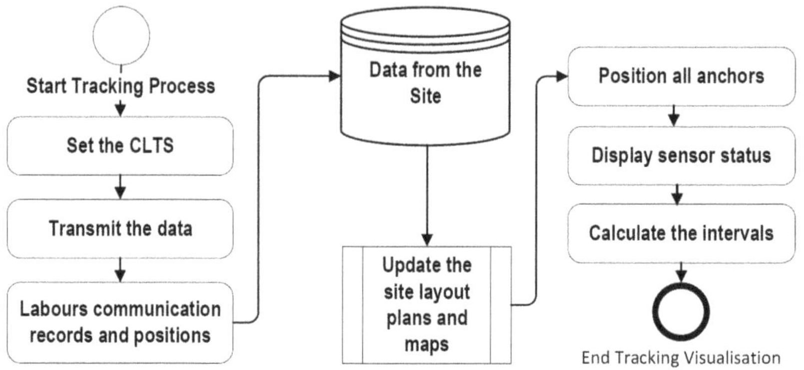

Fig. 1.   The flowchart of the implementation process.

Jiang *et al.* (2015) used GPS and GIS to formulate a three-layered system aimed at solving several labour computational and usability challenges in dam construction. The output software presents real-time site state visualization, and an accurate analysis of on-site labour consumption is run by GPS-equipped smartphones, wireless on-site base stations and servers. The proposed system, when applied to hydropower project and tested, demonstrated the ability to log entire on-site trajectories of labourers and calculate the accurate labour consumption of each dam monolith.

Figure 1 shows that the data were collected and processed in two phases: utilizing sensors and connecting to GIS. The installation took 15 min each time, and the data was collected based on the number of workers.

4. Implementation

The study is based on a series of field experimentations in an education building. A Wireless Local Area Network (WLAN) was utilized for localization of the system and for high accuracy in indoor environments. They emit signals of frequency ranging from ISM 2.4 to 2.48 GHz. The devices are assigned to several individuals to track who was walking into the building. The tags can emit its identification information at regular intervals, making it possible to get real-time locations of the tag. A PC with a LINUX operating system was installed to run the required software to collect and process data.

The CLTS is used for data acquisition and transmission, data processing and real-time visualization. The data acquisition includes intercom technology inducing a built-in tag. The CLTS is configured via a web-client system. The web-client system enabled us to browse the position of

Fig. 2.   CLTS utilization in the selected field.

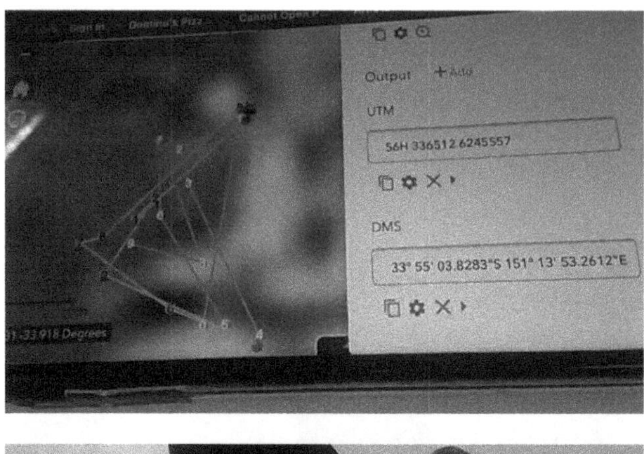

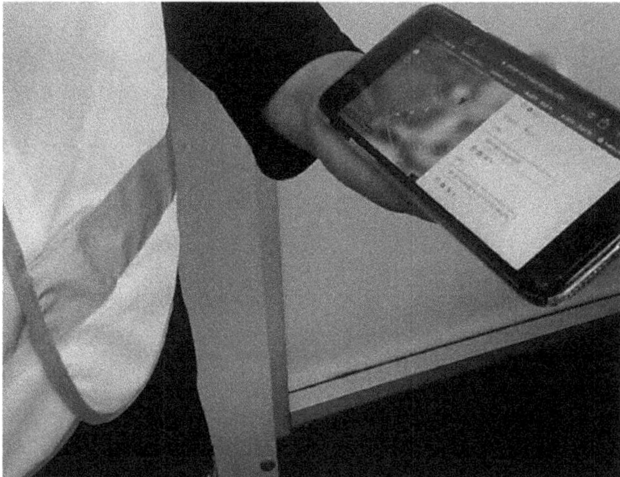

Fig. 3.   Monitoring other labourers' movements on an iPhone 7, including their position coordinates.

labourers on the map, assess device information and analyze all required data provided by the system. The instruments include indoor access points, intercom devices, analyzer, portable screen and GIS. The system was assembled in the selected area in the field as shown in Fig. 2.

As Fig. 3 illustrates, the collected data are imported into the GIS and used in a smartphone. The location of each labourer and the corresponding route paths are stored in different layers. These layers can be overlaid to compare the locations of the labourers at each time point. In addition, the distance of these labourers at each time point can be measured.

Also, an analysis of finding hot spots can be run to identify the locations with statistically significant number of workers. This information is crucial for supervisors for managing hazards and increasing the level of safety by finding the distribution pattern of the labourers in risky zones, identifying the labourers crossing the hazardous zones or moving towards such zones in construction sites.

5. Conclusion

This study aimed to develop and utilize a novel tracking system for monitoring objects' (e.g. labourer and/or vehicles) locations using radio communication systems. The system is called the Construction Labour Tracking System (CLTS). This system employs a customized GIS to visualize the paths of labourers. The CLTS is used for monitoring labour in the working area and examining its capability for visualization of moving objects in construction. The CLTS implementation procedure includes determination of the functionalities of the system components and the interrelationships between them and the GIS. A series of field experimentations were carried out at a selected space covered by WiFi. The results of this ongoing study show that CLTS has the capability to provide constant and reliable data and visualize them on mobile devices in detail. This system is different from the previous attempts in the literature, since it provides an individual's two-way communication (e.g. between workers and supervisors) in conjunction with tracking them in a GIS environment for productivity purposes and daily performance evaluation. The developed CLTS tool is highly valuable to construction project managers because it increases the quality of the collected data and enables them to measure labour productivity in construction sites. The system will be used for job-site equipment and workforce monitoring during construction and will be critical for improving productivity.

References

CII (2008). *Leveraging Technology to Improve Construction Productivity*, New York, NY: Construction Industry Institute.

Ding, L., Zhou, C., Deng, Q., Luo, H., Ye, X., Ni, Y. and Guo, P. (2013). "Real-Time Safety Early Warning System for Cross Passage Construction in

Yangtze Riverbed Metro Tunnel Based on the Internet of Things", *Automation in Construction*, 36, 25–37.

Grau, D., Caldas, C. H., Haas, C. T., Goodrum, P. M. and Gong, J. (2009). "Assessing the Impact of Materials Tracking Technologies on Construction Craft Productivity", *Automation in Construction*, 18(7), 903–911.

Guo, H., Yu, Y. and Skitmore, M. (2017). "Visualization Technology-Based Construction Safety Management: A Review", *Automation in Construction*, 73, 135–144.

Hooper, B. and Haris, M. (2010). *2020 Vision*, London, UK: Royal Institution of Chartered Surveyors.

Jang, W.-S. and Skibniewski, M. J. (2009). "Embedded System for Construction Asset Tracking Combining Radio and Ultrasound Signals", *Journal of Computing in Civil Engineering*, 23(4), 221–229.

Jiang, H., Lin, P., Qiang, M. and Fan, Q. (2015). "A Labor Consumption Measurement System Based on Real-Time Tracking Technology for Dam Construction Site", *Automation in Construction*, 52, 1–15.

Li, H., Chan, G., Wong, J. K. W. and Skitmore, M. (2016). "Real-Time Locating Systems Applications in Construction", *Automation in Construction*, 63, 37–47.

Omar, T. and Nehdi, M. L. (2016). "Data Acquisition Technologies for Construction Progress Tracking", *Automation in Construction*, 70, 143–155.

Sepasgozar, S., Lim, S., Shirowzhan, S. and Kim, Y. (2014a). *Implementation of As-built Information Modelling Using Mobile and Terrestrial LiDAR Systems*, translated by Vilnius Gediminas Technical University, Vilinius: Department of Construction Economics & Property, 1.

Sepasgozar, S. M., Lim, S. and Shirowzhan, S. (2014b). Implementation of Rapid as-Built Building Information Modeling Using Mobile LiDAR, in *The 31st International Symposium on Automation and Robotics in Construction and Mining*, pp. 209–218.

Sepasgozar, S., Lim, S., Shirowzhan, S., Kim, Y. and Nadoushani, Z. M. (2015). *Utilisation of a New Terrestrial Scanner for Reconstruction of as-Built Models: A Comparative Study*, translated by Vilnius Gediminas Technical University, Vilinius: Department of Construction Economics & Property, 1–10.

Sepasgozar, S. M., Loosemore, M. and Davis, S. R. (2016). "Conceptualising Information and Equipment Technology Adoption in Construction: A Critical Review of Existing Research", *Engineering, Construction and Architectural Management*, 23(2), 158–176.

Shirowzhan, S. and Trinder, J. (2017). "Building Classification from LiDAR Data for Spatio-Temporal Assessment of 3D Urban Developments", *Procedia Engineering*, 180, 1453–1461.

Vähä, P., Heikkilä, T., Kilpeläinen, P., Järviluoma, M. and Gambao, E. (2013). "Extending Automation of Building Construction — Survey on Potential Sensor Technologies and Robotic Applications", *Automation in Construction*, 36, 168–178.

Zhou, C. and Ding, L. (2017). "Safety Barrier Warning System for Underground Construction Sites Using Internet-of-Things Technologies", *Automation in Construction*, 83, 372–389.

Chapter 2

Towards Conceptualization of SocioBIM-based Post-occupancy Evaluation Framework for Learning Spaces

Olatunji Abisuga*, Changxin Wang and Imriyas Kamardeen

Faculty of Built Environment, The University of New South Wales, Sydney, NSW 2052, Australia
**o.abisuga@unsw.edu.au*

Abstract

Higher education institutions (HEIs) produce critical human resources and innovations needed to facilitate economic development. In today's knowledge-driven economy, constructed spaces at HEIs are critical components, which support learning activities. However, performance of learning environment tends to decline due to design, age, use and maintenance. Several investigations have used post-occupancy evaluation (POE) to examine performance of learning spaces using diverse criteria. But, there is a deficit of research in higher education (HE) facilities assessment and building information modelling (BIM) and facility management (FM) integration. To address this gap, this study aims to develop a conceptual framework for the advancement of a holistic socioBIM-based POE criteria model to assess HE facilities

---

*Corresponding author.

performance. An exploratory research approach was adopted. The study reviews previous studies on BIM–POE integration, POE frameworks and HE facilities evaluations to facilitate criteria extraction, with the input of an expert panel. This enables the formation of an integrated framework with spectrum across diverse facility performance attributes. This integrated framework can scaffold the development of a holistic socioBIM-based POE assessment model for buildings in HE considering its peculiar purpose and for future POE studies.

Keywords: Building Information Modelling; Facilities Performance; Higher Education; Learning Space; Post-occupancy Evaluation.

1. Introduction

Higher education institution (HEI) is one of the places where knowledge can be acquired. The acquired knowledge culminates in the advancement of human capital (Asteriou and Agiomirgianakis, 2001) and innovations (Mowery, 2004). HEIs disseminate knowledge within their precinct with the support of auxiliary facilities in the learning space. Learning spaces reflect the image of the pedagogy of HEIs and can support or hinder teaching and learning methods (Oblinger, 2005). Also, the effective functioning of these facilities determines the level of productivity of the staff and students (Tucker and Smith, 2008). Hence, learning space facilities must be well managed to be at optimum performance to meet institutional mission through continuous evaluation and monitoring.

Buildings performance is assessed by conducting post-occupancy evaluation (POE). Various methods have been designed and used for POE of HE buildings (Riley *et al.*, 2010). POE is use to assess the physical environment of learning spaces based on end user perception of the building attributes during the operational life (Yang *et al.*, 2013). Hence, it is essential to solicit user's cooperation on a collaborative platform to achieve continuous user-generated facilities information through POE for facility management (FM) decision-making.

Different technologies have been developed to facilitate collaborative efforts in the architecture, engineering, construction and operation (AECO) industry. Building information modelling (BIM) technologies in recent times have been identified as a collaborative platform within the

AECO industry. Kassem *et al.* (2015) attest that BIM utilization improves information handover, accuracy and accessibility of FM data and work efficiency in FM practice. In addition, BIM due to its functional characteristics foster multi-disciplinary integration than the previous fragmented conventional approach experienced in FM practice (Olatunji and Akanmu, 2015). According to Motawa and Almarshad (2015), BIM's purpose is to facilitate stakeholder input and participation throughout the entire project life cycle.

During the operational life of a building, POE is paramount to investigate the building performance and how it meets user's requirements. Many studies on HEI POE have adopted various performance attributes (e.g. Hassanain and Mudhei, 2006). This necessitates the need to develop a holistic POE framework for educational buildings that can capture diverse facilities components. In addition, facility users' requirements differ and change over time (Price *et al.*, 2003; Kwun *et al.*, 2013), so there is a need to continuously evaluate facilities performance (Lavy, 2008). This can only be achieved through an automated process of the traditional POE approaches and the participation of facility end users in building evaluation. However, facility user's participation in building evaluation can only be improved further on a collaborative platform. Unfortunately, BIM development in the past only facilitates professional collaboration (Azhar, 2011; Becerik-Gerber *et al.*, 2012; Miettinen and Paavola, 2014; Ilter and Ergen, 2015), but in recent advancement, it has been enabled to include client's and users' participation (Adamu *et al.*, 2015; Das *et al.*, 2015; Grover *et al.*, 2015; Shoolestani *et al.*, 2015). However, this advancement is still underdeveloped and needs further investigation. Therefore, this study is aimed at developing a conceptual framework for socioBIM-based POE of buildings, which can facilitate collaborative practice among facility users and built environment professionals.

1.1. *Existing frameworks for post-occupancy evaluation of higher education buildings*

There are various types of POE frameworks for building evaluation. Riley *et al.* (2010) highlighted six POE methods for HE as post-occupancy review of buildings and their engineering (PROBE), higher education design quality forum (HEDQF), soft landing, overall liking score, construction industry council design quality indicators (CIC DQIs), and building use studies (BUS) occupancy survey. Other identified frameworks for HE include the technology, architecture and furniture (TAF) model, evaluation of quality in educational spaces (EQES) and pedagogy-space technology (PST) and learning space rating system (LSRS). Also, most HE buildings utilized intelligent technologies and their assessment frameworks were also considered in the study.

2. Literature Review of Existing Higher Education Post-occupancy Evaluation

Various researchers have conducted POE for HE buildings using diverse performance criteria for assessment. The performance criteria used differ in accordance to the type of buildings surveyed. Most importantly, the assessment of the performance of the individual educational building is paramount because each building has its peculiar characteristics. Moreover, two different locations in the same building may differ in contents and functions, thereby exhibiting different performance levels. Hence, POE researches have been conducted in different learning spaces, such as library, accommodation, cafeteria, laboratories/workshop, lecture theatres, offices and classrooms (Steiner, 2005; Adewunmi *et al.*, 2011; McGrath and Horton, 2011; Sawyerr and Yusof, 2013; Yang *et al.*, 2013; Sanni-Anibire and Hassanain, 2016; Teeroovengadam *et al.*, 2016). For instance, Hassanain and Mudhei (2006) focus on POE of academic and research library facilities at King Fahd University of Petroleum and Minerals (KFUPM), Dhahran, Saudi Arabia. A walk-through and user questionnaire survey approach was used for data collection. The surveyed sample included 27 students and three staff. The respondents were requested to assess the performance of 22 functional and technical elements. Douglas *et al.* (2006) used a questionnaire survey to evaluate student satisfaction at Liverpool John Moore University's Faculty of Business and Law in United Kingdom. They adopted the principle of service–product bundle to inform the questionnaire. The attributes measured were ancillary facilities, facilitating goods, lecture and tutorial facilities, implicit and explicit services. From the sample size of 3,800 students, 865 were surveyed. Further, a focus group discussion was employed to discover vital issues that were not captured in the questionnaire survey. McGrath and Horton (2011) conducted POE study of student accommodation in a modern method of construction (MMC)/modular building in Trent University, Nottingham, UK. Furthermore, an investigative POE of postgraduate hostel facilities at the University of Lagos, Nigeria was also conducted, considering main functional and technical criteria performance (Adewunmi *et al.*, 2011). For the research, a self-administered questionnaire survey was adopted for data collection, and organizational data were also collected via an interview with the coordinator of the hostels of the university and the facility manager. Sanni-Anibire and Hassanain (2016) carried out a quality assessment of student housing facilities through POE of a university campus in Saudi Arabia. Three assessment approaches were employed for data collection, which include walk-throughs, questionnaire surveys and focus group meetings. Similarly, a research on student perceptions of HE classroom attributes was conducted by Yang *et al.* (2013).

A general view of student satisfaction of the importance of facilities at a Norwegian University, Norway, was explored and the University of Nordland (UoN) was used as a case study (Hanssen and Solvoll, 2015). The population sample of the institution was 6,000 students and 600 staff. The student perception of campus life and the facilities of the university was collected with a purpose-made web-based questionnaire. The total number of Norwegian students surveyed was 5,232, with 1,457 valid feedbacks. User satisfaction of the indoor environmental quality of HE facilities was investigated (El Asmar *et al.*, 2014). The indoor environmental quality attributes measured were space furniture, space layout, thermal comfort, indoor air quality, lighting level, water efficiency, acoustic quality, cleanliness and maintenance and overall satisfaction. An IEQ satisfaction was compared between two universities: the Arizona State University (ASU) in Tempe, Arizona, USA and the American University of Beirut (AUB) in Beirut, Lebanon. In addition, Mustafa (2017) studied the performance assessment of educational buildings via POE using the department of architecture and software engineering departments in Salahaddin University-Erbil, Iraq. Design quality, indoor environmental quality and quality of building support services were used as POE indicators in the study, and a questionnaire survey approach was adopted.

Table 1.   The proposed POE indicators for HE buildings.

Attributes	Performance Criteria (Major Groups)		References
Functional (C01)	C1	Space layout	Winnicka-Jasłowska (2015), El Asmar *et al.* (2014), Farrenkopf *et al.* (1980)
	C2	Flexibility	Fianchini (2007), OECD (2006)
	C3	Functionality	van der Voordt *et al.* (2008)
	C4	Ergonomic	Hughes (2002), Farrenkopf *et al.* (1980), Radcliffe (2008)
	C5	System integration	Sanni-Anibire *et al.* (2016), Kahraman *et al.* (2012)
	C6	Accessibility	van der Voordt *et al.* (2008)
Technical (C02)	C7	Acoustic comfort	Winnicka-Jasłowska *et al.* (2015), Hassanain *et al.* (2006), OECD (2006)
	C8	Thermal comfort	Tookaloo *et al.* (2015), El Asmar *et al.* (2014), McGrath *et al.* (2011)
	C9	Visual comfort	El Asmar *et al.* (2014), Hebert and Chaney (2012), Yang *et al.* (2013), Adewunmi *et al.* (2011)
Environmental (C03)	C10	Indoor environmental quality	Tookaloo and Smith (2015), Radcliffe (2008)
	C11	Site planning and land use	Alborz *et al.* (2015), Sanni-Anibire *et al.* (2016), van der Voordt *et al.* (2008)
	C12	Energy consumption	Alborz and Berardi (2015)
	C13	Water consumption	El Asmar *et al.* (2014)
	C14	Waste management	Sawyerr and Yusof (2013)
	C15	Sustainable methods and materials	Kahraman and Kaya (2012), OECD (2006)

Technological (C05)	C16	Security monitoring and access control system	Sawyerr and Yusof (2013)
	C17	Emergency preparedness	Sanni-Anibire *et al.* (2016), Hassanain and Mudhei (2006)
	C18	Addressable fire detection and system	OECD (2006)
	C19	Use of high-tech system	Kahraman and Kaya (2012), Radcliffe (2008)
	C20	Telecom and data system connectibility	Adewunmi *et al.* (2011)
	C21	Work efficiency	van der Voordt *et al.* (2008)
	C22	Use of advanced artificial intelligence	Alborz and Berardi (2015)
	C23	New technologies	Yang *et al.* (2013), Hughes (2002)
Socio-cultural (C06)	C24	Human comfort	Adewunmi *et al.* (2011)
	C25	Health and sanitation	Kahraman and Kaya (2012)
	C26	Functionality and Aesthetic aspects	van der Voordt *et al.* (2008), McGrath *et al.* (2011), Farrenkopf and Roth (1980)
	C27	Architectural consideration and Symbolic	Winnicka-Jasłowska (2015)
	C28	Usability (community use)	OECD (2006)
	C29	Social spaces	Hanssen and Solvoll (2015)
Operational (C07)	C30	Management and operation	El Asmar *et al.* (2014), Sanni-Anibire *et al.* (2016)
	C31	Life-cycle costing	Kahraman and Kaya (2012)
	C32	Feedback loops	van der Voordt *et al.* (2008)
	C33	Design selection	OECD (2006)

The ongoing research debate in HE POE investigation indicates the need for the development of an automated POE process to improve on the traditional POE process. Previous studies have shown the weakness of traditional POE, such as lack of implementation of its outcome, tedious, not regular and uniform as required (Lavy and Bilbo, 2009; Germany, 2014; Göçer *et al.*, 2015).

2.1. *Development of a holistic (POE) indicator framework*

POE is seen as the logical process feedback loop, where lessons learned from occupants' experience in building space is utilized to inform future design and continuous improvement (Zimmerman and Martin, 2001). But most of the POE frameworks are not comprehensive to cover the essential performance attributes in an educational building. A holistic POE for HE buildings is the one that indicates in detail what is affecting the viability and performance of the precinct. According to Sanni-Anibire *et al.* (2016), different approaches from the previous studies can be consolidated to achieve a holistic POE indicator (see Table 1). Therefore, we propose holistic indicators for POE of learning spaces that will be configured into a BIM user's interface platform.

3. Building Information Modelling for Post-occupancy Evaluation of Buildings

POE of buildings is to ascertain whether the design functions meet the expected functions. POE of educational buildings has been conducted with several approaches. According to Coates *et al.* (2012) and Ozturk *et al.* (2012), the data utilized for POE are customarily from occupants' input, bills and metrics, and measurements and readings. In addition, to be able to meet sustainability objectives, continuous evaluation of buildings is paramount (Coates *et al.*, 2012). For sustainability objectives to be achieved in buildings, there is need for improvement on the traditional methods of POE, with the advantages of BIM and its possible integrated technologies.

BIM has been adopted for collaborative practice in the architecture, engineering, construction industry, but its integration to facilities management functions at the operational phase is underdeveloped and unexplored (Arayici *et al.*, 2012; Liao *et al.*, 2012; Miettinen and Paavola, 2014; Ilter and Ergen, 2015; Liu and Issa, 2015; Motawa and Almarshad, 2015; Olatunji and Akanmu, 2015; Pärn *et al.*, 2017). According to Motawa and Almarshad (2015), the primary purpose of BIM is to scaffold the administration of stakeholder participation throughout the facility life cycle. However, FM practices have less benefits in this collaborative innovation (Kassem *et al.*, 2015; Olatunji and Akanmu, 2015). BIM, which is a digital visual representation of building facilities, holds a promising added value in FM. According to Naghshbandi (2016), BIM will support facility personnel's speedy retrieval of facility information, while a longer time will be expended without BIM. Additionally, the other FM benefit is that BIM provides an accurate geometrical representation of facility, fosters faster information-sharing, quality production of documentation, accurate prediction of life-cycle costing and environmental performance, and reduces lead time of non-productive activities, locating building components, checking maintainability, emergency and space management (Arayici *et al.*, 2012; Shalabi and Turkan, 2016). However, few studies have attempted to implement BIM technologies to conduct POE in learning spaces.

BIM application to POE is growing gradually and of interest to researchers in recent times. For instance, Coates *et al.* (2012) proposed a new framework for POE utilizing BIM technologies and sensors. The framework enables the ability to capture real-time facilities information (Coates *et al.*, 2012). In a similar study, Ozturk *et al.* (2012) explored the utilization of BIM and sensing devices for POE of a residential dwelling, where energy usage aspects, such as temperature, humidity and energy consumption, was investigated. They believed that the integration of BIM-POE is a developing field that needs further exploration as technology advances. Sustainable BIM-driven POE for buildings was also investigated (Motawa and Corrigan, 2012). Motawa and Corrigan proposed a framework for BIM-based system for building performance evaluation. In 2013, Motawa and Carter proposed a conceptual model for sustainable BIM-based evaluation of buildings focusing on energy management. Furthermore, BIM simulation tool was integrated in a proposed POE model for intelligent buildings (Göçer, 2014). Additionally, Göçer *et al.* (2015) integrated BIM with geographical information systems (GIS) to enhanced POE process for effective feedback for FM decision-making on building performance. However, most of these proposed frameworks lack a collaborative platform that includes the facility owners and the end users.

Most of the BIM's solutions are basically inclusive of built environment professional collaboration, but exclude client/end user involvement. This creates a limitation for client and end user contribution to the product service delivery. However, BIM's purpose is to facilitate stakeholder input and participation throughout the entire project life cycle (Motawa and Almarshad, 2015). Therefore, the engagement of facility users in HE on BIM platform to interact with the FM department can possibly create a collaborative practice to improve stakeholder satisfaction. This situation leads to the advancement of a BIM concept that can facilitate the participation of all stakeholders in the management of facility throughout its life cycle.

3.1. *SocioBIM concept*

Earlier professionals in the construction industry were only engaged on BIM platform, while the clients and end users were excluded. Operation of facilities is complex and is difficult for FM to handle it alone. This necessitated the involvement of end users' input on a collaborative platform to assist FM in continuous monitoring of facilities performance. Furthermore, to improve the POE process, facilities performance and user's satisfaction, the end user's integration to BIM platform is needful. The advancement of BIM technologies to accommodate the end users and the public is termed as socioBIM (Grover *et al.*, 2015; Shoolestani *et al.*, 2015).

SocioBIM is a platform that facilitates the participation of building users to interact with their building and give useful comments and feedback to the facility manager that can positively impact decision on asset management and sustainable facility operations (Shoolestani *et al.*, 2015). SocioBIM technologies are not just meant for BIM data availability to the end users for utilization but also used to collect input from users by connecting BIM innovation with technologies of social interaction network (Shoolestani *et al.*, 2015). Furthermore, Grover *et al.* (2015) proposed the need for a BIM-to-public platform for integrating user's participation in design and facilities operation. Furthermore, Adamu *et al.* (2015)

developed and validated a framework for social BIM for situational awareness among selected remote users. In addition, a cloud-based social BIM framework to capture, manage and exchange project information over the life cycle of a building was proposed (Das *et al.*, 2015).

All these previous proposed frameworks lack a holistic view of POE criteria which are supposed to feature comprehensive facility performance indicators to be measured over time. Moreover, the information provided by the public will assist in facilities operation efficiency. According to Shoolestani *et al.* (2015), socioBIM provides the users capacity to do the following:

- give feedback on the condition and maintenance of the building;
- give feedback on the design and management levels;
- give feedback on the indoor and socio-environmental performance;
- express their level of satisfaction on working conditions (health issues and productivity);
- advice on solutions to facilities problems;
- give feedback on aesthetic and appearance;
- give their apparent value of sustainability highlights;
- give suggestions for upgrades and improvement, attaching significant videos and voice updates, photos or any needful documents to the feedback;
- gives feedback on functionality, performance and usability of any building elements.

To accomplish this, we propose a conceptual framework for a socioBIM-based POE using learning spaces as a specific case study because the facility performance indicator requirements of building differ on its functions.

4. Research Method

To achieve a comprehensive examination and review of existing research studies and frameworks related to a particular field, a precise assessment of publications in academic journals is fundamental (Tsai and Wen, 2005). This study adopted an exploratory research approach. A content analysis of the POE of HE researches and frameworks relevant to the study in the span of 1980–2017 was conducted. The content analysis process involves the identification, selection and examination of the target papers (see Yi and Wang, 2013). The papers were searched using Scopus and Google Scholar. The content analysis focuses majorly on the extraction of facilities performance indicators. The extracted indicators were further subjected to a two-member expert panel judgement to identify suitable performance indicators and their grouping. The extracted variables were consolidated to form holistic POE indicators for HE buildings to achieve the socioBIM-based POE.

5. The Proposed Conceptual Framework for SocioBIM-based POE

The conceptual framework we develop for the BIM-based POE for existing buildings comprises three stages. Initially, we will build up a theoretical procedure model from the literature and physical observation of the case area. Used cases will be utilized to outline the prerequisites and procedures that the application will have in this stage. Secondly, we use POE indicators identified to drive the improvement of a conceptual information framework. This stage will be utilized to decide how to gather the required information and how the application collaborates with the information. Thirdly, software developers will turn the conceptual framework information into a prototype system for implementation. In the remainder of this section, we provide some brief explanations on the conceptualization process.

POE is paramount to investigate the building performance and how it meets the user's requirements. Many studies on HEI POE have adopted various performance attributes.

The concept of the socioBIM-based POE focusses on how to advance the development of POE API plug-in on BIM interface. Within the framework architecture, the graphical user interface (GUI) takes into consideration the users to characterize essential data needs to express their level of satisfaction/perception of the performance of the buildings components/facilities. It also allows for query on satisfaction towards FM services from the users. The GUI platform in the module comprises the POE criteria template where stakeholders can assess the facilities tag and input their comments, needs and suggestions as it concerns a specific facility in the building. In addition, the facility personnel also access the platform to receive the input and output for decision-making. The developed socio-BIM-based POE system demonstrates as a compartment for post-occupancy information storage. Associated with that compartment is an ontology of information/data exchange systems that consider filtering, querying and getting to various data as input by the users. The information exchange system includes the operation bit where data are analyzed and turned to useful decision-making information to address FM services and facility information update (as shown in Fig. 1).

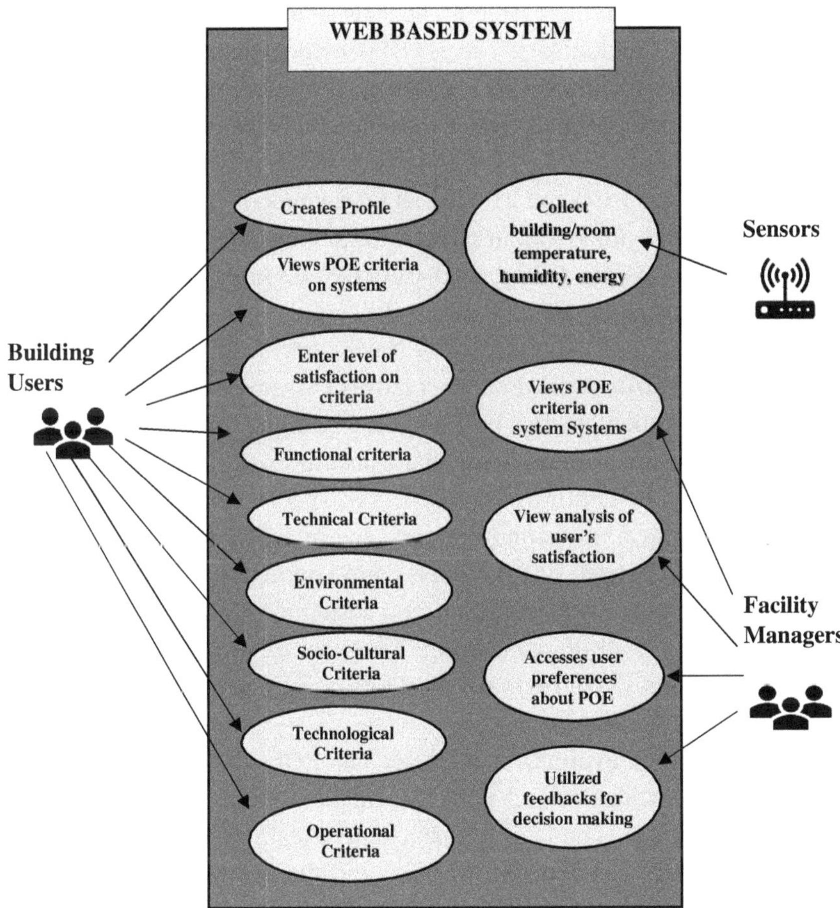

Fig. 1.   Platform for end user interaction during the operational stage.

The system architecture is sub-divided into three basic modules: user's information module, POE module and the building information module (see Fig. 2). The activities and the purpose of each module differ but complement each other. The integration of all the modules formulates the BIM-based POE model. The content of each module, basic activities at each stage and necessary technologies to facilitate the functions are described as follows:

(1) **Building information module:** The BIM represents the interface where the stakeholders interact more with the 3D visualization of the as-built facilities in their systems, such as laptops, desktop computers, mobile phones and tablets. The 3D BIM model enhances vertical pictorial view of facilities and location in the building, which the 2D drawing cannot facilitate. The BIM platform is developed and maintained using Autodesk Revit Architecture and Revit MEP. The as-built BIM is created from the existing 2D drawings of the building. The Revit drawings are keyed in directly, where original computer-aided design (CAD) is not available. Better still, 3D laser scanning can be employed to capture the existing facility conditions. The facilities are integrated and read using Autodesk Navisworks from the BIM model.

(2) **User information module:** This platform assembles the description of each apartment in the building. It highlights and indicates the purpose of each compartment that constitutes the entire structure. For instance, it identifies the classroom spaces and indicates their characteristics, fixtures and features, the intended activities to take place in it, location, access levels, and capacity. How each space facilities support or influences users' activities due to its conditions determine its level of satisfaction. The state of the condition of the space facilities is keyed in on the POE questionnaire integrated on the web-enabled BIM.

(3) **Post-occupancy evaluation module:** The POE module will provide the FM staff with detailed POE information on facilities, condition and performance. The facility management team (FMT) will be able to make analysis and assess the state and the perception/satisfaction level of facility users. The POE survey questionnaire developed (Fig. 2) will be integrated in the Autodesk Revit application programming

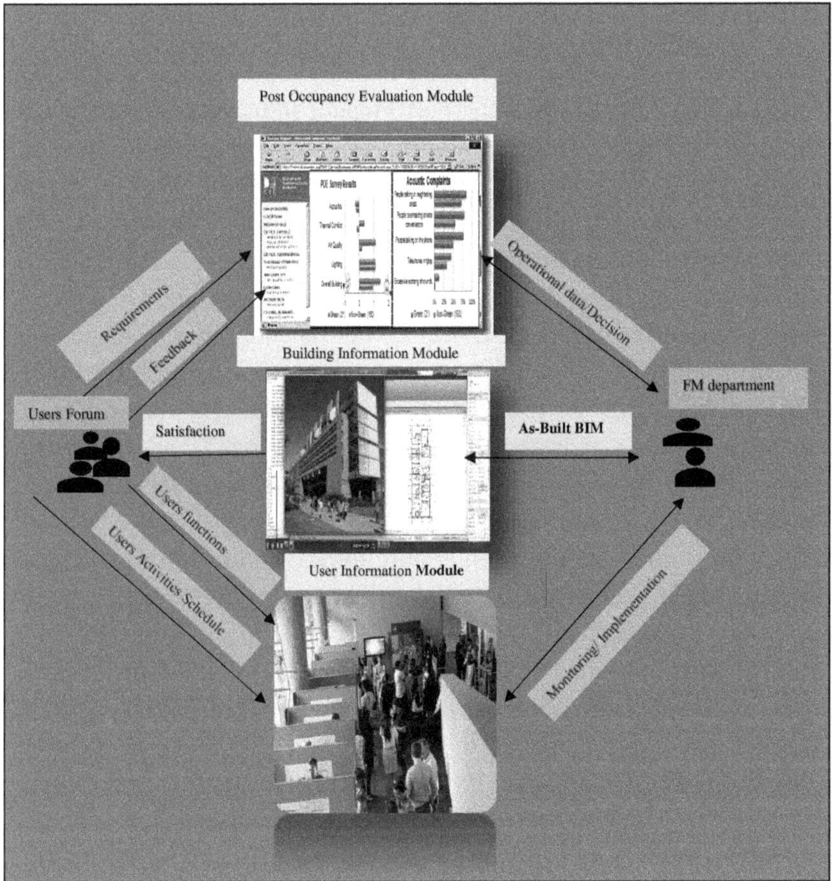

Fig. 2.   BIM-based post occupancy evaluation conceptual architecture platform.

interface (API) using C# programming language. From the platform, the users can input their opinions and needs with comments on a web-enabled interface. The information input will be analyzed as POE data to generate a readable outcome in the form of mean score, weight average, graphs and charts for decision-making by the FMT. The platform will indicate the moving average of POE criteria (i.e. functional, technical, environmental, socio-cultural, technological and operational).

6. Conclusion

POE is the method of ensuring that the constructed facilities satisfied the intended design purpose. Building performance indicators are paramount to discover how users and the building attributes integrate to enhance productivity and sustainability goals. POE result can be used to inform FM decision-making, facility improvement and new development. Ability to effectively capture post-occupancy facility information is essential, but is still difficult for FM to accomplish. In this study, we sought to conceptualize a socioBIM-based POE framework anchor on user-generated experiences and expert systems through the integration of traditional POE and BIM technologies. We propose a conceptual framework that relates to educational building performance attributes categorized under six main groups, namely functional, technical, environmental, technological, socio-cultural and operational factors. The framework consolidates all existing POE frameworks in the literature for HE building evaluation and 33 attributes have been extracted as sub-criteria under the six main categories, based on the literature and expert panel opinions. These attributes will be adopted to evaluate building spaces, such as classrooms, offices, studios, workshops/labs, libraries and other services on a BIM-based POE platform.

Basically, BIM technologies have the potential to empower stakeholder participation in learning space facilities management. Hence, these potentials need exploration to facilitate collaborative practice in facilities management that foster effective decision-making. The application of socioBIM in HE will approximately improve FM performance and increase user satisfaction. The socioBIM-based POE platform will also facilitate the process of continuous evaluation of buildings and improvement of the traditional POE approaches. Additionally, the platform will enhance the capacity to capture the continuous end users dynamic needs and expectations over time, which can prompt FM response to meet user requirements in the built environment. Nevertheless, this study is not comprehensive enough to highlight the benefits or challenges of socioBIM-based POE. It must be noted that the POE criteria extracted may not cross all building components which may be essential for a facility user. Furthermore, the development of a comprehensive socioBIM-based POE may be building-specific because the building information of each building differ. However, it will be beneficial to employ the advantages of BIM technologies for the advancement of traditional POE. According to Pärn and Edwards (2017), there is a need to integrate POE with BIM API plug-in that is informed from user's experience.

Further investigation of the major criteria and their sub-criteria is needed for the allotment of weighting scale based on established rules and expert's perspectives. The developed framework is applicable to any buildings and can inform future automated POE of constructed facilities and new project development and procedures.

References

Adamu, Z. A., Emmitt, S. and Soetanto, R. (2015). "Social BIM: Co-creation with Shared Situational Awareness", *Journal of Information Technology in Construction*, 20, 230–252.

Adewunmi, Y. *et al.* (2011). "Post-occupancy Evaluation of Postgraduate Hostel Facilities", *Facilities*, 29, 149–168. doi: 10.1108/02632771111109270.

Alborz, N. and Berardi, U. (2015). "A Post-Occupancy Evaluation Framework for LEED Certified U.S. Higher Education Residence Halls", *Procedia Engineering*, 118, 19–27. doi: 10.1016/j.proeng.2015.08.399.

Arayici, Y., Onyenobi, T. and Egbu, C. (2012). "Building Information Modelling (BIM) for Facilities Management (FM): The Mediacity Case Study Approach", *International Journal of 3-D Information Modeling*, 1(1), 55–73. doi: 10.4018/ij3dim.2012010104.

Asteriou, D. and Agiomirgianakis, G. M. (2001). "Human Capital and Economic Growth Time Series Evidence from Greece", *Journal of Policy Modeling*, 23, 481–489. doi: PII: S0161-8938(01)00054-0.

Azhar, S. (2011). "Building Information Modeling (BIM): Trends, Benefits, Risks, and Challenges for the AEC Industry", *Leadership and Management in Engineering*, 11(3), 241–252. doi: 10.1061/(ASCE)LM.1943-5630.0000127.

Becerik-Gerber, B. *et al.* (2012). "Application Areas and Data Requirements for BIM-enabled Facilities Management", *Journal of Construction Engineering and Management*, 138(3), 431–442. doi: 10.1061/(ASCE)CO.1943-7862. 0000433.

Coates, P., Arayici, Y. and Ozturk, Z. (2012). "New Concepts of Post Occupancy Evaluation (POE) Utilizing BIM Benchmarking Techniques and Sensing Devices", in *Sustainability in Energy and Buildings*, N. M'Sirdi and A. Et, eds., Berlin: Springer-Verlag Berlin Heidelberg, pp. 319–329.

Das, M., Cheng, J. C. and Kumar, S. S. (2015). "Social BIMCloud: A Distributed Cloud-based BIM Platform for Object-Based Lifecycle Information Exchange", *Visualization in Engineering*, 3(1), 8. doi: 10.1186/s40327-015-0022-6.

Douglas, J., Douglas, A. and Barnes, B. (2006). "Measuring Student Satisfaction at a UK University", *Quality Assurance in Education*, 14(3), 251–267. doi: 10.1108/09684880610678568.

El Asmar, M., Chokor, A. and Srour, I. (2014). "Are Building Occupants Satisfied With Indoor Environmental Quality of Higher Education Facilities?" in *The International Conference on Technologies and Materials for Renewable Energy, Environment and Sustainability, TMREES14, Energy Procedia*, pp. 751–760. doi: 10.1016/j.egypro.2014.06.093.

Farrenkopf, T. and Roth, V. (1980). "The University Faculty Office as an Environment", *Environment and Behavior*, 12(4), 467–477.

Fianchini, M. (2007). "Fitness for Purpose: A Performance Evaluation Methodology for the Management of University Buildings", *Facilities*, 25(3/4), 137–146. doi: DOI 10.1108/02632770710729728.

Germany, L. (2014). "Learning Space Evaluations — Timing, Team, Techniques", in *The Future of Learning and Teaching in Next Generation Learning Spaces*, pp. 267–288.

Göçer, Ö. (2014). "A Post-Occupancy Evaluation Model for Intelligent Buildings", *İstanbul Ticaret Üniversitesi Fen Bilimleri Dergisi*, 13(26), 125–139.

Göçer, Ö., Hua, Y. and Göçer, K. (2015). "Completing the Missing Link in Building Design Process: Enhancing Post-Occupancy Evaluation Method for Effective Feedback for Building Performance", *Building and Environment*, 89, 14–27. doi: 10.1016/j.buildenv.2015.02.011.

Grover, R., Li, P. and Froese, T. M. (2015). The Interface Between Building Information Models and the Public, in *5th International/11th Construction Speciality Conference*, pp. 1–7.

Hanssen, T.-E. S. and Solvoll, G. (2015). "The Importance of University Facilities for Student Satisfaction at a Norwegian University", *Facilities*, 33(13/14), pp. 744–759. doi: http://dx.doi.org/10.1108/F-11-2014-0081.

Hassanain, M. A. and Mudhei, A. A. (2006). "Post-occupancy Evaluation of Academic and Research Library Facilities", *Structural Survey*, 24(3), 230–239. doi: 10.1108/02630800610678878.

Hebert, P. R. and Chaney, S. (2012). "Using End-user Surveys to Enhance Facilities Design and Management", *Facilities*, 30(11/12), 458–471. doi: 10.1108/02632771211252306.

Hughes, J. C. (2002). "Developing a Classroom Vision and Implementation Plan", in *New Directions for Teaching and Learning,* Wiley Periodicals, Inc., pp. 63–72. doi: 10.1002/tl.80.

Ilter, D. and Ergen, E. (2015). "BIM for Building Refurbishment and Maintenance: Current Status and Research Directions", *Structural Survey*, 33(3), 228–256. doi: http://dx.doi.org/10.1108/SS-02-2015-0008.

Kahraman, C. and Kaya, İ. (2012). "A Fuzzy Multiple Attribute Utility Model for Intelligent Building Assessment", *Journal of Civil Engineering and Management*, 18(6), 811–820. doi: 10.3846/13923730.2012.720932.

Kassem, M. *et al.* (2015). "BIM in Facilities Management Applications: A Case study of a Large University Complex", *Built Environment Project and Asset Management*, 5(3), 261–277. doi: 10.1108/BEPAM-02-2014-0011.

Kwun, D. J.-W., Ellyn, E. and Choi, Y. (2013). "Campus Foodservice Attributes and their Effects on Customer Satisfaction, Image, and Word-of-mouth", *Journal of Foodservice Business Research*, 16(3), 276–297. doi: 10.1080/15378020.2013.810534.

Lavy, S. (2008). "Facility Management Practices in Higher Education Buildings : A Case Study", *Journal of Facilities Management*, 6(4), 303–315. doi: http://dx.doi.org/10.1108/14725960810908163.

Lavy, S. and Bilbo, D. L. (2009). "Facilities Maintenance Management Practices in Large Public Schools, Texas", *Facilities*, 27, 5–20. doi: 10.1108/02632770910923054.

Liao, C., Tan, D. and Li, Y. (2012). "Research on the Application of BIM in the Operation Stage of Green Building", *Applied Mechanics and Materials*, 174–177, 2111–2114. doi: 10.4028/www.scientific.net/AMM.174-177.2111.

Liu, R. and Issa, R. (2015). "Survey: Common Knowledge in BIM for Facility Maintenance", *Journal of Performance of Constructed Facilities*, 30(2010), 1–8. doi: 10.1061/(ASCE)CF.1943-5509.0000778.

McGrath, P. T. and Horton, M. (2011). "A Post-occupancy Evaluation (POE) Study of Student Accommodation in an MMC/modular Building", *Structural Survey*, 29(3), 244–252. doi: 10.1108/02630801111148211.

Miettinen, R. and Paavola, S. (2014). "Beyond the BIM Utopia: Approaches to the Development and Implementation of Building Information Modeling", *Automation in Construction*, 43, 84–91. doi: 10.1016/j.autcon.2014.03.009.

Motawa, I. and Almarshad, A. (2015). "Case-based Reasoning and BIM Systems for Asset Management", *Built Environment Project and Asset Management*, 5(3), 233–247. doi: http://dx.doi.org/10.1108/BEPAM-02-2014-0006.

Motawa, I. and Carter, K. (2013). "Sustainable BIM-based Evaluation of Buildings", *Procedia — Social and Behavioral Sciences*, 74, 419–428. doi: 10.1016/j.sbspro.2013.03.015.

Motawa, I. and Corrigan, W. (2012). "Sustainable BIM-Driven Postoccupancy Evaluation for Buildings". Available at: http://www.cicstart.org/userfiles/file/FS-49-REPORT.PDF.

Mowery, C. D. (2004). "Universities in National Innovation Systems". doi: 10.1093/oxfordhb/9780199286805.003.0008.

Mustafa, F. A. (2017). "Performance Assessment of Buildings via Post-Occupancy Evaluation: A Case Study of the Building of the Architecture and Software Engineering Departments in Salahaddin University-Erbil, Iraq", *Frontiers of Architectural Research*, 6(3), 412–429. doi: 10.1016/j.foar.2017.06.004.

Naghshbandi, S. N. (2016). "BIM for Facility Management: Challenges and Research Gaps", *Civil Engineering Journal*, 2(12), 679–684.

Oblinger, D. (2005). "Leading the Transition from Classrooms to Learning Spaces". *Educause Quarterly*, 28(1), 14–18.

OECD (2006) (Revised). *CELE Organising Framework on Evaluating Quality in Educational Spaces*, Paris: OECD.

Olatunji, O. A. and Akanmu, A. (2015). "BIM-FM and Consequential Loss: How Consequential Can Design Models Be?", *Built Environment Project and Asset Management*, 5(3), 304–317. doi: http://dx.doi.org/10.1108/BEPAM-03-2014-0021.

Ozturk, Z., Arayici, Y. and Coates, P. (2012). "Post-Occupancy Evaluation (POE) in Residential Buildings Utilizing BIM and Sensing Devices: Salford Energy House Example". doi: http://usir.salford.ac.uk/20697/1/105_Ozturk.pdf.

Pärn, E. A. and Edwards, D. J. (2017). "Conceptualising the FinDD API Plug-in: A Study of BIM-FM Integration", *Automation in Construction*, 80, 11–21. doi: 10.1016/j.autcon.2017.03.015.

Pärn, E. A., Edwards, D. J. and Sing, M. C. P. (2017). "The Building Information Modelling Trajectory in Facilities Management: A Review", *Automation in Construction*, 75, 45–55. doi: 10.1016/j.autcon.2016.12.003.

Price, I. *et al.* (2003). "The Impact of Facilities on Student Choice of University", *Facilities*, 21(10), 212–222. doi: 10.1108/02632770310493580.

Radcliffe, D. (2008). "A Pedagogy-Space-Technology (PST) Framework for Designing and Evaluating Learning Places", in *Learning Spaces in Higher Education: Positive Outcomes*, D. Radcliffe, H. Wilson and D. Powell, eds., St. Lucia: The University of Queensland, pp. 10–16.

Riley, M., Kokkarinen, N. and Pitt, M. (2010). "Assessing Post Occupancy Evaluation in Higher Education Facilities", *Journal of Facilities Management*, 8(3), 202–213. doi: 10.1108/14725961011058839.

Sanni-Anibire, M. O. and Hassanain, M. A. (2016). "Quality Assessment of Student Housing Facilities Through Post-Occupancy Evaluation", *Architectural Engineering and Design Management*, 2007(May), 1–14. doi: 10.1080/17452007.2016.1176553.

Sanni-Anibire, M. O., Hassanain, M. A. and Al-Hammad, A.-M. (2016). "Holistic Post-Occupancy Evaluation Framework for Campus Residential Housing Facilities", *Journal of Performance of Constructed Facilities*, 30(5), 4016026. doi: 10.1061/(ASCE)CF.1943-5509.0000875.

Sawyerr, T. P. and Yusof, N. (2013). "Student Satisfaction With Hostel Facilities in Nigerian Polytechnics", *Journal of Facilities Management*, 11(4), 306–322. doi: 10.1108/JFM-08-2012-0041.

Shalabi, F. and Turkan, Y. (2016). "IFC BIM-Based Facility Management Approach to Optimize Data Collection for Corrective Maintenance", *Journal of Performance of Constructed Facilities*, 040160811–13. doi: 10.1061/(ASCE) CF.1943-5509.0000941.

Shoolestani, A. *et al.* (2015). SocioBIM: BIM-to-end User Interaction for Sustainable Building Operations and Facility Asset Management, in *Proceedings of ICSC15: The Canadian Society for Civil Engineering 5th International/11th Construction Speciality Conference*, University of British Columbia, Vancouver, Canada. June 7–10. Vancouver, British Columbia, pp. 3261–3266. doi: 10.14288/1.0076332.

Steiner, J. (2005). "The Art of Space Management: Planning Flexible Workspaces for People", *Journal of Facilities Management*, 4(1), 6–22. doi: 10.1108/14725960610644195.

Teeroovengadam, V., Kamalanabhan, T. J. and Seebaluck, A. K. (2016). "Measuring Service Quality in Higher Education", *Quality Assurance in Education*, 24(2), 244–258. doi: 10.1108/QAE-06-2014-0028.

Tookaloo, A. and Smith, R. (2015). "Post Occupancy Evaluation in Higher Education", in *International Conference on Sustainable Design, Engineering and Construction, Procedia Engineering*, pp. 515–521. doi: 10.1016/j.proeng.2015.08.470.

Tsai, C. and Lydia Wen, M. (2005). "Research and Trends in Science Education from 1998 to 2002: A Content Analysis of Publication in Selected Journals", *International Journal of Science Education*, 27(1), 3–14. doi: 10.1080/0950069042000243727.

Tucker, M. and Smith, A. (2008). "User Perceptions in Workplace Productivity and Strategic FM Delivery", *Facilities*, 26(5/6), 196–212. doi: 10.1108/02632770810864989.

van der Voordt, T. J. and van der Klooster, W. (2008). "Post-occupancy Evaluation of a New Office Concept in an Educational Setting", in *CIB W70 International Conference in Facilities Management*, Heriot Watt University, Edinburgh, June 16–18.

Winnicka-Jasłowska, D. (2015). "Quality Analysis of Polish Universities Based on POE Method — Description of Research Experiences", in *UAHCI 2015, Part III, LNCS 9177*, M. Antona and C. Stephanidis, eds., pp. 236–242. doi: 10.1007/978-3-319-20684-4.

Yang, Z., Becerik-Gerber, B. and Mino, L. (2013). "A Study on Student Perceptions of Higher Education Classrooms: Impact of Classroom Attributes on Student Satisfaction and Performance", *Building and Environment*, 70, 171–188. doi: 10.1016/j.buildenv.2013.08.030.

Yi, H. and Wang, Y. (2013). "Trend of the Research on Public Funded Projects", *Open Construction and Building Technology Journal*, 7, 51–62. doi: 10.217 4/1874836820130716002.

Zimmerman, A. and Martin, M. (2001). "Post-occupancy Evaluation: Benefits and Barriers", *Building Research & Information*, 29(2), 168–174.

Chapter 3

Critical Barriers to BIM Adoption in the Chinese Architectural, Engineering and Construction Industry

Xianbo Zhao*

*Central Queensland University,
400 Kent Street, Sydney, NSW 2000, Australia
b.zhao@cqu.edu.au*

Josua Pienaar

*Central Queensland University,
44 Greenhill Road, Wayville, SA 5034, Australia*

Abstract

The Chinese government has recently promoted the use of building information modelling (BIM) in the architecture, engineering and construction (AEC) industry to improve construction innovation, working efficiency and project performance. The objective of this chapter is to identify the critical barriers to BIM adoption. A questionnaire survey was performed with 95 professionals. A total of 14 barriers were recognized as critical barriers to BIM adoption and are worth

---

* Corresponding author.

more attention from the government and industry practitioners. The top barriers included "lack of BIM implementation data in construction phase", "reluctance to openly share information", "lack of industry standards", "lack of tangible benefits", "lack of demand for BIM use" and "poor interoperability among BIM software". In addition, the influence of barriers was not correlated with the years of BIM experience. Some barriers would persist after several years of BIM adoption. This chapter allows industry practitioners to take measures to overcome the potential barriers and enhance the implementation level of BIM and finally assure the far-reaching benefits of BIM.

Keywords: Building Information Modelling; Barriers; China.

1. Introduction

In recent years, building information modelling (BIM) has been transforming the architecture, engineering and construction (AEC) industry in many countries (Azhar, 2011; Zhao *et al.*, 2017) and has thus received increasing attention from researchers and practitioners. BIM is a process of representing building and infrastructure over its whole life cycle from planning, design, construction, operations, maintenance to recycling (Building SMART Australasia, 2012). The expansion of this technology may be attributed to its potential benefits to users. Multiple previous studies have shown that adopting BIM could bring about better design solutions, fewer design errors and omissions, better organization's image, reduced rework, fewer field problems, repeat business, reduced project duration and reduced construction cost (Cao *et al.*, 2014; Dodge Data & Analytics, 2015). In industry practice, BIM has been applied in diverse areas, such as site analysis, design option analysis, 3D presentation, design coordination, cost estimation, energy simulation, clash detection, construction system design, schedule simulation, quantity takeoff, site resource management, and offsite fabrication (Azhar, 2011; Eastman *et al.*, 2011; Li *et al.*, 2012; Bynum *et al.*, 2013; Cao *et al.*, 2014).

The Chinese construction market is huge in terms of its size. In 2015, the total value of this market reached CNY 18,075.747 billion, approximately USD 2,702.9 billion. Hence, there is a great potential for BIM adoption in the Chinese AEC industry. The Chinese government has pushed the adoption of BIM for the purpose of innovation and development. The 12th National Five-Year Plan of the Ministry of Housing and Urban–Rural Development (MOHURD) has recognized industrialization, informatization, urbanization and agricultural modernization as the key points of the national building industry development, and BIM is expected to play an important role in each of them. Additionally, the MOHURD issued a regulation to push the widespread adoption of BIM in order to enhance the working efficiency of construction project stakeholders and assure construction quality, safety and environment friendliness. Therefore, the regional governments follow this policy and create the regional policies relating to BIM adoption. For example, in 2014, the Shanghai municipal government issued various policies, including a detailed three-year action plan to push the BIM adoption in the AEC companies located in Shanghai; similarly, Guangdong province plans to push the BIM adoption and sets a goal that BIM will be used in all the new building projects with gross floor area above 20,000 m^2 by the end of 2020. Because of the "Belt and Road" strategy of the Chinese government, an increasing number of Chinese AEC organizations have ventured into the international market. Thus, the BIM adoption in the Chinese AEC organizations would also influence the BIM adoption in the host countries' AEC industry.

To expand the BIM adoption in the Chinese AEC industry, it is necessary to overcome the barriers to BIM adoption. The objective of this chapter is to identify the critical barriers to BIM adoption. Thus, this chapter allows industry practitioners to take measures to overcome the potential barriers and enhance the implementation level of BIM and finally assure the far-reaching benefits of BIM.

2. Background

The studies of BIM have been diverse and more emerging technologies have been integrated into BIM. For instance, BIM can facilitate 3D printing implementation (Arayici *et al.*, 2012) and has been applied in the 3D printing of small-scale models and large-scale buildings, respectively. Wu *et al.* (2016) and Mahdjoubi *et al.* (2013) developed a model to facilitate the delivery of real-estate services by integrating 3D laser scanning and BIM and Wang *et al.* (2013) attempted to integrate BIM with augmented reality (AR) to enable the real-time visualization of the physical context of each construction activity or task. Zhao (2017) summarized that the hot topics of BIM research were mobile and cloud computing, laser scan, AR, ontology, safety rule and code checking, semantic web technology, and automated generation. Indeed, BIM is not just a technology, but also a project management tool and process (Succar, 2009). Hence, in addition to the studies focused on the relevant technologies, there are studies focused on the management issues in BIM adoption.

Various studies have reported that BIM adoption has been slower than anticipated (Bernstein and Pittman, 2004). Hence, studies have been performed to identify the factors influencing BIM adoption, most of which focused on the barriers to BIM adoption. Gu and London (2010) found that the lack of experience in BIM because of a limited understanding of the industry needs and technical requirements was a major barrier to the adoption of BIM-related technologies within the Australian AEC industry. Coates *et al.* (2010) argued that the existing BIM tools do not allow "personal nuances in the design process" and "ambiguities of early design" because BIM provides ways to represent the final design (Coates *et al.*, 2010). Additionally, Gerrard *et al.* (2010) reported that lack of BIM expertise, lack of awareness and resistance to change were the main barriers to BIM adoption. According to Hartmann *et al.* (2008), the low-level BIM acceptance can be attributed to the highly reactive and action-oriented nature of the construction industry, which in turn hinders the adoption of sophisticated technologies by practitioners. Some practitioners simply believe that the current technologies are enough and it is unnecessary to adopt BIM; as shown by the study of Alabdulqader *et al.* (2013), BIM were not well adopted mainly because practitioners perceived the existing CAD system can fulfil the need and BIM was expensive to operate and maintain. Furthermore, Khosrowshahi and Arayici (2012) hold a perspective of change management and consider BIM adoption as a major change management task, which involves various challenges and barriers, such as resistance to cultural change, inadequate tangible benefits, lack of demand for BIM adoption within a company and lack of relevant training. Through a literature review, 20 barriers to BIM adoption were identified, as shown in Table 1.

Table 1.   Identification of barriers to BIM adoption.

Barriers to BIM Adoption	References
Cost of investment	Tse *et al.* (2005), D'Agostino *et al.* (2007), Gilligan and Kunz (2007), Yan and Damian (2008), Young *et al.* (2009), Eastman *et al.* (2011), Ku and Taiebat (2011), Bernstein *et al.* (2012), Khosrowshahi and Arayici (2012), Eadie *et al.* (2013), Won *et al.* (2013)
Learning curve for BIM technologies	D'Agostino *et al.* (2007), Gilligan and Kunz (2007), Krygiel *et al.* (2008), Yan and Damian (2008), Young *et al.* (2009), Eastman *et al.* (2011), Ku and Taiebat (2011), Bernstein *et al.* (2012), Demian and Walters (2013), Won *et al.* (2013)
Lack of support from senior management	D'Agostino *et al.* (2007), Gilligan and Kunz (2007), Bernstein *et al.* (2012), Won *et al.* (2013)
Design liability issues	D'Agostino *et al.* (2007), Gilligan and Kunz (2007), Eastman *et al.* (2011), Bernstein *et al.* (2012), Khosrowshahi and Arayici (2012), Won *et al.* (2013)
Data ownership issues	D'Agostino *et al.* (2007), Becerik-Gerber and Kensek (2010), Dossick and Neff (2010), Gu and London (2010), Azhar (2011), Eadie *et al.* (2013)
Poor collaboration among participants	D'Agostino *et al.* (2007), Eastman *et al.* (2011), Ku and Taiebat (2011), Won *et al.* (2013)
Poor interoperability among BIM software	D'Agostino *et al.* (2007), Young *et al.* (2009), Becerik-Gerber and Kensek (2010), Grilo and Jardim-Goncalves (2010), Gu and London (2010), Eastman *et al.* (2011), Ku and Taiebat (2011), Demian and Walters (2013), Won *et al.* (2013)
Reluctance to openly share information	Gilligan and Kunz (2007), Young *et al.* (2009), Becerik-Gerber and Kensek (2010), Eadie *et al.* (2013), Won *et al.* (2013)
Lack of relevant training	Becerik-Gerber and Kensek (2010), Gu and London (2010), Khosrowshahi and Arayici (2012)
Unsupportive organizational structure	Eastman *et al.* (2011), Won *et al.* (2013)

(Continued)

Table 1.   (*Continued*)

Barriers to BIM Adoption	References
Unsupportive organizational culture	Becerik-Gerber and Kensek (2010), Khosrowshahi and Arayici (2012), Eadie *et al.* (2013), Jensen and Jóhannesson (2013)
Lack of sub-contractors who can use BIM technology	Young *et al.* (2009), Won *et al.* (2013)
Data security issues	D'Agostino *et al.* (2007), Gu and London (2010), Won *et al.* (2013)
Lack of relevant expertise and knowledge	Tse *et al.* (2005), Gilligan and Kunz (2007), Young *et al.* (2009), Becerik-Gerber and Kensek (2010), Khosrowshahi and Arayici (2012), Eadie *et al.* (2013), Won *et al.* (2013)
Limitation of current BIM applications	Gilligan and Kunz (2007), Young *et al.* (2009), Won *et al.* (2013)
Lack of industry standards	D'Agostino *et al.* (2007), Young *et al.* (2009), Eastman *et al.* (2011)
Lack of BIM implementation data in construction phase	Young *et al.* (2009), Eastman *et al.* (2011), Won *et al.* (2013)
Perception that current technology is enough	Yan and Damian (2008)
Lack of tangible benefits	D'Agostino *et al.* (2007), Becerik-Gerber and Kensek (2010), Gu and London (2010), Khosrowshahi and Arayici (2012), Won *et al.* (2013)
Lack of demand for BIM use	Khosrowshahi and Arayici (2012), Eadie *et al.* (2013)

3. Method

A preliminary questionnaire survey was performed to investigate the critical barriers to BIM adoption in China. In the survey, the respondents were asked to rate the influence of the 20 barriers to BIM adoption using a five-point scale (1 = very low, 2 = low, 3 = neutral, 4 = high and 5 = very high). In addition, they were requested to provide general information, such as their firm type, years of industry experience, designation and years of BIM experience of their companies (see Table 2). The population consisted of all the industry practitioners in the Chinese AEC industry. As the list of practitioners was unavailable, there was no sampling framework and the non-probability sampling method was used in this study. Wilkins (2011) considered it appropriate to use a non-probability sampling plan when respondents were selected based on their willingness to participate in the study, and Patton (2001) argued that this method could still produce a representative sample (Patton, 2001).

A total of 400 questionnaires were sent out and 95 completed questionnaires were received, representing a response rate of 24%, which was acceptable compared with the norm of 20–30% with most questionnaire surveys in the construction industry (Akintoye, 2000; Hwang *et al.*, 2015). Out of the 95 respondents, 72% of them were from design companies, and 19% were from contractors. Around 66% had 5–10 years of industry experience and 18% had over 15 years of experience. In terms of BIM experience, 47% had used BIM for over three years while 11% had no BIM experience.

4. Results and Discussion

To improve the level of BIM adoption, critical barriers to BIM adoption were identified by ranking them by their mean scores. The one-sample *t*-test was used to check whether the 20 barriers were significantly influential to BIM adoption in the Chinese AEC industry. As shown in Table 3, the mean scores ranged from 3.60 to 3.03. A total of 14 barriers received mean values above 3 and significance value below 0.05, implying that they were deemed as critical barriers to BIM adoption and worth further attention. This method has been widely used in various construction management studies that identified critical factors (e.g. Low and Chuan, 2006; Hwang *et al.*, 2014; Zhao *et al.*, 2015).

Table 2.   Profile of respondents and their firms.

Organization Type	Count	Percentage (%)	Industry Experience	Count	Percentage (%)	BIM Experience	Count	Percentage (%)
Contractor v	18	19	5–10 years	63	66	None	10	11
Client	9	9	11–15 years	15	16	1–2 years	40	42
Designer	68	72	16–20 years	14	15	3–4 years	22	23
			Over 20 years	3	3	≥ 5 years	23	24

Table 3. Ranking of barriers to BIM adoption in China.

Barriers to BIM Adoption	Mean	Sig.	Rank	Barrier–BIM Experience	
				Correlation	Sig.
Lack of BIM implementation data in construction phase	3.60	0.000	1	0.072	0.487
Reluctance to openly share information	3.54	0.000	2	0.126	0.224
Lack of industry standards	3.52	0.000	3	0.05	0.629
Lack of tangible benefits	3.51	0.000	4	−0.057	0.582
Cost of investment	3.43	0.000	5	0.163	0.115
Poor interoperability among BIM software	3.43	0.000	5	−0.013	0.898
Lack of demand for BIM use	3.43	0.000	5	0.058	0.576
Poor collaboration among participants	3.41	0.000	8	0.09	0.388
Limitation of current BIM applications	3.38	0.002	9	−0.004	0.968
Data ownership issues	3.37	0.002	10	−0.084	0.419
Lack of relevant training	3.36	0.006	11	−0.123	0.236
Lack of sub-contractors who can use BIM technology	3.36	0.001	11	0.043	0.679
Design liability issues	3.32	0.009	13	0.038	0.714
Lack of support from senior management	3.26	0.047	14	−0.147	0.156
Lack of relevant expertise and knowledge	3.21	0.068	15	0.004	0.97
Learning curve for BIM technologies	3.09	0.445	16	0.101	0.328
Data security issues	3.09	0.435	16	−0.034	0.742
Unsupportive organizational culture	3.05	0.667	18	−0.048	0.646
Perception that current technology is enough	3.05	0.665	18	0.026	0.802
Unsupportive organizational structure	3.03	0.799	20	0.025	0.809

"Lack of BIM implementation data in construction phase" was ranked top, suggesting that project stakeholders required more BIM implementation data in construction. Won *et al.* (2013) believed this factor can hinder both company-specific and inter-organizational innovation. Eastman *et al.* (2011) indicated that shortage of BIM implementation data posed challenges in adopting BIM in the UK construction practice. This is because such BIM implementation data can help managers demonstrate the tangible benefits relating to project performance to the senior management.

"Reluctance to openly share information" received the second position in the ranking. BIM should transform the ways that architects, engineers, contractors, and owners share information, models and data (Won *et al.*, 2013). The willingness to share information has been seen as a critical success factor of the BIM adoption (Young *et al.*, 2008; Won *et al.*, 2013) while the reluctance to share information would hinder the use of BIM within a project (Gilligan and Kunz, 2007; Young *et al.*, 2009; Eadie *et al.*, 2013; Won *et al.*, 2013). Dossick and Neff (2010) found that the reluctance to share information tended to result from the issues of ownership, intellectual property and design liability.

"Lack of industry standards" was the third most critical barrier, indicating that practitioners need industry standards to adopt BIM in the AEC industry. This result echoed the finding of D'Agostino *et al.* (2007) that lack of industry standards was one of the greatest hurdles to collaborative construction processes and BIM adoption. In China, some regional governments, such as Guangdong and Liaoning provinces, have issued the regional industry standards. However, in most parts of China, such industry standards were still lacking or not well known to the practitioners.

"Lack of tangible benefits" was ranked fourth, indicating that tangible benefits should be demonstrated to all the managers. This seems to be a prerequisite for BIM adoption because the construction industry is cost-sensitive. Khosrowshahi and Arayici (2012) argued that if benefits were not tangible enough to warrant its use, the implementation of BIM would face challenges. Hanna *et al.* (2013) recognized the costs of using BIM in a project as a main factor that would influence the current state of BIM practice. Therefore, it is necessary to demonstrate that the benefits of BIM exceed the cost associated with BIM adoption.

"Lack of demand for BIM use" occupied the fifth position. This suggested that the actual demand for BIM in the industry was not as great as the government thought. As revealed by Khosrowshahi and Arayici (2012), no demand for BIM use would significantly hinder the expansion of BIM adoption. Indeed, the Chinese government recognizes BIM adoption as an innovative way to enhance working efficiency and productivity in the AEC industry. To initiate this change, the demand for change should be demonstrated to the industry practitioners.

"Poor interoperability among BIM software" was also ranked fifth. Interoperability refers to the ability to exchange data between applications to facilitate automation and avoidance of data re-entry (Azhar *et al.*, 2012). This suggested that the interoperability issues significantly hindered BIM adoption in China, consistent with the findings of previous studies that poor interoperability among BIM software could serve as a barrier to the BIM adoption in project management (D'Agostino *et al.*, 2007; Grilo and Jardim-Goncalves, 2010; Eastman *et al.*, 2011; Ku and Taiebat, 2011; Demian and Walters, 2013).

The correlation between barriers' influence and respondents' BIM experience was assessed. As shown in Table 3, eight barriers seem to be negatively correlated with the years of BIM experience. However, none of the 20 barriers received a significance score below 0.05, suggesting that no barrier's influence was significantly correlated with BIM experience. This result further indicated that longer BIM experience did not necessarily lead to less negative influence of barriers in the Chinese AEC industry. Some barriers would persist after several years of BIM adoption.

5. Conclusion

The Chinese government has promoted the BIM adoption in the AEC industry to improve construction innovation, working efficiency and project performance. The objective of this chapter is to identify the barriers to BIM adoption in China. A questionnaire survey was performed with 95 professionals. A total of 14 barriers were recognized as critical barriers to BIM adoption and are worth more attention from the government and industry practitioners. The top barriers included "lack of BIM implementation data in construction phase", "reluctance to openly share information", "lack of industry standards", "lack of tangible benefits", "lack of demand for BIM use" and "poor interoperability among BIM software". Longer BIM experience did not necessarily bring about less negative influence of barriers in the Chinese AEC industry. Some barriers would persist even though BIM has been adopted for several years.

Future research would examine the differences in barriers among different project stakeholders and investigate the correlations among these barriers. Thus, the root causes of the low-level BIM implementation would be identified and addressed. Additionally, as BIM adoption involves changes in companies, organizational change in theories would be used to interpret the critical barriers and provide the theoretical rationale behind the interrelationships among these barriers.

Acknowledgement

This research is funded by the Central Queensland University New Staff Research Grant.

References

Akintoye, A. (2000). "Analysis of Factors Influencing Project Cost Estimating Practice", *Construction Management and Economics*, 18(1), 77–89.

Alabdulqader, A., Panuwatwanich, K. and Doh, J.-H. (2013). "Current Use of Building Information Modelling Within Australian AEC Industry", *The 13th East Asia-Pacific Conference on Structural Engineering and Construction*, Sapporo, Japan.

Arayici, Y., Egbu, C. and Coates, P. (2012). "Building Information Modelling (BIM) Implementation and Remote Construction Projects: Issues, Challenges, and Critiques", *Journal of Information Technology in Construction*, 17, 75–92.

Azhar, S. (2011). "Building Information Modeling (BIM): Trends, Benefits, Risks, and Challenges for the AEC Industry", *Leadership and Management in Engineering*, 11(3), 241–252.

Azhar, S., Khalfan, M. and Maqsood, T. (2012). "Building Information Modeling (BIM): Now and Beyond", *Australasian Journal of Construction Economics and Building*, 12(4), 15–28.

Becerik-Gerber, B. and Kensek, K. (2010). "Building Information Modeling in Architecture, Engineering, and Construction: Emerging Research Directions and Trends", *Journal of Professional Issues in Engineering Education and Practice*, 136(3), 139–147.

Bernstein, H. M., Jones, S. A. and Russo, M. A. (2012). *The Business Value of BIM in North America: Multi-year Trend Analysis and User Ratings (2007–2012)*, Bedford, MA: McGraw-Hill Construction.

Bernstein, P. G. and Pittman, J. H. (2004). *Barriers to the Adoption of Building Information Modeling in the Building Industry*, Autodesk Inc.

Building SMART Australasia. (2012). *National Building Information Modelling Initiative Volume 1: Strategy*, buildingSMART Australasia.

Bynum, P., Issa, R. R. A. and Olbina, S. (2013). "Building Information Modeling in Support of Sustainable Design and Construction", *Journal of Construction Engineering and Management*, 139(1), 24–34.

Cao, D., Li, H. and Wang, G. (2014). "Impacts of Isomorphic Pressures on BIM Adoption in Construction Projects", *Journal of Construction Engineering and Management*, 140(12), 04014056.

Coates, P., Arayici, Y., Koskela, L. J., Kagioglou, M., Usher, C. and O'Reilly, K. (2010). "The Limitations of BIM in the Architectural Process", *First International Conference on Sustainable Urbanization (ICSU 2010)*, Hong Kong, China.

D'Agostino, B., Mikulis, M. and Bridgers, M. (2007). *FMI/CMAA Eighth Annual Survey of Owners: The Perfect Storm-Construction Style*, Raleigh, NC: FMI Corp./Construction Management Association of America.

Demian, P. and Walters, D. (2013). "The Advantages of Information Management Through Building Information Modelling", *Construction Management and Economics*, 32(12), 1153–1165.

Dodge Data & Analytics. (2015). *The Business Value of BIM in China*, Bedford, MA: Dodge Data & Analytics.

Dossick, C. S. and Neff, G. (2010). "Organizational Divisions in BIM-enabled Commercial Construction", *Journal of Construction Engineering and Management*, 136(4), 459–467.

Eadie, R., Browne, M., Odeyinka, H., McKeown, C. and McNiff, S. (2013). "BIM Implementation Throughout the UK Construction Project Lifecycle: An Analysis", *Automation in Construction*, 36, 145–151.

Eastman, C., Teicholz, P., Sacks, R. and Liston, K. (2011). *BIM Handbook — A Guide to Building Information Modeling for Owners, Managers, Designers, Engineers, and Contractors*, Hoboken, NJ: John Wiley & Sons.

Gerrard, A. J., Zuo, J., Zillante, G. and Skitmore, M. (2010). "Building Information Modeling in the Australian Architecture Engineering and Construction Industry", *Handbook of Research on Building Information Modeling and Construction Informatics: Concepts and Technologies*, J. Underwood and U. Isikdag, eds., IGI Global, Hershey, PA, 521–544.

Gilligan, B. and Kunz, J. (2007). "VDC Use in 2007: Significant Value, Dramatic Growth, and Apparent Business Opportunity", *Construction Users Roundtable (CURT) National Meeting*, Stanford, CA.

Grilo, A. and Jardim-Goncalves, R. (2010). "Value Proposition on Interoperability of BIM and Collaborative Working Environments", *Automation in Construction*, 19(5), 522–530.

Gu, N. and London, K. (2010). "Understanding and Facilitating BIM Adoption in the AEC Industry", *Automation in Construction*, 19(8), 988–999.

Hanna, A., Boodai, F. and El Asmar, M. (2013). "State of Practice of Building Information Modeling in Mechanical and Electrical Construction Industries", *Journal of Construction Engineering and Management*, 139(10), 04013009.

Hartmann, T., Gao, J. and Fischer, M. (2008). "Areas of Application for 3D and 4D Models on Construction Projects", *Journal of Construction Engineering and Management*, 134(10), 776–785.

Hwang, B. G., Zhao, X. and Do, T. H. V. (2014). "Influence of Trade-level Coordination Problems on Project Productivity", *Project Management Journal*, 45(5), 5–14.

Hwang, B. G., Zhao, X. and Ong, S. (2015). "Value Management in Singaporean Building Projects: Implementation Status, Critical Success Factors, and Risk Factors", *Journal of Management in Engineering*, 31(6), 04014094.

Jensen, P. A. and Jóhannesson, E. I. (2013). "Building Information Modelling in Denmark and Iceland", *Engineering, Construction and Architectural Management*, 20(1), 99–110.

Khosrowshahi, F. and Arayici, Y. (2012). "Roadmap for Implementation of BIM in the UK Construction Industry", *Engineering, Construction and Architectural Management*, 19(6), 610–635.

Krygiel, E., Nies, B. and McDowell, S. (2008). *Green BIM: Successful Sustainable Design with Building Information Modeling*, Indianapolis, IN: John Wiley & Sons.

Ku, K. and Taiebat, M. (2011). "BIM Experiences and Expectations: The Constructors' Perspective", *International Journal of Construction Education and Research*, 7(3), 175–197.

Li, H., Chan, N. K., Huang, T., Skitmore, M. and Yang, J. (2012). "Virtual Prototyping for Planning Bridge Construction", *Automation in Construction*, 27, 1–10.

Low, S. P. and Chuan, Q. T. (2006). "Environmental Factors and Work Performance of Project Managers in the Construction Industry", *International Journal of Project Management*, 24(1), 24–37.

Mahdjoubi, L., Moobela, C. and Laing, R. (2013). "Providing Real-Estate Services Through the Integration of 3D Laser Scanning and Building Information Modelling", *Computers in Industry*, 64(9), 1272–1281.

Patton, M. Q. (2001). *Qualitative Research & Evaluation Components*, Thousand Oaks, CA: Sage.

Succar, B. (2009). "Building Information Modelling Framework: A Research and Delivery Foundation for Industry Stakeholders", *Automation in Construction*, 18(3), 357–375.

Tse, T.-C. K., Wong, K. A. and Wong, K. F. (2005). "The Utilisation of Building Information Models in nD Modelling: A Study of Data Interfacing and Adoption Barriers", *Journal of Information Technology in Construction*, 10(Special Issue From 3D to nD modelling), 85–110.

Wang, X., Love, P. E., Kim, M. J., Park, C.-S., Sing, C.-P. and Hou, L. (2013). "A Conceptual Framework for Integrating Building Information Modeling With Augmented Reality", *Automation in Construction*, 34, 37–44.

Wilkins, J. R. (2011). "Construction Workers' Perceptions of Health and Safety Training Programmes", *Construction Management and Economics*, 29(10), 1017–1026.

Won, J., Lee, G., Dossick, C. and Messner, J. (2013). "Where to Focus for Successful Adoption of Building Information Modeling Within Organization", *Journal of Construction Engineering and Management*, 139(11), 04013014.

Wu, P., Wang, J. and Wang, X. (2016). "A Critical Review of the Use of 3-D Printing in the Construction Industry", *Automation in Construction*, 68, 21–31.

Yan, H. and Damian, P. (2008). "Benefits and Barriers of Building Information Modelling", in *The 12th International Conference on Computing in Civil and Building Engineering*, Beijing, 1–5.

Young, N. W., Jones, S. A. and Bernstein, H. M. (2008). *Building Information Modeling (BIM)-Transforming Design and Construction to Achieve Greater Industry Productivity*, Bedford, MA: McGraw-Hill Construction.

Young, N. W., Jones, S. A., Bernstein, H. M. and Gudgel, J. (2009). *The Business Value of BIM-Getting Building Information Modeling to the Bottom Line*, Bedford, MA: McGraw-Hill Construction.

Zhao, X. (2017). "A Scientometric Review of Global BIM Research: Analysis and Visualization", *Automation in Construction*, 80, 37–47.

Zhao, X., Feng, Y., Pienaar, J. and O'Brien, D. (2017). "Modelling Paths of Risks Associated With BIM Implementation in Architectural, Engineering and Construction Projects", *Architectural Science Review*, 60(6), 472–482. doi: 10.1080/00038628.2017.1373628.

Zhao, X., Hwang, B. G. and Low, S. P. (2015). "Enterprise Risk Management in International Construction Firms: Drivers And Hindrances", *Engineering, Construction and Architectural Management*, 22(3), 347–366.

Chapter 4

Automatic Detection of Positions and Shapes of Various Objects at Construction Sites from Digital Images Using Deep Learning

Naoto Nishimura*, Nobuyoshi Yabuki[†] and Tomohiro Fukuda[‡]

*Division of Sustainable Energy and Environmental Engineering,
Graduate School of Engineering, Osaka University,
Suita, Osaka, Japan*

**nishimura@it.see.eng.osaka-u.ac.jp*
[†]yabuki@see.eng.osaka-u.ac.jp
[‡]fukuda@see.eng.osaka-u.ac.jp

Abstract

At construction sites, many pictures are taken for inspection and management of construction. Objects such as construction machinery, signages, signboard, construction workers, etc., are captured, but their existence and locations must be manually detected by humans. Thus, it is desirable to automate the object detection process in order to improve the efficiency. For detection of objects from digital images, machine

---

*Corresponding author.

learning with feature values of images has generally been employed. However, this method requires determination of features and contents, which is learned by humans, taking much labour and making it difficult to achieve satisfactory accuracy. On the other hand, deep learning can automatically determine these features and contents of various objects from digital images so that it can reduce labour and can increase accuracy compared to the conventional machine learning. Therefore, this research aims to automatically detect positions and shapes of objects from digital images by using deep learning. First, a dataset is created and Single Shot Multibox Detector (SSD) is employed for an object detection algorithm to detect positions. Next, re-detection of positions is performed by fine-tuning the weights of SSD. In addition, using detected object information can improve the efficiency of filing images. The shape of an object can be assumed using fully convolutional network (FCN). In this research, construction machinery, workers and signages were detected from digital images taken at construction sites by the proposed method. The proposed method has shown better performance compared to the conventional machine learning method. Finally, object position detection and shape detection are overlapped and the result shows the visually detailed object detection.

Keywords: Deep Learning; Construction Machinery; Detection; Artificial Intelligence; Single Shot Multibox Detector (SSD); Fine-tuning.

1. Introduction

Many pictures are taken at construction sites, and they are used for inspection and management of construction. These pictures include objects such as construction machinery, signages, signboard, construction workers, etc. In order to use these pictures for inspection and management of construction, these objects must be manually detected by humans. In order to improve the efficiency, the detection process must be automatically conducted.

In object detection on an image, machine learning using feature values of the image has often been employed. Feature values used for object detection include local feature values, such as histograms of oriented gradients (HOGs) feature (Dalal and Triggs, 2005). The accuracy of object detection using HOG feature is high. Also, the detection results for specific objects such as people and faces are accurate. Recently, object detection by artificial intelligence has been attracting attention, whereas improvement on the accuracy using HOG feature to detect objects is actively executed. The initial purpose of HOG feature was detection of people (Dalal and Triggs, 2005). Studies on object detection using HOG feature have been actively conducted. By combining HOG feature and Support Vector Machine (SVM), the object detection accuracy increased (Pang *et al.*, 2011). Recognition of road marks using HOG feature and SVM was conducted and the recognition accuracy was more than 80% (Ouerhani *et al.*, 2017). Cascaded AdaBoost was used in addition to HOG feature and SVM to reduce the failure rate of object detection and to improve the accuracy (Liu *et al.*, 2013). However, in the object detection of conventional machine learning using the feature values, only the designated feature values can be used which are determined by humans and misdetection may occur due to a single feature value. Therefore, it takes much labour and it is difficult to achieve satisfactory accuracy. On the other hand, deep learning using neural networks has attracted attention in recent years. Deep learning is an extension of machine learning, and features used for learning are automatically determined. Thus, labour can be reduced and accuracy can be improved. Pedestrian face detection was conducted using deep learning (Rowley *et al.*, 1998). In addition, recognition of human behaviour was successfully done by using 3D convolution (Ji *et al.*, 2013). In the field of deep learning, many datasets for object recognition and detection, including people, animals, vehicles, and objects in living environment, such as TVs, tables and chairs, etc., are provided from datasets such as PASCAL VOC, COCO, and ILSVRC, etc. Proposals for improving detection accuracy using these provided datasets have actively progressed, e.g. Regions with CNN features (R-CNN) (Girshick *et al.*, 2015), Fast Region-based Convolutional Network (Fast r-cnn) (Girshick, 2015), Faster R-CNN (Ren *et al.*, 2015), You Only Look Once (YOLO) (Redmon *et al.*, 2016) and Single Shot Multibox Detector (SSD) (Liu *et al.*, 2016). However, object detection using deep learning has not been heard much, where a new dataset is created and utilized in a specific field. As for the object shape detection, it has been widely used in the medical field to grasp the shape of the cell and the present condition of caries, such as U-Net (Ronneberger *et al.*, 2015). Seg-Net is used for dividing (segmenting) the shapes of objects such as cars and persons for automatic driving (Badrinarayanan *et al.*, 2015). However, in the object detection in the civil engineering field, especially construction machinery, deep learning has not been employed so far.

Therefore, this research aims to detect objects from digital images by using deep learning. In this research, object position detection and object shape detection are executed. In the object position detection, construction machinery, workers and signages were detected from digital images, and position and type of the object were detected using a rectangle. The proposed method of position detection has shown better performance compared to the conventional machine learning method. Since a rectangle is used for only object position detection, the detailed shape cannot be detected. Therefore, object shape detection was also cavered in this research. Only the construction machine is detected in object shape detection. Finally, by overlapping the results of object position detection and shape detection, a detailed object detection is executed.

2. Method

2.1. *Overview of the method*

The proposed method is to detect an object by overlapping both results of object position and shape detection. The flow of object position detection is shown in Fig. 1. SSD is employed for an object detection algorithm to detect positions. Next, the correct answer labels to re-learn (fine-tune) SSD are created. ImageNet-Utils is used to create learning data labels (Lin and Kao, 2015–2016). Finally, the created dataset is used to re-learn the SSD. By using the new weight acquired by fine-tuning, the object position on the digital image is determined. The flow of object shape detection is

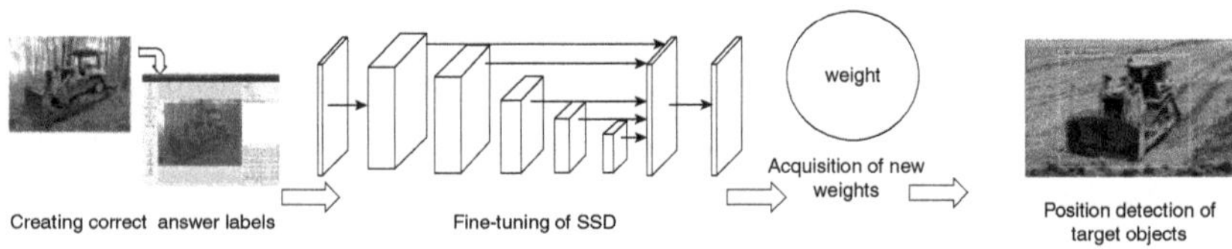

Fig. 1.   Overview of object position detection.

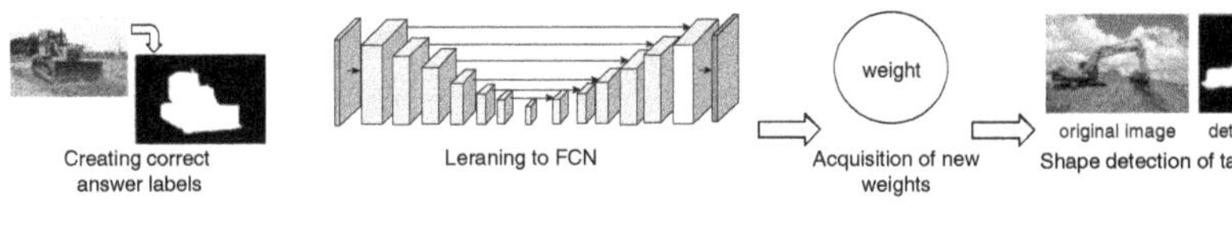

Fig. 2.   Overview of object shape detection.

shown in Fig. 2. For shape detection, the dedicated neural network is built, and the object shape is detected by carrying out segmentation of the object on the digital image. First, a mask image as a correct answer label is created for each image of a dataset. Next, the dataset and correct answer labels are included in the neural network. It outputs the object shape as a binary image by using the learned result. Finally, the object position detection result and the object shape detection result are overlapped, and the detection result is confirmed visually.

In this research, Keras is used as a machine learning library. It is one of machine learning libraries written in Python. Both TensorFlow and Theano can be used as the backend with Keras. TensorFlow, developed and provided by Google Inc., is used in object position detection as the backend. Theano, developed and provided by Montreal Institute for Learning Algorithms (MILA) at University of Montreal, is used in object shape detection as the backend.

2.2. *Object position detection*

2.2.1. *Single shot multiBox detector (SSD)*

SSD is an object detection algorithm using a simple network built in 2016 by Liu *et al.* SSD is faster than conventional object detection algorithms. A simplified network model of SSD is shown in Fig. 3. VGG-16 (Fig. 3(a)) (Simonyan and Zisserman, 2015), which is a learning model of image recognition, is used for the base network of SSD. After this base network, by using hierarchical feature maps (Fig. 3(b)), various scale objects can be dealt with, and the accuracy of detection can achieve high rate. The reason is that it identifies each aspect ratio of the object. For these reasons, the SSD can detect a target object even with a relatively low-resolution image. Even if digital images taken at a construction site are of high resolution, a

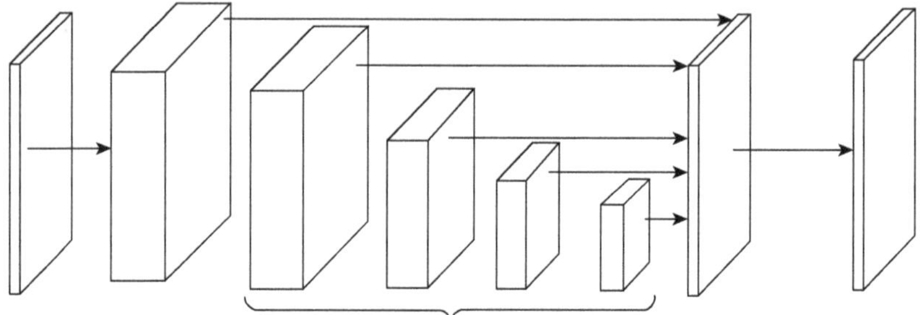

Fig. 3.   Simplified network image model of SSD (created by author referring to Fig. 2 of Liu *et al.*, 2016).

low-resolution object exists in the distance of images. The accuracy of object detection of SSD is better than the accuracy of R-CNN, Fast R-CNN, Faster-RCNN and YOLO. Therefore, it is considered that SSD is optimal for object detection on digital images taken at a construction site.

2.2.2.   Creation of correct answer label for object position detection

In order to detect an object from a digital image using machine learning, it is necessary to know where and what is there. Hence, a bounding box and its coordinate are required to create the correct answer labels. In addition, correct answer labels must exist with the original digital images. They are created by using ImageNet-Utils. This tool was built by Lin and Kao to create a correct label called annotation data. The annotation data includes information such as the name of the original image and coordinates of a bounding box (Horizontal and vertical coordinates on the image). First, image data to be learned is opened. After that, the coordinates of an upper left corner and a lower right corner of the object are determined, which would be detected. It is possible to create bounding boxes of a plurality of objects for one image (Fig. 4). The created annotation data is saved as XML format as shown in Fig. 5.

2.2.3. Fine-tuning and object position detection

The created correct label and the original image are learned and stored in SSD. When learning is completed, a new set of weights is acquired.

Fig. 4.   Example of creating correct answer labels (Annotation data). It also shows that bounding boxes of multiple objects can be created.

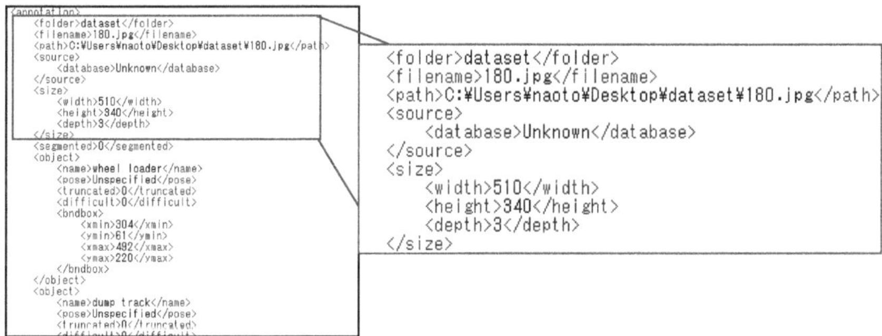

Fig. 5.   Example of contents of annotation data.

An appropriate set of weights is selected from the newly acquired weights and used for object detection. The selection of an appropriate set of weights is determined from the transition of the loss coefficient acquired by fine-tuning (Fig. 6). In the case of Fig. 6, since the increase in the loss coefficient (validation loss coefficient) in the test data at Epoch 27–28 is the largest and the training data coefficient is in a stable state, the weight at that point is selected.

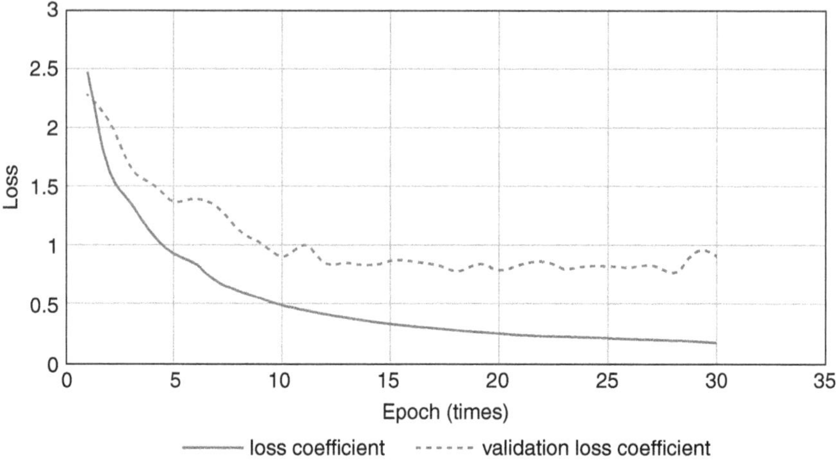

Fig. 6. Example of transition of loss coefficient (loss coefficient means the training data loss, validation coefficient means testing data's loss).

2.3. *Object shape detection*

2.3.1. *Building of N-layer convolution neural network*

For object shape detection, fully convolutional network (FCN) was used. In FCN, a fully connected layer is not prepared, and a final prediction result is generated as an image. FCN is frequently used in medical and automobile fields. U-Net has been successful in grasping cell movement and the present condition of caries and Seg-Net has been successful in segmentation by an object such as people, automobiles, buildings and sky. In this research, a parallel type FCN was built. The network model is shown in Fig. 7. The network has shortcut structures which improve the prediction accuracy by using the feature map in the upper layer and the lower layer. In the upper layer (Fig. 7 (a)), a pooling layer is provided after the convolution layer, i.e. layer for extracting the representative value of a feature map. In the lower layer (Fig. 7(b)), an upsampling layer (a layer for enlarging the feature map and having a reduced feature map by the pooling layer) is provided after the convolution layer. Dice coefficient was used as loss function (Eq. (1)). The dice coefficient expresses the similarity between two sets of an input value (X) and an output value (Y). If the similarity

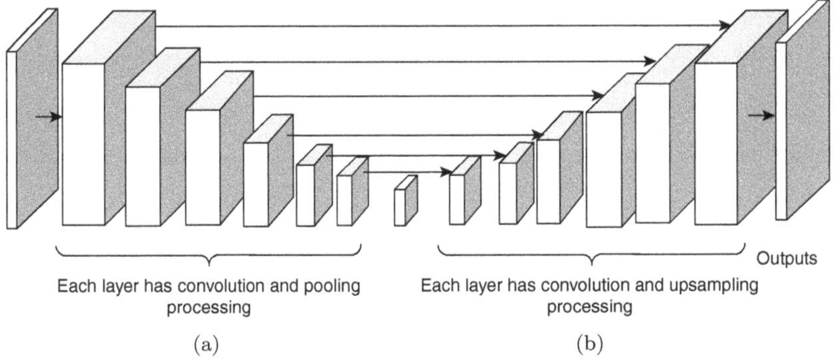

Fig. 7.   Network image model of FCN.

Fig. 8.   Example of original digital image and correct label image for shape detection.

between X and Y is high, it approaches 1, else it approaches 0. The sigmoid function is used for the activation function.

2.3.2. *Creating correct answer labels for object shape detection*

First, an image as a correct label was manually created using an image editing software. It is called a mask image which is binary. In the images, the object to be detected is white and the other objects are black (Fig. 8). It is created for each dataset. In general, the dataset is randomly extended with the processing of rotation, horizontal/vertical movement, shearing, etc. Then, the created dataset is learned and stored in the network. When learning is finished, the weight for shape detection is selected similar to that meathared in Section 2.3.1.

3. Experiment

Construction machinery (backhoes, bulldozers, dump trucks and wheel loaders), workers and signages were selected as detection objects, and the accuracy of object position detection was verified. Construction machinery was selected for object shape detection as well. Since construction machineries are similar in shape, the verification was conducted with the construction machinery in one unit. Detection accuracy is evaluated based on the criteria shown in Table 1. Learning environment in this research is shown in Table 2. The number of images for verification is shown in Table 3. Each learning dataset for object, the position detection and the object shape detection is different.

Table 1.   Evaluation criteria for detection accuracy.

Types of Detection	Position Detection	Shape Detection
Positive detection	• Both position and object type are correct	• The outline of the object can be recognized correctly (perfect) • Shape of the object is roughly recognized (good)
Negative detection	• The position is correct, but the type of object being detected is different • The type of the object being detected is correct, but the position is different	• Outline of the object cannot be recognized correctly • Detecting the shape of things that are not related at all
Not detected	• It has not been detected at all	• It has not been detected at all

Table 2.   Learning environment.

Item	Type	Note
GPU	GeForce GTX 1080 Ti	Memory speed: 11 Gbps Memory: 11 GB
CPU	Core i7-7700	TDP: 3.60 GHz
OS	Ubuntu 14.04	Linux based operating system. GPGPU becomes possible.
Machine learning library	Keras ver.1.2.2	Both TensorFlow and Theano can be used as backend.

Table 3.   Number of digital images collected for verification.

Types of Construction Machinery	Number of Images
Backhoe	165
Bulldozer	67
Dump truck	94
Wheel loader	174

Table 4.   Number of digital images for learning (SSD).

Types of Construction Machinery	Number of Images
Backhoe	557
Bulldozer	340
Dump truck	507
Wheel loader	349
Signage	242
Worker	217

3.1. *Experiment of object position detection*

3.1.1. *Detection target and number of images*

Table 4 shows the number of images used for learning. These images were created from movies and downloaded from pixabay (Braxmeier and Steinberger, 2017) and Flicker (Ludicorp, 2002). The number of images indicates the number of objects in which the target object is present since it contains objects with multiple objects in one image.

3.1.2. *Result of object position detection*

Detection accuracy verification for each object was executed using an image in which the object is single. Image example of detection results of each object (Fig. 9), the object detection result (Table 5) and the accuracy using test data (Fig. 10) are shown. The image shown as an example can be used for commercial purposes. Detection accuracy is 82.4% for backhoes, 68.7% for bulldozers, 83.0% for dump trucks and 68.4% for wheel

Fig. 9.   Digital image example of position detection result (backhoe, bulldozer, dump truck, wheel loader, signage and worker are shown. Positive means correct detection and Negative means misdetection).

Table 5.   Result of object position detection.

Types of Construction Machinery	Number of Images	Not Detected	Positive Detection	Negative Detection
Backhoe	165	4	136	25
Bulldozer	67	6	46	15
Dump truck	94	4	78	12
Wheel loader	174	6	119	49

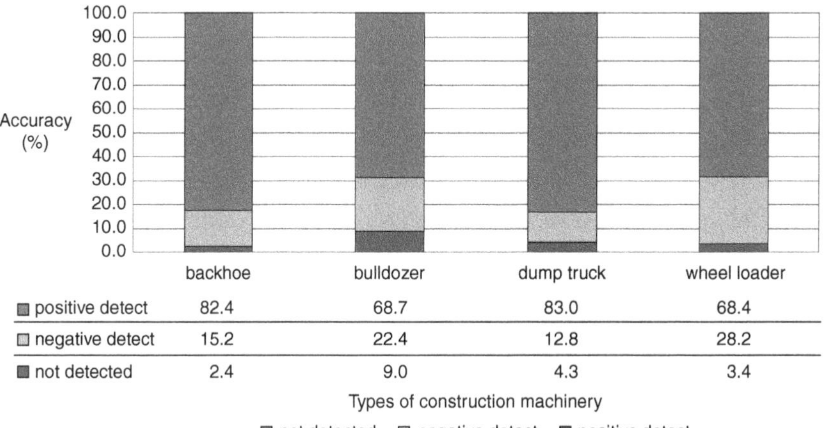

Fig. 10.   Detection accuracy of object position detection.

loaders. Since it was difficult to collect images of workers and signages which include only a single object, the accuracy is not shown. In addition, the number of digital images of workers and signages in which objects were detected was low. As for image arrangement, it depends on the image detection result in Fig. 10.

3.2. *Experiment of object shape detection*

3.2.1. *Detection target and the dataset*

In this research, the object shape detection was executed only for construction machinery. The shape was not detected by each construction machinery, but shape detection was executed as a unit of one construction

Table 6. Number of digital images used for learning (FCN).

Types of Construction Machinery	Number of Images	Total (Number of Images × 40)
Backhoe	25	1000
Bulldozer	38	1520
Dump truck	38	1520
Wheel loader	48	1920
Total	151	5960

machine. The number of images for learning is shown in Table 6. In this research, the number of datasets is increased. Processing such as rotation, horizontal/vertical movement, and shearing, etc., was done randomly, and the dataset was extended by 40 times.

3.2.2. *Result of object shape detection*

Examples of images and accuracy of object shape detection results are as follows. Examples of original images and positive and negative detection results are shown in Fig. 11. Detection result and the accuracy are shown in Table 7 and Fig. 12. The accuracy of positive detection was high, 67.9% for backhoes, 76.1% for bulldozers, 71.3% for dump trucks and 72.4% for wheel loaders in total. On the other hand, among the positive detections, the result of judgment of "perfect" was low.

3.3. *Overlapping experiment of position and shape detection*

Overlapped images with the results of Sections 3.1 and 3.2 created manu-ally by using image editing software are shown in Fig. 13. Images used for superimposition are all determined to be positive detection in Sections 3.1 and 3.2. As for Section 3.2, images determined as "perfect" in positive detection were used. The results in Section 3.1 were detected only with a rectangle, but by combining the results in Section 3.2, it is possible to

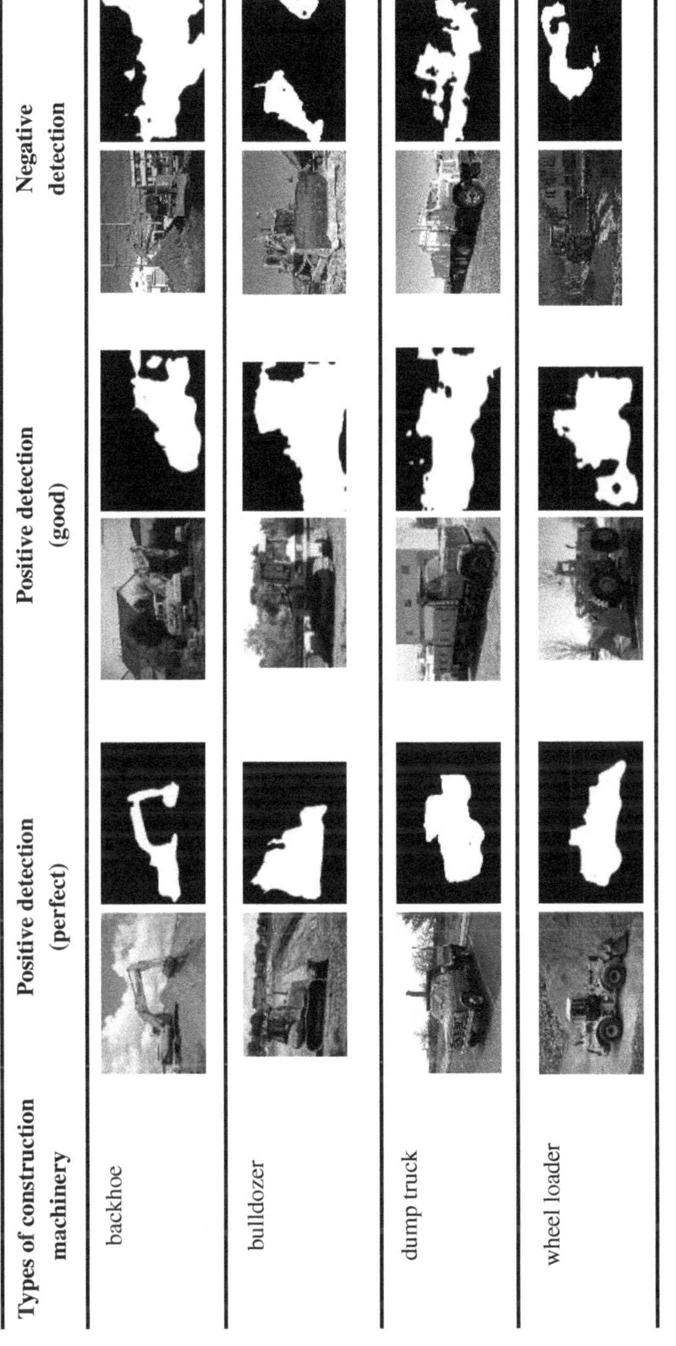

Fig. 11. Digital image example of object shape detection result (backhoe, bulldozer, dump truck and wheel loader, are shown. Positive means correct detection and Negative means misdetection).

Table 7. Result of object shape detection.

Types of Construction Machinery	Number of Images	Not Detected	Positive Detection (Perfect)	Positive Detection (Good)	Negative Detection
Backhoe	165	None	47	65	53
Bulldozer	67	None	17	34	16
Dump truck	94	None	32	35	27
Wheel loader	174	None	45	81	48

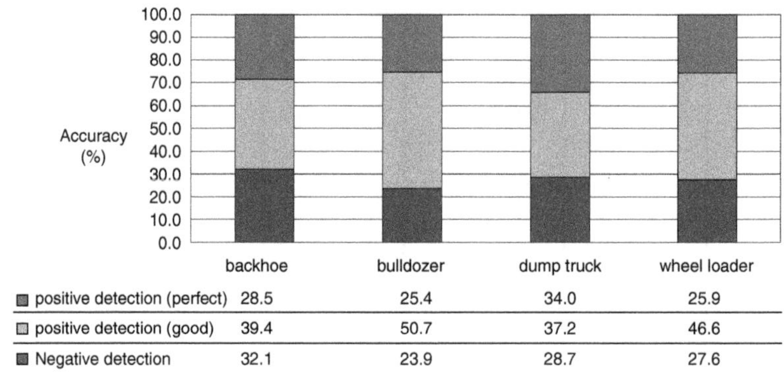

Fig. 12. Detection accuracy of object shape detection.

simultaneously determine the position, type and shape of the construction machinery. It is considered that the accuracy of appearance has improved.

4. Comparative Experiment

4.1. *Comparison of object position detection with conventional machine learning*

4.1.1. *Conventional machine learning implementation and number of images*

In this research, object detection by HOG feature and SVM was executed using dlib, which is a machine learning library, in order to compare with

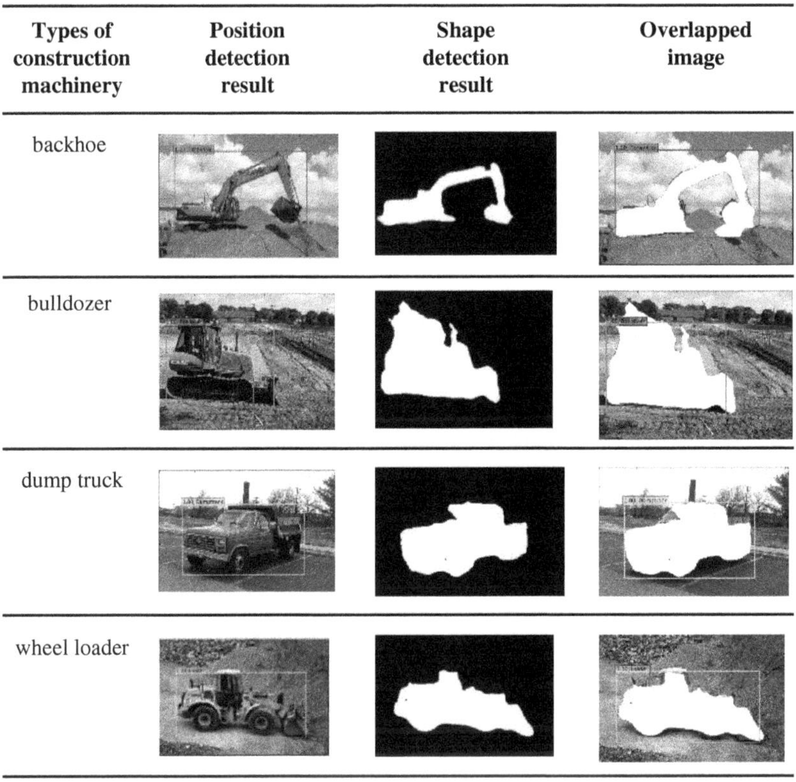

Types of construction machinery	Position detection result	Shape detection result	Overlapped image
backhoe			
bulldozer			
dump truck			
wheel loader			

Fig. 13.   Result of overlapping with position detection result and shape detection result.

the proposed method. The reason for using dlib is that objects can be detected with the same creation method of the correct labels as SSD. In object detection using dlib, if the aspect ratio of the bounding box created as a correct label is greatly different as shown in Fig. 14, it cannot be used for learning. Therefore, the number of data was reduced. This result was compared with the construction machines (backhoes, bulldozers, dump trucks and wheel loaders), and the data used are shown in Table 8.

4.1.2. *Comparison of detection results of conventional machine learning and depth learning*

The detection result image examples by using convolutional machine learning (CML) and deep learning (DL) are shown in Fig. 15. In CML,

Fig. 14.   Example of difference in aspect ratio of correct label (shown with backhoe).

Table 8.   Number of digital images for comparative verification between conventional machine learning and deep learning (reduced from Table 3 to use dlib).

Types of Construction Machinery	Number of Images
Backhoe	223
Bulldozer	306
Dump truck	295
Wheel loader	289

dlib with HOG and SVM was used for detection, whereas in DL, SSD was used for detection as in Section 3.1. Table 9 shows the comparison of detection results between CML and DL, and Fig. 16 shows the comparison of the accuracy. The accuracy of the method using DL in the construction machine excluding wheel loaders was higher than that using the CML method. In particular, there was a detection accuracy difference of 44.8% in backhoes, 49.3% in bulldozers and 47.9% in dump trucks. On the other hand, in the result of wheel loaders in which detection accuracy in DL was lower than CML, many cases were erroneously detected as dump trucks and bulldozers.

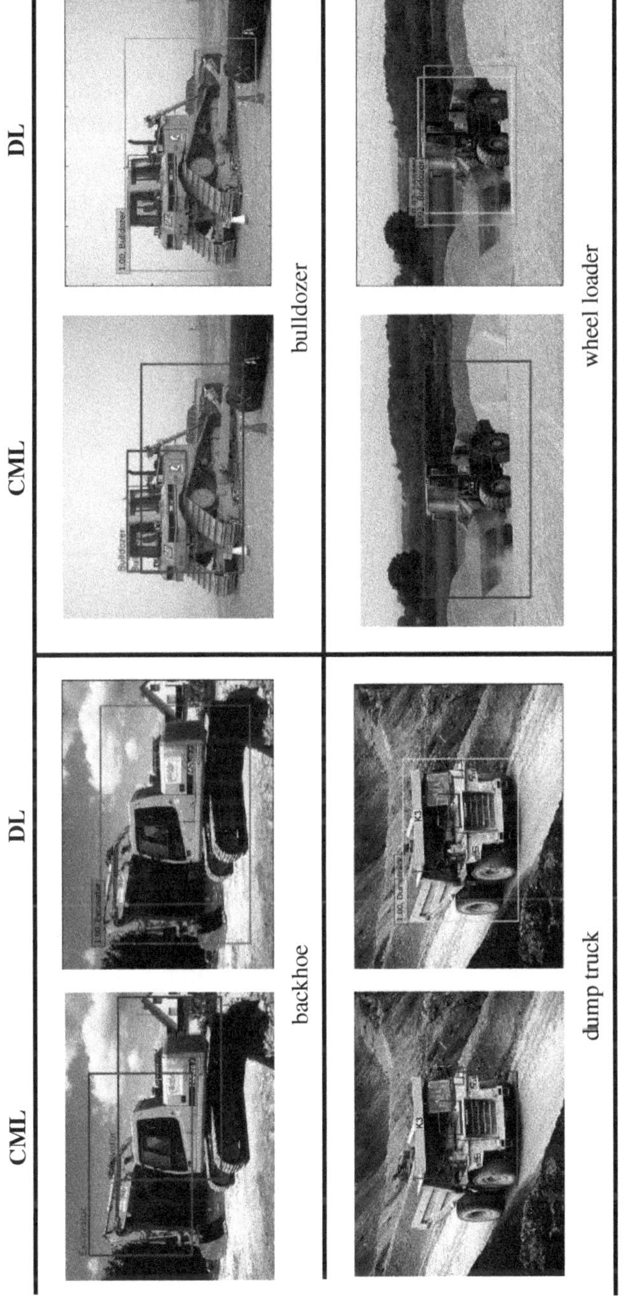

Fig. 15.   Digital image example of object position detection of CML and DL.

Table 9. Result of object position detection of CML and DL.

Types of Construction Machinery	Method	Number of Images	Number of Images Which Objects Were Detected	Not Detected	Positive Detection	Negative Detection
Backhoe	CML	165	112	53	56	56
	DL	165	161	4	130	31
Bulldozer	CML	67	46	21	17	29
	DL	67	66	1	50	16
Dump truck	CML	94	77	17	29	48
	DL	94	91	3	74	17
Wheel loader	CML	174	134	40	114	20
	DL	174	172	2	59	113

5. Conclusion

In this research, construction machines (backhoes, bulldozers, dump trucks and wheel loaders) were detected by using deep learning. Also, workers and signages were detected, but the number of digital images of workers and signages in which objects were detected was low. So the accuracy and the comparison detection between CML and DL of workers and signages were not conducted. In the detection of construction machines using deep learning, the detection accuracy improved over conventional machine learning. Specifically, this research has shown that the detection accuracy improved by 47.3% as a whole excluding the wheel loader in which the accuracy was not so good. Moreover, it was shown that anyone can easily detect the target object by using the deep learning algorithm. In the proposed method, it is possible to detect objects easily by creating a new dataset. However, the detection accuracy depends on the data amount and its quality. Even if the amount of learning data is large, if the quality is not appropriate, the detection accuracy will not be high. On the other hand, if the amount of learning data is small, the detection accuracy will not improve even if the quality is appropriate. In this

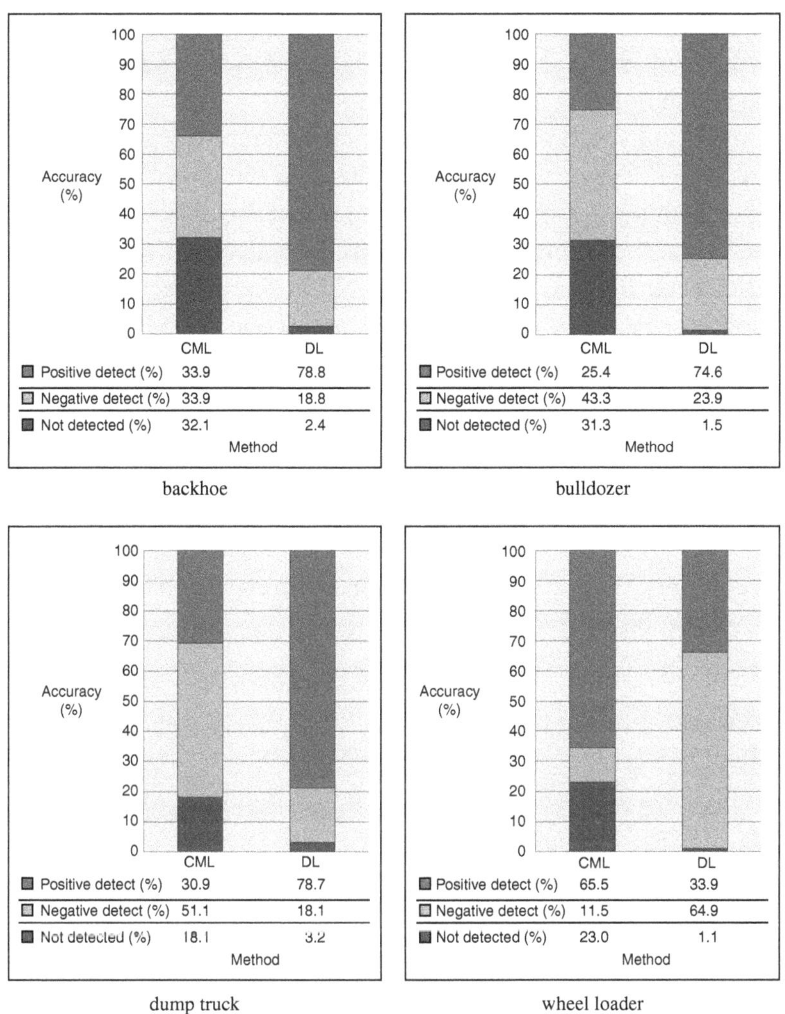

Fig. 16. Comparison of detection accuracy between CML and DL.

research, the accuracy of the detection result in the wheel loaders can be cited as an example. The reason is that the data used for learning with wheel loader were similar to dump trucks and bulldozers in many cases. It is considered that the results of this research can contribute to improving the efficiency of filing of data at the construction site in the future. In addition, because SSD can detect objects on moving images, it helps in the monitoring of construction sites. Moreover, if combined with the monitoring of sensors, etc., it is possible that this result will lead to safer and smoother operation of a construction site.

In the result of object shape detection different from the result of object position detection, the detailed shape of the object was comprehended. In addition, it was shown that a more detailed object detection can be executed by combining the result of this shape detection with the result of position detection. It is considered that object shape detection is applied to photographs taken for use in structure from motion (SfM). If the result is used as a mask image, it will be useful for improving the accuracy of model generation.

Future works include detection of other types of construction machinery and improving the detection accuracy by enhancing the quality of data and increasing the amount of data as well as site verification. In terms of workers and signages, increasing the number of digital images for verification, clarifying the accuracy of detection and comparison between CML and DL should be executed.

References

Badrinarayanan, V., Alex, K. and Roberto, C. (2015). "SegNet: A Deep Convolutional Encoder-Decoder Architecture for Image Segmentation", preprint, arXiv:1511.00561.

Braxmeier, H. and Steinberger, S. (2017). "Pixabay". Available at: https://pixabay.com/ (Accessed on September 10, 2017).

Dalal, N. and Triggs, B. (2005). "Histograms of Oriented Gradients for Human Detection", in *Proceedings of Computer Vision and Pattern Recognition,* San Diego, CA, pp. 886–893.

Girshick, R., Donahue, J., Darrell, T. and Malik, J. (2014). "Rich Feature Hierarchies for Accurate Object Detection and Semantic Segmentation", in *Proceedings of the IEEE Conference on Computer vision and pattern recognition*, Columbus, OH, pp. 580–587.

Girshick, R. (2015). "Fast r-cnn", in *Proceedings of the IEEE International Conference on Computer Vision,* Santiago, Chile, pp. 1440–1448.

Ji, S., Xu, W., Yang, M. and Yu, K. (2013). "3D Convolutional Neural Networks for Human Action Recognition", *IEEE Transactions on Pattern Analysis and Machine Intelligence*, 35(1), 221–231.

Lin, T. and Kao, W. (2015–2016). "ImageNet-Utils", Github. Available at: https://github.com/tzutalin/ImageNet_Utils (Accessed on September 12, 2017).

Liu, H., Xu, T., Wang, X. and Qian, Y. (2013). "Related HOG Features for Human Detection Using Cascaded Adaboost and SVM Classifiers", in *Proceedings of International Conference on Multimedia Modeling*, pp. 345–355.

Liu, W., Anguelov, D., Erhan, D., Szegedy, C., Reed, S., Fu, C.Y. and Breg, A.C. (2016). "SSD: Single Shot MultiBox Detector", in *European Conference on Computer Vision*, Amsterdam, Netherlands: Springer, pp. 21–37.

Ludicorp (2002). "CC Photo Search in Flicker", *Flicker*. Available at: https:// ccphotosearch.com/ (Accessed on September 10, 2017).

Ouerhani, Y., Ayman, A. and Christian, B. (2017). "Road Mark Recognition Using HOG-SVM and Correlation", in *Proceedings of Optics and Photonics for Information Processing XI*, San Diego, CA, Vol. 10395. https://doi. org/10.1117/12.2273304.

Pang, Y., Yuan, Y., Li, X. and Pan, J. (2011). "Efficient HOG Human Detection", *Signal Processing*, 91(4), 773–781.

Redmon, J., Divvala, S., Girshick, R. and Farhadi, A. (2016). "You Only Look Once: Unified, Real-time Object Detection", in *Proceedings of the IEEE Conference on Computer Vision and Pattern Recognition*, Las Vegas, NV, pp. 779–788.

Ren, S., He, K., Girshick, R. and Sun, J. (2015). "Faster R-CNN: Towards Real-time Object Detection with Region Proposal Networks", in *Proceedings of Advances in Neural Information Processing Systems*, Montreal, Canada, pp. 91–99.

Ronneberger, O., Fischer, P. and Brox, T. (2015). "U-Net: Convolutional Networks for Biomedical Image Segmentation", *Medical Image Computing and Computer-Assisted Intervention 2015*, Munich, Germany: Springer, pp. 234–241.

Rowley, H, A., Baluja, S. and Kanade, T. (1998). "Neural Network-based Face Detection", *IEEE Transactions on Pattern Analysis and Machine Intelligence*, 20(1), 23–38.

Simonyan, K. and Zisserman, A. (2015). "Very Deep Convolutional Networks for Large-Scale Image Recognition", in *Proceedings of International Conference on Learning Representation*, San Diego, CA, CoRR, arXiv: 1409.1556v6 [cs.cv].

Chapter 5

Daylight Assessment and Energy Consumption Analysis at an Early Stage of Designing Residential Buildings Integrating BIM and LCA

Mohammad Najjar and Assed Haddad*

Programa de Engenharia Ambiental,
Universidade Federal do Rio de Janeiro, Rio de Janeiro, Brazil
**assed@poli.ufrj.br*

Karoline Figueiredo

Departamento de Construção Civil,
Universidade Federal do Rio de Janeiro, Rio de Janeiro, Brazil

Abstract

Building information modelling (BIM) tool provides a distinctive way of observing and estimating energy consumption and daylight analysis in buildings. Integrating this tool with life-cycle assessment (LCA) empowers the decision-making process and sustainable building design in the construction sector. This chapter presents the interests of BIM–LCA integration in examining different design alternatives and orientations in

*Corresponding author.

order to increase daylight efficiency and energy performance in buildings at an early designing phase. This work encourages the improvement of construction options by showing the methodology and tools capable of facilitating decision-making in the designing phase of sustainable projects. It evaluates the LCA methodology of residential buildings based on ISO 14040 guidelines within the available database. The methodology of this chapter aims to conduct a conceptual energy consumption analysis using Autodesk Green Building Studio, a plug-in that allows designers to perform building performance simulations in a cloud-based service to optimize energy efficiency, testing different possibilities for the construction. Besides, this work uses the advantages of Autodesk Revit software to assess the impact of natural daylight analysis using LEED daylight plug-in. The results show that BIM–LCA integration is considered as an optimistic course in terms of sustainable development and decision-making process in the construction sector. Furthermore, it encourages reviewing some critical factors, such as building orientation, HVAC systems and the construction of external walls and roofs in the construction projects at an early stage of design in order to increase energy efficiency and capture natural daylight in buildings.

Keywords: Building Information Modelling; Sustainability in Construction; Life-cycle Assessment; Daylight Analysis; Energy Consumption.

1. Introduction

The construction industry involves activities that consume energy and natural resources. It is known as "the industry of 40%" (Lasvaux, 2010). This comes back to the fact that the life cycle of buildings produces nearly 40% of CO_2 emissions, 40% of waste generation and consumes 40% of natural resources (Lassio *et al.*, 2016; Kwok Wai Wong and Zhou, 2015). The world is witnessing an increasing concern in the field of energy efficiency, particularly non-renewable energy. Thus, advanced solutions are required to achieve the sustainability standards in this field, particularly in serious circumstances such as deteriorating of natural resources and insufficient criteria to protect the environment (Šaparauskas and Turskis, 2006). Such circumstances are affecting the surrounding environment and energy consumption over the entire life-cycle assessment (LCA) of construction materials (Buyle *et al.*, 2013; Gustavsson *et al.*, 2010).

Several tools and methods have been assessed to support the implementation of sustainable strategies in the built environment

(Kang, 2015; Azhar *et al.*, 2011; Alwan *et al.*, 2017). One of these methods is LCA methodology, which is considered as a complete process to evaluate the sustainability of buildings over their entire lifespan (Asdrubali *et al.*, 2013). It is highly important to declare that the number of publications in the field of environmental LCA studies have increased significantly after the release of ISO 14040 series. The number of publications that could be found in Scopus up to 2011 was only 88 whereas this figure increased up to 264 publications in 2015 (Anand and Amor, 2017). Some works highlighted the importance of improving the application of LCA in the construction sector (Martínez-Rocamora and Solís-Guzmán, 2016; Huang *et al.*, 2015; Soust-Verdaguer *et al.*, 2016). However, applying LCA in the construction sector requires an integration with building tools (Anand and Amor, 2017). In this discipline, building information modelling (BIM) is being discussed as a building tool that optimizes the application of LCA (Soust-Verdaguer *et al.*, 2016). It provides opportunities to estimate the energy consumption of buildings in the designing stage (Jrade and Jalaei, 2014; Antón and Díaz, 2014) and empowers the decision-making process (Shafiq *et al.*, 2015; Peng, 2015). Despite these publications, it can be recognized that a gap lies in the insufficient methodological framework in the field of BIM and LCA integration. This part of the study needs to be addressed in a comprehensive way to support the decision-making process in the construction sector.

The novelty of this work is to present the interests of integrating BIM tools with LCA methodology at early design stages in order to empower the decision-making process and sustainable design procedure in the construction sector. This work reviews LCA from a building perspective and examines the integration process at designing residential buildings. It aims to evaluate the benefits of such integration to conduct a daylight analysis using LEED daylight plug-in in Revit and estimate the consumption of energy at the operating phase based on different design alternatives and orientations using Autodesk Revit and Autodesk Green Building Studio. In this discipline, design alternatives mean modifying parameters of HVAC systems, and construction of external walls and roofs, while design orientations mean rotating buildings in different directions with the intention of achieving the objectives. This work applies to one of the typical multi-storied residential buildings in Brazil, recognized as a low- profile construction building, as a case study to validate the methodological framework of BIM–LCA integration and achieve the objectives of the research.

2. Methodological Framework of BIM–LCA Integration

2.1. *Decision support analysis*

Integrating BIM and LCA is an urgent process that achieves sustainability standards (Antón and Díaz, 2014) and protects the environment (Eleftheriadis *et al.*, 2017). This process depends on exchanging data between BIM software and LCA application (Soust-Verdaguer *et al.*, 2017). It gives the opportunity to estimate energy performance, evaluate environmental impacts and empower the decision-making process in the construction sector (Shadram *et al.*, 2016). This work applies the methodological framework of LCA based on ISO 14040 as shown in Fig. 1 (UNEP, 2012): Goal and Scope, Life-Cycle Inventory (LCI), Life-Cycle Impact Assessment (LCIA), and Interpretations. Besides, it considers the different models, taxonomies and classifications of BIM framework (Porwal and Hewage, 2013; Succar, 2009), based on the main axes of BIM domain as shown in Fig. 1 (Succar, 2009). However, Fig. 1 shows that the first step at an early designing stage of a construction project is to determine design parameters. This means to recognize BIM fields by conducting the required clusters, interactions and overlaps between the different players of the applied technologies, processes and policies. Moreover, it means to define the goal and scope of the construction project by stating the functional unit, system boundary and the set of building materials (UNEP, 2012).

The next step is to state the performance parameters of the study by classifying BIM stages that reflect the maturity level of BIM implementation (Succar, 2009). BIM stages are divided into different steps, such as inventory database (LCI), 3D modelling, collaboration and integration of design, considering the local environmental data of the construction project. In this discipline, LCI is the most challenging step of LCA methodology due to the difficulty of accumulating reliable and relevant data (UNEP, 2012). However, the performance parameters step considers the

Lifecycle Stages

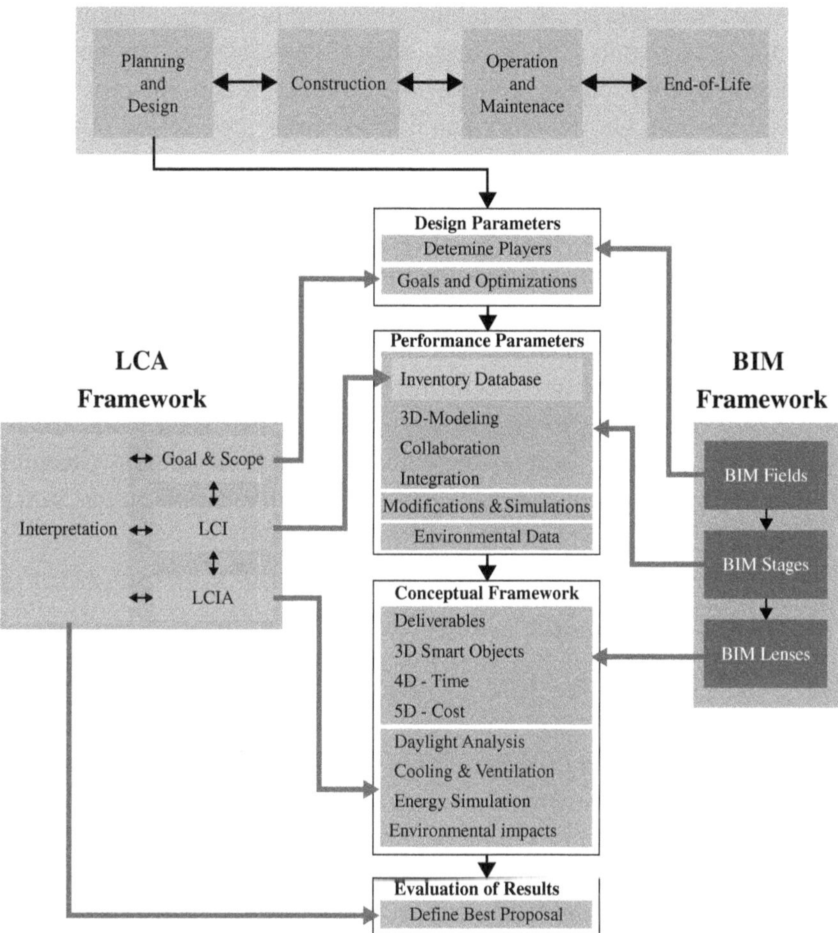

Fig. 1.   Decision support analysis.

possible modifications and simulations, orientations and materials of building design (Østergård *et al.*, 2016; Barrios *et al.*, 2017; Harish and Kumar, 2016; Nguyen *et al.*, 2014). In this study, design alternatives mean adjusting parameters of HVAC systems and the construction of exterior walls and roofs.

The following step is clarifying the conceptual framework, as shown in Fig. 1. This is an important step in the integration process where the outputs of the LCIA analysis integrate with BIM lenses. In this term, LCIA is the total sum of the quantities of materials, energy consumption and resulting emissions (UNEP, 2012), such as daylight study, cooling and ventilation analysis, energy performance estimation and environmental impact evaluation. On the other hand, BIM lenses provide the required indicators to classify BIM fields and BIM stages (Succar, 2009), such as the deliverables of BIM, 3D smart objects, estimation of energy simulation and time and cost schedules. The last step in the integration process is the evaluation of results and interpretation in order to define the best proposal at the early design stage of buildings, as seen in Fig. 1. Interpretation identifies and evaluates the results obtained from LCI and LCIA (UNEP, 2012). However, there is an interconnected relationship between the performance parameters and conceptual framework. This allows observation of several modifications and simulations within different modelling types and deliverables in a way to figure out the best proposal that serves the objectives of the project.

2.2. *Application of tools*

This work illustrates the BIM–LCA integration within the available database in order to conduct a daylight assessment and energy consumption analysis based on different design alternatives and orientations of residential buildings in Brazil. It targets the early design phase of construction projects when the model encloses only basic geometry and spaces, construction cost is characteristically low with an excess of alternative designs and there are prospects to reduce the consumption of operating energy and improve the daylight efficiency in buildings (Liu *et al.*, 2011; Azhar *et al.*, 2011). This work uses the construction of a multi-storied residential building in Brazil as a case study to examine the validity and usability of BIM–LCA integration in evaluating the daylight analysis, estimating energy performance and clarifying the development of design concepts generated by BIM and LCA tools.

The first step of this work is to state the goal and scope of the work and define BIM fields. Next, the performance parameters are to be identified by defining BIM stages, and the possible simulations and environmental issues are considered. This includes defining the parameters, conducting the 3D modelling of the building, recognizing the various design alternatives, building up of the inventory database of construction materials and considering the local climate data of the case study using Autodesk Revit as a BIM software. The next step is to integrate the benefit indicators and outputs of Revit, BIM lenses with the impact assessment (LCIA). At this step, LEED daylight plug-in is applied to assess the potential impact of natural daylight analysis, and Autodesk Green Building Studio is used to conduct a conceptual energy consumption analysis of the building. Results at this step are required to build up life-cycle energy analysis at the operation phase of buildings (KT Innovations, Thinkstep and Autodesk, 2016). Such results are evaluated and meet the criteria under ANSI/ASHRAE Standard 140 (Autodesk, 2014). However, Green Building Studio could produce various simulations for buildings, taking into consideration the building model, various specifications and orientation of the building and properties of construction materials. After that, this work assesses the impacts by evaluating the significance of the outputs with the elementary flows in order to make simpler understanding results (UNEP, 2012). The next step is interpretation. This includes presenting the results, classifying the sources of the impacts, comparing solutions, and suggesting recommendations (UNEP, 2012). The last step is to discuss the results, review the work and present the conclusion.

3. Case Study

This work examines the structure of a multi-storied residential building recognized as a low-profile construction based on the regulations of the Union of the Civil Construction Industry to model the case study building. The building consists of 10 levels with a total floor area of (1558 m^2), including 36 residential apartments, as shown in Fig. 2. The construction materials that are applied in the case study building as mentioned in SINDUSCON (SINDUSCON-MG, 2007) are ceramic masonry block for exterior walls, concrete floor with ceramic tiles for floors, plate and single mass for ceilings, sheet metal rail (120 cm × 120 cm) with smooth glass for windows, and solid concrete flat slab for roof. However, Autodesk

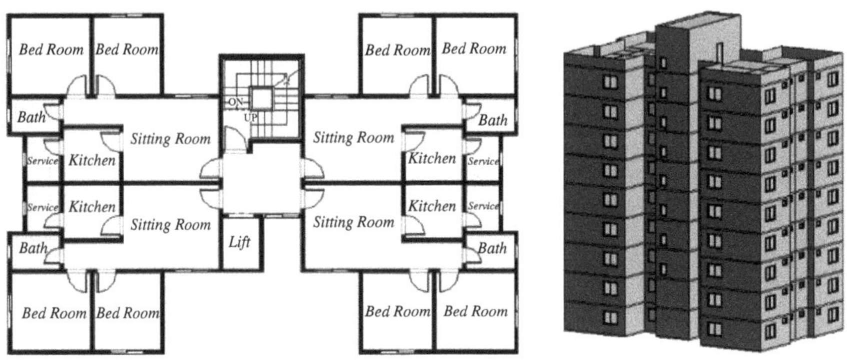

Fig. 2.   2D plan and 3D modelling of the multi-storied residential building.

Revit generates all graphs of energy consumption analysis and daylight assessment presented in this work.

3.1. *Goal and scope*

This work targets the early design phase of construction projects. Despite this, the system boundary that refers to the size of LCA focuses on the operation phase only, neglecting all other phases of LCA. In this chapter, the entire building is considered as a single unit and the analysis of the functional unit concentrated on the parameters of the design alternatives and orientations. According to the default weather stations in Autodesk Green Building Studio, the average lifespan of energy consumption analysis is limited in the range of 30 years of weather data (Autodesk, 2011). The scope of this chapter is to examine different design alternatives and orientations in order to increase daylight efficiency and reduce consumption of energy in such types of buildings at the early design phase.

3.2. *Life-cycle inventory (LCI)*

The model of the case study building and specific properties of the building material are constructed in Autodesk Revit software, which uses Green Building Studio application as an intelligent energy analysis engine

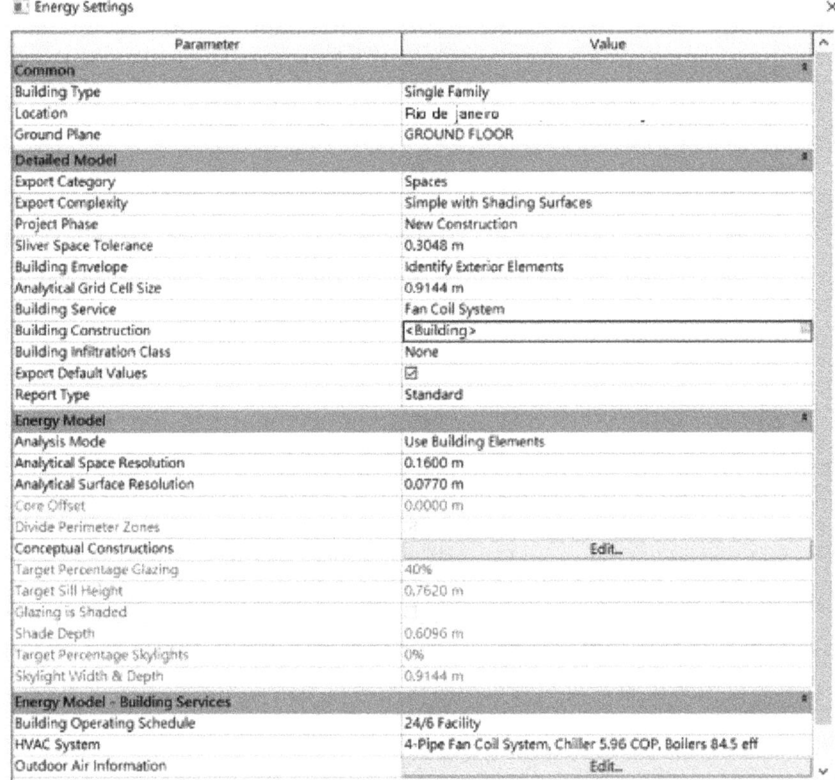

Fig. 3.   Energy consumption analysis in Autodesk Revit.

in order to estimate the energy performance in buildings (Autodesk Revit, 2017). Different assumptions and parameters are required to be filled in precisely in this application, such as building type, location, thermal properties, project phase, building envelope, analysis mode, concept of construction, building operating schedule, Heating, Ventilation and Air Conditioning (HVAC) system, and outdoor air information, etc., as shown in Fig. 3.

Such assumptions and parameters give the opportunity to examine the suggested design alternatives of this work by adjusting the parameters of HVAC systems and the construction of exterior walls and roofs as shown in Table 1. However, the alternative HVAC systems chosen in this work

Table 1.   Adjusting parameters of the case study building.

HVAC System		Construction of External Walls		Construction of Roofs	
No.	Type	No.	Type	No.	Type
1	Residential 14 SEER/ 0.9 AFUE Split/ Packaged Gas <5.5 tonne	8	Insulated Concrete Form Wall, 10″ thick form	15	Wood Frame Roof with Super High Insulation
2	Residential 14 SEER/ 8.3 HSPF Split Packaged Heat Pump	9	Insulated Concrete Form Wall, 12″ thick form	16	Wood Frame Roof without Insulation
3	HP, 13 SEER, Electric Heat, Residential	10	Insulated Concrete Form Wall, 14″ thick form	17	Metal Frame Roof with Super High Insulation
4	Residential 17 SEER/ 9.6 HSPF Split HP <5.5 tonne	11	Metal Frame Wall with Super High Insulation	18	Metal Frame Roof without Insulation
5	Residential 14 SEER/ 0.9 AFUE Split/ Packaged Gas <5.5 tonne	12	Metal Frame Wall without Insulation	19	Continuous Deck Roof with Super High Insulation
6	Residential 17 SEER/ 0.85 AFUE Split/ Packaged <5.5 tonne	13	Massive Wall with Super High Insulation	20	Continuous Deck Roof without Insulation
7	PSZ, ASHRAE 90.1-2007, 11 EER, 78% AFUE, Residential	14	Massive Wall without Insulation	—	—

are provided as residential options in Green Building Studio application. Besides, the alternative construction materials of external walls and roofs chosen for this work aim to show the important role of the thickness form and the insulation process using different types of materials.

Autodesk Revit provides the opportunity to rotate the model in different directions. Besides, it uses LEED daylight plug-in to perform daylighting simulations in the cloud (Autodesk, 2017). In this

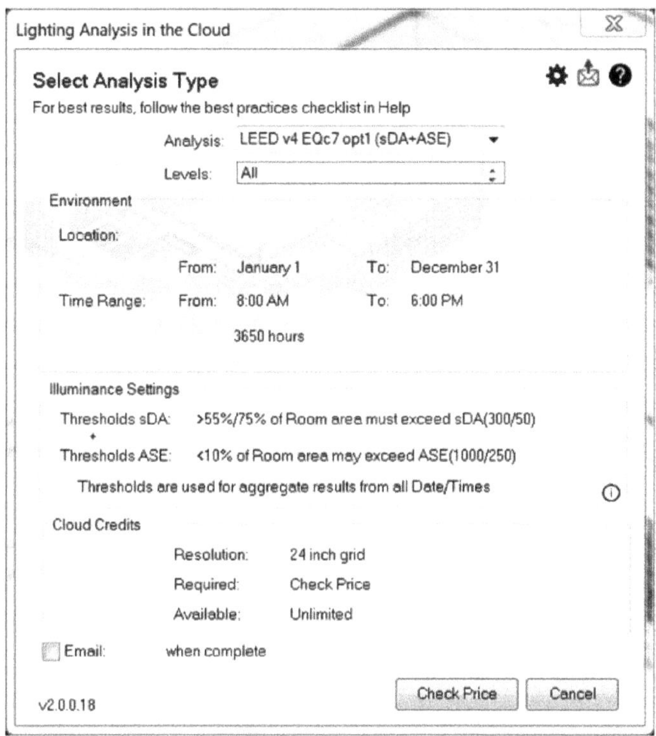

Fig. 4.   Lighting analysis in Autodesk Revit.

discipline, LEED v4 plug-in in Revit has adopted two important metrics, which are spatial daylight autonomy (sDA) and annual sunlight exposure (ASE) in order to help designers and engineers in understanding annual daylight availability and quality, as well as glare and overheating potential within their construction projects (Autodesk, 2016). The assumptions at this level of study are based on the local environmental issues, illumination settings and cloud credits, as shown in Fig. 4. Such assumptions give the opportunity to automate daylight simulations and improve design decisions. Figure 4 clarifies that the time range is automatically simulated for 10 h per day, from 8 am to 6 pm, covering 3650 h over a full annual simulation. LEED requires that DA achieves at least 55% or 75% of regularly occupied floor area within ASE of no more than 10% for the occupied day-lit floor area per sDA.

However, both metrics require a resolution of at least 24-inch analysis grid at the cloud credit level.

3.3. *Life-cycle impact assessment (LCIA)*

The presented results at this level of analysis are based on the evaluation of construction materials used in the functional unit of this work that considered the whole building as a single unit. The conceptual energy analysis of the case study building shows that the consumption of electricity and fuel accounts for 97% and 3%, respectively. Results show that the case study building consumes 96,293 kWh of electricity per year, however, Table 2 presents the values of electricity consumption based on the adjustment of parameters.

According to the LEED daylight analysis, the results presented in this research are based on the building orientation within 180°. This comes back to the symmetric design of the case study building. Hence, this analysis considers seven basic cases to rotate the building between 0° and 180° within a variation of 30°, as shown in Table 3. Case (A) assumes that the longer side of the building is oriented to the east–west axis directly,

Table 2.   Energy consumption based on adjustment of parameters.

	HVAC System		Construction of External Walls		Construction of Roofs
No.	Electricity Consumption (kWh/Year)	No.	Electricity Consumption (kWh/Year)	No.	Electricity Consumption (kWh/Year)
1	121,801	8	97,747	15	95,108
2	117,347	9	97,340	16	101,684
3	87,509	10	95,820	17	95,624
4	76,550	11	99,104	18	102,488
5	121,801	12	122,977	19	95,708
6	77,234	13	99,189	20	101,343
7	121,801	14	104,938	—	—

Table 3.   LEED daylight analysis of the case study.

	Case (A)	Case (B)	Case (C)	Case (D)	Case (E)	Case (F)	Case (G)
Building orientation							
sDA + ASE	33%	30%	15%	15%	21%	25%	33%

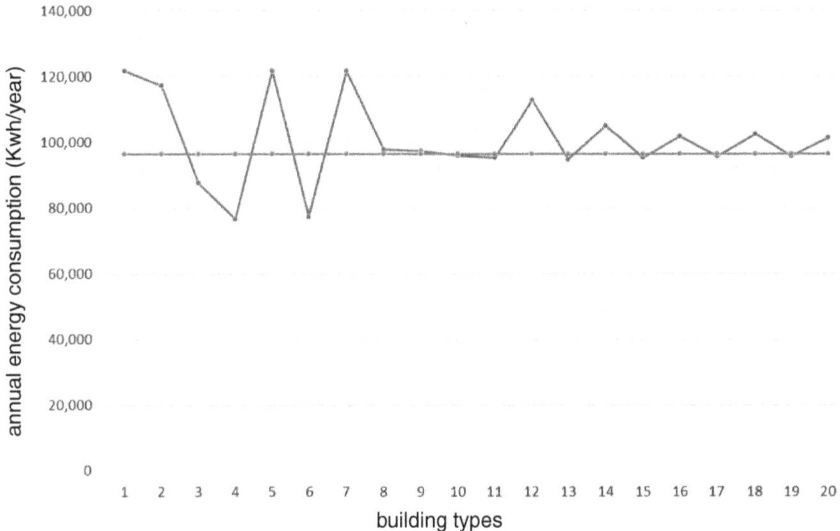

Fig. 5.   Annual electricity consumption of the building and adjusting parameters.

while Cases (B)–(G) reflect a rotation of the main axis within 30°, 60°, 90°, 120°, 150° and 180°, respectively.

3.4. *Interpretation*

Comparing the energy performance in the case study building and adjustment parameters shows that using particular HVAC systems or some alternative construction materials for exterior walls and roofs would influence the energy efficiency in buildings, as shown in Fig. 5. In terms of HVAC systems, it can be seen that type numbers (3, 4 and 6) could reduce the consumption of electricity whereas type numbers (1, 2, 5 and 7) would dramatically increase this issue. In terms of construction materials of external walls and roofs, it is clear that applying super-high insulation process on the different construction materials such as type numbers (11, 13, 15, 17 and 19) would increase the energy efficiency in buildings. Furthermore, reducing the thickness of insulated concrete walls would result in more energy consumption and vice versa. This point is approved in type numbers (8, 9 and 10).

On the other hand, the collected outcomes of LEED daylight analysis as shown in Table 2 illustrate that Cases (A and G) would achieve the best results ever in terms of sDA and ASE, clarifying that 33% of building area would meet the required percentage hours of sDA in rooms within less than 10% of area above ASE. However, Case (B) presents the second best standard of the building that meets sDA and ASE requirements with a value of 30%. Besides, Cases (E) and (F) illustrate lower standards in terms of sDA and ASE with values of 21% and 25%, respectively while Cases (C) and (D) demonstrate the worst standards for such types of buildings with a value of 15% each.

4. Discussion

This chapter highlights the growing interest in the field of BIM–LCA integration and evaluates the natural daylight assessment and energy consumption in buildings based on different design alternatives and orientations at the early design stage of residential buildings. It clarifies that BIM models allow using various construction materials within different performance parameters at the design phase in order to empower the decision-making process. Besides, this chapter shows that LCA methodology aims to evaluate the influence of such estimations over the entire lifespan of the construction project. This chapter presents a BIM–LCA integration framework in order to analyze construction projects from a sustainable perspective, applying tools that allow the creation of different simulations in a short period. It presents the difficulties in comparing the different scenarios of energy and daylight analysis in such integration course (Anand and Amor, 2017).

The analyzed case study shows that there are different factors influencing energy efficiency in buildings, such as the type of HVAC system, the application of the insulation process and the thickness form of the construction of external walls and roofs. The results show that the appropriate choice of such factors would reduce waste generation and environmental impacts. The applied tools at this step of analysis provide a wider vision of examining various alternatives to construction materials at an early designing stage in order to increase energy efficiency in the construction sector. However, climate data constitute a critical input to validate building performance analysis. This chapter presents an uncertainty issue facing this study in terms of the applied climate database in Green Building Studio that dates back to the year 2006. This means that the data might be outdated or do not reflect the reality of the region today.

Furthermore, this chapter shows that the building design and orientation factors play a fundamental role in controlling the capacity of natural daylight, glare and overheating potential in buildings. In other words, such factors are largely influencing the design decisions in such types of construction projects. The results show that orienting the long axis of the building towards the east–west direction, as shown in Cases (A) and (G), would achieve the highest inducing results in capturing the natural daylight potential. On the other hand, orienting this axis towards the north–south direction, as shown in Case (D), would achieve the lowest inducing results in these terms. However, rotating this axis with 30°, 120° or 150°, as shown in Cases (B), (E) and (F), would expose the building to less natural daylight compared with Cases (A) and (G).

5. Conclusion

This chapter motivates the sustainability in terms of BIM–LCA integration at the early design stage of construction projects. BIM models allow using various conceptual parameters, while LCA methodology evaluates the impact of these parameters on the construction sector and the environment. Thus, this chapter presents a methodological framework for the proposed integration between LCA methodology and BIM tools, and analyzes a case study building in a way to achieve the objectives of this research by assessing the energy consumption and daylight analysis based on different design alternatives and orientations at the early design stage of residential buildings. Besides, the tools presented in this work help architects, designers and engineers in making more conscious choices in terms of natural daylight analysis and energy efficiency.

This chapter presents some critical points such as the difficulties of comparing the different scenarios of BIM tools in such integration course. On the other hand, it encourages reviewing the building orientation, the application of HVAC systems and the construction of external walls and roofs in the construction projects at the initial stages of design in order to increase energy efficiency and ventilation in buildings. In light of the results, integrating BIM models with the LCA methodology is considered as an optimal procedure towards achieving a sustainable development and empowering the decision-making process in the construction sector. The proposed methodology helps increase both energy efficiency and capturing of natural daylight over the entire lifespan of buildings.

Acknowledgements

The authors want to acknowledge the financial support from CAPES Foundation (Coordination for the Improvement of Higher Education Personnel) and CNPq (Brazilian National Council for Scientific and Technological Development).

References

Alwan, Z., Jones, P. and Holgate, P. (2017). "Strategic Sustainable Development in the UK Construction Industry, through the Framework for Strategic Sustainable Development, Using Building Information Modelling", *Journal of Cleaner Production*, 140, 349–358. Available at: http://dx.doi.org/10.1016/j.jclepro.2015.12.085.

Anand, C. K. and Amor, B. (2017). "Recent Developments, Future Challenges and New Research Directions in LCA of Buildings : A Critical Review", *Renewable and Sustainable Energy Reviews*, 67, 408–416.

Antón, L. Á. and Díaz, J. (2014). "Integration of LCA and BIM for Sustainable Construction", *World Academy of Science, Engineering and Technology International Journal of Social, Education, Economics and Management Engineering*, 8(5), 1356–1360.

Asdrubali, F., Baldassarri, C. and Fthenakis, V. (2013). "Life Cycle Analysis in the Construction Sector: Guiding the Optimization of Conventional Italian Buildings", *Energy and Buildings*, 64, 73–74.

Autodesk. (2011). "Getting Started with Autodesk Green Building Studio", *Autodesk® Ecotect*[TM] *Analysis*, 17. Available at: www.greenbuildingstudio. com/default.aspx.

Autodesk. (2014). *Green Building Studio Validation*. Available at: https://googleweblight.com/?lite_url=https://knowledge.autodesk.com/search-result/caas/ CloudHelp/cloudhelp/ENU/BPA-GBSWebService/files/GUID-EF68E7D5- C0A5-4805-BFE5-7C74C57B712E-htm.html&lc=pt-BR&s= 1&m=654&host=www.google.com.br&ts=1506512870&sig=ANTY_ L3qfNFfK2vQbBsqOZ_AdfH34TKMog (Accessed on September 12, 2017).

Autodesk. (2016). *LEED sDA & ASE Studies with Insight Revit 2017 Plugin* [online]. INSIGHT BLOG. Available at: http://blogs.autodesk.com/insight/ leed-sda-ase-studies-with-insight-revit-2017-plugin/ (Accessed on September 12, 2017).

Autodesk. (2017). *Autodesk Apps for LEED Automation*. Available at: https://www.usgbc.org/resources/autodesk-apps-leed-automation (Accessed on August 14, 2017).

Autodesk Revit. (2017). *About Green Building Studio and Energy Analysis*. AUTODESK KNOWLEDGE NETWORK. Available at: https://knowledge. autodesk.com/support/revit-products/learn-explore/caas/CloudHelp/cloud- help/2016/ENU/Revit-Analyze/files/GUID-7948A714-1B97-4176-A942- D99A8ECA4786-htm.html (Accessed on August 27, 2017).

Azhar, S., Carlton, W. A., Olsen, D. and Ahmad, I. (2011). "Building Information Modeling for Sustainable Design and LEED Rating Analysis", *Automation in Construction*, 20, 217–224.

Barrios, G., Rojas, J., Huelsz, G., Tovar, R. and Jalife, S. (2017). "Implementation of the Equivalent-Homogeneous-layers-set Method in Whole-building Simulations: Experimental Validation", *Applied Thermal Engineering Journal*, 125, 35–40.

Bayer, C., Gamble, M., Gentry, R. and Joshi, S. (2010). *AIA Guide to Building Life Cycle Assessment in Practice*, USA: The American Institute of Architects.

Buyle, M., Braet, J. and Audenaert, A. (2013). "Life Cycle Assessment in the Construction Sector: A Review", *Renewable and Sustainable Energy Reviews*, 26, 379–388. Available at: http://dx.doi.org/10.1016/j.rser.2013.05.001.

Eleftheriadis, S., Mumovic, D. and Greening, P. (2017). "Life Cycle Energy Efficiency in Building Structures: A Review of Current Developments and

Future Outlooks Based on BIM Capabilities", *Renewable and Sustainable Energy Reviews*, 67, 811–825.

Gustavsson, L., Joelsson, A. and Sathre, R. (2010). "Life Cycle Primary Energy Use and Carbon Emission of an Eight-story Wood-framed Apartment Building", *Energy and Buildings*, 42(2), 230–242.

Harish, V. S. K. and Kumar, A. (2016). "A Review on Modeling and Simulation of Building Energy Systems", *Renewable and Sustainable Energy Reviews*, 56, 1272–1292.

Huang, B., Xing, K. and Pullen, S. (2015). "Energy and Carbon Performance Evaluation for Buildings and Urban Precincts: Review and a New Modelling Concept", *Cleaner Production*. Available at: http://dx.doi.org/10.1016/j.jclepro.2015.12.008.

Jrade, A. and Jalaei, F. (2014). "Integrating Building Information Modeling (BIM) and Energy Analysis Tools With Green Building Certification System to Conceptually Design Sustainable Buildings", *Journal of Information Technology in Construction*, 19, 494–519.

Kang, H. J. (2015). "Development of a Systematic Model for an Assessment Tool for Sustainable Buildings Based on a Structural Framework", *Energy and Buildings*, 104, 287–301. Available at: http://linkinghub.elsevier.com/retrieve/pii/S0378778815301316.

KT Innovations, Thinkstep and Autodesk. (2016). *Tally*. Available at: http://choosetally.com/methods/ (Accessed August 6, 2017).

Kwok Wai Wong, J. and Zhou, J. (2015). "Enhancing Environmental Sustainability Over Building Life Cycles Through Green BIM: A Review", *Automation in Construction*, 57, 156–165. Available at: http://dx.doi.org/10.1016/j.autcon.2015.06.003.

Lassio, J. De, França, J., Santo, K. E. and Haddad, A. (2016). *Case Study: LCA Methodology Applied to Materials Management in a Brazilian Residential Construction Site*. Cairo: Hindawi Publishing Corporation.

Lasvaux, S. (2010). *Study of a Simplified Model for the Life Cycle Analysis of Buildings*, Paris: Paris Institute de Technologie.

Liu, Z., Osmani, M., Demian, P. and Baldwin, A. N. (2011). "The Potential Use of BIM to Aid Construction Waste Minimalisation", in *The CIB International Conference*, Sophia Antipolis, France.

Martínez-Rocamora, A. and Solís-Guzmán, J. (2016). "LCA Databases Focused on Construction Materials: A Review", *Renewable and Sustainable Energy Reviews*, 58, 565–573.

Nguyen, A., Reiter, S. and Rigo, P. (2014). "A Review on Simulation-Based Optimization Methods Applied to Building Performance Analysis", *Applied Energy*, 113, 1043–1058. Available at: http://dx.doi.org/10.1016/j.apenergy.2013.08.061.

Østergård, T., Jensen, R. L. and Maagaard, S. E. (2016). "Building Simulations Supporting Decision Making in Early Design — A Review", *Renewable and Sustainable Energy Reviews*, 61, 187–201.

Peng, C. (2015). "Calculation of a Building's Life Cycle Carbon Emissions Based on Ecotect and Building Information Modeling", *Journal of Cleaner Production*, 1, 453–465.

Porwal, A. and Hewage, K. N. (2013). "Automation in Construction Building Information Modeling (BIM) Partnering Framework For Public Construction Projects", *Automation in Construction*, 31, 204–214. Available at: http://dx.doi.org/10.1016/j.autcon.2012.12.004.

Šaparauskas, J. and Turskis, Z. (2006). "Evaluation of Construction Sustainability by Multiple Criteria Methods", *Journal Ukio Technologinis ir Ekonominis Vystymas ISSN*, 12(4), 321–326.

Schlueter, A. and Thesseling, F. (2008). "Building Information Model Based Energy/Exergy Performance Assessment in Early Design Stages", *Automation in Construction*, 18, 153–163.

Shadram, F., Johansson, T. D., Lu, W., Schade, J. and Olofsson, T. (2016). "An Integrated BIM-Based Framework for Minimizing Embodied Energy During Building Design", *Energy and Buildings*, 128, 592–604. Available at: http://dx.doi.org/10.1016/j.enbuild.2016.07.007.

Shafiq, N., Nuruddin, M., Gardezi, S. and Kamaruzzaman, A. (2015). "Carbon Footprint Assessment of a Typical Low Rise Office Building in Malaysia Using Building Information Modelling (BIM)", *International Journal of Sustainable Building Technology and Urban Development*, 6(3), 157–172.

SINDUSCON-MG. (2007). *Custo Unitário Básico (Cub/M^2): Principais Aspectos*, Belo, Horizonte: Estado de Minas Gerais.

Soust-Verdaguer, B., Llatas, C. and García-Martínez, A. (2016). "Simplification in Life Cycle Assessment of Single-Family Houses: A Review of Recent Developments", *Building and Environment*, 103(May), 215–227.

Soust-Verdaguer, B., Llatas, C. and García-Martínez, A. (2017). "Critical Review of BIM-Based LCA Method to Buildings Critical Review of BIM-Based LCA Method to Buildings", *Energy and Buildings*, 136, 110–120. Available at: http://dx.doi.org/10.1016/j.enbuild.2016.12.009.

Succar, B. (2009). "Automation in Construction Building Information Modelling Framework: A Research and Delivery Foundation for Industry Stakeholders", *Automation in Construction*, 18(3), 357–375. Available at: http://dx.doi.org/10.1016/j.autcon.2008.10.003.
UNEP. (2012). *Towards a Life Cycle Sustainability Assessment*, Nairobi: UNEP.

Chapter 6

Ex post Impact Evaluation of PPP Projects — The Project Success Evaluation Pyramid Model

Jose O. Romero* and Ajibade A. Aibinu

*Faculty of Architecture, Building and Planning,
The University of Melbourne, Australia*

**joseo@student.unimelb.edu.au*

Abstract

Public–private partnership (PPP) model of procurement has been employed extensively and promoted internationally. However, their performance and their real capacity to achieve public welfare have been questioned. To address this criticism, there is no specific method to assess the impact of PPP projects after its operation phase; an *ex post* impact evaluation tool needs to be designed. The aim of the research is to develop an *ex post* evaluation procedure that can address the complexities of PPPs, namely (1) large size and technically complex projects, (2) multiple perceptions of the impacts, (3) vague and uncertain understanding of "public interest", (4) long time horizon for the evaluation, which in some cases is more than 20 years, and (5) political

---

* Corresponding author.

and ideological drivers that are relevant and difficult to address. A Design Science approach is adopted with four stages. Currently, the research is in its first stage, in which a conceptual framework called Project Success Evaluation Pyramid Model (PSEPM) has been designed and presented to experts. This chapter describes this model and the theory that supports it. The research contributes to the project management and public policy body of knowledge and, consequently, to practice by (1) improving future PPP developments and (2) providing a basis for discussion about the use of PPPs.

Keywords: Innovative Construction Management; Impact Evaluation; Project Evaluation; Project Success; Public–private Partnership; Stakeholders.

1. Introduction

Public–private partnership (PPP) model of procurement is a way to deliver public infrastructure using private funding and managing risk for public purposes. Investment in infrastructure using this method includes many sectors, such as health, education, water supply, transport, electric power, etc. Studies show that PPPs have good results, especially in terms of efficiency, delivering on average more projects on time and on budget (National Audit Office, 2013; Raisbeck *et al.*, 2010). PPPs have been promoted worldwide with a large number of projects in the pipeline in many countries.

Currently, the PPP model has been criticized in the media, social media and the academia (Pollock, 2012; Rogers, 2013). For example, the international organization of work unions (PSI) has launched a report completely against PPPs stating that they are expensive and inefficient compared to a public alternative (Hall, 2015). In the UK, the restricted access to actual PPP contracts (PFI) due to "commercial confidentiality" has given rise to criticism about the transparency of the PPPs. An article in The Guardian states, "… show the true cost of PFI" (Pollock, 2012). In NSW Australia, the bankruptcy of the private company that was building social housing PPP projects opened the discussion about how reliable are private companies that deliver public welfare (Rogers, 2013). Similarly, the operator of Brisbane's AU\$ 4.8 billion Airport Link toll road went out of administration in 2013 because the average daily traffic and revenue was half of what was predicted (Han, 2017).

As with any kind of public investment, the taxpayer interest could be protected with an *ex post* impact evaluation. However, there is no *ex post* evaluation method in the public domain that can address the complexity of PPP projects as (1) they are large and technically complex, (2) there are multiple perceptions of the impacts, (3) they deal with a vague and uncertain understanding of what is called "public interest", (4) they have long time horizon for evaluation, which in some cases is more than 20 years and (5) they are affected by political and ideological drivers that are relevant and difficult to address. Citizens at large should know if public projects are making the difference (Owen and Alkin, 2007), and the accountability of public decisions must be evaluated in the best possible way. Existing evaluations follow the pattern of considering only a limited view of the project without a systematic process for addressing the multiple stakeholder perspectives present in large projects such as PPPs. Additionally, technical performance measures are usually employed to the detriment of perceptive indicators, which in many cases are the real reasons for accepting, investing or changing a PPP.

The aim of the research is to design an *ex post* impact evaluation tool that can assess PPP projects. For doing so, the project management concept "project success (PS)" has been employed as a starting point to develop an approach based on the multiple stakeholders that are interrelated in a PPP. The outcome of the research is a designed PPP *ex post* evaluation tool that can be applied for (1) improving future PPP developments and (2) providing arguments for the discussion about the use of PPPs.

2. Methodology and Structure

To address the research problem, a Design Science approach is adopted (Hevner *et al.*, 2004) with the following four main stages: (1) development stage, (2) prototype stage, (3) testing stage and (4) validation stage. The research is currently at the development stage, in which the conceptual framework called "Project Success Evaluation Pyramid Model (PSEPM)" was designed and discussed with experts. The preliminary design was generated through a starting literature and document review, in which the problem was confirmed. Then, a reflexive process and a design synthesis were done, creating the conceptual framework.

The problem, the PSEPM, and the tool architecture were presented and discussed with 13 experts and PPP practitioners from the public and private sectors in Chile and Australia by interviewing them in one-hour sessions: three former senior public sector authorities in infrastructure and public evaluation, one senior public sector authority in infrastructure, two junior public sector executives, two senior PPP consultants, one mega contractor project manager, and four junior private developers in PPP social housing projects. Further details about the experts cannot be provided due to confidentiality reasons. Thirteen experts are considered a robust group for testing a new evaluation approach because as a group, they are able to represent the diversity of perspectives, all in the context of PPP projects. Also, there are limited highly knowledgeable experts in this area and the experience of the 13 experts is considered adequate. Each session started with a 20 min presentation for presenting the PPP context, the PSEPM, and the tool architecture. Then, 40 min of discussion were separated into (i) general comments about PPP evaluations, (ii) general comments about the proposed tool, (iii) specific review of each component of the presentation (tool components) and (iv) significance, applicability and limitations of the tool. The interviews were transcribed verbatim and then studied by employing theme analysis. Employing MS Excel, 170 pieces of information were generated, and the coding included more than 40 key concepts.

This chapter is structured in three main sections excluding the introduction and conclusion. The literature review section confirms the lack of evaluation procedures that can be applied to PPPs. The next section explains the PSPEM in four sub-sections which represent the different components of the PSEPM. The last section discusses the inputs and insights that were gathered through the experts' interviews.

3. Literature Review and Gap

There is no specific procedure to assess (in any way) the impact of PPP projects on stakeholders, and the multiple stakeholder perspective approach is something new and a gap in the existing literature.

However, many bodies of knowledge could be employed to support the development of an *ex post* evaluation.

3.1. *PPP ex post evaluation guideline*

The PPP guidelines of the European PPP Expertise Centre from the European Investment Bank (EIB) underscore the need for an *ex post* evaluation. In order for a PPP to be well procured and delivered, the guideline considered *ex post* evaluation as a part of the steps to be undertaken (European PPP Expertise Centre, 2011). However, the evaluation is presented as a general guideline with no clear instruction on how to really assess a PPP.

3.2. *PPP whole life assessment*

The PPP whole life assessment presented by Liu and Love has the aim of improving the performance of the PPPs by measuring the performance at every stage of the project (i.e. formative evaluation) (Liu *et al.*, 2014; Liu *et al.*, 2015). However, this assessment is still a conceptual approach without addressing the impacts of the project on every stakeholder (summative evaluation).

3.3. *PPP performance*

PPP performance has been studied by many researchers in construction management (Hardcastle *et al.*, 2005; Jingfeng *et al.*, 2009; Liu *et al.*, 2015; Mladenovic *et al.*, 2013; Raisbeck *et al.*, 2010; Xiong *et al.*, 2015). However, they focus mostly on (1) comparing PPP KPIs and traditional procurement KPIs, (2) success factors for PPPs and (3) mathematical models to manage stakeholder's satisfaction.

3.4. *Value for money*

In deciding to proceed with PPPs, many public sector guides recommend the use of value for money (VfM) indicator to measure efficiency in public investment by comparing the PPP option with a public-sector comparator.

This method does not address the complexity of multiple stakeholders and it is essentially used to provide financial support for investment decision-making. Moreover, studies have concluded that this indicator presents biases towards the PPPs in cases where there is no funding for developing the project with public resources (Heald, 2003).

3.5. *Project audit*

Project audit is another body of knowledge to address a PPP *ex post* evaluation. Originally, an audit is conceptualized as a financial tool to assess compliance rather than the quality and impacts of the processes and outcomes. Nevertheless, there are approaches such as Nalewaik and Mills (2017) that address the question: "Was the right thing done?" In contrast to the traditional, "Was the thing done right?" The problem with this approach is that it is based on a single stakeholder perspective (the auditor), receiving only inputs from the rest of the stakeholders.

3.6. *Project success concept*

To assess the impact of a PPP on stakeholders, the Project Success (PS) concept from the Project Management discipline could be employed. Existing literature has focused on (1) developing frameworks to analyze PS, (2) success factors, which are key areas that can improve the possibility of having a successful project and (3) success criteria (SC), which are the measurements that can be assessed to determine if a project is successful or not. However, the existing literature (Müller and Jugdev, 2006) in this area lack details. For example, there is not yet an accepted method to assess the success of a project, and development in this area is still limited. They are theoretical conceptualizations with limited application in practice. Moreover, the definition of PS is ambiguous and difficult to apply because it disregards the differences that could arise in the judgement of PS by different stakeholders. Nevertheless, the theory of PS is the only approach that supports a broad view of the project, involving all the stakeholders.

3.7. *Performance measurement systems (PMS)*

Performance measurement systems (PMS), such as the Balance scorecard (Kaplan and Norton, 2007) and the Performance Prism (Neely *et al.*, 2001), could also be an approach to perform an *ex post* evaluation. The problem is that a PMS is good for measuring performance, and an evaluation has the aim for not only measuring but also making a judgement out of that measurement for decision-making purposes.

3.8. *Program evaluation (public policy)*

A PPP is a model of procurement for building an infrastructure and providing a public service. In fact, a PPP contract includes the whole project from the design stage to the operational stage. Therefore, Program Evaluation theory could be used as an approach for evaluating PPPs using, what is called, an impact evaluation (Langbein and Felbinger, 2006; Owen and Alkin, 2007; Quade and Carter, 1989). However, the problem with this approach is that all the different types of impact evaluation are mostly performed by a team of experts using their unique perspective without considering each stakeholder as an object of study. Thus, the impacts of the project on each stakeholder are disregarded and are not clearly assessed and explained so that the lessons learned can be used for improving future PPPs.

There is no specific procedure to assess (in any way) the impact of PPP projects on stakeholders, and the multiple stakeholder perspective approach is something new and disruptive in the existing literature. However, many bodies of knowledge could be employed to support the development of an *ex post* evaluation.

4. Project Success Evaluation Pyramid Model (PSEPM)

4.1. *The problem of existing evaluation pattern*

Existing evaluation methods have the pattern of including only the criteria that concerns the evaluation team without addressing, in a systematic way, the complexity of having multiple perspectives. With almost no exception, evaluations and audits have a common strategy.

In a broad sense (not only *ex post*), existing evaluations can be synthesized as a procedure of two steps.

The first step is the selection of a set of criteria, which is done by an "expert" team. The criteria can have a big spectrum of possibilities, from the most technical (cost, time, budget, profit, waste, CO_2 emission, etc.) to the most social/perceptive criterion (compliance, political revenue, VfM, social impact, satisfaction, good or bad, etc.). The criteria selection depends on the type of evaluation and on the client/user/requester of the evaluation.

The second step is to assess the selected criteria by employing one or more of the following strategies:

- comparison before–after (i.e. for social interventions),
- comparison expected–actual (i.e. contract compliance),
- comparison project A–project B (performance comparison),
- comparison project A–standard (i.e. minimum requirements),
- comparison project A–expert judgement (deep qualitative evaluation), etc.

The problem with this pattern is that it only addresses the perspective of the evaluation team, which leads to a legitimacy problem. Legitimacy refers to the acceptance of an authority for taking decisions. Using Max Weber's classification, the rational–legal legitimacy is the one involved in this context, where the rationality is assumed to be behind the public interest. Legitimacy is not an issue when there is only one stakeholder who concentrates the power over the rest of the stakeholders. However, this issue is more complex when a project works with a series of stakeholders that have powers and interests that can generate conflict. This is the case of a PPP, which has a broad public sector (different agencies and ministries), a private consortium with several companies within the entity, an aggregated private sector (lobbies and guild groups), a significant role of the users, and an even more determinant extended community (citizens) that can vote for an overall strategy in favour or against PPPs.

Another problem is that usually existing evaluations address technical issues to the detriment of "non-technical" or "perceptive" issues, which in some cases are the real decision-making triggers (i.e. political discussion).

Perceptions are critical for the development of PPPs. Whether they are correct or incorrect, stakeholders' perceptions are often the basis of disinterest and rejection of PPP use and outcomes. Heald (2003) and Bauld and McGuinness (2006) refer to the subjectivity of the value for money indicator, suggesting that the concept "needs" a non-technical component.

4.2. *Project success concept*

PS concept has been employed to inform the evaluation tool development. Even though there is not yet an agreement in the literature about the definition of PS, many authors agree that "meeting of stakeholders' expectations" is central to any assessment of PS (Davis, 2013, 2016; de Wit, 1988; Koops *et al.*, 2015). The existing literature suggest that success is a judgement. Shenhar *et al.* (2001) stated that the success of a project should be more accurately expressed as "perceived success" due to the personal validation of the concept and a subsequent agreement of success between the different judgements.

With the agreed definition of success, the question to be asked is "How is it possible to meet all stakeholders' expectations?" Or even, "How do we know when stakeholders' expectations are met?" Asking them directly seems to be a reasonable option, but in practice, it is neither that simple nor that useful; a satisfaction survey from all the stakeholders does not solve the problem. A more robust structure is needed to implement this theoretical approach.

4.3. *Project success evaluation pyramid model general structure*

This study proposed the PSEPM as essentially a procedure that identifies a big spectrum of SC from each of the involved stakeholders and then employs a set of principles to assess the validity of all those criteria. For each stakeholder, a unique perspective is generated, which contains the relevant information that supports the stakeholder judgement towards the project. The outcome of the framework is the process for reaching a judgement of PS rather than the measurement of PS itself.

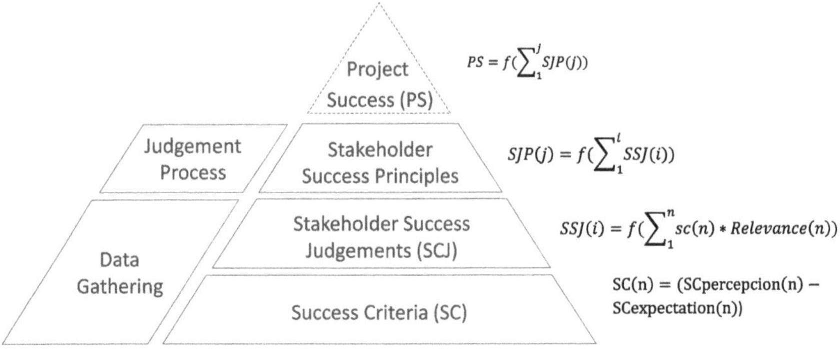

$$PS = f(\sum_1^J SJP(j))$$

$$SJP(j) = f(\sum_1^l SSJ(i))$$

$$SSJ(i) = f(\sum_1^n sc(n) * Relevance(n))$$

$$SC(n) = (SCpercepcion(n) - SCexpectation(n))$$

Fig. 1.   Project success evaluation pyramid model.

The model (Fig. 1) analyzes the success using the following three components: (1) SC, (2) stakeholders' success judgement, and (3) stakeholders' judgement principles. Each component/level of the pyramid informs and supports the next level in order to achieve a theoretical judgement of PS. The model illustrates that a set of SC informs a single stakeholder success judgement (SSJ); then, the combination of all the stakeholders success judgements are tested against a single "stakeholder judgement principle" (SJP); finally, all the stakeholder judgement principles inform the judgement of PS.

4.4. *Success criteria (SC)*

The bottom of the pyramid represents all the existing SC that can be employed to measure PS, including time, cost, quality, safety, profit, environmental sustainability, value for money, etc. A specific subset of this pool of SC is what stakeholders use to inform a success judgement. The model establishes that a stakeholder must recognize and be accountable for the way they judge the project. In other words, they must provide the set of criteria that will represent their perspective.

Existing literature suggest that the success depends on each stakeholder (Baccarini, 1999; Bourne and Walker, 2005; Davis, 2013; Müller and Jugdev, 2006; Shenhar *et al.*, 2001; Turner, 2009; Westerveld, 2003), which means that every stakeholder has a different judgement of success. According to de Wit (1988), this is the main complexity for addressing PS; it represents a different thing to everybody.

4.5. *Stakeholder success judgement (SSJ)*

The second level of the pyramid represents the judgement that every stakeholder must generate employing their particular set of criteria. They must declare if the project is meeting their expectations in every criterion. However, the following needs to be clarified: (i) What are stakeholders' expectations and how can they be measured? and (ii) How do we ensure the validity of the expectations?

Based on Xiong *et al.* (2015), a gap in expectation can be expressed as the difference between what stakeholders expect and what they perceive in a specific success criterion. Every gap in expectation must be expressed in terms of quantitative and qualitative information to understand the gap and assess its relevance and validity. A *j* gap in expectations is defined as follows:

$$\text{GAP}_i = (\text{Stakeholder perception}_i - \text{Stakeholder expectation}_i). \qquad (1)$$

A constraint must be considered to limit the expectations of the stakeholders. The addition of all the expectations should make the project feasible. For example, if a hospital is expected to have the newest technology, but it is also expected to be within a maximum budget that makes it the technology impossible to buy, then the project is not "feasible". Consequently, project success will never be reached even theoretically speaking. The overall purpose of the project (i.e. business case, contract or approved scope) as constructed by the stakeholders is the source for validating the expectations. This constraint is expressed in the following equation for *n* stakeholder with *j* expectations:

$$\sum_{1}^{n} \sum_{1}^{j} \text{Stakeholder expectation}_{nj} \leq \text{Feasible project}. \qquad (2)$$

The judgement of the stakeholder about the success of the project is called a "perspective", which is supported on the criteria that contains specific expectations and traceable perceptions. This judgement is no more than what it has been conceptually stated in the literature; a stakeholder has its own judgement of success, and that judgement is based on the expectations over certain SC.

4.6. *Stakeholder judgement principles (SJP)*

The assumption that a project can objectively be declared successful means that it is necessary to unify the different stakeholder success judgements into one overall statement. For doing so, the concept of "stakeholder judgement principles" is proposed. These principles are the key characteristics that projects should have in order to be considered successful from the perspective of multiple stakeholders. They have a "high level of abstraction" and they differ from the traditional SC such as cost, time and quality because the focus is on the stakeholders' acceptance rather than the technical measurements. This research has developed the following three judgement principles:

- **Willingness to repeat the project:** Are the stakeholders willing to repeat the same project and receive the same outcomes assuming that they could choose from the same original options? This willingness to repeat the project supports the assumption that the stakeholder can accept the gaps in expectations, when they are compared with the benefits they received.
- **No excessive negative outcome:** Has any stakeholder received an evident negative outcome that could have been avoided? An excessive negative outcome can be judged by the other stakeholders in order to preserve the context of that outcome. Then, are the stakeholders able to accept an exchange of roles with any of the other stakeholders (including exchanging impacts)? This requirement provides a minimum of "fairness" in which an excessive and evident negative outcome cannot be accepted for considering a project as successful.

- **Best option:** In the case that the project does not want to be repeated and it has evident negative outcomes, do the stakeholders think that the project is still the best option from the original alternatives that were available at the conceptualization of the project (including the "do nothing" option)? Not having a better alternative is a strong argument for judging a project as successful. It helps in adjusting (calibrating) the expectations with the real available options and the estimated results.

The following three limitations need to be considered for the principles: (i) they are guidelines to inform a judgement and not formulas, so they do not fit perfectly in every case, (ii) they are generic and should be adapted for each project and stakeholder and (iii) it is also possible to add other principles (with a similar level of abstraction) for specific cases. For the PSEPM, the concept of "judgement principle" is the relevant component rather than the specific principles developed by this study.

The underlying theory for each of the proposed principles is beyond the scope of this chapter. The theory has been developed through a new philosophical approach of the PS concept employing the Alan Turing Test Approach. This approach addresses the theoretical question, "How do we know when a project ends being a failure and starts being successful? This question is answered by understanding a project as the "group of stakeholders that are interrelated for creating an endeavour", which differs from the traditional definition of the project that separates the stakeholders from the project (from the endeavour).

4.7. *Data gathering and judgement process*

The goal of an infrastructure PPP project is to provide a public service using private funding. Consequently, Program Evaluation from public policy can be employed to assess a PPP project.

Program Evaluation, as a discipline, establishes that "evaluating" involves two processes: a measuring process (gathering data) and a decision-making process (a judgement) (Langbein and Felbinger, 2006). The PSEPM is consistent with the Program Evaluation theory including the two steps that operationalize the pyramid. The first step represents all the information that needs to be gathered from each of the multiple perspectives, and the second step represents the judgement process that analyzes the information and generates a decision, which is guided by the judgement principles.

4.8. *PSEPM outcome*

The outcome for determining the impact of the PPP on stakeholders is a judgement employing the selected stakeholder judgement principles (SJP), which are based on the set of gaps that were previously gathered and assessed. This outcome specifies the impacts on each stakeholder and the connection to the overall view of the project.

Specifically for PPP projects, it is important to take into account the fact that there is a difference between (i) the impact of the project on a stakeholder and (ii) the impact of the procurement strategy (PPP) on that same stakeholder. This separation is relevant for avoiding the problem of assigning a negative or positive impact as a result of the PPP, whereas in reality, it can be an impact that would have occurred with any kind of procurement method.

The impact of a PPP project on the stakeholders involves not only the final outcome of the project but also the process of delivering the outcome. This is why the gathering process is the most relevant component of the model, as it generates a broad spectrum of information.

The model's contribution is to provide a solution to the main problem of PS, which is the difference between stakeholders' perspectives. Consequently, it is possible to assess the impact of a PPP on stakeholders by judging if the project was successful for each of them.

5. Expert Discussion

When presenting the PSEPM to experts, there a was general agreement about the need for a tool to assess complex projects such as PPPs. The general structure of the pyramid was accepted as a feasible way to reach a PPP *ex post* impact evaluation tool. However, there are certain issues that need to be considered, may be not at the theoretical level (conceptual framework) but at the practical level, where the tool will be designed and applied.

- **Subjectivity of success criteria:** There was no consensus about the subjectivity of SC. There is a sense that objective criteria must be included even if no relevant stakeholder is accountable for them. This perspective was discussed, but mostly agreed by the inter-viewees that SC are subjective and specific for a single project.
- **Change in expectations:** The expectations must have support and references; they should not be just a statement at the moment of the evaluation. In theory, it should be connected or based on an *ex ante* evaluation.
- **Weighting the relevance of the stakeholder:** When addressing the different perspectives, the relevance of the stakeholder must be con-sidered when prioritizing a specific conflict. This must be explicit in the pyramid.
- **Real accountability of the set of criteria:** Not all stakeholders can agree on a set of criteria. This process will have certain biases in favour of who (in practice) is selected as representing the stakeholder.

The discussed issues generated some changes in the model, which are included in the currently presented version. The issues are consid-ered as a part of the application phase, meaning that the theoretical model (PSEPM) is a good support for the subsequent stages of tool design and tool testing.

6. Conclusion

For evaluating the impact of PPP projects, an original approach is needed. To address the complexity of having multiple relevant perspectives inside a PPP, the PS concept is employed as a starting point to design a PSEPM conceptual framework.

The final aim is to obtain a clear procedure to evaluate PPP projects based on the PSEPM. The conceptual model presents the general rules to drive a multiple perspective evaluation. Subsequently, the PPP evaluation tool needs to be based on the PSEPM, specifying all the issues and concerns that were discussed with the experts plus the ones that can arise when prototyping and testing.

Even though the PSPEM is developed specifically for PPP projects, it can be extrapolated for general projects that need to be analyzed from different stakeholder perspectives. In fact, it is possible to suggest that any project should consider a multiple perspective evaluation whenever the process or the outcomes generate conflicts between different stakeholders.

The stakeholder judgement principles are described generally in this chapter and a larger development is not part of the scope. The presented principles (repetition, fairness and best possible option) are currently being developed for another academic paper.

In the next stages of the research, the PPP evaluation tool will be designed, and it will be presented and discussed in future papers. The PSPEM is a contribution in the theoretical domain and it is the conceptual base for an applicable tool.

References

Baccarini, D. (1999). "The Logical Framework Method for Defining Project Success", *Project Management Journal*, 30(4), 25.

Bauld, S. and McGuinness, K. (2006). *Value for Money*. Paper presented at the Summit.

Bourne, L. and Walker, D. H. (2005). "Visualising and Mapping Stakeholder Influence", *Management Decision*, 43(5), 649–660.

Davis, K. (2013). "Different Stakeholder Groups and their Perceptions of Project Success", *International Journal of Project Management*, 32(2), 189–201. doi:http://dx.doi.org/10.1016/j.ijproman.2013.02.006.

Davis, K. (2016). "A Method to Measure Success Dimensions Relating to Individual Stakeholder Groups", *International Journal of Project Management*, 34(3), 480–493. doi:http://dx.doi.org/10.1016/j.ijproman.2015.12.009.

de Wit, A. (1988). "Measurement of Project Success", *International Journal of Project Management*, 6(3), 164–170. doi:http://dx.doi.org/10.1016/0263-7863(88)90043-9.

European PPP expertise center. (2011). *Guide to Guidance: How to Prepare Procure and Deliver*, PPP Projects, European Investment Bank.

Hall, D. (2015). *Why Public–Private Partnerships Don't Work*. Retrieved from University of Greenwich.

Han, M. (2017). "Arup to Face Court over BrisConnections Collapse", *Financial Review*. Available at: http://www.afr.com/business/infrastructure/arup-to-face-court-over-brisconnections-collapse-20170929-gyr5lu.

Hardcastle, C., Edwards, P. J., Akintoye, A. and Li, B. (2005). "Critical Success Factors for PPP/PFI Projects in the U.K. Construction Industry", *Construction Management and Economics*, 23(5), 459–471.

Heald, D. (2003). "Value for Money Tests and Accounting Treatment in PFI Schemes", *Accounting, Auditing & Accountability Journal*, 16(3), 342–371.

Hevner, A. R., March, S. T., Park, J. and Ram, S. (2004). "Design Science in Information Systems Research", *MIS Quarterly*, 28(1), 75–105.

Jingfeng, Y., Yajun Zeng, A., Skibniewski, M. J. and Qiming, L. I. (2009). "Selection of Performance Objectives and Key Performance Indicators in Public–Private Partnership Projects to Achieve Value for Money", *Construction Management & Economics*, 27(3), 253–270. doi:10.1080/01446190902748705.

Kaplan, R. S. and Norton, D. P. (2007). "Using the Balanced Scorecard as a Strategic Management System", *Harvard Business Review*, 85(7/8), 150–161.

Koops, L., Coman, L., Bosch-Rekveldt, M., Hertogh, M. and Bakker, H. (2015). "Public Perspectives on Project Success — Influenced by National Culture?" *Procedia — Social and Behavioral Sciences*, 194, 115–124. doi:http://dx.doi.org/10.1016/j.sbspro.2015.06.126.

Langbein, L. I. and Felbinger, C. L. (2006). *Public Program Evaluation [Electronic Resource]: A Statistical Guide*, Armonk, N.Y.: M.E. Sharpe.

Liu, J., Love, P. E. D., Davis, P. R., Smith, J. and Regan, M. (2014). "Conceptual Framework for the Performance Measurement of Public–Private Partnerships", *Journal of Infrastructure Systems*, 21(1), 04014023.

Liu, J., Love, P. E. D., Smith, J., Regan, M. and Palaneeswaran, E. (2015). "Review of Performance Measurement: Implications for Public–Private Partnerships", *Built Environment Project and Asset Management*, 5(1), 35–51. doi:10.1108/BEPAM-12-2013-0070.

Mladenovic, G., Vajdic, N., Wündsch, B. and Temeljotov-Salaj, A. (2013). "Use of Key Performance Indicators for PPP Transport Projects to Meet Stakeholders' Performance Objectives", *Built Environment Project & Asset Management*, 3(2), 228.

Müller, R. A. and Jugdev, K. A. (2006). "A Retrospective Look at our Evolving Understanding of Project Success", *IEEE Engineering Management Review*, 110.

Nalewaik, A. and Mills, A. (2017). *Project Performance Review [Electronic Resource]: Capturing the Value of Audit, Oversight, and Compliance for Project Success*, London: Routledge.

National Audit Office. (2013). *Savings from Operational PFI Contracts*, UK: NAO Communications.

Neely, A., Adams, C. and Crowe, P. (2001). "The Performance Prism in Practice", *Measuring Business Excellence*, 5(2), 6–13.

Owen, J. M. and Alkin, M. C. (2007). *Program Evaluation: Forms and Approaches*, 3rd Edn., New York: Guilford Press.

Pollock, A. (2012). "How PFI is Crippling the NHS", *The Guardian*. Retrieved from http://www.theguardian.com/commentisfree/2012/jun/29/pfi-crippling-nhs.

Quade, E. S. and Carter, G. M. (1989). *Analysis for Public Decisions*, New York: North-Holland.

Raisbeck, P., Duffield, C. and Xu, M. (2010). "Comparative Performance of PPPs and Traditional Procurement in Australia", *Construction Management and Economics*, 28(4), 345–359. doi:10.1080/01446190903582731.

Rogers, D. (2013). "Funding the Future After the Demise of PPPs", *The Conversation*. Retrieved from http://theconversation.com/funding-the-future-after-the-demise-of-ppps-18869.

Shenhar, A. J., Dvir, D., Levy, O. and Maltz, A. C. (2001). "Project Success: A Multidimensional Strategic Concept", *Long Range Planning*, 34(6), 699–725. doi:http://dx.doi.org/10.1016/S0024-6301(01)00097-8.

Turner, J. R. (2009). *The Handbook of Project-based Management [Electronic Resource]: Leading Strategic Change in Organizations*, 3rd edn., New York: McGraw-Hill.

Westerveld, E. (2003). "The Project Excellence Model®: Linking Success Criteria and Critical Success Factors", *International Journal of Project Management*, 21(6), 411–418. doi:http://dx.doi.org/10.1016/S0263-7863(02)00112-6.

Xiong, W., Yuan, J.-F., Li, Q. and Skibniewski, M. J. (2015). "Performance Objective-based Dynamic Adjustment Model to Balance the Stakeholders' Satisfaction in PPP Projects", *Journal of Civil Engineering & Management*, 21(5), 539–547. doi:10.3846/13923730.2014.895409.

Chapter 7

Psychological Contracts in Construction: Two Case Studies

Yongjian Ke

Senior Lecturer, University of Technology Sydney,
School of Built Environment, Ultimo NSW 2007, Australia

Yongjian.Ke@uts.edu.au

Abstract

A major contributing factor to the inefficiency in the construction industry is the adversarial relationships that exist within the industry. The strategic importance of relationship style contracting is hence getting recognized. Both public and private sector clients are stipulating more integrated and collaborative forms of procurement. The avoidance of traditional business-as-usual thinking and effective participation in relationship development have the potential to significantly enhance project outcomes. This chapter is limited to non-contractual agreement and behaviours, which are not defined in the contractual documents. Because contracting parties can practice to a different extent according to their will, non-contractual agreements and behaviours can better demonstrate the shared perceptions regarding practices that affect relationships in construction projects. This chapter adopts the concept of psychological contract (PC), which is defined as "individual beliefs in reciprocal obligations between employees and employers" and, refers to unwritten agreements and behaviours between a construction firm and its upstream or downstream procurement partners. This paper aims to investigate the impacts of contracting parties' relationships arising from PC in two distinct procurement systems, including the upgrading project of Beijing Subway Lines 1 and 2 and the improvement project of Fuk Man Road Nullah in Sai Kung, Hong Kong. The Beijing project was procured traditionally while the Hong Kong project adopted the Engineering and Construction Contract (ECC) Option C of New Engineering Contract (i.e. a collaborative contracting approach). Regardless of the procurement methods, findings suggest PCs are present in construction delivery teams.

Keywords: Innovative Construction Management; Psychological Contract; Relational Contracting; Relationship.

1. Introduction

The construction industry is one of the most important industries in the Australian economy in terms of its contribution to the growth of the economy. According to the Australian Bureau of Statistics (2015), the construction industry was the second largest contributor to the Australian Gross Value Added in the year 2014–2015. However, the Australian construction industry is inefficient (Fulford and Standing, 2014). Using simple labour productivity measures, value added per worker in construction ranks 9th out of 18 industries (Richardson, 2014). One of the major contributing factors to this inefficiency is the adversarial relationships that exist within the construction industry (Fulford and Standing, 2014; Smiley *et al.*, 2014). Effective participation in relationship development has the potential to significantly reduce transaction costs in projects (Ke *et al.*, 2015). In practice, both public and private sector clients are calling for more integrated and collaborative forms of procurement, such as Alliance or Public–Private Partnerships (Department of Infrastructure and Regional Development, 2015). However, despite relationship and integrated procurement opportunities being available for some time, some Australian firms hold the traditional business-as-usual thinking and have been slow to adopt them, and those that have participated have problems in articulating significant benefits (Walker *et al.*, 2015).

To argue over the traditional business-as-usual thinking, this chapter aims to examine the existence of the potential impact of relationship style contracting through two case studies, including the upgrading project of Beijing Subway Lines 1 and 2 and the improvement project of Fuk Man Road Nullah in Sai Kung, Hong Kong. The Beijing project was procured traditionally while the Hong Kong project was delivered through an official relational contracting approach. To focus the findings, this research is limited to non-contractual agreement and behaviours, excluding contractual agreement and behaviours. Non-contractual behaviours are those that are not defined in the contractual documents and contracting parties can practice to a different extent according to their will (Sariola and Martinsuo, 2016). The reason is that non-contractual agreements and behaviours can better demonstrate the shared perceptions of contracting parties regarding practices that affect relationships in construction projects. In context, this research adopts the concept of psychological contract (PC) mainly developed by Rousseau (1990) to explain non-contractual agreement and behaviours, which is defined as "individual beliefs in reciprocal obligations between employees and employers". Accordingly, this research aims to investigate whether the PC can be adopted as a suitable framework to explore organizational relationships, thereby responding to the research aim, i.e. to investigate the impacts of contracting parties' relationships arising from PC in two distinct procurement systems.

2. Psychological Contracts in Construction

The term 'PC' refers to the unwritten expectations of an employment relationship as distinct from the formal codified employment contract. It first gained popularity in human resource studies during the 1990s (Rousseau, 1990, 1995; Sparrow and Marchington, 1998). The basis of the PC relationship within a firm is reciprocity between the organization and employees on the perceived obligations and expectations from one another (Guest and Conway, 2002). What is clear is that as the PC evolves around individual beliefs and perceptions, it is highly subjective and can be particular to each employee (Rousseau, 1995). In essence, the PC constitutes an unwritten agreement between the organization and employees based on mutually accepted promises and obligations among the organization and the employees (Sparrow and Marchington, 1998). The PC theory has been widely adopted in many different sectors, such as education (Bordia *et al.*, 2015), hospitality (Bashir and Nasir, 2013), and banking (Yang and Fu, 2013).

In 2004, a team of British academics first wrote about PC displayed by construction project managers (Dainty *et al.*, 2004). Dainty *et al.* (2004) investigated the dynamics that govern the PC between construction project managers and employees to understand its influence on employee turnover. Interestingly, in the context of this research, they identified that understanding based on trust in the PC was being undermined at the time by the organizational change that encompassed flattened organizational structures and overall expansion (Raiden *et al.*, 2009). Studying employee resourcing in construction, Raiden *et al.* (2009) observed inconsistencies due to a reliance on line management personal assessments. These decisions potentially disillusion employees through a violation of the PC. Thus, a more flexible approach to employee resourcing in construction organizations was suggested, where individuals' preferences and expectations are considered throughout resourcing processes (Raiden *et al.*, 2009). Most recently, Dainty and Loosemore (2013) reinforced the link between the individualization of the employment relationship, productivity maximization and the PC. They also noted that intrinsic rewards are becoming more important in the job market, citing an example from their case study where individuals involved in a construction project suggested that the PC offers scope for creativity, innovation and a feeling of long-term impact on their environment (Dainty and Loosemore, 2013).

Beyond the abovementioned limited body of scholarly writing, studies centred on the application of the PC in the construction industry have been largely unexplored, with the exception of recent interest specifically related to PC and workplace safety (Walker, 2010, 2013; Walker and Hutton, 2006). Consequently, it was not known whether the PC can be applicable to the upstream (for example, a client) or downstream (for example, a specialist contractor) organizational relationships among procurement partners in the construction industry until the recent publications by Ke *et al.* (2016) and Davis *et al.* (2017). Ke *et al.* (2016) developed a conceptual model of the PC that was validated using the results from a questionnaire survey administered to construction professionals in Australia. Davis *et al.* (2017) further uncovered the relationships that existed among relational conditions and relational benefits, the PC and the partners' satisfaction. Following Ke *et al.* (2016) and Davis *et al.* (2017), this chapter adopted the concept of PC to describe unwritten agreement and behaviours and used two case studies to investigate the impacts of contracting parties' relationships arising from PC in two distinct procurement systems.

3. Case One: Upgrading Project of Beijing Subway Lines 1 and 2

3.1. *Project background*

Beijing Subway Line 1 is the oldest line in Beijing's mass transit rail network. It is also one of the most heavily used lines from the time the subway opened in 1969. Beijing Subway Line 2 runs in a rectangular loop around the city centre that opened in 1984 and is the second oldest subway line and the only one to serve the Beijing Railway Station. Due to the limitation of design and technology at that time, a comprehensive upgrade was required for both lines to improve the carrying capacity and eliminate hidden safety defects.

The feasibility study of the upgrading project was initially approved in 2003. Preliminary design of the project was completed in 2005. Thereafter, the project was completely approved and the procurement process started. 2007 was the busiest year when the construction work of all specialities involved in the project was carried out at the same time. The main part of the upgrading project had to be completed before the commencement of the 2008 Olympic Games. Other periphery works were completed by the middle of 2009.

The scope of the upgrading project included vehicle renewal, power supply, mechatronics, line network, civil engineering, and the system for communication, signalling and automated fare collection. The total investment hit 8.58 billion RMB (about AU$ 1.64 billion). It involved seven design firms, 77 contractors and 115 equipment suppliers from different specialities.

The upgrading project of Beijing Subway Lines 1 and 2 was the first upgrading project in China's mass transit rail system. There were no strict or standardized guidelines in place at that time. A summary of the characteristics of this project is as follows:

(1) The upgrading project must be completed before the commencement of the 2008 Olympic Games.
(2) The daily operation of Lines 1 and 2 must not be affected by the upgrading works, which meant shorter daily working hours and limited effective working space for the various parties involved in the project.
(3) Because the project involved many participants from different specialities, it increased the difficulty of cooperation and project management.
(4) The scope of the upgrading project covered many different specialities such as civil, mechatronics and electronic information engineering.
(5) The upgrading project had to take into consideration not only the function and durability of the new equipment but also its compatibility with the existing equipment.
(6) Because of the location of Lines 1 and 2, the upgrading project was of vast concern and had a great impact on the community.

Considering the above project characteristics, project management was naturally very complex in terms of overall planning, communication and the management of relationships, schedules, interfaces and changes.

3.2. *Impacts of psychological contract*

The client directly negotiated the overall design contract with the Beijing Urban Engineering Design & Research Institute Co., Ltd. (BUEDRI), who was the original designer of Lines 1 and 2. With the client's agreement, BUEDRI further transferred some design packages to sub-consultants via open bidding. Thereafter, three-way contracts were signed between the client, BUEDRI and sub-consultants. The reason behind the procurement method selection is the historical trustworthy collaboration between the client and the BUEDRI in Beijing Subway Lines 1 and 2. In the procurement of other work packages, past experiences could improve the chance of winning a bid in this project. Bidders with experience in urban rail projects were given bonus marks, and bidders who had previously worked with the client were allotted a higher bonus score. Although the bonus score was not dominant, it may become the key to competitiveness, especially when two bidders offered similar prices. In addition, given a close past relationship with the client, private partners could be able to better understand the call for bidding and thereby prepare more competitive bidding documents. It was also found that the collaboration with the client in this project could increase the chance of undertaking more work at a later stage. This is because the client would expand the scope of a current project if the additional work was not too substantial. Contractors were then paid for the extra work as a change order. If the additional work was substantial and an open bidding was required, the contractors could still have a greater chance to win the bid if they performed well in the earlier stage. The foregoing echoes the findings of Ke *et al.* (2016) and Wong *et al.* (2008) that trust develops from the confidence built upon understanding and knowledge revealing the cognitive bearings of an organization. The managerial implication is hence that positive past relationships have a significant and positive effect on procurement partners.

Reputation and track record in this project helped in clinching contracts in the future. As this project was the first upgrading project in China's mass transit rail system, a proven track record in this project could possibly give the contracting parties a significant advantage in their future participation in similar projects. Future tangible relational outcomes may be correlated with commitment and contracting behaviours (Ke *et al.*, 2016). Knowing the promise of future tangible relational outcomes, contracting parties are likely to present commitment and participation to receive a tied future gain (Scholl, 1981). In the upgrading project, contracting parties focused more on the relationship than profit/cost. Contractors and consultants were keen to build and maintain relationships with the client who was responsible for Beijing's mass transit rail system. The impact of relationships on procurement discussed above could be a convincing proof.

Contracting parties were also found to rely more on relationships to guard against trouble instead of contractual or institutional arrangements. The contracts for the project were not drawn up comprehensively enough as this was the first urban rail upgrading project in China. As such, contracting parties without any direct experiences in such projects were forced to deal with disputes using relationships as a platform. For example, one of the civil work packages had a significant cost overrun and schedule delay due to shorter daily working hours and more limited working space than expected. Unfortunately, the contract signed between the client and the contractor did not provide a standard procedure to deal with such a situation. They soon achieved an agreement to continue the collaboration and leave the dispute to be solved at the end of the project to avoid further delay. Since the client was a source of potential work opportunities in the future, private partners were more willing to compromise on disputes so as to build and maintain a good relationship with the client.

There was no significant difference in project performance in this project between companies with past collaboration and those without past relationships. However, the project management styles were different. A higher level of mutual trust, understanding and communication existed among the private partners who had past good relationships with clients. They made behaviours of each other more predictable and eradicated fears that create difficulties among project participants. Moreover, there were no weak links between the client and private partners, thereby helping in achieving a relationally integrated team. However, the private partners with good past relationship were inclined to assign new staff to work on the project. Public clients also found it embarrassing to criticize the private partners whom they had a prior relationship with. As new staff did not participate in the bidding stage, it took a longer time for them to be familiar with the project and make a real contribution. Interpersonal clashes may occur as well at the start resulting from this issue. On the other hand, staff from those private partners without past relationship with clients were found to be relatively more experienced and proactive to make a good impression on the client. However, there was a lack of mutual understanding and trust at the start. The cost and time involved for communication were also much greater. For instance, staff from contractors without past relationship with the client did not understand the security requirements of Beijing mass transit rail system and were greatly averse to the strict daily identity checks at first. Another example was that some contractors and consultants neglected to attend the frequent meetings at the start of the project. Thereafter, more meetings had to be held during the project to achieve a common understanding.

Interpersonal relationship harmony is also vital to the success of the project, as the development and maintenance of relationship are largely at the interpersonal level. The quality of relationship in this project was good because there was an effective communication in place. This good relationship existed both at the individual and organizational levels. However, since the client usually employs most of the staff, the good relationship at the individual level that was established in a past project would not continue into a future project if the staff assigned to the new project are different from those in the previous project. It was also found that contractors were more concerned about interpersonal clashes. Being placed lower than clients and consultants on the value chain, contractors cannot ask for replacement of client's representative or a consultant should any interpersonal clashes occur between them. Interpersonal barriers, originating from an individual's working attitude or a previous contractual relationship, were the hardest to overcome. This was more critical at the start of the project with those consultants who had a good past relationship with the client, as they assigned many new staff to work on the project.

3.3. *Project outcomes*

As explained above, time performance was the most critical factor in this project. The bulk of the upgrading project was completed before the 2008 Olympic Games. All other works done by the various specialities were finished by the middle of 2009. The client was satisfied with the output quality and service quality of consultants and contractors. No significant safety mishap happened during the project while the operation of Lines 1 and 2 continued normally. Although the relational contracting approach was not formally adopted in this project, relationship played an important role. The quality of relationship among contracting parties was enhanced.

4. Case Two: Improvement Project of Fuk Man Road Nullah

4.1. *Project background*

Early in 2001, the Construction Industry Review Committee (CIRC), Hong Kong, published a report stating that key participants need to work together more effectively and efficiently to complete the project in a satisfactory manner through a partnering approach (CIRC, 2001). Partnering has since been used in various forms, most of which have been non-contractual. Hong Kong has had some unsuccessful partnering experiences as is to be expected when the industry is pioneering a new procurement strategy (CIC, 2010). Contractual partnering has also been adopted in Hong Kong. Some private employers, for example, the Hong Kong Jockey Club, undertook several New Engineering Contract (NEC) projects around 1995 (Kumaraswamy, 2012). Given the positive experiences reported on the use of NEC in the UK and in other regions, the Development Bureau was keen to gain experience using NEC on public works contracts in Hong Kong. Improvement project of Fuk Man Road Nullah in Sai Kung was carried out via NEC Option C with a view to gaining experience and assessing the suitability of NEC for future public works projects.

The project comprises decking of about 180 m long and 12 m wide open *nullah* at Fuk Man Road in Sai Kung, landscaping works and local road junction improvement. Upon completion of the project, the quality of local environment would be enhanced because of the landscaping works on the decked *nullah* and its adjacent open space. Environmental problems posed by the open *nullah* will be substantially reduced.

Drainage Services Department (DSD), responsible for the wastewater and stormwater drainage services enabling the sustainable development of Hong Kong, was the public client of the project. Black & Veatch Hong Kong Ltd. was employed as the supervisor. The contractor was Chun Wo Construction & Engineering Co. Ltd.

4.2. *Impacts of psychological contract*

Engineering and Construction Contract (ECC) Option C "Target Contract with Activity Schedule" was selected for this project, which should be run by cooperation developed among the client, the consultant and the contractor in the interests of the project. Instead of carrying out the duties and exercising the power specified in the contract in a traditional public project, contracting parties would team up and provide their support to the execution of day-to-day project management with a unified objective and a long-term commitment because of the gain/pain sharing mechanism in the contract. The provisions of gain/pain sharing mechanism defined in the contract denote a common purpose for the contracting parties, thereby providing the incentives for the contracting parties to work collaboratively and obtain the future relational outcomes. This is consistent with the findings of Ke *et al.* (2016) and Davis *et al.* (2017) who perceived that future tangible relational benefits are one of the important enabling factors of the PC.

In the ECC Option C, if the risk is a compensation event or is a risk that remains with the client, its cost and time implications are borne by the client. If the event is not one of these, the costs will initially be borne by the client but will eventually be shared through the contractor's share mechanism. In addition, there was a single and unified procedure in place for all issues related to time and cost change, which was designed with an aim to a quick settlement. Such arrangements to protect the fairness and interests of different parties would to some extent increase their readiness to compromise on unclear issues. Ke *et al.* (2016) indicated that there is a significant correlation between good faith and fair dealing and unwritten contracting behaviours such as signalling and disclosing behaviours with regard to technical arguments, disputes and changes. This suggests that when circumstances fall outside existing contractual principles, good faith and fair dealing may make a material difference. Jackson (2007) claimed that good faith and fair dealing is particularly important to result in an acceptable level of initiating and disclosing behaviours at the stage of pre-contractual negotiations when there is no contract imposing obligations on the contracting parties.

To establish a partnering spirit among contracting parties, a start-up partnering workshop, joint NEC training for team members and a joint site office were arranged in this project. Half-yearly partnering reviews and bi-monthly champion group meetings were also set up to maintain and improve the relationship. There were many other team building events such as the joint dragon boat team in 2001 to improve the team working attitude, mutual trust and mutual understanding. Similar to the Beijing case, all these team building events led to a strong interpersonal relationship harmony that is vital for a project's success. However, this good relationship existed mainly at the individual level. Since Hong Kong government departments usually employ a large number of engineers, the good relationship that was established in a past project would not continue into a future project if the engineers assigned to the new project are different from those in the previous project. In the tender evaluation of public projects, a small part of the total score takes into account the performance of a contracting party in past projects. But the quality of a past relationship with the client is not one of the factors. Therefore, the impact of relationship on procurement design and management is not significant in this project either.

The effective share of project information is required as well by the ECC Option C. For example, near transparency of the real cost of the work was seen through the open book accounting. It provides the client with visibility of what they are going to pay and when they are going to pay it. In addition, the NEC also requires the parties to each notify the other of any matter that could increase the total cost, delay in completion or impair the quality. The parties are motivated to identify problems as early as possible and have a proactive approach to jointly finding a solution, rather than putting off decisions or ignoring their resolution. The definite timeframe for both the client and contractor to reply to the early warning notice were also provided. Such contractual arrangements further enhanced the contracting parties' willingness of disclosing and trusting behaviours and expected it to be reciprocated in their business dealings (Davis and Love, 2011). This is also solid evidence to indicate how the formal contract could have an impact on the PC.

4.3. *Project outcomes*

According to DSD, the project was started on August 31, 2009, and completed on May 18, 2012, with a 6-months saving. There was also a 5% of cost saving, which was shared between client and contractor. DSD was satisfied with the quality of both output and service.

The positive management approach encourages full cooperation and provides a good basis for the use of partnering arrangements, leading to a reduction in disputes and a significantly good relationship. For example, all the 15 early warning notices were resolved and 81 out of 93 quotations of compensation events submitted by the contractor were agreed. In addition, there were zero dispute occurrences.

5. Conclusion

The aim of this chapter is to examine the existence of the potential impact of relationship style contracting through two case studies. This research is limited to non-contractual agreement and behaviours, which are not defined in the contractual documents and presented in the concept of PCs. The two cases studied are the upgrading project of Beijing Subway Lines 1 and 2 and the improvement project of Fuk Man Road Nullah in Sai Kung, Hong Kong. The Beijing project was procured traditionally while the Hong Kong project adopted the ECC Option C of NEC3. It was found that regardless of the procurement methods, PCs are present in construction delivery teams. In the Beijing project, the platform of relationship worked very well, because contracting parties shared the common objective, and were then willing to compromise to resolve disputes rapidly. Another important driving force is the perceived future tangible relational benefits. However, it may be costly and time-consuming in other projects. While in the Hong Kong case, the functions of PCs in the project were driven by the contractual arrangements such as gain/pain sharing and open book communication. It is hence likely to replicate the delivery in other projects, which was also the original purpose of the government, i.e. to adopt the NEC3 contract in the project.

This chapter provides a greater understanding of the significant implications and managerial decisions leading to collaborative models of the firm–customer relationship. The taxonomies of PCs developed can be used as a management tool to guide and support the adoption of relationship style contractual arrangements that reduce the likelihood of firms, together with their clients, experiencing reduced productivity and enhanced project success.

Several limitations of the chapter are acknowledged. It could be argued that the case studies only represent a limited breadth of construction delivery. The focus of the research is novel and considers PCs from a construction perspective and this essentially provides limited empirical data on the construction sector to draw from. Possible future research may focus on exploring the internal function mechanisms in the model of the PC as well as quantitatively assessing the impacts on project performance.

References

Australian Bureau of Statistics (2015). *Australian System of National Accounts, 2014–15*, ABS Catalogue No. 5204.0, Australian Bureau of Statistics, Canberra.

Bashir, S. and Nasir, M. (2013). "Breach of Psychological Contract, Organizational Cynicism and Union Commitment: A Study of Hospitality Industry in Pakistan", *International Journal of Hospitality Management*, 34, 61–65.

Bordia, S., Bordia, P. and Restubog, S. L. D. (2015). "Promises from Afar: A Model of International Student Psychological Contract in Business Education", *Studies in Higher Education*, 40(2), 212–232.

Construction Industry Council (CIC) (2010). *Guidelines on Partnering.* Hong Kong, August 2010.

Construction Industry Review Committee (CIRC) (2001). Construct for Excellence, Report of the Construction Industry Review Committee, HKSAR Government.

Dainty, A. R. J., Raiden, A. B. and Neale, R. H. (2004). "Psychological Contract Expectations of Construction Project Managers", *Engineering, Construction and Architectural Management*, 11(1), 33–44.

Davis, P., Jefferies, M. and Ke, Y. (2017). "Psychological Contracts: A Framework for Relationships in Construction Procurement", *Journal of Construction Engineering and Management*, 143(8), 04017028.

Davis, P. R. and Love, P. E. D. (2011). "Alliance Contracting: Adding Value through Relationship Development", *Engineering Construction and Architectural Management*, 18(5), 444–461.

Department of Infrastructure and Regional Development (2015). *National Alliance Contracting Guidelines: Policy Principles*, Australian Government.

Fulford, R. and Standing, C. (2014). "Construction Industry Productivity and the Potential for Collaborative Practice", *International Journal of Project Management*, 32(2), 315–326.

Guest, D. E. and Conway, N. (2002). "Communicating the Psychological Contract: An Employer Perspective", *Human Resource Management Journal*, 12(2), 22–38.

Jackson, S. (2007). "Good Faith in Construction-Will It Make a Difference and is it Worth the Trouble?", *Construction Law Journal*, 23(6), 420–435.

Ke, Y., Davis, P. and Jefferies, M. (2016). "A Conceptual Model of Psychological Contracts in Construction Projects", *Construction Economics and Building*, 16(3), 20–37.

Ke, Y., Ling, F. Y. Y. and Zou, P. X. W. (2015). "Effects of Contract Strategy on Interpersonal Relations and Project Outcomes of Public-Sector Construction Contracts in Australia", *Journal of Management in Engineering*, 31(4), 04014062.

Kumaraswamy, M. M. (2012). Relational Contracting and Principles Underpinning Partnering Practices. Available at: http://www.civil.hku.hk/cicid/3_events/ 117/117_PPT_4_HKU.pdf (Accessed on January 10, 2017).

Raiden, A. B., Dainty, A. R. J. and Neale, R. H. (2009). *Employee Resourcing in the Construction Industry: Strategic Considerations and Operational Practice*, London: Routledge.

Richardson, D. (2014). *Productivity in the Construction Industry*, Canberra: The Australia Institute.

Rousseau, D. M. (1990). "New Hire Perceptions of their Own and their Employer's Obligations: A Study of Psychological Contracts", *Journal of Organizational Behavior*, 11(5), 389–400.

Rousseau, D. M. (1995). *Psychological Contracts in Organizations: Understanding Written and Unwritten Agreements*, Thousand Oaks, California: Sage Publications Inc.

Sariola, R. and Martinsuo, M. (2016). "Enhancing the Supplier's Non-Contractual Project Relationships with Designers", *International Journal of Project Management*, 34(6), 923–936.

Scholl, R. W. (1981). "Differentiating Organizational Commitment from Expectancy as a Motivating Force", *Academy of Management Review*, 6(4), 589–599.

Smiley, J. P., Fernie, S. and Dainty, A. (2014). "Understanding Construction Reform Discourses", *Construction Management and Economics*, 32(7–8), 804–815.

Sparrow, P. and Marchington, M. (1998). *Human Resource Management: The New Agenda*, Upper Saddle River: Financial Times Pitman Pub.

Walker, A. (2010). "The Development and Validation of a Psychological Contract of Safety Scale", *Journal of Safety Research*, 41(4), 315–321.

Walker, A. (2013). "Outcomes Associated with Breach and Fulfillment of the Psychological Contract of Safety", *Journal of Safety Research*, 47, 31–37.

Walker, A. and Hutton, D. M. (2006). "The Application of the Psychological Contract to Workplace Safety", *Journal of Safety Research*, 37(5), 433–441.

Walker, D. H. T., Harley, J. and Mills, A. (2015). "Performance of Project Alliancing in Australasia: A Digest of Infrastructure Development from 2008 to 2013", *Construction Economics and Building*, 15(1), 1–18.

Wong, W. K., Cheung, S. O., Yiu, T. W. and Pang, H. Y. (2008). "A Framework for Trust in Construction Contracting", *International Journal of Project Management*, 26(8), 821–829.

Yang, H. and Fu, H. (2013). "Psychological Contract Structure of Banks and Credit Customers Based on the Empirical Study of Banking Industry in Jiangxi Province", *Journal of Applied Sciences*, 13(17), 3431–3434.

Chapter 8

An Exploratory Model on Greenhouse-Gas Emissions and Costing Assessment in Building Operation Stage

Vivian W. Y. Tam*, Khoa N. Le, Cuong N. N. Tran
and I. M. Chethana S. Illankoon

*School of Computing Engineering and Mathematics,
Western Sydney University, Locked Bag 1797, Penrith,
NSW 2751, Australia*
**College of Civil Engineering, Shenzhen University,
Shenzhen, China*
**vivianwytam@gmail.com*

Abstract

One of the several reasons that lead to global warming appears to be the large contributions of greenhouse-gas (GHG) emissions. International researchers are trying to simulate life-cycle GHG emissions and cost by using a variety of building optimization models. A simple automated model which can be run in a friendly and popular software such as Microsoft Excel is needed in the industry.

---

*Corresponding Author.

This chapter presents a comprehensive energy consumption-optimization model, analyzing different types of buildings, which comply with the energy-efficiency requirements of the National Construction Building Code of Australia. This model has been developed in Microsoft Excel and Visual Basic in order to determine GHG emissions associated with energy usage in building operations for 50 years of building life as well as energy costs. In conjunction with this, one of the five points in the GHG emission reduction Credit 15A of Green Star Design & As Built rating system assesses project sustainability at all life-cycle stages by selecting 13 roof types, 16 wall types and 3 floor types according to Australia and New Zealand's *Insulation Handbook, Part I: Thermal Performance*. By considering eight climate zones in Australia, developers and designers may use this model to choose the optimum building types with the highest total R-value including roof, wall and floor for their design.

Keywords: Australia; Commercial Building; Construction; Environment; Green Star; Green Building; Greenhouse-gas Emissions.

1. Introduction

The total energy usage in Australian commercial buildings has been projected from 134.6 PJ in 2009 to 169.6 PJ in 2020, which will produce about 41 Mt CO_2-e, which is about 1.2 times the volume of greenhouse-gas (GHG) emissions in 2009. This tremendous critical enigma will create a significant impact on the global climate and should be resolved by comprehensive solutions. In fact, solutions to this problem lie in the hands of architects and designers who play an important role in energy conservation in the whole building operation. GHG emissions from energy usage in buildings can originate from (1) direct emissions from fuel combustion for cooking and heating and (2) emissions from electricity usage for heating, cooling, and providing energy for buildings (Center for Climate and Energy Solutions, 2016). Energy usage for heating and cooling accounts for about 18% of the office tenancies' average energy usage from 1999 to 2012 in Australia (Council of Australian Governments (COAG), 2012).

According to Pitt and Sherry (2012), a standalone office consumes 35.6 PJ amount of energy, out of which 43% is used for heating and ventilation. Further, there is 40,911,000 m^2 total area of offices in Australia (Pitt and Sherry, 2012). Therefore, approximately a standalone office consumes around 395 kWh of energy annually per square metre. An up-to-date literature review has shown that a building envelope consisting of exterior walls, windows, external doors, roof and floor could have a significant impact on energy consumption during the operation stage (Lawania and Biswas, 2016). In order to assist designers, especially in a design phase which is the period when the premature selection of components for the building envelope is necessitated, the Insulation Council of Australia and New Zealand has published a handbook that shows the actual structure system complying with the Building Code of Australia, AS/NZS 4859. Nevertheless, with particular geographic and climatic characteristics of Australia (Australia's total land area is 7,692,024 km^2, divided into eight climate zones (Ga.gov.au, 2016)) and the number of combinations among the systems of roof (13 types), wall (11 types) and floor (3 types), it is a huge task to determine which is the clear optimum choice.

This chapter aims to develop a computer-aided model focusing on various building envelopes, required commercial building areas and climate in different zones in Australia in order to achieve the required points for Credit 15A: In Green Star. The proposed model can then be used to eliminate difficulties while green-building design can be automated, user-friendly and effective. Companies can readily use this model to implement their green-building strategies without the lengthy design processes. The objectives of the model are to (1) choose the optimum building envelope system for each climate zone in Australia, (2) calculate the average energy usage in building operations in 50 years of the building's lifespan and (3) estimate the average GHG emissions during the building operations as well as the median energy expenditure in the building's lifetime.

2. Research Methodologies

2.1. *Green Star energy credit*

The Green Building Council of Australia operates as Australia's only national voluntary comprehensive environmental rating system for buildings and communities known as Green Star. Criteria on "Energy" in Green Star account for 24 out of 100 points. To encourage the reduction of GHG emissions from building operation, up to 20 points can be achieved, which is declared in Credit 15 (GBCA, 2015). Specifically, "Energy" contains all energy types used in building's operation as per the definitions used by the National Greenhouse and Energy Reporting (NGER) scheme. In order to achieve points for credit 15 in Green Star, a condition must be met for each of the five pathways, which is verified using a modelling approach or an authoritative approach as explained in this credit. The project uses Credit 15A prescriptive pathway that can be awarded up to a maximum of five points where eight credit elements are (building envelope: one point; glazing: one point; lighting: one point; ventilation and air-conditioning: one point; domestic hot water systems: one point; building sealing: one point; Accredited Greenpower®: one to two (1–2) points) of the building, which is classified in NCC's requirements.

2.2. *Model development*

2.2.1. *Assumptions*

Because of the unique architectural features of each building, data may be difficult to collect and be compared with other available datasets (Cattarin *et al.*, 2016). The typical lifespan of a building has been estimated as at least 50 years (Lawania and Biswas, 2016). Therefore, the temperature that is used for calculation in this period of time will be randomly selected using 10 temperature combinations in the last 11 years from the Australian Bureau of Meteorology for seven major cities: Adelaide, Brisbane, Canberra, Hobart, Melbourne, Perth and Sydney. A foundation temperature is closely connected to the model. The Floor is thus assumed to be cooler than a mean difference of +3°C compared with the air temperature (Gerner and Budd, 2015).

The appropriate bounds on office temperatures help improve working quality and reduce negative effects on employee's attitude. The optimum temperature for office work depends on seasons and employee's attire and ranges from 20°C to 26°C. The internal

Table 1. Office and room areas in Queensland, Australia.

Category (AusHFG)	Work Space m^2 (AusHFG)/Features (m^2)
Office type A	18
Office type B	15
Office type C	12
Office type D	9
Shared office type A	12
Shared office type B	15
Shared office type C	20
Meeting room A	9
Meeting room B	12
Meeting room C	20
Meeting room D	30
Meeting room E	55

temperature inside the building used for the model is 25°C. The number of working hours is 12 h per day including overtime work for five days a week.

In order to adapt a work space design to meet future agency requirements within the existing premises and allow owners to achieve GHG emission credits in Green Star, office design should be based on a modular model. According to the Queensland Health Work Place and Office Accommodation Guideline, there are four office types (types A, B, C and D) with work spaces of 9, 12, 15 and 18 m^2, respectively (Table 1), whereas New South Wales Government Office Design Requirements stated that the work area benchmark is not to exceed 15 m^2 per person in metropolitan and not to exceed 14 m^2 per person in regional centres. Similarly, in the Government Office Accommodation Framework of South Australia Government, space standard for an individual office area is less than 12.6 m^2.

To simplify the calculation of the roof upper surface solar absorbance value, the model uses three roof colour categories: light, medium or dark (Vorobyev and Guo, 2010; Yacouby *et al.*, 2011). Based on this value, the model will be able to define the minimum total R-value for each climate zone. Hence, the model is developed based on the modular design and is be classified into four types of working space in order to cover all requirements in all states in Australia. Four office-room types (types A, B, C and D) will have the areas of 9, 12, 15 and 18 m^2, respectively. The minimum room clearance ceiling height has been set at 2.4 m for the calculation in the model (Australian Building Codes Board (ABCB), 2016).

2.2.2. *Electricity consumption, greenhouse-gas emissions and energy costs analyses*

(a) Electricity consumption calculation

The electricity consumed for heating and cooling in the building equals the conductive heat loss rate, which is proportional to the U-value, surface area and temperature difference, as shown in Eq. (1) (Jankovic, 2013):

$$Q_c = U \times A \times (T - T_o), \tag{1}$$

where Q_c denotes conductive heat loss rate (W), A is the surface area (m), T is the internal air temperature (K or °C), To is the external air temperature (K or °C), U is the thermal conductance ($W.m^{-2} \cdot K^{-1}$), which can be calculated by the thermal resistance of a material, and R-value can be shown in Eq. (2) (Butcher, 2015).

$$U = \frac{1}{R}. \tag{2}$$

The total thermal resistance for each of the above building structures is calculated by Eq. (3) (Butcher, 2015):

$$R = R_i + R_1 + R_2 + R_3 + R_4 + R_o, \tag{3}$$

where R denotes the total thermal resistance of the structure, R_i and R_o are the thermal resistances of internal and external surface of structure, R_1, R_2, R_3, R_4 are the thermal resistances of physical layers of structure.

The data for thermal resistances of the building elements such as roof, wall or floor are collected from the Insulation Council of Australia and

New Zealand's handbook on the model which is developed in Microsoft Excel and Visual Basic.

(b) Greenhouse-gas emission calculation

GHG emissions often include the following: (a) GHG emission scope 1 is defined as the distribution of GHG into the air as a direct product of an activity or series of activities (including supplementary activities) that constitute the facility or (b) GHG emission scope 2 is defined as the GHG distribution in the air as a direct product of one or more activities that generate electricity, heating, cooling or steam that is consumed by the building, though that do not constitute the facility's parts. In this research, the amount of GHG emissions associated with the energy usage in building operations, or the amount of GHG emission scope 2 is calculated by Eq. (4) (Office of Parliamentary Counsel, 2016):

$$Y = \frac{Q \times E}{1000}, \tag{4}$$

where Y denotes the scope 2 emissions (CO_2-e tonnes), Q is the purchased electricity quantity consumed from the building's operation (kW $\times$ h), E is the GHG emission scope 2's emission factors for Australian state or territory (Kg CO_2-e/kW $\times$ h), which is presented in Table 2.

Table 2. Indirect (scope 2) emission factors from consumption of purchased electricity from grid.

State, Territory or Grid Description	Emission Factor kg CO_2-e/kWh
Australian Capital Territory	0.84
New South Wales	0.84
Victoria	1.09
Queensland	0.78
South Australia	0.53
South West Interconnected System in Western Australia	0.72
Tasmania	0.12
Northern Territory	0.67

(c) Energy costs

From the previous sections, the energy consumption for the changes in the insulation material is considered. However, generally in offices, the energy consumption is fulfilled through the electricity supplied from the national grid. Therefore, in order to calculate the cost, the prices relating to the electricity supply from the national grid are considered. Herein, the "energy cost" implies the cost of electricity consumed by each of these options. Further, for the purpose of this research, it is assumed that the entire energy requirement is fulfilled through the electricity supplier and renewable energy sources such as solar panels are not considered.

After calculating the energy consumption, the cost of energy is also calculated for the entire lifespan of 50 years. For this calculation, it is assumed that entire energy requirement is satisfied through electricity and renewable energy is not used. The electricity cost per unit of energy (AUD per kWh) varies with the different states within Australia. To calculate the total energy costs, the Net Present Value (NPV) calculation is used, which is given in the following equation (Dell'Isola and Kirk, 2003):

$$\mathrm{NPV}(i,N) = \sum_{t=0}^{N} \frac{R_t}{(1+i)^t}, \tag{5}$$

where i denotes the discount rate, t denotes the time of cash flow, R^t denotes the net cash flow and N is the total number of periods.

The discount rate is established considering the time value of money and the associated risk. The minimum attractive rate of return is commonly used as the discounting rate (Dell'Isola and Kirk, 2003). The rate of interest on a 25-year Treasury bond in Australia is 3.25% per annum (Australia Building Codes Board, 2016). Further, the return of assets for a non-residential construction firm is around 3.30% in Australia (Deloitte Access Economics, 2016). Therefore, considering these rates, the real discounting rate is taken as 3.25% for this calculation.

The energy costs occur quarterly per annum. In this calculation, a flat rate for cost is identified annually. Commencing from the current year, the energy costs is forecast for a period for 50 years. Australia Energy Market

Commission (AEMC) (2015) report forecasts the energy price increasing up by 2018. According to this report, there will be a 2.45% increase in energy prices in Queensland, 0.2% increase in New South Wales, 0.8% increase in South Australia and 6.2% increase in Western Australia in 2018. However, there is no change expected in Victoria. The price projections for the 50-year period from 2018 is considered based on the report on retail electricity price projections. Further, Australia Energy Market Commission (AEMC) (2015) report projects electricity prices for a period till 2037, considering economic scenarios and past trends from 1981. According to this report, the average electricity growth rates are as follows: 0.35% in Queensland, 1.18% in New South Wales, 0.78% in Victoria, 0.85% in South Australia and 0.85% in Tasmania (Parisot and Nidras, 2016). In Perth, the electricity prices are regulated by the state government (Australia Energy Market Commission (AEMC), 2015). Therefore, the average of the price growth is taken for Perth in this calculation. These growth rates of electricity prices are considered to forecast the energy prices in the next 50 years.

The current energy costs per kWh vary depending on retailers. Therefore, quotations for energy prices are obtained from 10 retailers. There is a minor price range change for the quotations from these retailers and therefore, the average flat rate is established for each state. The unit of measurement is AUD per kWh. After establishing the current energy costs per unit, it is compounded based on the identified growth rates for electricity prices. Finally, the energy costs per unit are multiplied by the energy consumption to obtain the energy costs for each year and the summation of the annual energy costs gives the total energy costs for each option.

2.2.3. *Types of wall, floor and roof structures*

According to *Insulation Council of Australia and New Zealand* (2014), there are mainly 13 types of roofing arrangements, 11 types of wall arrangements and three types of floor arrangements. However, for each of these arrangements, there are several options and combinations for insulation material. For roofing solutions, the required *R*-value can be obtained by changing different types of sarking material and ceiling insulation.

For walls, there are also wall insulation and membranes. There are mainly nine types of insulations used in this model: (1) single-sided foil, (2) double-sided anti-glare foil, (3) vapour permeable membrane, (4) bubble/foam foils, (5) ceiling and wall batts, (6) foil-based blanket, (7) foil-faced board, (8) anti-glare expanded polystyrene (EPS) board and (9) reflective rigid polyisocyanurate (PIR) sheets.

3. Data Analysis

This model provides data on GHG emissions and energy costs for different insulated floor, wall and roof types. Further, it selects the effective solutions for a selected element structure (R0100: pitched tiled roof with flat ceiling, R0200: pitched metal roof with flat ceiling) for each state considering GHG emissions and energy costs. Table 3 reports structures with the lowest GHG emissions and energy costs in Canberra. Analyses on GHG emissions and energy costs for walls, floors and roofs are given in Sections 3.1–3.3.

3.1. *Analyses on roof structures*

The pattern of GHG emissions and the energy costs appear similar. There is a slight increment in GHG emissions and energy costs for roof type R0450NVW: foil-faced R1.8 blanket, ceiling insulation R2.0 (75 mm), rafter remaining air space: 75 mm and R0410NVS: single-sided foil, ceiling insulation R2.5 (90 mm), rafter remaining air space: 100 mm. Both these roof types use ceiling insulation with R-value of 3.0. For roof type R0410NVS: pitched metal roof with cathedral ceiling below rafters (concealed rafters), the sarking material is single-sided foil. However, there is no contribution from this type of foil for the insulation. Roof type R0440NVW: pitched metal roof, foil faced R1.3 blanket, ceiling insulation R2.0 (75mm), rafter remaining air space: 95 mm is the lowest GHG emitting as well the roof with least energy costs. This includes foil-faced R1.3 blanket with the R-value of 1.38 and ceiling insulation R4.0. The total R-value decreases with the increase in GHG emissions and energy costs. There is an inverse relationship between the total R-value and the GHG emissions.

Table 3.   Optimum cost and GHG emissions for each type of structure in Canberra.

	Roof Types			Floor Types			Wall Types	
Type of Roof	GHG Emissions (CO_2-e Tonnes)	Energy Costs (AUD)	Type of Floor	GHG Emissions (CO_2-e Tonnes)	Energy Costs (AUD)	Type of Wall	GHG Emissions (CO_2-e Tonnes)	Energy Costs (AUD)
R0230VS	0.029	5.709	F0230NVW	0.0012	7.515	W1490NVW	0.106	20.486
R0660NVW	0.029	5.657	F0220NVW	0.0032	8.271	W0140NVW	0.133	27.612
R0860NVW	0.029	5.656	F0350NVW	0.0032	8.271	W1640NVW	0.133	27.103
R0740NVW	0.032	6.227	F0150NVW	0.0048	9.026	W0340NVW	0.143	28.884
R1270NVS	0.033	6.382	F0300NVW	0.009	9.941	W1240NVW	0.143	28.884
R0130VS	0.035	6.865	F0310NVW	0.009	9.941	W1210NVW	0.156	30.284
R1090NVS	0.035	6.849	F0240NVW	0.0158	10.736	R0940NVW	0.163	31.033
R0530NVS	0.04	7.889	F0100NVW	0.021	11.054	W1100NVW	0.184	35.374
R0450NVW	0.044	8.668	F0110NVW	0.021	11.054	W0850NVW	0.186	36.391
R0320NVW	0.047	9.291	F0270NVW	0.0448	14.712	W0240NVW	0.188	37.409
						W1550NVW	0.215	42.372

The lowest carbon emitting roof structures for the climatic conditions of Canberra are R0230VS: pitched metal roof with flat ceiling, double-sided bubble foam foil, Rm0.2, R0660NVW: pitched metal roof with cathedral ceiling above rafters (exposed rafters), foil-faced R2.5 blanket and R860NVW: pitched metal roof with plasterboard ceiling (exposed rafters), foil-faced R2.5. Further, the relative energy cost is the lowest for R0660NVW: pitched metal roof with cathedral ceiling above rafters (exposed rafters), foil-faced R2.5 blanket. The energy costs and the GHG emissions have a positive relationship whereas with the increase of GHG emissions, energy costs also tend to increase.

3.2. *Analysis on floor structures*

The lowest GHG-emitting floor combination for Type F0200: suspended concrete slab is floor type F0220NVW: suspended concrete slab with fibrous insulation boards or blankets, foil-faced board/blanket, enclosed subfloor, R1.4 board or blanket. This floor type includes a non-faced board/blanket and vapour permeable membrane.

In addition, insulation is basically provided by R3.0 boards or blankets. The highest GHG-emitting material is F0220NVS: suspended concrete slab with insulation material R1.4 board or blanket. Similar to roof structures, in floor also, there are GHG emissions and the energy costs follow a similar pattern. When the GHG emissions increase, so do the energy costs.

The lowest GHG-emitting floor structure is F0230NVW: suspended concrete slab with rigid board system, PIR insulation board, enclosed subfloor, R1.4 Reflective PIR (30 mm). The lowest energy costs are obtained from the same floor type F0230NVW: suspended concrete slab with rigid board system, PIR insulation board, enclosed subfloor, R1.4 Reflective PIR (30 mm). Both the highest energy costs and the GHG emissions are obtained from floor type F0270NVW: suspended concrete slab with foil membrane system with spacer, double-sided bubble/foam Rm0.2, enclosed subfloor, 25 mm air gap.

3.3. *Analysis on wall structures*

In wall structure type W0300: light weight cladding (fixed to battens) in Perth, the lowest GHG emissions are for wall type

W0340NVW: double-sided antiglare EPS board R_m 0.37, 15 mm, stud wall batts R2.0 (90 mm). This type of wall includes double-sided antiglare EPS board R_m 0.37, 15 mm for sarking and stud wall batts R2.7 for insulation. The energy costs are also recorded as the lowest for this wall type. The highest GHG emissions and the energy costs are reported from wall type W0390NVW: vapour permeable, stud wall batts R2.5 (90 mm). The lowest GHG-emitting wall type is W0340NVW: double-sided antiglare EPS board R_m 0.37, 15 mm, stud wall batts R2.7 (90 mm). Similar to floor and roof, in wall structures also, the energy costs are lower for the lowest GHG-emitting wall type.

In this model, all the three structures can be analysed in terms of GHG emissions and energy costs. Further, this can be applied for each climatic zone in Australia. Therefore, developers can analyze the most suited type of insulation, sarking and other material once the structure of the roof, wall or the floor is finalized. Based on the electricity consumption data, average electricity consumption for this type of office building (with 9 m² floor area) is approximated at 300 kWh. This figure is in line with the electricity consumption data given in the assumption made in this chapter.

4. Conclusion

This chapter has developed the computer-aided model using Microsoft Excel and Visual Basic platforms to automate the numerical calculations for life-cycle GHG emissions to achieve optimal points in Credit 15A: GHG emissions reduction in Green Star. Different aspects of building envelopes (roof, wall and floor) and various climate zone scenarios can be varied in the model including weather condition and climate zone to better fit individual organizations and designers.

The model is designed to be flexible and is one of the simplest ways of choosing the optimal building envelope for achieving the Green Star's Energy Credit (Credit 15A) focusing on insulation. This model can identify the most suitable type of insulation for the selected structure in the selected climate zones.

Apart from that, this provides the GHG-emission calculation for the selected option and the energy costs for the selection. Therefore, this can be used to compare and contrast different options for insulation.

This model could become an independent application and consequently can be applied for different building configurations.

Acknowledgement

The authors wish to acknowledge the financial support from the Australian Research Council (ARC) Discovery Project under grant number DP150101015.

References

Australia Building Codes Board (2016). *National Construction Code 2016: Building Code of Australia*. Canberra: The Australian Building Codes Board.

Australia Energy Market Commission (AEMC) (2015). *2015 Residential Electricity Price Trends*. Sydney, Australia: Australian Energy Market Commission.

Australian Building Codes Board (ABCB) (2016). *NCC 2016*. Canberra: National Building Code of Australia.

Butcher, K. J. (2015). *CIBSE Guide A — Environmental Design*. 7th edn., CIBSE.

Cattarin, G., Causone, F., Kindinis, A. and Pagliano, L. (2016). "Outdoor Test Cells for Building Envelope Experimental Characterisation — A Literature Review", *Renewable and Sustainable Energy Reviews*, 54, 606–625.

Center for Climate and Energy Solutions (2016). "Buildings and Emissions: Making the Connection", viewed 30.6.2017. Available at: http://www.c2es. org/technology/overview/buildings.

Council of Australian Governments (2012). *Baseline Energy Consumption and Greenhouse Gas Emissions in Commercial Buildings in Australia, Part 1*, Council of Australian Governments, Australia.

Dell'isola, A. J. and Kirk, S. J. (2003). *Life Cycle Costing for Facilities: Economic Analysis for Owners and Professionals in Planning, Programming, and Real Estate Development: Designing, Specifying, And Construction, Maintenance, Operations, And Procurement*. Kingston, MA: Reed Construction Data.

Deloitte Access Economics. (2016). *Construction Sector — Outlook, Labour Costs and Productivity*. Retrieved from https://www2.deloitte.com/content/ dam/Deloitte/au/Documents/Economics/deloitte-au-economics%E2%80%93 outlook-labour-costs-productivity-310316.pdf.

Ga.Gov.Au (2016). Maps of Australia — Geoscience Australia. Available at: http://www.ga.gov.au/data-pubs/maps (accessed on September 15, 2016).

GBCA (2015). *Green Star — Design & As Built.* Sydney: Green Building Council of Australia.

Gerner, Ed. and Budd, A. (2015). "Australian Surface Temperature Corrections for Thermal Modelling", in *Proceedings of the World Geothermal Congress 2015*, Pleiades Publishing, Melbourne, Australia.

Insulation Council of Australia and New Zealand (2014). Insulation handbook, Insulation Council of Australia and New Zealand, Melbourne, Australia, viewed 30.6.2017. Available at: https://www.qld.gov.au/environment/assets/documents/climate/climate-action-submissions/insulation-council-aus-nz-icanz.pdf.

Jankovic, L. (2013). *Designing Zero Carbon Buildings Using Dynamic Simulation Methods.* New York: Taylor and Francis.

Lawania, K. K. and Biswas, W. K. (2016). "Achieving Environmentally Friendly Building Envelope for Western Australia's Housing Sector: A Life Cycle Assessment Approach", *International Journal of Sustainable Built Environment*, 5(2), 210–224.

Office of Parliamentary Counsel (2016). *National Greenhouse and Energy Reporting* (Measurement Determination 2008).

Parisot, L. and Nidras, P. (2016). Retail electricity price history and projections. In *Retail Price Series Development.* Melbourne, Australia: Australian Energy Market Operator (AEMO).

Pitt and Sherry (2012). *Baseline Energy Consumption and Greenhouse Gas Emissions in Commercial Buildings in Australia.*

Vorobyev, A. Y. and Guo, C. (2010). "Solar Absorber Surfaces Treated by Femto-second Laser", International Conference on Biosciences (Biosciencesworld), March 7–13, 2010, pp. 135–138.

Yacouby, A. M. A., Khamidi, M. F., Nuruddin, M. F., Farhan, S. A. and Razali, A. E. (2011). "Study on Roof Tile's Colors in Malaysia for Development of New Anti-Warming Roof Tiles With Higher Solar Reflectance Index (SRI)." *National Postgraduate Conference (NPC)*, September 19–20, 2011, pp. 1–6.

Chapter 9

Applicability of Casa Azul Certification as Sustainability Tool Available in Brazil — Case Study: Condo Neo Niterói

Vivian W. Y. Tam*

School of Computing Engineering and Mathematics, Western Sydney University, Locked Bag 1797, Penrith, NSW 2751, Australia
College of Civil Engineering, Shenzhen University, Shenzhen, China
vivianwytam@gmail.com

Camila Dinamarco, Assed Haddad and Ana Evangelista

Universidade Federal do Rio de Janeiro, Brazil

Abstract

The aim of this chapter is to assess the feasibility of acquiring the Brazilian sustainability guide tool named "Selo Casa Azul Environmental Certification — CEF (Brazilian Federal Bank)" by the condominium

---

* Corresponding author.

Neo (Rio de Janeiro, Niterói). It is important to consider the impact of human activities in the environment on the construction industry. The Casa Azul Seal is the first classification system of sustainability of projects developed based on the characteristics of the Brazilian constructions. The sustainability guide proposes solutions appropriate to the local reality, so as to improve the use of natural resources and social benefits. The main mission of the seal is to recognize projects that adopt efficient solutions in the construction, use, occupation and maintenance of the buildings. There are 53 evaluation criteria, which are divided into six categories: urban quality, design and comfort, energy efficiency, material resource conservation, water management and social practices. The construction industry and even the built environment generally have a dominant impact on the economic, social and environmental aspects. Concerns regarding environmental issues are increasing worldwide and it is necessary to investigate new environment-friendly alternatives for the construction industry. The methodology used for the development of this chapter consists of a literature review and case study to validate the theoretical data available in the current technical literature. The literature review was through scientific articles and national and international publications, and the case study was conducted in a multi-dwelling building condominium in the city of Niterói (Rio de Janeiro, Brazil). It was verified that the building would be eligible for a possible acquisition of the Casa Azul in the silver-level grading if it met the mandatory criteria 5.1 (individualized water measurement), 6.1 (education for construction and demolition waste (CDW) management) and 6.2 (environmental education of employees). However, this condominium did not meet these mandatory requirements. The mandatory criteria are indispensable to promote water savings, reuse and final disposal of CDW. Additionally, these motivate construction professionals to embrace environmental management and education causes.

Keywords: Casa Azul; Green Building; Environmental Certification; Sustainability.

1. Introduction

The Green Building Council (GBC), Brazil, was created in 2007 as a non-governmental organization that promotes the training of companies to acquire certifications in the area of sustainable buildings. Leadership in

Energy and Environmental Design (LEED) is another organization, created in the US through the US GBC, which is a green label for buildings that follow international sustainability models (GBC Brazil, 2017). The LEED green label has been implemented by the GBC Brazil organization. LEED is the most important label in the field of sustainable construction around the world. Another certificate was created by a Brazilian governmental bank, Caixa Econômica Federal, in the year 2010 under the name of the "Guia de Sustentabilidade Ambiental do Selo Casa Azul". The purpose of this guide is to instruct professionals, students and companies focused on the construction sector to develop sustainable projects.

One of the first concepts of sustainability that came in 1987 was titled "Our Common Future through the Brundtland Report". This document was prepared by the World Commission on Environment and Development and was signed by the United Nations (UN). The term "sustainable development" has been introduced in this document and it refers to the development that meets the present needs without compromising the ability of future generations to meet their own needs (Brundtland, 1987).

Environmental issues are gaining importance and are increasingly being discussed worldwide. With this, architecture and construction needed to adapt to this new reality. Several countries have developed systems for evaluating sustainable buildings. Systems for environmental assessment of buildings emerged in the 1990s in Europe, US and Canada to motivate the construction market to achieve better levels of environmental performance (Matos and Wagner, 1998).

As environmental agendas vary from country to country, methods used in other countries should not be applied without appropriate adaptations, including the definition of sustainability requirements that should be met by buildings established in Brazil. Some European countries, in addition to the United States, Canada, Australia, Japan and Hong Kong, have a building certification system. In Brazil, the most well-known environmental assessment system is the GBC LEED Certification or Green rating, created by the USGBC, in addition to the recently created Selo Casa Azul (Dinamarco, Haddad and Evangelista, 2016).

According to the Caixa Econômica Federal (CEF), the Casa Azul certification seeks to recognize projects that prove their contributions towards reducing environmental impacts. Criteria related to urban quality, comfort design, energy efficiency, conservation of material resources, water management and social practices are evaluated. According to the guide, Boas Práticas para Habitação mais Sustentável (John and Prada, 2010), in creating the Casa Azul, CEF intends to encourage the rational use of natural resources in the construction of housing developments, reduce the cost of maintaining buildings and the monthly expenses of its users as well as promote awareness among entrepreneurs and residents about the advantages of sustainable buildings.

This is the very first certification on sustainability classification system for civil construction projects offered in Brazil and developed for the particularities of the country. Its methodology was developed by a team of technicians from the state bank CEF having extensive experience in housing developments projects. Its methodology focuses on ascertaining, through the technical feasibility analysis of the enterprise, the fulfilment of the criteria established by the instrument. It also instigates the adoption of practices aimed at the sustainability of housing developments. Adherence to the certification is voluntary. The applicant must show interest in acquiring it in order for the project to be analyzed under the criteria of this certification (Caixa, 2010).

In Brazil, according to Schenini *et al.* (2004), lack of ecological awareness in the construction industry resulted in irreparable environmental damages, which were aggravated by the massive migration process that occurred in the second half of the last century, which caused a great demand for new housing.

Currently, the civil construction model practised in Brazil, throughout its production chain, causes several environmental damages, since, besides using non-renewable raw material and consuming high amounts of energy, both in the extraction and in the transportation and processing of inputs, it is also wasteful in the use of materials and is considered a major source of waste within society (Roth and Garcias, 2009).

Civil construction needs to adopt more sustainable practices avoiding excessive and indiscriminate consumption of non-renewable raw materials. It is also important to observe the life cycle of the materials used and reuse construction and demolition waste (CDW) and enhance techniques of clean energy sources and especially if they are engaged in developing sustainable projects that contribute to the recovery of the environment (Dinamarco, 2016).

Through the proposal of Casa Azul Certification in housing developments in Brazil, it will be possible to contribute to the awareness of the implementation of sustainable housing developments that does not result in the environmental degradation. Each criteria adopted by Casa Azul to determine the possible rating acquisition is through the implementation of more sustainable practices during construction and also more sustainable techniques to reduce energy consumption.

2. Methodology

The methodology of the present study is initially a literature review. In addition to this, a screening of all the material that specifically addresses the topic under study was performed. This study based in a case study t, employing the analysis of qualitative data. The case study was carried out in the Neo Niterói condominium, in the city of Niterói (Brazil). As data collection instruments, photographic cameras, owner's manual, apartment plans, descriptive memorial of the Neo Niterói condominium and notebooks were used. In Brazil, the most well-known environmental assessment system is the LEED or USGBC. However, the Brazilian certification system was created by a Brazilian governmental bank, Caixa Econômica Federal, in the year 2010 under the name of the "Guia de Sustentabilidade Ambiental do Selo Casa Azul" aiming the Brazil reality to improve the use of natural resources including the social benefits. The aim of this chapter is to investigate the suitability of the criteria in the Brazilian sustainability rating tool, contributing to the dissemination of the sustainable and social benefits achieved by the implementation of a local rating tool.

2.1. *Purpose of the case study*

The main purpose of the case study is to evaluate the viability of the real-estate development Neo Niterói before the acquisition of the Casa Azul certification. The main goal is to assess the residential building attainment of the Casa Azul rating: bronze, silver or gold. The case study was randomly chosen to answer the main questions, "Would any development be able to reach the criteria?" and "What type of improvements could be implemented to achieve a higher rating in a constructed development?"

2.2. *Description of the real-estate development — Neo Niterói*

The project consists of four towers, exclusively residential, under the condominium regime regulated by Federal Law No. 4,591 of December 16, 1964, and subsequent pertinent legislation. Approval was obtained within the City Council of Niterói and the project was constructed according to the Brazilian Association of Technical Standard (ABNT) standards. The Neo Niterói condominium is composed of the following:

(1) **External area:** It has exclusive common parts, containing internal circulation of vehicles, sidewalks, gardens, open parking for 408 vehicles, leisure area, adult pool, children's pool, deck, party room with kitchen, toilet and two bathrooms and two toilet/dressing rooms, a guardhouse with bathroom, four edicts with barbecue grills and sports court.

(2) **Towers 1–4:** The ground floor of each tower is composed of common and private parts. Common parts consist of a hall with an access to two lifts, storage ladder, pump house, two compartments for metres and garbage compartment. Its private part consists of eight units numbered from 101 to 108. Numbers 101, 102, 105 and 106 consist of hall, living room, circulation, two bedrooms, bathroom, kitchen and a service area. The units numbered 103,104, 107 and 108 are made up of living room, bedroom, bathroom, kitchen and a service area.

(3) **Typical pavements (2–13 floors in Towers 1–3 and 2–12 floors in Tower 4):** All of them are made up of common and private parts. Common parts include entrance hall for two elevators, ladder, stairway, antechamber and private parts include each of the eight units, the units numbered as 201/1301, 202/1302, 205/1305 and 206/1306 of Towers 1–3 and 201/1201, 202/1202, 205/1205 and 206/1206 of Tower 4, and are composed of living room, circulation, two bedrooms, bathroom, kitchen and a service area.

(4) **Penthouse:** It consists only of common parts, engine room for elevators, fire system room, stairs, fire chamber, water tank and roof.

3. Results

The analysis of the results was done through surveys, photos, plant verification, owner's manual and descriptive memorial. Casa Azul's Caixa Certification guide was used throughout the case study.

As described in the Casa Azul Seal guide, the certification is obtained from a detailed review of the pre-established criteria. Criteria attainment will enable the grant at one of the three rating levels presented in Table 1.

The case study presents the evaluation of 53 criteria established by the Casa Azul aiming to identify which criteria were met by the real-estate development in Neo Niterói. Table 2 presents the summary of whether these criteria were met or not by Neo Niterói.

Table 1.   Rating: Casa Azul (Caixa, 2010).

Rating	Minimum Criteria
BRONZE	The project development must meet the 19 mandatory criteria
SILVER	The project development must meet the 19 mandatory criteria and 6 free choice criteria
GOLD	The project development must meet the 19 mandatory criteria and 12 free choice criteria

Table 2.   Casa Azul criteria vs. Neo Niterói development (Caixa, 2010).

Category/Criteria	Details	Observations Regarding the Neo Niterói Real-estate Development
URBAN QUALITY/INFRASTRUCUTRE		
1.1 Quality of the environment — Infrastructure	Mandatory	Successfully met
1.2 Quality of the environment — Impacts	Mandatory	Successfully met
1.3 Improvements in the environment	Elective	Not met
1.4 Recovery of degraded areas	Elective	Not met
1.5 Rehabilitation of real estate	Elective	Not met

(Continued)

Table 2.   (*Continued*)

Category/Criteria	Details	Observations Regarding the Neo Niterói Real-estate Development
DESIGN AND COMFORT		
2.1 Landscaping	Mandatory	Successfully met
2.2 Design flexibility	Elective	Not met
2.3 Relationship with the neighbourhood	Elective	Not met
2.4 Alternative transport solution	Elective	Successfully met
2.5 Location for selective collection	Mandatory	Successfully met
2.6 Leisure, social and sporting equipment	Mandatory	Successfully met
2.7 Thermal performance — Sections	Mandatory	Successfully met
2.8 Thermal performance — Orientations to the sun and wind	Mandatory	Successfully met
2.9 Natural lighting of common areas	Elective	Successfully met
2.10 Ventilation and natural lighting for bathrooms	Elective	Successfully met
2.11 Suitability to the physical conditions of the terrain	Elective	Successfully met
ENERGY EFFICIENCY		
3.1 Low consumption light bulbs — Private areas	Mandatory	Partially met
3.2 Saving devices — Common areas	Mandatory	Partially met
3.3 Solar heating system	Elective	Not met
3.4 Gas heating system	Elective	Not met
3.5 Individual measurement — Gas	Mandatory	Successfully met
3.6 Efficient lifts	Elective	Successfully met
3.7 Efficient appliances	Elective	Successfully met
3.8 Alternative sources of energy	Elective	Not met
RESOURCE MATERIALS CONSERVATION		
4.1 Modular design	Elective	Not met
4.2 Quality of materials and components	Mandatory	Successfully met
4.3 Industrialized or pre-fabricated components	Elective	Successfully met

(*Continued*)

Table 2.   (*Continued*)

Category/Criteria	Details	Observations Regarding the Neo Niterói Real-estate Development
4.4 Reusable molds and scaffold	Mandatory	Not met
4.5 Management of CDW	Mandatory	Not met
4.6 Concrete with optimized dosage	Elective	Not met
4.7 Blast furnace cement and Pozzolanic cement	Elective	Not met
4.8 Paving with CDW	Elective	Not met
4.9 Ease of façade maintenance	Elective	Successfully met
4.10 Wood planted or certified	Elective	Not met
WATER MANAGEMENT		
5.1 Individualized measurement — Water	Mandatory	Not met
5.2 Saving devices — Discharge system	Mandatory	Successfully met
5.3 Saving devices — Aerators	Elective	Successfully met
5.4 Saving devices — Flow control register	Elective	Successfully met
5.5 Rainwater harvesting	Elective	Not met
5.6 Stormwater retention	Elective	Partially met
5.7 Rainwater infiltration	Voluntary	Not met
5.8 Permeable areas	Mandatory	Successfully met
SOCIAL PRACTICES		
6.1 Education for CDW management	Mandatory	Not met
6.2 Employee environmental education	Mandatory	Not met
6.3 Personal development of employees	Elective	Not met
6.4 Professional training of employees	Elective	Not met
6.5 Inclusion of local workers	Elective	Not met
6.6 Community involvement in project design	Elective	Not met
6.7 Orientation to residents	Mandatory	Successfully met
6.8 Environmental education of residents	Elective	Partially met
6.9 Training for management of the enterprise	Elective	Partially met
6.10 Actions to mitigate social risks	Elective	Not met
6.11 Actions to generate employment and income	Elective	Not met

According to Table 2, Neo Niterói condominium meets 16 mandatory criteria and also meets 14 elective criteria. In summary, the Neo Niterói achieved the total of 30 criteria out of the 53 criteria specified. Considering this, to claim the Casa Azul certification, the Neo Niterói project needs to meet the following three mandatory criteria:

(1) individualized water measurement,
(2) education for CDW management,
(3) employee environmental education.

4. Discussion

Regarding Criterion 5.1, it would be necessary for the implantation of individual water metres in new projects. Considering that the buildings are already built, the implementation of a new project must go through a condominium/strata assembly to be approved. Individual water metres would lead to a significant reduction in the monthly condominium fee, since each resident would pay its own water separately. So, everyone would reduce their consumption, paying only for their individual usage.

In order to meet Criteria 6.1 and 6.2, educational actions aiming at the management of CDW as well as environmental educational actions focused on the employees should be implemented during the construction period. If Criteria 5.1, 6.1 and 6.2 were not mandatory, it would have been possible to claim the Casa Azul certification in the silver category based on the perspective of the real-estate project having met a total of 30 criteria. However, despite this, it is important to highlight the environmental benefits achieved by meeting at least 30 criteria.

4.1. *Benefits acquired with the criteria served by the real-estate development Neo Niterói*

Criterion 1: Quality of the environment — Infrastructure: Through the fulfilment of this criterion, it was possible to provide the residents of Neo Niterói an improvement in their life quality, since it is located in an area close to trade, churches, health centres, schools, park Palmier Silva (Horto do Barreto) and bus stops.

Criterion 2: Quality of the environment — Impacts: Meeting this criteria provided an improvement in life quality of residents since it was built far from factories, industries, water treatment plants, airports, highways, which characterizes the absence of noise, auditory, visual and air pollution in the environment.

Criterion 3: Landscaping: In accordance with this criterion, Neo Niterói has a well-designed landscaping project, has several trees scattered throughout the condominium, including fruit trees, which provide shade for the residents and keep the environment cool.

Criterion 4: Relationship with the neighbourhood: Neo Niterói was strategically constructed aiming to have no negative impacts on the neighbourhood. The condominium does not obstruct the luminosity or the incidence of winds in the neighbouring buildings. The buildings are positioned at a standard distance in order to preserve individual privacy.

Criterion 5: Alternative transportation solution: In accordance with this criterion, Neo Niterói was built near bicycle paths and also has a bicycle track for the residents. This is a significant difference, since it allows the residents to move easily through the neighbourhood using a bicycle without polluting the environment.

Criterion 6: Selective waste collection site: In compliance with these criteria, the condominium has a space outside, where all the organic waste, plastic, paper, glass and metals are separated from the garbage generated by the residents. After the separation, some of the separated garbage is sent for collection and then directed to the landfill by the local cleaning company and all separate cardboard is sold to a recycling cooperative. The profit obtained through this sale is reversed in improvements to the condominium. The separation of waste allows recycling, favouring the environment in general.

Criterion 7: Leisure, social and sports equipment: In compliance with this criterion, the condominium counts on a multi-sport court, an adult and children's pool, a ballroom, a children's playground, an outdoor gym and a living space. All of these structures allow residents to have a lot of entertainment and encourage them to lead a healthier life by practising exercise and sports and maintaining good interpersonal relationships with neighbours.

Criterion 8: Thermal performance — Walls: In compliance with this criterion, the condominium has a good thermal performance in relation to the walls. The walls have been built according to the local climate, enabling energy efficiency. It is not always necessary to use a fan or an air conditioner. The walls of the apartments do not dissipate the heat to the interior of the apartment.

Criterion 9: Thermal performance — Orientation to the sun and wind: In compliance with this criterion, the Neo Niterói condominium has a good thermal performance in relation to the incidence of sun and wind. The rooms remain pleasant throughout the year, since the sun is only a part of the day and the apartments receive natural ventilation due to their position near Guanabara Bay. It is not always necessary to use a fan or air conditioner, allowing savings in the residents' energy bill.

Criterion 10: Natural lighting of common areas: Neo Niterói has well-positioned apartments with natural lighting. The apartments receive natural light during the whole day, this is due to the fact that they have windows with adequate ventilation for lighting. For this reason, it is not necessary to light bulbs during the day resulting in savings to the residents' energy bill.

Criterion 11: Ventilation and natural lighting of bathrooms: Apartments have a 30 × 30 cm window that allows the incidence of luminosity and ventilation of the same. During the day, it is possible to save electricity, as the incidence of luminosity inside the bathroom allows the use of the same without the need of artificial light.

Criterion 12: Adequacy of the physical conditions of the land: Neo Niterói was built after a landscaping project in which several stones were preserved in it. Even the pool was built on a large stone. This initiative made it possible not to damage the pre-existing natural space on the land, keeping part of it originally preserved.

Criterion 13: Low-energy bulbs — Private areas: Apartments using low-energy bulbs, such as fluorescent and LED, have a reduction in energy consumption and consequent reduction in the energy bills.

Criterion 14: Energy-saving devices — Common areas: Availability of presence sensors provide significant energy savings for the condominium.

Criteria 15: Individualized measurement — Gas: Apartment has its own individualized gas measurement. The gas facilities are only for use of stoves. Individual gas metres give residents control over spending and end up encouraging consumption reduction.

Criterion 16: Efficient lifts: The condominium has efficient traffic control of tower elevators. Each tower has two elevators during the day and part of the night. During the dawn, each tower has only one elevator, since the other is disconnected from 24 h to 6 h, where there is less demand for their use. This initiative allows a reduction in energy consumption, as four elevators remain off during the whole dawn.

Criterion 17: Efficient Appliances: The condominium has certified (Procel energy rating label/Brazilian energy rating label) appliances installed in common areas such as a ballroom and living space for employees. The option for these energy-efficient appliances provides a significant reduction in monthly energy consumption.

Criterion 18: Quality of materials and components: The project had standardized quality materials, manufactured by qualified companies, according to PBQP-H5 (Brazilian quality sectoral programmes). By using standardized material quality during construction, it was possible to avoid material wastage owing to defects, avoiding repairs and early maintenance. As a result, environmental impacts were reduced since it was not necessary to produce new replacement materials.

Criterion 19: Industrialized or prefabricated components: The towers of Neo Niterói were designed and built using prefabricated structural elements. A prefabricated structural project is more sustainable, since it reduces the loss of materials and generation of waste during construction.

Criterion 20: Reusable molds and scaffold: Neo Niterói used reusable molds and scaffold during the construction. The use of molds for prefabricated structural components and scaffold allowed the reduction of material wastes, including wood, since the molds and scaffold could be reused on other constructions.

Criterion 21: Management of CDW: In compliance with this criterion, during the construction of Neo Niterói, CDW was properly delivered to the local landfill. The CDW management contributed to the mitigation of the impacts on urban environment.

Criterion 22: Façade maintenance: Standardized quality materials were used in the façades resulting in a significant reduction of the regular maintenance. The increase in the service life of the façade due to the material quality generated a cost reduction with regular maintenance, avoided the waste of materials and consequent reduction of environmental impacts.

Criterion 23: Saving devices — Discharge system: The bathroom in each apartment has a discharge with water-saving devices. These devices have two water-operated valves, allowing more or less volume of water according to need. This system allows a significant saving of water because the largest amount of water in the reservoir is only triggered when necessary.

Criterion 24: Water-saving devices — Aerators: The kitchen and bathroom of each apartment present aerators in the taps. These faucet aerators provide a reduction in water consumption, allowing a better dispersion of the water jet in the taps.

Criterion 25: Water-saving devices — Flow control register: The kitchen and bathroom of each apartment present flow control registers. These flow records allow residents to regulate the amount of water used in the apartment and consequently save water. Therefore, the amount of sewage waste generated by each apartment also decreases.

Criterion 26: Rainwater retention: Neo Niterói has gutters that allow the drainage of rainwater, avoiding possible flooding. This system avoids flooding in the condominium and surroundings, and also reduces the flow of rainwater, thus contributing to the urban drainage system.

Criterion 27: Permeable areas: Neo Niterói condominium has permeable areas according to the established standards. These permeable areas allow rainwater to penetrate the soil, thus renewing the water cycle of nature as the water returns to the water table.

Criterion 28: Orientation to the residents: The residents of Neo Niterói were properly advised on the most diverse issues related to the proper functioning of the condominium. Even the owner's manual has been distributed with detailed information about Neo Niterói. This initiative provided a space for discussion among residents as well as the search for more sustainable alternatives for the proper functioning of the real-estate project.

Criterion 29: Environmental education of the residents: The residents received the owner's manual and a guide, presenting a social project aimed at the sustainability of the condominium. This action made possible the implementation of more sustainable practices by the residents contributing significantly to the environment.

Criterion 30: Training for management of the enterprise: The residents were previously advised about the need for administrative management of the condominium. This orientation was given through meetings and through the owner's manual and the booklet distributed to all residents. This initiative has made the residents aware of the use and maintenance of buildings in a way that does not harm the environment.

5. Conclusion

The analysis of Neo Niterói case study was carried out through on-site inspection, verification of documents and photographic records. It is understood that the condominium would be eligible for the possible acquisition of the Casa Azul certification in the silver rating if it met the mandatory criteria: 5.1 (individualized water measurement), 6.1 (education for RCD management) and 6.2 (environmental education of employees), however the development does not meet these mandatory criteria.

The mandatory criteria Neo Niterói met are indispensable because they would generate significant water savings, reuse and final disposal of CDW, and would also train all construction professionals in relation to environmental management and education.

In a real situation of litigation against Caixa's sustainable certification, the project studied would be impossible to achieve the certification, as it is not in the design stage, not under construction, nor is it undergoing any work of addition or modification. But if the project was in the construction phase, it would be possible to adapt it easily in order to meet the mandatory criteria 5.1, 6.1 and 6.2 and submit to the Casa Azul.

The criteria determined by the Casa Azul guideline are easy to be met by a housing developer, whereas most of them seek to meet the basic levels of environmental sustainability. By meeting the criteria established, it is possible to achieve financial savings during construction, reduction of the consumption of raw materials, reuse of diverse materials, as well as provide future residents with water savings and energy efficiency. It was possible to realize that a large part of the housing constructions could apply for the acquisition of a sustainable Casa Azul certification. This initiative would benefit construction companies, buyers and especially the environment.

References

Bruntland, G. H. (1987). *Our Common Future the World Commission on Environment and Development*. Oxford: Oxford University Press.

Caixa Econômica Federal (CEF). (2010). Guia Selo Casa Azul. Available at: http://www.caixa.gov.br/Downloads/selo_azul/Selo_Casa_Azul.pdf (accessed on August 5, 2015) (in Portuguese).

Dinamarco, C. (2016). Selo Casa Azul Certificação Ambiental: Estudo De Caso. Thesis, Brazil: Federal University of Rio de Janeiro, pp. 1–165 (in Portuguese).

Dinamarco, C. P. G., Haddad, A. and Evangelista, A. (2016). "Selo Casa Azul Certificação Ambiental Estudo De Caso: Condomínio Neo Niterói", *Revista Sustinere*, 4(1) (in Portuguese).

Green Building Council (GBC) Brasil. (2017). Available at: http://www.gbcbrasil. org.br/?p=certificacao (accessed on June 29, 2017).

John, V. M. and Prado, R. T. A. (2010). *Boas Práticas Para Habitação Mais Sustentável*, São Paulo, Brazil: Editora e Gráfica (in Portuguese).

Matos, G. and Wagner, L. (1998). "Consumption of Materials in the United States, 1990–1995", *Annual Review of Energy and the Environment*, 23(1), 107–122.

Roth, C. D. G. and Garcias, C. M. (2009). "Construção Civil e a Degradação Ambiental", *Desenvolvimento em Questão*, 7(13), 111–128 (in Portuguese).

Schenini, P. C., Bagnati, A. M. B. and Cardoso, A. C. F. (2004). Gestão de resíduos da construção civil. In *Cobrac — Congresso Brasileiro de Cadastro Técnico Multifinalitário*, (ed.). Jorge Campos de Costa, Florianópolis: UFSC (in Portuguese).

Chapter 10

Review on Green Building Rating Tools Used in Australia

I. M. Chethana S. Illankoon, Vivian W. Y. Tam*,
Khoa N. Le and Cuong N. N. Tran

*School of Computing Engineering and Mathematics,
Western Sydney University, Locked Bag 1797, Penrith,
NSW 2751, Australia*
*College of Civil Engineering, Shenzhen University,
Shenzhen, China*
vivianwytam@gmail.com

Abstract

The construction industry is one of the main contributors towards climate change. Therefore, green buildings have become a requirement and are no longer a luxury. With the development of green buildings, there are many assessment tools and rating tools developed for evaluating the performance of different criteria of green buildings. Certain tools evaluate one criterion such as water efficiency or energy efficiency of green buildings. Similarly, other tools evaluate some criteria, finally providing a star rating for the overall performance. There are also tools

---

*Corresponding author.

used to evaluate different options used for green buildings as well. In Australia, many green building rating tools evaluate the performance of green buildings falling into these categories. Therefore, this chapter aims to review the various green rating tools and incentives in Australia. There are different uses of various green building rating tools, and the state governments in Australia support certain tools such as building sustainability index (BASIX) and nationwide house energy rating scheme (NatHERS). However, these green building rating tools focus on different criteria. Apart from that, there are various green incentive schemes as well. After a thorough review of the current status of the green building rating tools in Australia, this chapter recommends developing government policies and regulations for promoting green buildings, government mandates on green building implementation, incentives to green building development, green building rating tools considering all criteria and including cost considerations for green building development in Australia.

Keywords: Australia; Green Buildings; Green Building Rating Tools.

1. Introduction

Climate change is a common and widely discussed topic of the decade. Stern (2007) indicated that if no action were taken to reduce emissions, the greenhouse gases (GHGs) in the atmosphere would rise double to the pre-industrial level by 2035, leading to a temperature increase of 2°C. There are many contributing factors to these adverse environmental impacts. Construction is one of the main contributors of these adverse climatic conditions, and many studies confirm the significant negative impacts towards the environment (Abidin, 2010; Tan *et al.*, 2011).

In the United Kingdom, out of the many sectors of the economy, construction process and buildings use the most energy and emit the highest amount of CO_2 emissions, which is altogether about 57% (European Information Service Commission, 2012). The building sector is one of the sectors that significantly emit GHGs in Australia (Reidy *et al.*, 2011), globally leading to detrimental environmental impacts. The report of Evolution 2013 (Green Building Council Australia, 2013) illustrated that minimizing the carbon footprint of the building can lead to a significant positive impact on the global environment. Therefore, considering the projected environmental hazards in future, it is necessary to regulate the construction of buildings in an environment-friendly manner.

There was a strong requirement to develop a yardstick to measure the buildings regarding environmental friendliness. Therefore, the primary role of a building rating tool is to provide a comprehensive assessment of the environmental characteristics of a building using a common and verifiable set of criteria and targets for building owners and designers to achieve higher environmental standards (Cole and Larsson, 1999). However, at present *status quo*, there are many green building rating tools used worldwide for different purposes. Software are developed to assist in the selection of environment-friendly material. Further, there are separate assessment tools developed for assessing buildings by many organizations.

In the Australian context, there are many rules and regulations put into practice to regulate the adverse environmental practices. Government regulations support certain tools, such as building sustainability index (BASIX) and nationwide house energy rating scheme (NatHERS). Green Star is also widely used. However, it is interesting to note that irrespective of the developments in these rating tools, Australia remains as one of the top greenhouse-gas emitting countries (Organisation for Economic Co-operation and Development (OECD), 2015). Therefore, this chapter aims to review the green building rating tools used in Australia and recommend essential developments in the evaluation and regulation process of green buildings in the Australian context.

2. Classification of Green Building Rating Tools

There are tools developed by various organizations to provide assessments so as to ensure each building is environmentally sustainable. According to Baumann *et al.* (2002), these tools can be categorized mainly into six categories: frameworks, analytical tools, checklists and guidelines, software and expert systems, rating and ranking tools and organizing tools.

According to the classification, frameworks and checklists are primary tools providing basic knowledge to the users. However, analytical tools, software and expert systems, and rating and ranking tools are complex. Further, these rating tools included three stages: (1) classification, referring to environmental change expectations that determine the impact category based on various inputs and outputs; (2) characterization, referring to identification of the impact of each input and output with relation to their categories; (3) valuation by weighting each category in comparison to other categories (Awadh, 2017; Fenner and Ryce, 2008). According to Seo *et al.* (2005b), Australian tools can be categorized as assessment tools providing quantitative performance indicators in assisting decision-making on design alternatives and rating tools assessing the performance levels of buildings against agreed standards often measured in stars.

Many researches have been carried out using analytical tools such as life-cycle assessment using various software like SimaPro (Geng *et al.*, 2017; Ness *et al.*, 2007; Pieragostini *et al.*, 2012; Rebitzer *et al.*, 2004; Saibuatrong *et al.*, 2017). In Australian classification, assessment tools give a quantitative output in supporting decision-making among alternatives. Further, rating tools illustrate the performance against agreed standards. Therefore, in summary, this research separately considers the "assessment tools" and "rating tools" as two separate sections (see Table 1).

3. Green Building Assessment Tools and Rating Tools Used in Australia

There are many tools used for buildings as a whole and certain other tools used for material. These tools are categorized into assessment tools and rating tools based on the classification given in Section 2. Assessment tools, such as AccuRate, FirstRate and Building Energy Rating Scheme (BERS), are used to assist rating tools such as NatHERS. Most of these tools are voluntary for use except for BASIX and NatHERS. BASIX is compulsory for New South Wales (NSW) residential building, and NatHERS is used as a regulation for thermal efficiency. Further, NABERS and BASIX ratings can be used to achieve certain credit points in Green Star.

Table 1. Green assessment tools, rating tools and specialized software in Australia.

Tool	Description	References
Green building assessment tools		
Athena EIE Athena's Environmental Impact Estimator	• Athena EIE is basically a software in which the user describes the primary building elements by inputting dimensions and choosing materials from the menu of options	Athena sustainable material institute (2017) Baek *et al.* (2013)
BEAT (Building Environmental Assessment Tool)	• BEAT is an inventory and assessment tool using Microsoft Access developed at the Danish Building and Urban Research	Dutch Green Building Council (2016) Eldridge and NSW Department of Public Works and Services Sydney Australia (2002)
BEES (Building for Environmental and Economic Sustainability)	• This software is used for selecting cost-effective, environmentally preferable building products using the environmental life-cycle assessment approach specified in the International Standards Organization (ISO) 14040 series of Standard	Erlandsson and Borg (2003) Forsberg and von Malmborg (2004)
Ecoprofile	• This is an assessment tool for buildings and their operations	Fossdal and Holm (1997)
Eco-quantum	• This is a simulation-based tool which supports the designers to calculate the environmental effects during the entire life cycle of the building	Graham (2003) Kortman *et al.* (1998)
ENVEST (Environmental impact estimating design software)	• This simplifies the design of buildings with low environmental impact and whole life costs	Lippiatt (2001) National Institute Of Standards and Technology (2016)
EQUER	• This performs yearly simulations of a building life cycle in order to provide mechanical, energy and architectural engineers or architects with environmental indicators	Pettersen (2000) Pettersen *et al.* (2000) Peuportier (2001) Peuportier (2002)

(*Continued*)

Table 1.   (*Continued*)

Tool	Description	References
GreenCalc	• This is a computer tool which can be used to calculate the environmental load of office buildings	Seo (2002)
LCADesign	• LCADesign is a software which provides life-cycle analysis of computer-aided drafting for design	Seo *et al.* (2005a)
LCAid	• This is used as a decision-making tool using life-cycle assessment (LCA) to evaluate the performance of design options	Seo *et al.* (2005b)
LISA Life cycle analysis in Sustainable Architecture	• This is an LCA decision support tool developed to assist in green design	Tae *et al.* (2011) Thiers and Peuportier (2008) Tucker *et al.* (2003) Tucker *et al.* (2005) Vijayan and Kumar (2005) Watson *et al.* (2004a)
Green rating tools		
AccuRate	• This provides the benchmark for accrediting other software for use with the Building Codes of Australia requirements	Alam and Ham (2014)
BASIX	• BASIX was introduced on July 1, 2004 by the NSW government as a sustainable planning measure to be undertaken in Australia • This was implemented under environmental planning and assessment regulations 2000 and environmental planning policy 2004 and applies to all residential dwelling types	Aldawi and Alam (2015) Anderson (2006) Baird (2009) Bannister (2005) Berry and Marker (2015) Crawford *et al.* (2015)
BERS	• This is a computer program and a powerful tool initially used to simulate and analyze the thermal performance of Australian houses in climates ranging from Alpine to tropical • This was accredited for use for energy ratings for NatHERS from May 1, 2016	Crawford *et al.* (2016) Department of the Environment and Energy Government of Australia (2017) EcoSpecifier Global (2017) Energy inspection (2017)

Ecospecifier	• This is an online database on environmentally preferable materials and aims to provide guidance and hard data in a form that is easy to use by designers and architects	Eroksuz and Rahman (2010) Graham (2003) Green Building Council Australia (2015) Hertzsch *et al.* (2011) Illankoon *et al.* (2016a) Illankoon *et al.* (2016b) Illankoon *et al.* (2017) Lewis and Ryan (2006) Mansoury and Tabatabaiefar (2014) Mitchell (2010) NatHERS (2017) NSW Department of Planing and Environment (2017) NSW Department of Planning and Environment (2013) O'Leary *et al.* (2016)
FirstRate	• FirstRate is an interactive tool with a graphic user interface that enables designers and thermal performance assessors to generate energy ratings for a home by tracing over floor plans • It is used by the majority of industries to rate the energy-efficiency compliance of residential dwellings to the 6-star standard under the National Construction Code of Australia (NCC) • This tool generates ratings based on the NatHERS 0–10 star scale for homes	
Green Star	• Green Building Council of Australia (GBCA) was launched in 2002 as a national not-for-profit organization focusing on the development sustainable property industry in Australia • In 2003, GBCA launched its green building rating system as "Green Star" • It includes four green building rating tools for Design & As-built, Interiors, Communities and Performances, and it also assess the sustainability of projects at all stages of life cycle	

(*Continued*)

Table 1.   (*Continued*)

Tool	Description	References
NABERS (National Australian Built Environment Rating System)	• NABERS is a collection of separate tools, each of which calculates and rates the performance of an existing building (or part of one) on a particular environmental indicator at a certain point in time • This can be undertaken individually (for example, a company may decide to rate its building only in energy and water), and they are not combined into an overall rating	Office of Environment and Heritage Government of Australia (2016) Seo (2002) Seo *et al.* (2005b) Shiel *et al.* (2017)
NatHERS	• NatHERS provides Australian homes with a star rating out of 10 based on an estimate of a home's potential (heating and cooling) energy use • Administered by the Department of the Environment and Energy on behalf of the states and territories of Australia • NatHERS accredits BERS and FirstRate software for the assessment of energy loads and rating houses	Van der Sterren *et al.* (2009) Verghese and Hes (2007) Victoria State Government (2017a) Victoria State Government (2017b) Vijayan and Kumar (2005) Watson *et al.* (2004b) Whaley *et al.* (2017)

These different tools identify the performance of buildings in various terms. As an example, LCAid focuses on energy, water and material consumption and excludes indoor environment quality. BASIX only focuses on residential buildings. Therefore, the use of these tools is specific. Table 2 reports the criteria addressed by each of these tools, and also the end use, application types and the phase of the life cycle for each tool can be used.

According to Table 2, most of the tools focus on the material, water, energy and indoor environmental quality for buildings throughout the life cycle. The assessment tools mostly focus on one criterion, and rating tools usually focus on more than one criterion. However, a tool such as NABERS focuses on the performance of the building and tools such as Green Star focus from the perspective of design. There are few assessment tools that focus on the cost of these buildings (see Table 2). All the rating tools except for Green Star can apply for rating in each criterion separately. In Green Star, the rating is given considering all the criteria together, and ratings for separate criteria are not possible. Noise levels in indoor environmental quality are not considered by majority of the tools. It is interesting to note that none of the rating tools consider cost in the respective ratings (see Table 2).

Apart from these green building rating tools, there was a suggestion to introduce tax breaks for green building programme in boosting green building development across Australia. However, this programme was scrapped in federal budget 2012–2013 (Green Building Council Australia, 2012). In Australia, there are tariff schemes such as gross-feed-in-tariff scheme, a net-feed-in-tariff scheme and a buy-back scheme for using renewable energy schemes (Tam *et al.*, 2017). There is a solar credit scheme initiated by the Australian government to provide support for household owners in receiving additional financial benefits for installing renewable energy systems (Tam *et al.*, 2017). Further, government funding of AU\$ 2.1 million (GST exclusive) was provided for the programme from 2014–2015 to 2015–2016 to support community organizations installing a renewable energy system (solar photovoltaic panels or a solar hot water system only) on an existing building that provides support to community groups (Department of the Environment and Energy Australian Government, 2017).

Table 2.   Detailed evaluation of green building assessment and rating tools used in Australia Development based on Seo (2002) and Seo *et al.* (2005b).

Tool	End Use	Applications	Phase*	Energy	Land	Water	Material	Environmental Loading	IEQ — Air	IEQ — Thermal	IEQ — Visual	IEQ — Noise	Cost
Green assessment tools													
ATHENA EIE	Building/material	Commercial & residential buildings	D, O&M	✓	—	✓	✓	✓	—	—	—	—	—
BEAT (Building Environmental Assessment Tool)	Building	Commercial & residential buildings	D, O&M, EoL	✓	—	—	✓	✓	—	—	—	—	—
BEES (Building for Environmental and Economic Sustainability)	Product	Building material	D, EoL	✓	—	✓	✓	✓	✓	—	—	—	✓
Ecoprofile	Building	Commercial building	D, O&M, EoL	✓	✓	✓	✓	✓	✓	✓	✓	—	—
Eco-quantum	Product/part	Building material	D, EoL	✓	✓	✓	✓	✓	✓	—	—	—	✓
ENVEST (Environmental impact estimating design software)	Building	Commercial building	D	✓		✓	✓	✓	—	—	—	—	—

EQUER	Building	Residential building	D, O&M, EoL	✓	✓	✓	—	✓	—	—	—	—	—
GreenCalc	Building	Commercial building	D, O&M, EoL	✓	—	✓	✓	✓	—	✓	—	—	—
LCADesign	Building/material	Commercial building	P, D and O&M	✓	✓	✓	✓	✓	✓	—	—	—	—
LCAid	Building/material	Commercial & residential buildings	P, D, O&M and EoL	✓	—	✓	✓	✓	—	—	—	—	✓
LISA	Building/material	Commercial & residential buildings	P, D, O&M and EoL	✓	—	✓	—	✓	✓	—	—	—	—
Green rating tools													
AccuRate	Building	Residential building	D, O&M	✓	—	—	—	—	—	✓	—	—	—
BASIX	Building	Residential building	P and D	✓	✓	✓	✓	✓	✓	✓	✓	—	—
BERS	Building	Residential building	D, O&M	✓	—	—	—	—	—	✓	—	—	—
Ecospecifier	Material	Building material	D, O&M	✓	✓	✓	✓	✓	—	✓	—	—	—
FirstRate	Building	Residential building	D, O&M	✓	—	—	—	—	—	✓	—	—	—
Green Star	Building	Buildings	D, O&M	✓	✓	✓	✓	✓	✓	✓	✓	✓	—
NABERS (National Australian Built Environment Rating System)	Building	Commercial & residential buildings	O&M	✓	—	✓	✓	✓	✓	✓	✓	✓	—
NatHERS	Building	Residential building	D, O&M	✓	—	—	—	—	—	✓	—	—	—

* P: Planning, D: Design, O&M: Operation and Maintenance, EoL: End-of-life.

4. Discussions

There are two major types of green building tools: (1) assessment tools and (2) rating tools. Assessment tools are used to choose among different alternative options, and rating tools give an indication to the public as to what extent a building or material is environment-friendly regarding the considered criteria. There are many criteria addressed by these rating tools, such as energy, water, indoor environment quality, land use and waste. Similarly, Green Star Australia also follow a similar pattern for the key criteria. However, the assessment tools focus on either one or two criteria separately (see Table 2). This is similar to most of the green building rating tools as well except for Green Star. None of the rating tools in Australia consider cost in evaluating buildings (see Table 2). According to the literature, the initial cost is one of the main concerns in green building development. Therefore, ignoring the cost considerations in the rating tools will hinder the development of green buildings.

In the Australian context, there are many assessment tools and rating tools. However, there is a clear lack of integration among these tools. Different state governments have different policies and various requirements in using these tools. Tools such as NatHERS and BASIX are supported by the state governments. However, these rating tools can obtain ratings for separate criteria. As an example in this context, a building can perform exceptionally well in energy efficiency and forgo water efficiency or indoor environmental quality. Therefore, the building needs to be evaluated in all criteria as a whole to obtain better environmentally friendlier buildings. This aspect is lacking in most of the Australian rating tools.

4.1. *Implications in findings in the Australian context*

According to the findings, there are various green assessment and rating tools utilized in Australia. However, as mentioned in Section 1, Australia is still within the top 10 GHG emitting countries. This shows that there is still need for clear development to regulate this adverse situation.

Further, separate consideration of different criteria, such as water efficiency and energy efficiency, will not address the Australian context effectively. There should be an approach to look into this issue as a whole. In the Australian context, different states support different policies. As an example, according to the findings, tools such as NatHERS and BASIX are supported by different state governments. GHG emission is a global issue. However, there is a clear lack of policies across Australia to address this issue.

Green buildings are promoted in Australia. In common terminology, green buildings are termed as environmentally, socially and economically friendly buildings. These three pillars are equally addressed in green buildings. However, according to the findings, there is only one rating tool assessing all these criteria in green buildings. As a result, there can be buildings with higher energy saving ratings but significant inefficiencies in water usage and indoor environmental quality. This is one of the major implications.

Taxation is one of the main concerns in the Australian context. However, there is a clear lack of support or financial incentive given to the green building developers. Based on the findings of this discussion, following recommendations can be made to boost the development of green buildings in Australia:

(1) *Government policies, regulations and mandates to promote green buildings*: As discussed earlier, there is a clear lack of government policies, regulations and mandates to promote green buildings. Initially, it is necessary to implement an Australian policy to integrate various policies and requirements of different state governments.

   The government can also include green building requirements to building codes as a mandatory requirement. Further, the government can mandate a minimum green certification for new development. Initially, the government building projects can meet these requirements and then extend it to private building projects as well.

(2) *Green building rating tool considering all the criteria*: All the rating tools except Green Star in Australia consider various criteria separately for evaluation. There is a clear need to develop a green rating

tool which covers all the criteria required for green building. Further, the separate rating tools and assessment tools can be used to verify the requirements. This method is available in Green Star rating tools to a certain extent by integrating with BASIX, NatHERS and NABERS in energy credit. However, this can be further improved.

(3) *Cost considerations in green building development*: Initial cost is one of the barriers to green building development. Therefore, when developing green rating tools, this parameter must be considered. It will be beneficial to develop a system to reward the developers certifying green buildings with a lower initial cost using innovative technologies and ideas.

5. Conclusions

This chapter initially discussed the difference between assessment tools and rating tools. Afterwards, it reviewed the green building assessment and rating tools in the Australian context. The criteria used to evaluate green buildings were similar to all the tools. Further, assessment tools and most of the green building rating tools in Australia focus on either one or two criteria separately for evaluating the green building features.

The incentives given to encourage green building development is also discussed in this chapter. Finally, based on the findings, it was concluded that Australia needs to develop government policies and regulations to promote green buildings, develop government mandates on green buildings, provide incentives to green building development, develop green building rating tool considering all the criteria and consider costs in green building development. Further research on different rating tools and practices worldwide and comparing it with the Australian context is also recommended.

Acknowledgements

The authors wish to acknowledge the financial support from the Australian Research Council (ARC), Australian Government (No. DP150101015).

References

Abidin, N. Z. (2010). "Investigating the Awareness and Application of Sustainable Construction Concept by Malaysian Developers", *Habitat International*, 34, 421–426.

Alam, J. and Ham, J. (2014). Towards a BIM-based Energy Rating System. *Proceedings of the 19th International Conference on Computer-Aided Architectural Design Research in Asia: Rethinking Comprehensive Design: Speculative Counterculture*, 285–294.

Aldawi, F. and Alam, F. (2015). "Residential Building Wall Systems: Energy Efficiency and Carbon Footprint", *Thermofluid Modeling for Energy Efficiency Applications*, 169–196.

Anderson, J. M. (2006). "Integrating Recycled Water into Urban Water Supply Solutions", *Desalination*, 187, 1–9.

Athena Sustainable Material Institute. (2017). Software and Data [Online]. Available: http://www.athenasmi.org/our-software-data/overview/ (Accessed on June 9, 2017).

Awadh, O. (2017). "Sustainability and Green Building Rating Systems: LEED, BREEAM, GSAS and Estidama critical analysis", *Journal of Building Engineering*, 11, 25–29.

Baek, C., Park, S.-H., Suzuki, M. and Lee, S.-H. (2013). "Life Cycle Carbon Dioxide Assessment Tool for Buildings in the Schematic Design Phase", *Energy and Buildings*, 61, 275–287.

Baird, G. (2009). "Incorporating User Performance Criteria into Building Sustainability Rating Tools (Bsrts) for Buildings in Operation", *Sustainability*, 1, 1069–1086.

Bannister, P. (2005). The ABGR Validation Protocol for Computer Simulations. *Proceedings of the Ninth International IBPSA Conference*, August 15–18, Montreal, Canada, 33–40.

Baumann, H., Boons, F. and Bragd, A. (2002). "Mapping the Green Product Development Field: Engineering, Policy and Business Perspectives", *Journal of Cleaner Production*, 10, 409–425.

Berry, S. and Marker, T. (2015). "Australia's Nationwide House Energy Rating Scheme: The Scientific Basis for the Next Generation of Tools", *International Journal of Sustainable Building Technology and Urban Development*, 6, 90–102.

Cole, R. J. and Larsson, N. K. (1999). "GBC'98 and GBTool: Background", *Building Research & Information*, 27, 221–229.

Crawford, R. H., Bartak, E., Stephan, A. and Jensen, C. A. (2015). Does Current Policy On Building Energy Efficiency Reduce A Building's Life Cycle

Energy Demand? in *Living and Learning: Research for a Better Built Environment: 49th International Conference of the Architectural Science Association*, Melbourne, Australia.

Crawford, R. H., Bartak, E. L., Stephan, A. and Jensen, C. A. (2016). "Evaluating the Life Cycle Energy Benefits of Energy Efficiency Regulations for Buildings", *Renewable and Sustainable Energy Reviews*, 63, 435–451.

Department Of The Environment And Energy Australian Government. (2017). Solar Towns [Online]. Available at: http://www.environment.gov.au/climate-change/renewable-energy/solar-towns (Accessed on June 12, 2017).

Department Of The Environment And Energy Government Of Australia. (2017). Nationwide House Energy Rating Scheme (NatHERS) [Online]. Available at: http://www.nathers.gov.au/ (Accessed on June 9, 2017).

Dutch Green Building Council. (2016). GreenCalc [Online]. Available at: https://www.dgbc.nl/ (Accessed on June 9, 2017).

Ecospecifier Global. (2017). Welcome to Ecospecifier Global [Online]. Available at: http://www.ecospecifier.com.au/ (Accessed on June 9, 2017).

Eldridge, C. and NSW Department Of Public Works And Services Sydney Australia (2002). Lcaid™ Software: Measuring Environmental Performance Of Buildings. in *9th International Conference on Durability of Building Materials and* Components, Netherlands.

Energy Inspection. (2017). About our Software [Online]. Available at: http://www.energyinspection.com.au/about/ (Accessed on June 9, 2017).

Erlandsson, M. and Borg, M. (2003). "Generic LCA-methodology Applicable for Buildings, Constructions and Operation Services — Today Practice and Development Needs", *Building and Environment*, 38, 919–938.

Eroksuz, E. and Rahman, A. (2010). "Rainwater Tanks in Multi-unit Buildings: A Case Study for Three Australian Cities", *Resources, Conservation and Recycling*, 54, 1449–1452.

European information service commission. (2012). Construction Sector Overview in the UK. Retrieved from http://www.prismenvironment.eu/reports_prism/UK_PRISM_Environment_Report_EN.pdf.

Fenner, R. A. and Ryce, T. (2008). "A Comparative Analysis of Two Building Rating Systems. Part 1: Evaluation", *Engineering Sustainability*, 161, 55.

Forsberg, A. and Von Malmborg, F. (2004). "Tools for Environmental Assessment of the Built Environment", *Building and Environment*, 39, 223–228.

Fossdal, S. and Holm, F. (1997). Ecoprofile for Buildings: A Method for Environmental Classification of Buildings, Buildings and the Environment, *International Conference*, 67–74.

Geng, S., Wang, Y., Zuo, J., Zhou, Z., Du, H. and Mao, G. (2017). "Building Life Cycle Assessment Research: A Review by Bibliometric Analysis", *Renewable and Sustainable Energy Reviews*, 76, 176–184.

Graham, P. (2003). "The Role of Environmental Performance Assessment in Australian Building Design", *International e-Journal of Construction*, 1–23.

Green Building Council of Australia. (2012). Financial Incentives [Online]. Available at: https://www.gbca.org.au/advocacy/federal/4-financial-incentives4/ (Accessed on June 12, 2017).

Green Building Council of Australia. (2013). Evolution: A Year in Sustainable Building. Retrieved from https://www.gbca.org.au/docs/evolution2013.pdf.

Green Building Council of Australia. (2015). About GBCA [Online]. Available at: http://www.gbca.org.au/about/ (Accessed on July 23, 2015).

Hertzsch, E., Heywood, C., Piechowski, M. and Rowe, A. (2011). Aspects of Life Cycle Investing for Sustainable Refurbishments in Australia. In Howlett, R. J., Jain, L. C., Lee, S. H. (eds.), *Sustainability in Energy and Buildings. Smart Innovation, Systems and Technologies*, vol. 7, Springer, Berlin, Heidelberg.

Illankoon, I. M. C. S., Tam, V. W. and Le, K. N. (2016a). "Environmental, Economic, and Social Parameters in International Green Building Rating Tools", *Journal of Professional Issues in Engineering Education and Practice*, 143.

Illankoon, I. M. C. S., Tam, V. W. Y., Le, K. N. and Shen, L. (2016b). "Cost Premium and the Life Cycle Cost of Green Building Implementation in Obtaining Green Star Rating in Australia", *Pacific Association of Quantity Surveyors Congress (PAQS)*. Christchursh, New Zealand.

Illankoon, I. M. C. S., Tam, V. W. Y., Le, K. N. and Shen, L. (2017). "Key Credit Criteria Among International Green Building Rating Tools", *Journal of Cleaner Production*, 164, 209–220.

Kortman, J., Van Ewijk, H., Mak, J., Anink, D. and Knapen, M. (1998). Presentation of Eco-Quantum, the LCA-based Computer Tool for the Quantitative Determination of the Environment Impact of Buildings. *Proceeding of the CIB World Building Congress*.

Lewis, H. and Ryan, C. (2006). *Imaging Sustainability,* Melbourne, Melbourne: RMIT Publishing Press.

Lippiatt, B. (2001). *Building for Environmental and Economic Sustainability (BEES)*. Construction and the Environment, CIB World Congress, Wellington, New Zealand.

Mansoury, B. and Tabatabaiefar, H. (2014). "Application of Sustainable Design Principles to Increase Energy Efficiency of Existing Buildings", *Building Research Journal*, 61, 167–177.

Mitchell, L. M. (2010). Green Star and Nabers: Learning from the Australian Experience with Green Building Rating Tools. In Bose, R., (ed.), *Energy Efficient Cities*, Washington: The World Bank.

National Institute of Standards And Technology. (2016). Bees [Online]. Available at: https://www.nist.gov/services-resources/software/bees (Accessed on June 9, 2017).

Nationwide House Energy Rating Scheme (Nathers). (2017). *BERS Pro Accreditation* [Online]. Available: http://www.nathers.gov.au/newsletters/issue-2-june-2016/bers-pro-accreditation (Accessed on June 9, 2017).

Ness, B., Urbel-piirsalu, E., Anderberg, S. and Olsson, L. (2007). "Categorising Tools for Sustainability Assessment", *Ecological Economics*, 60, 498–508.

NSW Department of Planning And Environment. (2017). About BASIX [Online]. Available at: https://www.planningportal.nsw.gov.au/planning-tools/basix (Accessed on June 9, 2017).

NSW Department Of Planning And Environment (2013). *Five Year Outcomes Summary*, Australia, BASIX.

O'leary, T., Belusko, M., Whaley, D. and Bruno, F. (2016). "Comparing the Energy Performance of Australian Houses Using NatHERS Modelling against Measured Household Energy Consumption for Heating and Cooling", *Energy and Buildings*, 119, 173–182.

Office of Environment and Heritage Government Of Australia. (2016). NABERS [Online]. Available at: https://nabers.gov.au/public/webpages/ContentStandard.aspx?module=0&template=3&include=homeIntro.htm (Accessed on June 9, 2017).

Organisation for Economic Co-Operation And Development [OECD]. (2015). Green House Gas Emmissions [Online]. Available at: https://stats.oecd.org/Index.aspx?DataSetCode=AIR_GHG# (Accessed on July 6, 2017).

Pettersen, T. D. (2000). *Ecoprofile for Commercial Buildings*. Oslo, Norway: Institute of Real Estate Studies.

Pettersen, T. D., Strand, S. M., Haagenrud, S. E. and Krigsvol, G. (2000). Ecoprofile — A Simplistic Environmental Assessment Method Experiences and New Challenges. *International Conference Sustainable Building*, Rotterdam, Netherlands.

Peuportier, B. (2002). Assessment and Design of a Renovation Project Using Life Cycle Analysis and GB Tool. *Proceedings of Sustainable Building Conference*, Oslo, Norway.

Peuportier, B. L. P. (2001). "Life Cycle Assessment Applied to the Comparative Evaluation of Single Family Houses in the French Context", *Energy and Buildings*, 33, 443–450.

Pieragostini, C., Mussati, M. C. and Aguirre, P. (2012). "On Process Optimization Considering LCA Methodology", *Journal of Environmental Management*, 96, 43–54.

Rebitzer, G., Ekvall, T., Frischknecht, R., Hunkeler, D., Norris, G., Rydberg, T., Schmidt, W.-P., Suh, S., Weidema, B. P. and Pennington, D. W. (2004). "Life Cycle Assessment: Part 1: Framework, Goal and Scope Definition, Inventory Analysis, and Applications", *Environment international*, 30, 701–720.

Reidy, C., Lederwasch, A. and Ison, N. (2011). Defining Zero Emission Buildings Review and Recommendations: Final Report, *Sustainability Victoria*. Institute for Sustainable Futures.

Saibuatrong, W., Cheroennet, N. and Suwanmanee, U. (2017). "Life Cycle Assessment Focusing on the Waste Management of Conventional and Bio-based Garbage Bags", *Journal of Cleaner Production*, 158, 319–334.

Seo, S. (2002). International Review of Environmental Assessment Tools And Databases. CRC Construction Innovation.

Seo, S., Mitchell, P., Watson, P. and Ambrose, M. (2005a). Analysing Environmental Impacts of Buildings Through Lcadesign: Approach and Case Study, *The 2005 World Sustainable Building Conference,* Tokyo, Japan.

Seo, S., Tucker, S., Ambrose, M., Mitchell, P. and Wang, C. H. (2005b). Technical Evaluation Of Environment Assessment Tools. Forest and Development Wood Product Research and Development Corporation.

Shiel, J. J., Aynsley, R., Moghtaderi, B. and Page, A. (2017). "The Importance of Air Movement in Warmer Temperatures: A Novel SET* House Case Study", *Architectural Science Review*, 60, 225–238.

Stern, N. (2007). *The Economics of Climate Change: The Stern Review*, Cambridge University press.

Tae, S., Shin, S., Woo, J. and Roh, S. (2011). "The Development of Apartment House Life Cycle CO_2 Simple Assessment System Using Standard Apartment Houses of South Korea", *Renewable and Sustainable Energy Reviews*, 15, 1454–1467.

Tam, V. W. Y., Le, K. N., Zeng, S. X., Wang, X. and Illankoon, I. M. C. S. (2017). "Regenerative Practice of Using Photovoltaic Solar Systems for Residential Dwellings: An Empirical Study in Australia", *Renewable and Sustainable Energy Reviews*, 75, 1–10.

Tan, Y., Shen, L. and Yao, H. (2011). "Sustainable Construction Practice and Contractors' Competitiveness: A Preliminary Study", *Habitat International*, 35, 225–230.

Thiers, S. and Peuportier, B. (2008). "Thermal and Environmental Assessment of a Passive Building Equipped With an Earth-To-Air Heat Exchanger in France", *Solar Energy*, 82, 820–831.

Tucker, S., Ambrose, M., Johnston, D., Newton, P., Seo, S. and Jones, D. (2003). LCADesign: An Integrated Approach to Automatic Eco-Efficiency Assessment of Commercial Buildings, CIB W078 Conference, Auckland, NewZealand.

Tucker, S. N., Seo, S., Ambrose, M. D., Johnston, D. R. and Newton, P. W. (2005). Eco-assessment of Commercial Buildings, *Proceedings of the Fourth Australian Conference on Life Cycle Assessment, Sustainability Measures for Decision Support*.

Van Der Sterren, M., Rahman, A., Shrestha, S., Barker, G. and Ryan, G. (2009). "An Overview Of On-Site Retention And Detention Policies For Urban Stormwater Management in the Greater Western Sydney Region in Australia", *Water International*, 34, 362–372.

Verghese, K. and Hes, D. (2007). "Qualitative and Quantitative Tool Development to Support Environmentally Responsible Decisions", *Journal of Cleaner Production*, 15, 814–818.

Victoria State Government. (2017a). About firstRate5 [Online]. Victoria State Government. Available at: https://www.fr5.com.au/aboutus (Accessed on June 9, 2017).

Victoria State Government. (2017b). Sustainability Victoria [Online]. Victoria State Government. Available at: http://www.sustainability.vic.gov.au/services-and-advice/business/firstrate5 (Accessed on June 9, 2017).

Vijayan, A. and Kumar, A. (2005). "A Review of Tools to Assess the Sustainability in Building Construction", *Environmental Progress*, 24, 125–132.

Watson, P., Jones, D. and Mitchell, P. (2004a). Are Australian Building Eco-Assessment Tools Meeting Stakeholder Decision-Making Needs, *Proceedings of the ANZSCA Conference*.

Watson, P., Mitchell, P. and Jones, D. (2004b). Environmental Assessment For Commercial Buildings: Stakeholder Requirements and Tool Characteristics, *Proceedings of the ANZSCA Conference*.

Whaley, D. M., O'Leary, T. and Al-Saedi, B. (2017). "Cost Benefit Analysis of Simulated Thermal Energy Improvements Made to Existing Older South Australian Houses", *Procedia Engineering*, 180, 272–281.

Chapter 11

Freeform Printing:
A Sustainable, Efficient Building Alternative

Kevin Sweet*, Armano Papageorge and Tim Miller

*Victoria University of Wellington,
Wellington 6011, New Zealand*

**kevin.sweet@vuw.ac.nz*

Abstract

Since the beginning of the 20th century, modernism introduced to the world an architectural composite that consists of concrete, steel and glass. Heading into the 21st century, the use of these three materials has only expanded as they continue to be the most economically efficient means of construction. While digital technology in design and construction continues to evolve, the materials with which we construct architecture has remained the same. Given the rapid growth of the human population, new and more sustainable approaches to construction methodologies and materials need to be explored and utilized.

This chapter will demonstrate the potential of freeform 3D printing or additive manufacturing (AM) as a sustainable and efficient alternative

---

*Corresponding author.

building method. It outlines contemporary digital design techniques including computation and simulation tools as a means to define and test this proposed building method including structural optimization tools to create the most structurally efficient form of AM. The computational methods described are then applied to a manufacturing process that includes a 6-axis robotic arm. The final result is a building methodology that supports a computational workflow from design conception to manufacture.

Keywords: Freeform 3D Printing; Mass Customization; Structural Optimization; Material Efficiency; Robotic Construction.

1. Introduction

Construction and demolition waste in New Zealand is estimated to represent up to 50% of all waste going to landfills (REBI, 2014). In the United States, in 2014 alone, there were 534 million tonnes of construction and demolition waste generated — more than twice the municipal solid waste (EPA, 2017). This amount of global waste is unnecessary, unsustainable and can be prevented. Given the extensive amount of waste generated by the building industry, an efficient alternative to typical construction methods is required. A potential solution defined in this chapter is a newly defined building methodology that incorporates several approaches to reduce waste. Together, these individual approaches could have a profound impact on the amount of construction waste produced globally and significantly change how buildings are constructed.

Building methodologies need to be updated to take advantage of contemporary processes we have for creating the built environment. Advancements in computational and manufacturing techniques as well as materials science have provided the methods necessary to reconceive how we build. Structural optimization software provides the means to tell us where material should be concentrated for structural performance and new techniques in material deposition allow material to be placed precisely with little to no waste. This chapter outlines a building methodology that has the potential to reshape the production of construction using these tools.

1.1. *New tools of construction*

Robotic arms have been in industrial use since the 1970s. Designed to perform the same task over and over with precision, they have become the tool of choice for most systems of automation. While initially complicated and time consuming to programme for a single task, recent advancements in software development have made this a fairly simple task. Because of this, the industrial robotic arm is now capable of running bespoke programmes with very little effort.

The robot arm operates around six axes, which is what gives it its advantage over more typical 3D printers and CNC machines. This is what provides the most potential freeform printing as it can reach any nearly any point in space with immense precision. With the customized extruder attachment, the potential of freeform 3D printing sees its full potential. The spatial precision of the robot arm combined with the flexibility of the extruder creates endless possibilities to what can be the output in the 3D space surrounding the robot.

With the rate at which these robot arms are being produced and replaced, they are becoming more readily available and affordable. This is extremely important for the accessibility of these machines, as a major reason why companies would be reluctant to adopt them is because they are too expensive and not readily available. Hence, if they were to be introduced to the construction industry in the masses, they could not be ignored.

1.2. *Additive manufacturing and freeform printing*

Typical additive manufacturing (AM) processes rely on the slicing of a 3D model to generate layers of material. Any part of the 3D model that is not sliced through for a particular layer requires more (sometimes different) material to be placed in order to support any parts of the model that may be above that part of the layer. This support material, which is removed and discarded after a part is complete, can at times use more material than the part itself. It is a wasteful process.

Freeform printing does not use any support material. Because the robot has six degrees of freedom, it is not limited to the typical layering system used in conventional 3D printing processes. This freedom, combined with a fast curing material, allows 3D prints to be made without the support material as the material extruded from the robot becomes self-supporting.

2. Robotic Control

The industrial robotic arm is an extremely valuable tool when combined with parametric software and one that is ideal for the freeform printing process. It is easily programmable, quickly re-programmable and extremely flexible with its six degrees of freedom. Parametric software such as Grasshopper and its plug-ins have provided a previously unseen accessibility to a complex tool. Combined with the customization of the end effector — the tool at the end of the robot — and the environment in which it interacts, complex fabrication methodologies can be developed. In general, there are three areas of knowledge that need to be considered in the use of the robotic arm, two of which will be discussed in detail related to freeform printing with biomaterials.

(1) Robot control and programming.
(2) End-effector: tool design and integration with robot.
(3) Cell: the environment the robot moves in and interacts with (not discussed in this chapter).

2.1. *Robot control and programming*

The ability to control a complex CNC machine, such as an industrial robotic arm, has been facilitated and simplified by advances made in software interfaces with commonly used 3D modelling programs. A machine once programmed to repeat the same task over and over can now be pro-grammed to produce one-off tasks utilizing parametric applications — a change in a design parameter will instantly generate a unique programme for the robot. The research in this chapter took advantage of these advancements and utilized HAL Robotics, a plug-in for the parametric plug-in Grasshopper for Rhino.

The process of learning Grasshopper and HAL Robotics first began with learning the required data inputs for Grasshopper that would allow the creation of a toolpath for the robot to follow. The mastering of this process utilized the HAL to produce a series of light drawings (Fig. 1).

These experiments highlighted the primary challenge with 3D printing toolpaths; that the extruder cannot overlap any existing toolpath. It must be

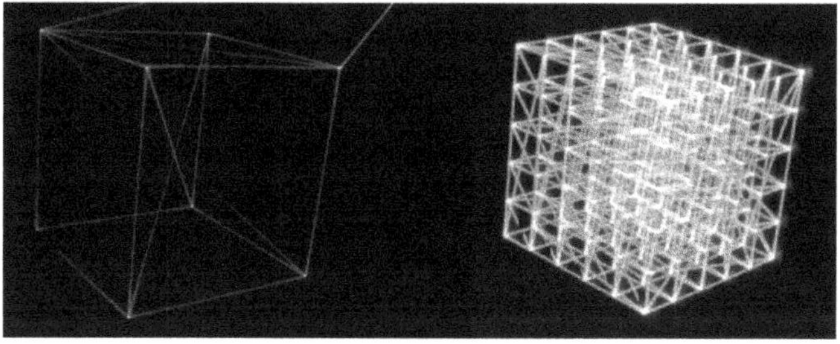

Fig. 1.   Example of LED-based toolpath used for learning and understanding Grasshopper and the robot.

a single continuous line without interruption throughout the entirety of the print. This led to redefining the programming of the toolpath, leading to rigorous experimentation with Grasshopper and its data handling. Various data inputs were tested to parametrically create a lattice structure with one single toolpath line; however, the majority of these attempts were failures because the toolpath was generated from a square grid input parameter from Grasshopper which had its own predefined interpretation of the toolpath. The point order could be manipulated to an extent; however, there was always a line overlapping in one or more instances.

It was determined that using Grasshopper's predefined data generators was not sufficient and toolpaths would have to be created from scratch. The new definitions would allow any toolpath to be parametrically created, forming any lattice structure with one single toolpath.

From this point, the progression of work to follow was to increase the complexity of the definition in order to provide mastery of Grasshopper to later have better control of the extrusion toolpath for freeform printing. The experimentation developed as follows:

- Parametrically manipulate an orthogonal lattice structure; adding more or less geometry in the X, Y, Z directions or changing the scale of the overall form.
- Create a logic that would enable the toolpath to adapt to more complex geometries.
- Create curvature in a single direction, which required reconfiguring input parameters. This resulted in an arch form being created from a curved surface.

- Create curvature in a second direction.
- Adding attractor points to be able to create curvature in the X and Y directions as well as in the Z. In achieving curvature in all three geometric directions, true parametric control was achieved.

Parallel to learning the methods to create toolpath definitions in the Grasshopper was the control and understanding of HAL inputs to create the logic for the robot arm. After learning the logic of Grasshopper, understanding the intricacies of HAL was easier. The process began with a simple base template with the plug-in and then components were added as needed to control robot speed, orientation and signal communication.

2.2. *Initial tests*

Control of the robot and toolpaths was achieved using an LED and light drawings. This was a simple process that gave visual feedback to the programming process. Toolpaths could be displayed through an image to compare the Grasshopper programming to the actual machine output. The process was invaluable to the learning process and did not rely on any physical material — making it a waste free and limitless process.

With basic programming and robot control mastered, the tests continued by adding a custom-made plastic extruder. The material used for these tests was an ABS plastic filament. This is a very effective material; it possesses excellent strength to weight ratio, and it adheres to itself well — an important trait for node connectivity.

The custom extruder (described below) had a limitation in that it could not be directly controlled through robot programming. Extruder instructions had to be set prior to the start of robot movement and run as parallel but separate processes. Because of the disconnect between robot and extruder, testing the system as a whole was tedious and resulted in multiple steps. The means of operating this system are as follows:

(1) Turning robot arm on and attaching the extruder to the gripper through the inputs/outputs command in the panel device.
(2) Turning on the extruder by plugging in all necessary cords, etc., and setting up the chosen filament.

(3) Loading of the desired print file into the robot arm.

(4) Start toolpath at robot.

(5) From a separate program (Grasshopper), send required extruder instructions.

(6) Supervise the process to ensure the output is correct.

This is the preliminary process that has been undergone for the initial tests thus far. This is not the ideal process of operations, as it is not fully automated. Steps 4 and 5 need to merge into one single streamlined system. Once the process has commenced, the robot arm and extruder must be able to communicate with one another. A logic must be created that informs the extruder when to print, how much to print and how fast to print. Then and only then can this system be considered as having potential for "mass manufacturing". Extensive analysis is being undertaken to ensure successful future development of this system (see Fig. 2).

2.3. *End effector: Extruder*

An end effector is the tool that is attached to the end of the robot to perform a defined task. In this case, it is a custom-made filament extruder that heats plastic filament to a set temperature and extrudes it in a softened state (see Fig. 3). The extruder also must be able to return the filament to a solid state after exiting the nozzle. The faster the material can be solidified, the better the plastic can be formed in 3D. Given this, there are key specifications in the creation of a basic extruder:

(1) Heat filament beyond its melting point should be softened for it to be extruded.

(2) Solidify the filament as it exits extruder's nozzle for it to maintain the desired form without sagging, etc.

(3) Fit all necessary components within a streamlined and ergonomic shape.

All preliminary tests were performed with a basic extruder that fit these specifications. After conducting many tests and experiments with

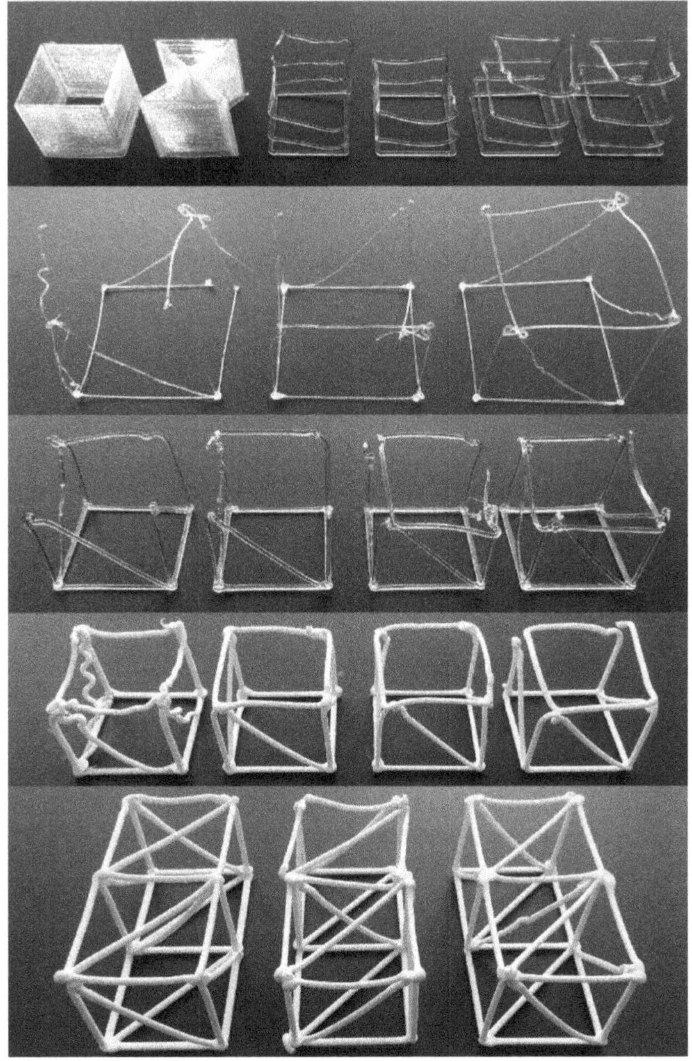

Fig. 2.   Initial tests using basic toolpaths and the first custom designed extruder.

the basic extruder, it has been determined that a second extruder will have to be created that will allow further control of the printing process and integrated control with the robot. The second extruder will have the added specifications:

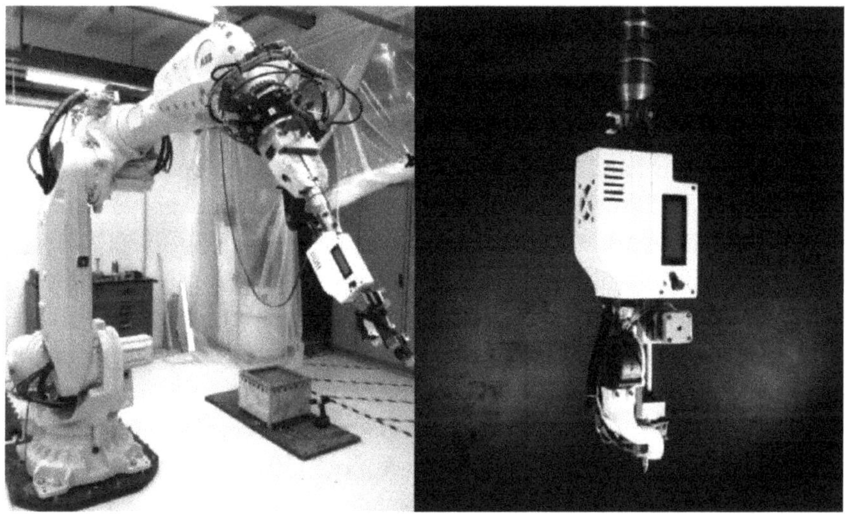

Fig. 3.   Custom-made extruder and its attachment to the robot.

(1) Remotely and succinctly coordinate with the robot programming to have full control of the extrusion rate and temperature of the filament.
(2) Print thicker (or various)-sized filament, thus producing stronger structures and providing expandability to use intended biomaterial still under development.
(3) Fit into an even smaller package to enhance the precision when plane reorientations are required. The current extruder is awkwardly sized and accuracy is lost because of this.

3. Additive Manufacturing as an Efficient Process

AM methods were first introduced in the 1980s as a promising approach towards construction automation and fabrication (Oxman, 2013). Until now, there has always been a certain degree of scepticism towards applying this technology to large-scale projects (Chalcraft, 2013). However, architects and engineers who have been researching this field for a number of years have gone from predicting that this may not be applicable for another 50 years, to now being convinced that this technology will very soon make the shift into the architectural realm (Krassenstein, 2015). AM differs from conventional processes such as subtractive processes (i.e. milling or drilling), formative processes (i.e. casting or forging) and joining processes (i.e. welding or fastening).

Freeform 3D Printing — a form of AM — can offer a new form of construction that utilizes contemporary digital technology to its fullest. AM traditionally involves manufacturing a part by depositing material layer by layer with the support material; however, in this case it would be used target by target. Freeform 3D printing involves moving to targeted points in 3D space as opposed to layering. Thus, it uses significantly less material. Combined with other computational tools such as optimization software and processes, it will provide a more efficient building system that will provide an innovative building technology.

3.1. *Structural optimization*

Structural optimization is a key component to a new efficient building methodology. Without it, there will be a tendency to use the AM process to reproduce methods that are currently used for building. The optimization processes offer another level of efficiency and waste control by dictating exactly where the material is required for performative reasons over aesthetic or conventional choices. Combined with AM, the material will be deposited where it is performatively required. In the case of using filament as a building material, lightweight structures will be considered.

A lightweight structure is defined by the optimal use of material to carry external loads or pre-stress. Material is used optimally within a structural member if the member is subjected to membrane forces rather than bending. The objective of an optimization procedure is to determine the layout and shape of a lightweight structure therefore minimizing bending or minimizing the strain energy rather than structural weight as the term "lightweight" may imply. This notion is extremely applicable in this case, as plastic structures are arguably the most lightweight structures available to the construction market today (Bletzinger, 2001).

Millipede is the primary tool for Grasshopper in achieving structurally optimized forms. The benefit in utilizing this tool as part of the design methodology is to computationally determine the most structurally efficient natural form to use for a building/building component (form finding). All that is required is a form to be analyzed, a series of force and supporting geometries which can vary in size and quantity, and a structurally optimized form will be generated based on these parameters (Fig. 4).

The experimental process of the project to date consists of progressively and logically manipulating all these parameters to test the output of workability of Millipede. Each iteration becomes increasingly complex and attempts to achieve a unique result. For example, various force and support geometries were added to the same initial diagram (a cube), however, eventually the output becomes relatively predictable; support members would appear between the force and support geometry (Figs. 4(a) and 4(b)). The next studies introduced a single uniform force/support that

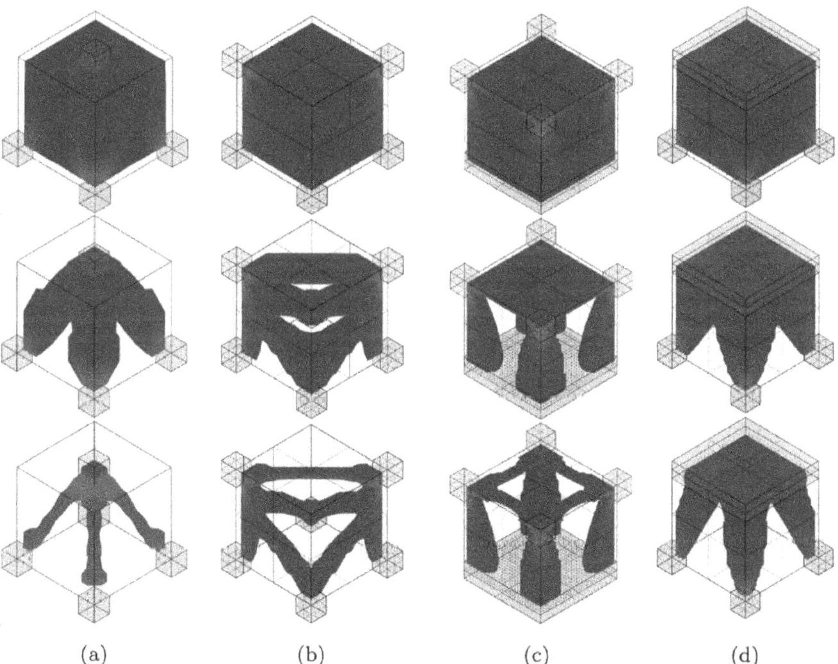

(a)   (b)   (c)   (d)

Fig. 4.   Simple cubes with different loadings and the resultant optimization graph using Millipede for Grasshopper. (a) Support members follow the force geometry; (b) Support members follow the support geometry; (c) Support members cover the top of the cube; (d) Support members cover the bottom of the cube.

covers the entirety of either the top or the bottom of the cube (Figs. 4(c) and (d)). This generated a much more unique output as it was not something that could be predicted as easily. From these experiments, a simple triangulated cuboid was developed to test the initial output of the freeform printing system. The simple tests on the cubes showed that forces moved directly from the point load to the support and this form was the most basic form to begin the freeform physical tests.

Future testing will evolve to develop a grasshopper definition that "fills" the resultant output from Millipede with an appropriate and efficient structural system that uses the least amount of material. These tests will be performed in the next phase of the project. The Grasshopper definition for the toolpath, i.e. the lattice structure, will have a logic that will enable it to instantly adapt to any surface or 3D form. If executed correctly, this design methodology can save time, money and resources due to the majority of the work being accomplished through computational calculations and form finding.

3.2. *Sustainable building system*

The freeform research is employing three strategies towards a more sustainable building system. The first is to employ spaceframe geometries determined by the initial Millipede studies, as these structures reduce the volume of material used while retaining structural integrity, as spaceframe geometries are highly efficient structures. Further evaluating these spaceframe structures, using Finite Element Analysis in phase two, will offer the ability to identify areas that require greater stiffness. Adding more material to an area using the freeform extruder system script can be achieved in a number of ways. The extrusion bead can be increased in diameter by either increasing the extrude rate or by slowing down the speed of movement. The third method is to simply extrude additional beads of material over the existing built spaceframe structure. Applying and testing these methods in phase two will add further hierarchy and control to the building system, making it even more efficient.

The second strategy is to utilize biopolymers, which are derived from living organisms and are biodegradable. The most prevalent biopolymer for 3D printing filament is Polylactic acid (PLA), which is derived from

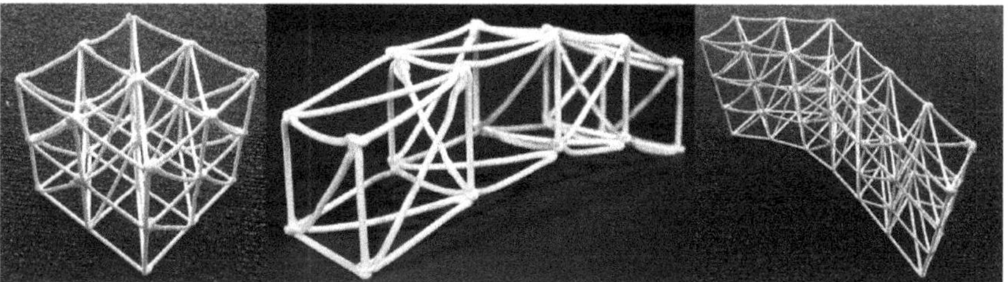

Fig. 5.   Current builds from the described building system.

organic materials such as corn starch, sugar cane and the like. It is commercially available in a range of filament diameters and has been tested through our freeform extruder. This study is also interested in biopolymers that offer greater structural efficiencies and is investigating long fibre biopolymers to enhance the structural qualities of freeform printing structures. The biopolymers are being developed by a research partner and should be ready for testing in phase two. The initial ABS studies have given them a datum to work from.

The third strategy is to investigate the opportunity of using recycling of plastic waste to form printing filament. This offers the most sustainable model as it reduces the landfill but costs associated with collecting, sorting, cleaning and filament production could make this economically unviable (Fig. 5).

4. Looking Forward

This chapter outlined the initial findings of a project to demonstrate the feasibility of a new sustainable building system using freeform printing. These initial steps were the foundation for an exploration that will continue using many of the strategies identified throughout the chapter. The key strategies to further this research into a viable system are categorized as follows:

(1) Robot and Extruder:
    (a) Redesign the extruder to integrate programmatic control with robot movement.

   (b) Alter building speed (robot movement and extruder speed) for specific building properties (more or less material for structural support).

   (c) Extruder cooler control through programming for faster printing or better adhesion of material.

   (d) Multiple materials for specific structural conditions (tension vs. compression).

   (e) Improved form factor of the extruder for tight building conditions.

(2) Structural Optimization:

   (a) Develop a Grasshopper definition that creates a toolpath with the most efficient form found through structural optimization.

   (b) Develop a Grasshopper algorithm to create a hierarchy of the building structure.

   (c) Test large-scale outputs based on the structural optimization.

(3) Materials:

   (a) Study alternative materials such as the biopolymers being developed by partner researcher.

   (b) Use recycled plastics to form the filament for building with plastic.

   (c) Test longevity and structural properties of materials once placed.

In the initial phases, this research has shown great promise and has created a solid foundation for further development of a sustainable building system using freeform 3D printing. Many roadblocks have been overcome and many more strategies for furthering the work have been identified. Construction waste management is a global issue that needs to be addressed soon and in innovative ways — the proposed building system is one of them.

References

Bletzinger, K.-U. and Ramm, E. (2001). "Structural Optimization and Form Finding of Light Weight Structures", *Computers & Structures*, 79(22–25), 2053–2062.

Chalcraft, E. (2013). In the Future We Might Print Not Only Buildings, But Entire Urban Sections. Available at: https://www.dezeen.com/2013/05/21/3d-printing-architecture-print-shift/ (Accessed on September 24, 2017).

D'Alessandro, N. (2014). 22 Facts about Plastic Pollution. Available at: https://www.ecowatch.com/22-facts-about-plastic-pollution-and-10-things-we-can-do-about-it-1881885971.html (Accessed on September 10, 2017).

EPA (2017). Sustainable Management of Construction and Demolition Materials. Available at: https://www.epa.gov/smm/sustainable-management-construction-and-demolition-materials (Accessed September 26, 2017).

Kishore, S. (2017). Only 9% of US Plastic Waste is Recycled, Total Waste Increasing at Alarming Rate. Available at: http://www.ibtimes.com/only-9-us-plastic-waste-recycled-total-waste-increasing-alarming-rate-2568325 (Accessed on September 24, 2017).

Krassenstein, E. (2015). Branch Technology 3D Prints Building Walls With World's Largest Freeform 3D Printer. Available at: https://3dprint.com/85215/branch-3d-printed-walls/ (Accessed on September 24, 2017).

Morten, J. (2015). More than 25,000 kg of Plastic Littered in NZ Daily. Available at: http://www.nzherald.co.nz/nz/news/article.cfm?c_id=1&objectid=11401696 (Accessed on September 24, 2017).

Oxman, N. Laucks, J., Kayser, M., Tsai, E. and Firstenberg, M. (2013). *Green Design, Materials and Manufacturing Processes*, Cambridge, Massachusetts, USA: CRC Press.

Rebri (2014). Waste Reduction — Construction. Available at: https://www.branz.co.nz/cms_show_download.php?id=5e8633f5234594b316612f186e49687aff5475dd (Accessed on September 22, 2017).

Chapter 12

A Reconceived Digital Workflow: A Case Study

Kevin Sweet* and Tonya Sweet

*Victoria University of Wellington,
Wellington 6011, New Zealand*
**kevin.sweet@vuw.ac.nz*

"Throughout the twentieth century, and into the twenty-first, the expectation of radical changes in the field of architecture as a result of technological innovation and the proliferation of new digital tools and techniques has prevailed. Professionals, academics, and students alike have shared in this general sense of anticipation. Still, there remains a great deal of mystery surrounding precisely how this revolution might reshape architectural practice and significantly impact the built environment."

— Corser, 2010

Abstract

Today's designers are inundated with a plethora of CAD/CAM software capable of meeting a variety of needs within the design process. In academia, it is commonplace for students to learn as many as 10 different

---

*Corresponding author.

software applications in order to develop their designs from conceptualization to digital model and, ultimately, to the production of a material output. With each shift from one application to another, the risk exists for valuable information to be lost in order to suit the limitations of the chosen software.

Emergent innovations in digital design tools are compressing the design workflow and allowing for greater flexibility and variability. Within the scope of these software options, parametric design tools have garnished increased validity and applicability across various disciplines of design: they afford the benefit of adaptability within the design process as well as highly customizable outcomes. Despite their recognized value, these tools remain secondary to their established counterparts. Designers are required to side-step through an inefficient series of procedures as they translate data between programs. In order to generate design outputs in a truly efficient manner, design practice (and, subsequently, design education) must capitalize on the potential of parametric software that supports a streamlined flow from design conception to production.

Through a case study, this chapter will explore how such a system was applied from design conceptualization to fabrication of a bespoke furniture object — the Earthquake Bench. The Earthquake Bench project utilizes a "simplified" digital workflow in which the entire process, from design to fabrication, was parametrically programmed and remained intact until the file was sent to the manufacturer, thus ensuring that all changes were consistently updated throughout the process. The case study identifies the shortcomings and benefits achieved from this process, as well as how this proposed system may begin to reshape design practice and education.

Keywords: Parametric Design; Digital Fabrication; Digital Workflow.

1. Introduction

Design has always been a parametric process. The designer e stablishes or is provided variables, or design parameters, that define the scope and limitations of the design. These variables are then explored over an iterative process to test multiple design solutions to a given problem towards the most suitable result. The application of digital parametric design tools and processes affords the designer control over the iterative exploration of a range of potential solutions as the design progresses from conception to resolution. Without this approach the development of a design requires repetitive editing, which is inefficient to say the least. With the technologization of contemporary design practice and the widespread availability of advanced software and tools in enabling complex design solutions, the potential for the development of innovative outcomes has grown exponentially. The reality of these potentials as they relate to design for manufacture, however, poses a number of unresolved challenges for the design industry.

The challenge may be understood as follows: Diversity in software tools and methodologies creates digital files that can be as varied as the designs and designers that produced them. Digital data created during the design development phase of a project are not always compatible with the systems employed in advanced manufacturing processes. Under the previous working methodologies, this was less of an issue as manufacturing processes were largely standardized. At the same time, however, these methodologies, were often inflexible to alternative and innovative design approaches, and did not support variable alterations in production processes. While contemporary manufacturing techniques have the benefit of utilizing computer numerically controlled (CNC) machines that allow for greater complexity as well as the efficient customization of bespoke or one-off pieces, the diverse software requirements for these tools remain an obstacle to achieving a flexible and optimal workflow. In order to evolve design practice and propel an economy rich in innovation, designers require an efficient and operational method of traversing the variances of our current technologized design/manufacture landscape.

This project is a case study that attempts to close the gap between contemporary design and manufacturing by devising a system that aims to streamline design-to-production processes and that enables feedback and consecutively link changes throughout the entirety of the process. It uses a small but complex piece of furniture as a proof of concept for the potential of a fully integrated process, and speculates on the ramifications that this process may have on the field of architecture.

2. Existing Parametric Process in Design and Manufacturing

2.1. *Design*

As design software has evolved and placed itself within mainstream practice, contemporary designers have learned to anticipate the application of a variety of programs within the development of a design. A design might be conceptualized in SketchUp, further developed in Rhino with a Grasshopper plug-in, and then remodelled in building information software (BIM) such as Revit. When the finished design finally goes to production, the data may then need to be translated or remodelled yet again using a manufacturer's proprietary software. For each move between software applications, the potential for loss or corruption of data is greatly increased. Each software is embedded with its own limitations and these often unintentionally drive the design rather than the design being driven according to the designer's intent. Further, the ability to make iterative alterations to the design is compromised as any geometrical relationship established in one software is lost in the translation when the data are uploaded to other software used in the process. Universally speaking, parametric connections are not compatible between software formats.

Today, parametric thinking, a component of advanced computational thinking, is changing how a design is conceived and produced. Sean Ahlquist and Achim Menges distinguish this form of thinking from standardized CAD methodologies by stating that "the distinction rests in the approach towards design, rather than in particular skill sets or knowledge. A computer-aided approach assumes an object-based strategy for encapsulating information into symbolic representations — methods of organising information. In contrast, a computational approach enables specific data to be realised out of initial abstraction — in the form of codes which encapsulate values and actions" (Menges, 2011). While not quite mainstream, the influx of young designers that are versed in this technology is rapidly influencing the profession. It is only through exposure to this type of software that the power of parametric thinking — especially in the system proposed by this chapter — becomes fully realized.

2.2. *Manufacture*

The application of computational tools is driving the need for new methods of fabrication in order to give material form to the complex designs developed with the aid of advanced digital software. While the potential to manufacture these forms exists, it is rare that manufacturers grasp the full capability of the technology to the degree necessary to produce the advanced and often enigmatic forms that designers desire. When a manufacturer does attempt to produce the model provided, they have a tendency to recreate the design data according to the non-progressive terms with which they are accustomed. As a result of the limitations of traditional modes of production, these terms are often far too simple to support the intended outcome. Manufacturers are largely limited to a predefined set of working methods and tools and, although the equipment may be capable of an expanded repertoire of toolpaths and capabilities, the standardization of mass production has left its imprint on the way these tools are used. This influence is similar to BIM software that has a pre-programmed library of parts that is often difficult to customize for an intended outcome: it tends to produce normative results. Occasionally, a forward-thinking designer has the opportunity to work with an experienced and open-minded manufacturer, and through their joint efforts a dialogue is facilitated regarding how the manufacturing process can be utilized to support the designer's vision. This type of exchange is ideal in positively impacting the evolution of the design industry, and it requires all parties to embrace an open mind and willingness to learn from and adapt to a wide net of parameters. It requires the broad application of parametric thinking and parametric affordances across all systems applied to the design.

In the effort to provide a more effective translation between the design process and the manufacture of the final output, there is a need for the entire process to be parametric — from design to production. Such a parametric system will enable designers to code or program the manufacturing limitations and manufacture-specific knowledge into the feedback loop. Likewise, this would enable manufacturers to program in the variables that will produce the complex design output that contemporary designers envision. Applied in this way, parametric design tools can be utilized to develop the digital model from start to finish with consideration to myriad parameters according to identified design needs and as manufacturing restrictions.

This strategy has the dual benefit of educating manufacturers regarding the potentiality of their tools in fabricating complex, innovative and specialized forms, and for designers to better understand the manufacturing processes used so that these variables can be considered at the initial phases of the design process. Ideally, this method of thinking and designing would be introduced to design and engineering students as the standard mode of operation, thus instigating a seamless workflow from the start of their careers. The application of such a system would function to minimize the knowledge and communication gap that is prevalent between designers and manufacturers and, if applied widely, would see tremendous benefits for the entire innovation economy.

3. Parametric System Reconceived

In order to achieve a more efficient workflow across the design and manufacturing sectors in architectural practice, the parametric process currently applied has to evolve to extend beyond solely focusing on design development to include the full spectrum entailed in the production and fabrication of material outputs. Parametric tools such as Grasshopper and Dynamo excel at allowing parametric processes to be applied during the conceptual phases of design, but these tools are limited in their scope availing generative capabilities as opposed to informative capacities. BIM software, on the other hand, is informative but cumbersome and inappropriate as a generative tool. None of these functions ideally supports the fabrication of complex forms and, as a result, designers rely on other non-parametric software that severs any link to the further manipulation of data. In other words, if the manufacturing process requires a dramatic design change, the model should be redeveloped by the designer and then provided again to the manufacturer, and in this process the data is likely to be translated back and forth between software platforms resulting in lost time and the inability to make additional changes as needed. The ideal parametric process will not sever the digital link to data information at any stage. Whether changes are required to be made by the designer, a consultant, or a manufacturer, the digital model should be robust enough for the changes to be made without hours of remodelling on anyone's part.

Designing with parameters is often perceived as a difficult process as it requires the designer to think about the design process in an all-encompassing manner. It requires a new set of tools that are not always provided in the educational environment. "It turns out these ideas are not easy, at least for those with typical design backgrounds. Mastering them requires us to be part designer, part computer scientist and part mathematician. It is hard enough to be an expert in one of these areas, let alone all. Yet, some of the best and brightest (and mostly young) designers are doing just that — they are developing stunning skill in evoking the new and surprising" (Woodbury, 2010) (see Fig. 1).

3.1. *Design process*

The Earthquake Bench project was conceived to test the proposed, holistic parametric system outlined above. A cohesive parametric process was applied across the entire development — from design to fabrication — and this was initiated with the creation of a digital model that could be altered or changed based on designer-defined parameters. These parameters allowed the model to undergo rapid adjustment in response to design

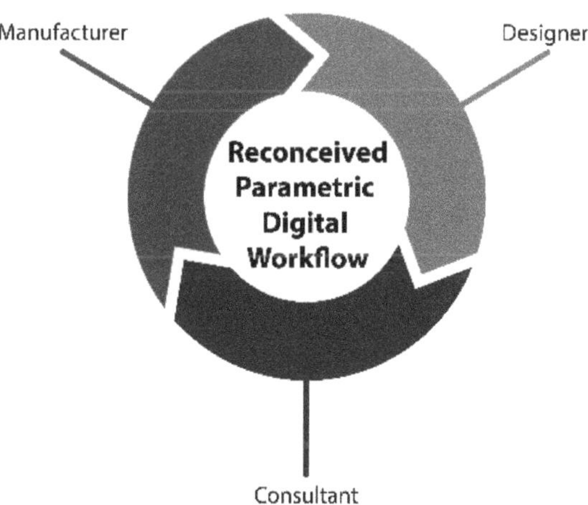

Fig. 1.   Diagram of the reconceived process.

and manufacturing needs and conditions as they arose. They also allowed multiple versions to be developed according to different sets of criteria. The data created across all phases of development remained intact while changes to the data were updated continuously thus resulting in an unbroken and efficient workflow.

Based on its reasonable scale and complexity, the Earthquake Bench was identified as an ideal case study to showcase the benefits of the proposed reconceived digital workflow. Although the project entails a piece of furniture as opposed to an architectural outcome, the principles considered within this model are equally applicable to architectural practice. Conceptually derived, the bench entails a monolithic glue-lam beam that is sculpted along every axis. While the bottom is shaped in a gentle arc, the sides are formed with compound curved surfaces. The "top" of the bench highlights the complex nature of the design where the 3D and multi-textured topographical expression of New Zealand is milled, complete with fault lines and the indication of major seismic events that have occurred throughout history. The designer's intention was to create numerous versions of the bench with topographical information that corresponds to specific regions within New Zealand. As a result of the linked digital data and live updating capability, select adjustments are easily applied to defined parameters and the design is therefore capable of supporting highly customizable, unique and efficiently manufactured outcomes.

In initiating a contemporary digital workflow, the design parameters are the first factor to be defined. These decisions must be made by the designer at the beginning of the process in regards the nature of the parameters and how they will be used to control the resulting geometry. Priority is given to some parameters over others based on the design intention and known limitations related to materials and manufacturing processes. In the case of the Earthquake Bench, there were several levels of design parameters that controlled different aspects of the final product. These formed a hierarchical structure that prioritized some parameters over others including the overall dimensions that dictated how all parameters react to one another, features that drove specific bench attributes such as the curvature and "shift" properties, base attributes, topographical attributes, earthquake attributes, and, lastly, fabrication parameters that were tied to manufacturing restrictions.

3.1.1. *Bench dimensions*

The bench dimension parameters controlled the overall dimensions of the object including length, width and height. These dominant parameters determined the overall size of the bench and functioned as the primary parameters to which all other parameters responded. With these parameters, the bench could easily be scaled to fit any proportion according to the specified limitations of the material. In this case, the material entailed a glue-lam beam that measured 2 m long by 0.42 m tall and 0.72 m wide (see Fig. 2).

3.1.2. *Bench features*

The bench-feature attributes controlled the various design features applied to the bench. For instance, parameters were created to alter the compound-curvature on each side of the bench so that the designer had aesthetic control over the final form. The seismic "shift" in the bench was also controlled with these parameters enabling the location, angle and offset of the shift to be controlled within the form (see Fig. 3).

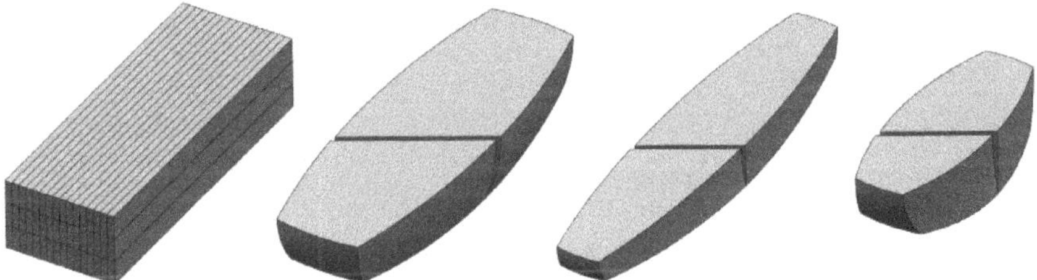

Fig. 2.   Final glue-lam beam and various dimensional configurations of the overall bench.

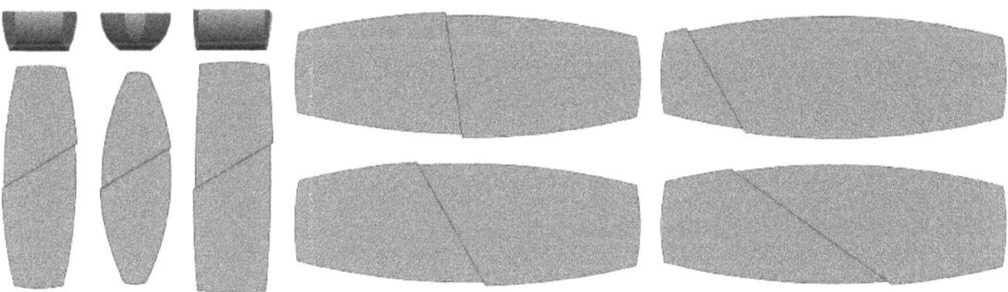

Fig. 3.   Various configurations of the bench curvature and split location.

Fig. 4.   Various configurations of the base.

3.1.3. *Base attributes*

The "rocking" arc base applied at the bottom of the bench was necessarily parametric in order to enable the design to adapt to the changing distribution of weight as the form underwent alterations. Other factors that drove the parameters of the base were in consideration to health and safety of the prospective users. For instance, as the object is of significant weight it was important to ensure that user's feet were protected. This factor was parametrized in the depth and height of the offset base surface. Likewise, the design required the rocking movement to be minimal, height appropriate and limited in range (see Fig. 4).

3.1.4. *Topological attributes*

The 3D topographical attributes of New Zealand were controlled by parameters that determine its location on bench and the scale and a rotational value in relationship to the overall form. This entailed the creation of built-in geometry that trimmed the islands to the edge of the bench in order to maintain a flush edge. The topography also responded to the "shift" of the overall form and changed location in space to align with this condition (see Fig. 5).

3.1.5. *Earthquake attributes*

Data regarding the dominant faults in New Zealand and a registry of major seismic events were collected from the Institute of Geological and Nuclear

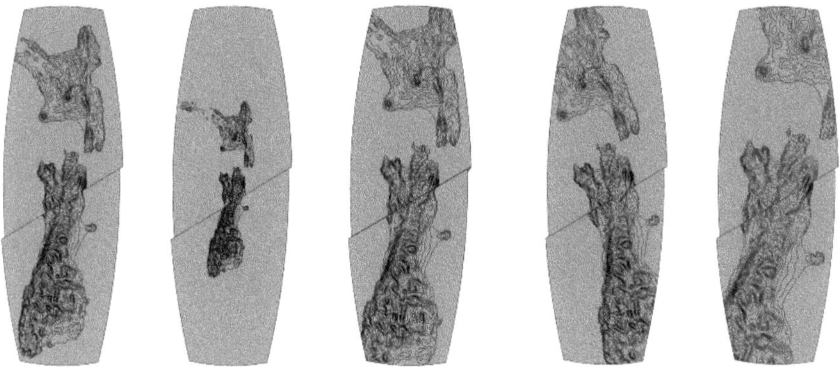

Fig. 5.   Various topological configurations.

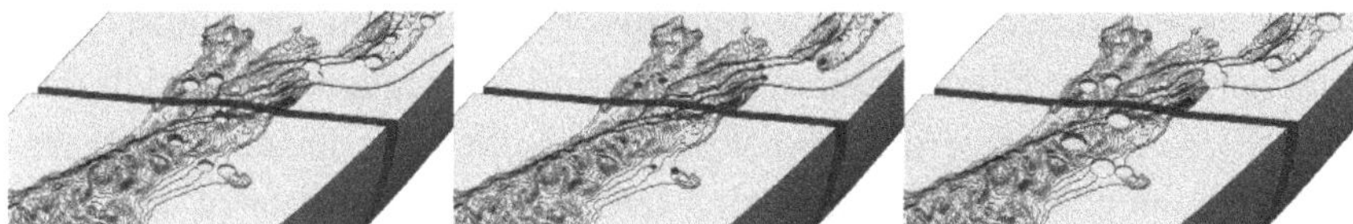

Fig. 6.   Various EQ data configurations.

Sciences (GNS Science) and applied to the earthquake attributes parameters. Although these data were fixed, the inclusion of this information within the parametric system enabled scaling and positioning according to the changes to the topography. The fault lines were parameterized to enable the designer to adjust the depth and width of the line to be cut, while the circular indentations that indicate major seismic events were parametric in their depth and diameter (see Fig. 6).

3.2. *Fabrication process*

Beyond its ability to showcase the advantages of a cohesive parametric system to the design of a material object, the Earthquake Bench effectively evidences the application of advanced parametric thinking to the fabrication process. Parametric considerations specific to manufacturing included the type of CNC tools to be employed and the subsequent limitations these tools have in dimensionality (X, Y and Z) and range of the tool

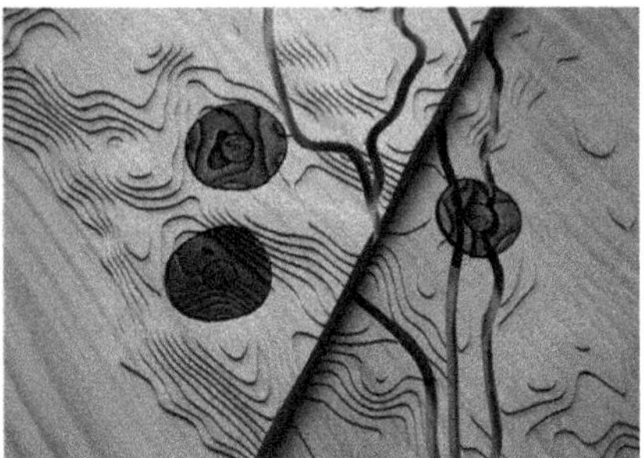

Fig. 7.   Final manufactured bench.

head. The shape, length and diameter of the tool head must also be taken into consideration, especially as this relates to the materiality and complexity of the form. Within this framework, the design data must also reflect an understanding of the tool paths and the sequence in which the tool will mill the material. With this knowledge, the designer is in an advantageous position in determining the aesthetic treatment of surfaces in accordance with the particular crafting enabled by the digital process.

In a non-parameterized "broken" system, the considerations noted above are typically navigated during the final stages of the design process. As discussed, the translation of data between the designer and manufacturer is severely challenged whereby a lack of fluidity of translatable data impedes the efficiency of the overall system (see Fig. 7). As applied to the Earthquake Bench, these considerations were implemented into the design programme in the early phase of the design process, thus enabling an unbroken workflow.

4. Conclusion

The Earthquake Bench Prototype attempted to use an existing parametric software application, Catia, to test a cohesive workflow that allowed for changes or alterations in the design at any time during the design process from conceptualization to fabrication. The goal of the project was to demonstrate the potential of digital design thinking using a fully parametric process. While the scope of the project was small enough for there to be success in the realization of a furniture piece, there were several limitations identified that would need further attention for the process to be scaled up for the production of architecture.

Designers, consultants and manufacturers would all need the capability to share parametric data. This goes beyond compatible file formats (which exist) and would have to extend to the designed parameters that are embedded into a parametric file. The parametric link would have to be robust enough to survive changes from any party involved in the process. The benefit of a fully parametric system that is applicable to contemporary architectural practice is that such a system would more effectively support the efficient production of advanced computational design outcomes. This system would be well situated to embrace the evolution of technologies that are currently shaping the field of design and, in turn, such a system has the potential to foster the growth of a robust innovation economy. With the recognition that this chapter implies that a reconceived parametric system has the power to change the nature of the industry, the seamlessness of the parametric workflow introduced depends upon and promotes a shared language between design practitioners and manufacturers with foreseeable advantages to everyone involved. Such a system has the potential of reshaping architecture by simply taking better advantage of all of the technologies and knowledge that we have at our disposal.

In moving forward, the first step to the accessibility and implementation of a comprehensive parametric digital workflow would entail the development of tools that expand the capabilities and emulate the processes currently used in programs such as Catia. Although Catia was specifically designed for the manufacturing sector and does not adequately support the scale of architectural projects, it provides a model from which to build. Ultimately, design practitioners need to be more proactive in defining the criteria they need to support a successful digital workflow rather than wait for software engineers to define it for them. In taking advantage of the technologies and cross-disciplinary skills that are currently giving shape to contemporary education and advanced practice, designers must consider the possibility of "designing" the tools in order to create a reconceived workflow and regain control of the design process.

References

Corser, R., ed. (2010). *Fabricating Architecture: Selected Readings in Digital Design and Manufacturing*, New York: Princeton Architectural Press.
Kolarevic, B., ed. (2003). *Architecture in the Digital Age: Design and Manufacturing,* New York: Spon Press.
Menges, A. and Ahlquist, S., eds. (2011). *Computational Design Thinking*, Hoboken, New Jersey: Wiley & Sons.
Woodbury, R. (2010). *Elements of Parametric Design*, New York: Routledge.

Chapter 13

Drivers and Barriers to the Adoption of Building Information Modelling (BIM) By Construction Firms in South Africa

Amanda Mtya* and Abimbola Windapo

Department of Construction Economics and Management, University of Cape Town, South Africa

**amanda.mtya@uct.ac.za*

Abstract

It remains a global challenge to complete construction projects predictably within the constraints of time, cost and quality. The Building Information Modelling (BIM) revolution has brought with it many interesting aspects in terms of changes in business processes and project practices in the construction sector. Construction firms play a crucial role in the delivery of projects; however, research shows that BIM adoption by construction organizations is still in its early stages in South Africa. Literature has shown that the South African Built Environment has not taken advantage of the BIM movement. This chapter examines the drivers and barriers to the adoption of BIM by construction companies and

---

**Corresponding author.*

whether BIM capability is a key driver of BIM adoption. The study adopts a comprehensive literature review approach in achieving the study objectives. The study found through literature that the key drivers and barriers to BIM adoption include organizational readiness and client request. The study aims to develop an assessment tool for BIM capabilities within construction organizations. Further empirical studies will be required to confirm this finding that the higher the grade of a construction company on the CIDB register of companies, the higher the level of its organizational readiness to implement and adopt BIM.

Keywords: Building Information Modelling; Adoption; Construction Firms.

1. Introduction

In South Africa over the period 2014–2016, the Construction Industry Indicators (CIIs) reflect a steady growth in the level of client dissatisfaction with the performance of contractors and consultants on construction projects (Construction Industry Development Board, 2016). Construction projects continue to record poor performance with time and cost overruns. A major cause of this problem is that the project environment is fragmented and uncollaborative. Despite the wide range of benefits for implementing BIM methodology documented in the literature (Windapo, 2017; Wang and Song, 2017; Barlish and Sullivan, 2012; Bryde *et al.*, 2013), Building Information Modelling (BIM) adoption in South Africa is still in its infancy (Smallwood *et al.*, 2012).

BIM is a term with evolving manifold definitions within the literature with multi-layered functions and applications. BIM is the interaction of processes, policies and technologies for planning, designing, constructing and maintaining facilities (Succar, 2010) that can be interpreted prototypically by computer-aided applications. BIM is the innovative approach that helps to reduce fragmentation and provides opportunities for enhanced collaboration and distributed project development (Arayici and Aouad, 2010). The implementation of BIM in the construction sector allows real-time rapid multi-directional data exchange. The key benefit of BIM is its accurate geometrical representation of the parts of a building in an integrated data environment (Azhar *et al.*, 2008).

For contracting organizations, the greater use of BIM is for clash detection (Smallwood *et al.*, 2012; Cao *et al.*, 2015). Other related benefits include faster and more effective processes through sharing and reusing information, better design through rigorous reviews and analysis using simulations and amendments to design, controlled whole life cycle costs and environmental data, automated assembly as a result of using digital data for fabricating and assembling, enhanced client service through visualization, and the use of life-cycle data in facilities management (Bernstein *et al.*, 2014, Smallwood *et al.*, 2012).

In a study conducted by Froise and Shakantu (2014), nearly three quarters (73%) of South African contractors were not aware of BIM and only 10% of contractors were familiar with, or had a fair understanding of BIM. The aim of this chapter is to identify the drivers and barriers to the adoption of BIM by construction firms in South Africa and whether there are differences in adoption based on the size of the construction organization.

2. Literature Review

2.1. *Drivers and barriers of BIM adoption*

In this section, drivers and barriers to BIM adoption are reviewed from the relevant literature. Table 1 lists a summary of core literature reviewed for this study and the focus area covered by each researcher. From Froise and Shakantu (2014) study, the level of South African contractors not being

Table 1.   Related literature for drivers and barriers to BIM adoption.

Author(s) (year)	Drivers and Barriers to BIM Adoption
Froise and Shakantu (2014)	Lack/awareness by: large clients, government and industry bodies, uncollaborative procurement
Hosseini *et al.* (2016)	Lack/evidence for BIM implementation (benefits)
Hong *et al.* (2016)	Adoption motivation, organizational competency and ease of implementation
Tsai *et al.* (2014)	Lack/support from top management, bureaucracy, competitive advantages, personnel training, BIM benefits

aware of BIM is nearly three quarters (73%). The main drivers identified from the literature are client request (United Kingdom) and organizational readiness (United States).

The key drivers and barriers of BIM adoption as identified by Hong *et al.* (2016) are adoption motivation, organizational competency and ease of implementation. All the other factors from the literature in Table 1 fit into these barriers. Adoption motivation refers to those that influence construction organizations' decision to adopt or reject BIM adoption, which are organizational innovativeness, subjective norms (client requirements and contractual obligations to share project information and awareness of BIM and its applications (Froise and Shakantu, 2014). Organizational competencies are measures of top management support (Tsai *et al.*, 2014), expertise and intention to BIM implementation efforts. The operational costs associated with BIM were found to be the main barrier to BIM adoption in this regard, where Hosseini *et al.* (2016) found that small- to medium-sized construction firms found it risky to invest in BIM with their limited resources and no evidence of a guaranteed returns on investment. Ease of implementation refers to technical issues such as ease of operation, maintenance and downtime of BIM operations and personnel training as identified by Tsai *et al.* (2014).

2.2. *Technology, capability, South African construction firms*

It is worth stating that BIM is more than a technological innovation as BIM includes processes and policies, and is thus better classified as an organizational innovation (Kassem *et al.*, 2015). For the purpose of this chapter, BIM is studied as an organizational innovation and technological resource as the primary indication of BIM capability of organizations. The Construction Industry Development Board (CIDB) categorizes construction organizations into nine grades of capabilities. Table 2 illustrates the composition of the construction firms by order of grades. Grade 2 contractors handle contracts of up to R 500,000.00 while grade 9 contractors handle contracts greater than R 100 million. The table further categorizes organizational firms, by virtue of their CIDB grade (resourcefulness), into technological capabilities (Rush *et al.*, 2007).

Table 2. Categories of South African construction firms.

CIDB Grade	Technological Capabilities	Average Percentage of Contractors (Civil and Building) (%)	Average Market Share by Value (Public Sector) (%)	Operations Characteristics
9	Creative	1.2	44	International/National
7–8	Strategic	12	40	Provincial/regional
5–6	Reactive	24.5	11	Local/Regional
2–4	Reactive	67.3	5	Local

Sources: CIDB (2016) and Windapo and Cattell (2013).

Grades 2–6 contractors are defined as "reactive", in that they recognize the challenge of change and the need for continuous improvement in their technological capabilities. Grades 7–8 contractors have a well-developed sense of the need for technological change and are referred to as strategic as they are capable of adopting a strategic approach to continuous innovation. Grade 9 contractors have well-developed sets of technological capabilities and are referred to as innovative. Grade 9 contractors are at ease with modern strategic frameworks for innovation and understand how to use technology, markets and organization to improve competitiveness (Rush *et al.*, 2007; Windapo and Cattell, 2013).

2.3. *Theoretical point of departure*

BIM adoption studies are aligned with innovation diffusion theories (Froise and Shakantu, 2014; Hosseini *et al.*, 2016). The diffusion of innovation is a theory of how, why and at what rate new ideas and technology spread through cultures at individual and firm levels (Rogers and Shoemaker, 1971).

The Diffusion of Innovation (DOI) theory at the firm level (Rogers, 1995) states that innovativeness is related to independent variables such as individual leader characteristics, internal organizational structural characteristics and external characteristics of the organization. Institutional

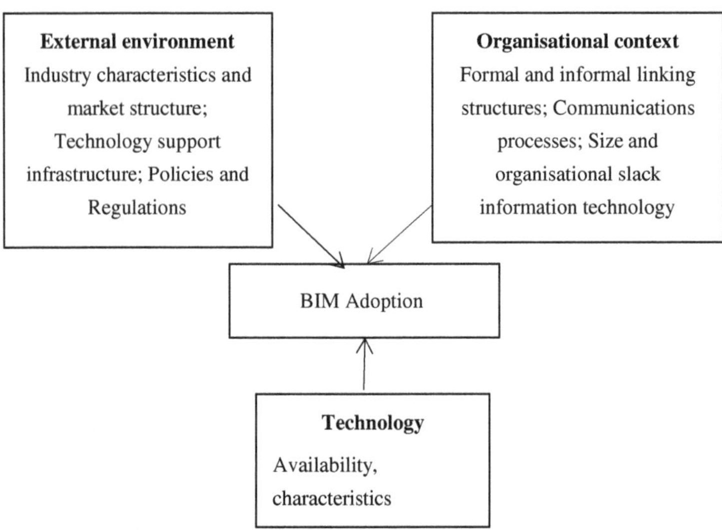

Fig. 1.   Technology, organization and environment framework (Tornatzky and Fleischer, 1990).

theory model (Iacovou *et al.*, 1995) explains the adoption of an innovation to be influenced by organizational readiness, perceived benefits and external pressure. Figure 1 is a technology, organization and environment (TOE) framework adopted from Tornatzky and Fleischer (1990). This framework synthesizes the DOI and institutional theory as this study recognizes that BIM is more than a technology and its adoption is influenced by organizational factors as well as the environment in which the construction organization operates within (Fig. 1). The organizational context and technology are both determinants of the organizations readiness to implement BIM.

3. Conceptual Framework

Inferring from institutional theory, organizational readiness is determined by the organizational context and availability of technology. By using Rush *et al.*'s (2007) study of the classification of organizations based on their technical capability and Windapo and Cattell's (2013) categorization

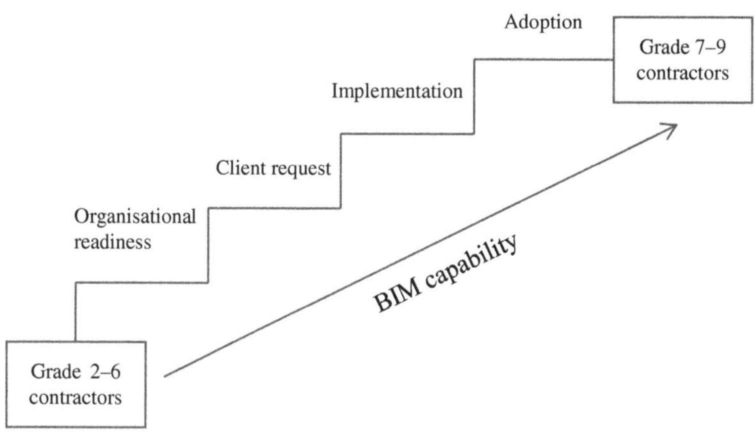

Fig. 2.   Concept of the drivers of BIM adoption within construction firms.

of the CIDB-registered contractor, the conceptual framework of the study is drawn (Fig. 2).

Hypothesis of the study: Level of BIM capability is directly related to the grade of the contractor. The higher the grade, the higher the level of BIM capability.

The conceptual framework (Fig. 2) shows that the main drivers for construction organizations in South Africa are requests by clients and organizational readiness. As the case with the United Kingdom mandating level 2 BIM for all public sector contracts within the construction industry, this in turn challenges organizations to be ready to implement BIM and this concept gains validity. Once firms implement BIM and are able to record evidence of perceived benefits then this can lead to firms adopting BIM as the new approach to business.

4. Conclusion and Further Research

This study looked at the main drivers and barriers. The study found the levels of BIM adoption among South African construction firms to be very low. The study aligned BIM adoption studies to Innovation Diffusion theories to establish the context and factors that influence the organizations innovativeness as well as adoption capabilities. Through the literature reviewed, the study found request by clients and organizational competency as the main drivers of BIM implementation, while the evidence of the perceived benefits of BIM implementation drives BIM adoption. Further empirical research needs to be conducted in this area to conclusively confirm these barriers and drivers of BIM adoption using case studies on firms that have implemented BIM in South Africa.

References

Arayici, Y. and Aouad, G. (2010). "Building Information Modelling (BIM) for Construction Lifecycle Management", *Construction and Building: Design, Materials, and Techniques*, 2010, 99–118.

Barlish, K. and Sullivan, K. (2012). "How to Measure the Benefits of BIM — A Case Study Approach", *Automation in Construction*, 24, 149–159.

Bernstein, H., Jones, S., Russo, M., Laquidara-Carr, D., Taylor, W., Ramos, J. and Pineda, R. (2014). "The Business Value of BIM in Australia and New Zealand", *SmartMarket Report*, 64.

Bryde, D., Broquetas, M. and Volm, J. M. (2013). "The Project Benefits of Building Information Modelling (BIM)", *International journal of project management*, 31, 971–980.

Cao, D., Wang, G., Li, H., Skitmore, M., Huang, T. and Zhang, W. (2015). "Practices and Effectiveness of Building Information Modelling in Construction Projects in China", *Automation in Construction*, 49, 113–122.

Froise, T. and Shakantu, W. (2014). "Diffusion of Innovations: An Assessment of Building Information Modelling Uptake Trends in South Africa", *Journal of Construction Project Management and Innovation*, 4, 895–911.

Hong, Y., Sepasgozar, S. M., Ahmadian, A. and Akbarnezhad, A. (2016). Factors influencing BIM adoption in small and medium sized construction organizations. *Proceedings of the International Symposium on Automation and Robotics in Construction, 2016*. Vilnius Gediminas Technical University, Department of Construction Economics & Property, 1.

Hosseini, M. R., Banihashemi, S., Chileshe, N., Namzadi, M. O., Udaeja, C., Rameezdeen, R. and Mccuen, T. (2016). "BIM Adoption within Australian Small and Medium-sized Enterprises (SMEs): An Innovation Diffusion Model", *Construction Economics and Building*, 16(3), 71–86.

Iacovou, C. L., Benbasat, I. and Dexter, A. S. (1995). "Electronic Data Interchange and Small Organizations: Adoption and Impact of Technology", *MIS Quarterly*, 19(4), 465–485.

Kassem, M., Succar, B. and Dawood, N. (2015). Building information modeling: analyzing noteworthy publications of eight countries using a knowledge content taxonomy. American Society of Civil Engineers.

Rogers, E. M. and Shoemaker, F. F. (1971). *Communication of Innovations: A Cross-Cultural Approach*, New York, USA, The Free Press.

Rush, H., Bessant, J. and Hobday, M. (2007). "Assessing the Technological Capabilities of Firms: Developing a Policy Tool", *R&D Management*, 37, 221–236.

Smallwood, J., Emuze, F. and Allen, C. (2012). "Building Information Modelling: South African Architects' and Contractors" Perceptions and Practices. First UK academic conference on BIM, pp. 141–151.

Succar, B. (2010). The Five Components of BIM Performance Measurement, CIB World Congress.

Tornatzky, L. and Fleischer, M. (1990). *The Process of Technology Innovation*, Lexington, MA. Lexington Books.

Trott, P. (2001). "The Role of Market Research in the Development of Discontinuous New Products", *European Journal of Innovation Management*, 4, 117–125.

Tsai, M.-H., Mom, M. and Hsieh, S.-H. (2014). "Developing Critical Success Factors for the Assessment of BIM Technology Adoption: Part I. Methodology and Survey", *Journal of the Chinese Institute of Engineers*, 37, 845–858.

Wang, G. and Song, J. (2017). "The Relation of Perceived Benefits and Organizational Supports to User Satisfaction With Building Information Model (BIM)", *Computers in Human Behavior*, 68, 493–500.

Windapo, A. O. (2017). *The Impact of Building Information Modelling on Quantity Surveying Practice and Project Performance. Integrated Building Information Modelling*, pp. 240 261, Sharjah, UAE, Bentham Science Publishers.

Windapo, A. O. and Cattell, K. (2013). "The South African Construction Industry: Perceptions of Key Challenges Facing its Performance, Development and Growth", *Journal of Construction in Developing Countries*, 18(2), 65–79.

Chapter 14

The Use of Geographic Information Systems (GIS) in Understanding Building Material Price Dispersion in South Africa

Abimbola Windapo* and Alireza Moghayedi

Department of Construction Economics and Management, University of Cape Town, Rondebosch, Cape Town 7700, South Africa

**abimbola.windapo@uct.ac.za*

Abstract

South Africa is a vast country, and there are significant variations in building material prices across the country. This chapter will use Geographic Information System (GIS) technology in explaining the role of distance in the differentiation in building material prices in South Africa. The research employs a quantitative research approach based on a cross-sectional survey research design and GIS technology in analyzing the data collected from each location. It emerged from the study that when distributed by provinces, the relationship between price and distance varies between inverse and direct. For example, while the

---

*Corresponding author.

price of clay bricks increases with distance in the Gauteng and KwaZulu-Natal provinces of South Africa, it decreases in the distance from the manufacturers in Western Cape province. The same ambivalent relationships are observed between timber, steel, bitumen, cement prices and distance. Based on these findings, the study concludes that the distance of a building material manufacturer from the retailer is not a predictor of price in the Western Cape province of South Africa, while the most uniform relationship is found in KwaZulu-Natal province. Further studies are therefore required to explain the other extraneous factors responsible for the irreconcilable retail prices of building materials in the Western Cape province when compared to other locations in South Africa.

Keywords: Building Material; GIS; Location Price; Retailers; Spatial Analysis.

1. Introduction

This chapter evaluates the role of distance and transportation cost in the variation in building material prices and whether there is a relationship between the distance of a retailer to the manufacturer and the price of building materials across different geographic locations in South Africa. The rationale for this study stems from reports by CIDB (2007), AECOM (2014), Compass (2016) and Windapo and Moghayedi (2017) that there are differences in the prices of building materials across South Africa. Akanni (2006) and Udosen and Akanni (2010) describe building materials as resources which are put together when erecting or constructing a structure. Arayela (2005) elucidated that the significant contribution of building materials to cost poses a substantial threat to the construction industry especially the affordability of newly proposed projects. This risk is important when added to the potential for the price of building materials to vary across locations. The construction industry has been the largest consumer of materials for almost a century with the building sector taking up two-fifths of the world's materials and energy flows while accounting for 40% of the total flow of raw materials worldwide each year (Horvath, 2004). Researchers show that building materials constitute between 50% and 65% of the total construction costs (Arayela, 2005; Oladipo and Oni, 2012; Windapo and Cattell, 2013; Akanni *et al.*, 2014).

The performance of projects and the growth of the construction industry are highly dependent on the price stability of building materials (Van Wyk, 2003), the increasing cost of building materials (Windapo and Cattell, 2010), and the differences in the prices of building materials, which is found to vary across different locations in most countries (Thisse, 2009). Compass (2016) highlighted that building material prices in Cape Town, Pretoria, and Durban are 1%, 2% and 2.5% lower than prices in Johannesburg, respectively, indicating that there is a building material price differential among the main cities in South Africa. This is further supported by a report by AECOM (2014) which posits that building costs vary across the provinces in South Africa. Price differential makes it difficult for the government and the private sector to have standard development plans across the country and discourages client investment. For example, in South Africa, the Reconstruction and Development Programme (RDP) houses cost ZAR 1,800, ZAR 2,070 and ZAR 1,980/m^2 in Johannesburg, Cape Town and Durban, respectively (AECOM, 2014).

A valid conclusion based on these studies cannot be drawn at this point as there has been no evidence and in-depth analysis using technology to verify whether there are significant differences between the prices of building materials based on location of manufacturers and retailers in South Africa. There is limited research into the nature of building material price differentiation across geographic areas in South Africa. Therefore, this chapter examines the phenomenon of building material price differentiation and whether there is a relationship between the distance of a retailer to the manufacturer using Geographic Information System (GIS) Technology.

To do this, this chapter firstly presents an overview of the nature and extent of the building material price variation across different geographic locations and the most common and price-volatile building materials in the South African construction industry. Secondly, the chapter outlines the research methodology used in the study. After that, it presents the findings generated from analysis using spatial models on ArcGIS, and the interpretation of the results. Finally, it delineates the conclusions and recommendations that are drawn from the study.

2. Overview of Price Differentiation and Location Theory

Location theory, as described by Beckman (1968), is the study of man's economic activity in different geographical areas. It creates a spatial configuration in which prices and costs of items can be compared and justified. Location theory has evolved over time, starting with von Thünen in 1875, who concentrated purely on the cost approach and was reformulated by Weber in 1906 (Greenhut, 1960). Weber applies freight rate of resources and finished goods, along with the finished good's production function, to develop an algorithm that identifies the optimal location for manufacturing plants.

Price differentiation is attributed to different location factors, namely transport cost, supply chain network, competition within an area, suitable labour, labour cost and income levels (Windapo and Cattell, 2014). This is because both manufacturers and suppliers need to include the cost of transportation in their selling price (ManuPrefab, 2013). According to Porter (1998) and Skitmore *et al.* (2006), the supply chain network and competition within a region contribute to the demand and supply of building materials within a location which puts pressure on prices of building materials in that area. Lalnunmawia (2010) citing Weber's location theory, acknowledges two primary causes related to the theory of location and regional factors as transport costs and labour. Canel *et al.* (2001) defined transportation cost as expenses involved in moving products to different places. They classified transportation costs in two types, inbound cost (from factories to facilities) and outbound cost (from facilities to demand points). Inbound transportation cost is added to the production cost to determine the production price at different location of manufacturers.

3. Review of Price Differentiation Within Key Building Materials Used in the Study

The price of building materials varies across locations in most countries. Research has shown that in India the prices of certain building materials including cement and paint differ across five of its cities. For example, the price per cubic metre of cement in Mumbai, Chennai, Bengaluru and

Kolkata is 10%, 17%, 26% and 22% higher, respectively (Reddy and Jagadish, 2003; Rangroo, 2014). The prices of clear glass, plywood, wood and steel bars showed a similar trend in India. Research done in the Netherlands by Vrijhoef and Koskela (2000) showed similar price variability. In that study, the prices of cement, brick and steel were found to be different across various locations and within the same areas in the country (Vrijhoef and Koskela, 2000). Brandt and Holz (2006) also found that urban prices in China are systematically higher that rural prices for similar products due to distance and transport costs.

Reports by Compass (2016), AECOM (2014) and Windapo and Moghayedi (2017) on the South African construction industry highlight the fact that there are differences in building material prices across South Africa. These reports by Compass (2016) and AECOM (2014) are aligned to Balcilar *et al.* (2013) who opined that price variation could be found across different locations in South Africa. An examination of the Lafarge Gypsum SA price list 2013 reveals that there are differences in the price of cement sold by the manufacturer in South Africa. The price list shows that while Lafarge plasterboards in Gauteng are priced at ZAR 76.70, the prices range from ZAR 80.54 in Durban to ZAR 82.07 in Cape Town (LaFarge Gypsum, 2013). The prices of cement ceiling boards, suspended ceiling and drywall partitioning showed a similar trend. The study by Windapo and Moghayedi (2017) found that there are significant differences in the prices of cement, timber, steel, bitumen and clay bricks across different locations (Cape Town, Johannesburg and KwaZulu-Natal) in South Africa, and that there are significant differences in the prices of clay bricks within all retailers.

This study focuses on the five most common construction materials used in civil engineering and building construction projects (Felix *et al.*, 2014; Horvath, 2004), namely cement, clay bricks, steel, timber and bitumen in three major South Africa provinces of Gauteng, Western Cape and KwaZulu-Natal. Also, these five key materials have the most volatile prices in the construction industry (CIDB, 2007; Laing *et al.*, 2011; Windapo and Cattell, 2014). Table 1 shows the distribution of the five key building materials manufacturers located in three major South African provinces.

Table 1 shows that all five key building materials have manufacturing plants in the three major South African provinces. Table 1 also shows that

Table 1. Distribution of key building materials by manufacturing plants in major provinces of South Africa.

Materials	Gauteng	Western Cape	KwaZulu-Natal	Total
Cement	3	4	1	8
Clay Bricks	27	36	6	60
Steel	3	1	1	5
Timber	5	2	9	16
Bitumen	1	1	2	4

the distribution of materials manufacturing plants in each province is diverse from others, for example, clay bricks have the highest number of plants across the three provinces; however, bitumen and steel are not as widespread across these three provinces which will influence the distribution, distance and transportation cost of building materials.

4. Analytical Framework

The theory of location acknowledges that transport costs and labour costs are the determining factors in price differentiation. The literature review suggests that the manufacturing plants for steel and bitumen are not as widespread across locations in South Africa and it is probable that there will be a significant differentiation in the transportation costs and accordingly price of steel and bitumen across South Africa. It can be deduced from the location theory and previous studies that building materials should cost more in provinces with a fewer representation of manufacturers in South Africa because of transportation costs.

The study will use GIS technology and spatial model in explaining the role of distribution distance in the differentiation of building material prices in South Africa using the location of the manufacturer and the distance between the retailer and the manufacturer. GIS is a system that enables one to visualize, question, analyze, and interpret data, to understand relationships, patterns, and trends. GIS geographically displays the information to give the user the ability to integrate, store, process and output geographic information (Ertug and Jacob, 2000; ESRI, 2014).

5. Research Methodology

The study adopts a quantitative research approach and a cross-sectional survey research design. Primary data were collected using telephonic survey and inquiries. The building material prices were sourced from three major cities, namely Johannesburg in Gauteng province, Cape Town in the Western Cape province and Durban in the KwaZulu-Natal province. The three top cities were selected to minimize the inefficiencies that might have resulted from insufficient or incomplete data collected from each province and not because all the major retailers had stores in all the nine provinces countrywide.

The population of the study comprised of all the retailers across South Africa although the data analyzed was limited to the information sourced from the five top retailers of building materials. The telephonic survey was conducted to collect the relevant data from 35% (120) of the 334 of the top five retail companies located in Cape Town, Durban and Johannesburg. The retailers contacted were randomly selected from the list of 334 retailers. At the end of the telephonic survey period from June to August 2016, a total of 75 positive responses were received.

The study forms part of a larger research whose questionnaire was set up in a way that would help determine the prices of the five building materials, the possible factors affecting the price variation of building materials, the location of the manufacturers from whom the retailers procured these materials and the freight costs involved in transporting the materials from the manufacturer to the retailers' stores.

6. Data Presentation

The study sought to know the prices of the five key building materials used in the study and the extent of their variation. The average prices for the five selected building materials obtained from the five selected retailers are presented in Table 2 and Fig. 1. The average prices for the five selected building materials for Gauteng were used as the base prices.

It can be seen from Table 2 and Fig. 1 that the prices in the Western Cape showed a greater price variation for the selected building materials

Table 2.   Variation of building material prices in 2016 by province in South Africa.

Province	Cement (Tonne)	Steel (Tonne)	Timber (Cubic Metre)	Clay Bricks (1,000 Bricks)	Bitumen (1,000 litres)
KwaZulu Natal					
Average Price (ZAR)	1,414.07	14,435.75	6,839.03	2,935.91	22,797.36
Variation (%)	0.94	6.77	26.31	87.74	−22.21
Gauteng					
Average Price (ZAR)	1,400.96	13,520.42	5,414.53	1,563.79	29,306.32
Variation (%)	0	0	0	0	0
Western Cape					
Average Price (ZAR)	1,989.26	16,954.21	5,545.19	3,730.16	16,828.52
Variation (%)	41.99	25.40	2.41	138.53	−42.58

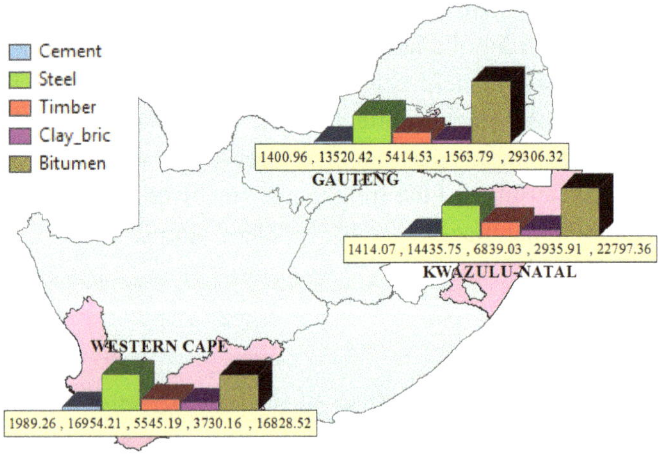

Fig. 1.   Average prices of five key building materials across the three provinces studied in South Africa.

compared to KwaZulu-Natal. The results further show a positive price change in cement at 41.99% and 0.94%, steel at 25.40% and 6.77%, timber at 2.41% and 26.31% and clay bricks at 138.53% and 87.74% between Gauteng, as the base Province, and Western Cape and KwaZulu-Natal, respectively. It also reveals a high negative average price variation for bitumen at 42.58% between Gauteng and Western Cape and 22.21% between Gauteng and KwaZulu-Natal.

6.1. *Analysis of price differentiation within provinces using GIS*

The study sought to explain the significant differences in building material prices across the three locations in South Africa using spatial analysis in the ArcGIS programme. To evaluate the price differences, the relationship between the distance from the suppliers, and price differentials of building materials within three distance buffers (20, 50 and further than 50 kilometres radius) are analyzed in five spatial models. The results of this inquiry are presented in Figs. 2–6 and Tables 3–7.

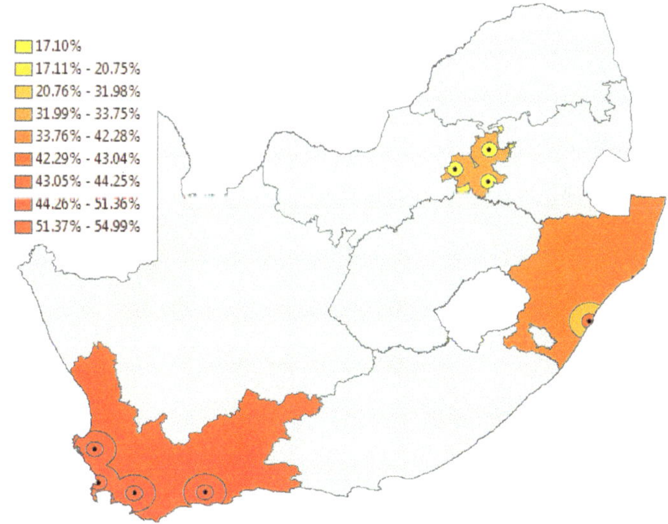

Fig. 2. Spatial model of cement prices across three provinces — KwaZulu-Natal, Guateng and Western Cape.

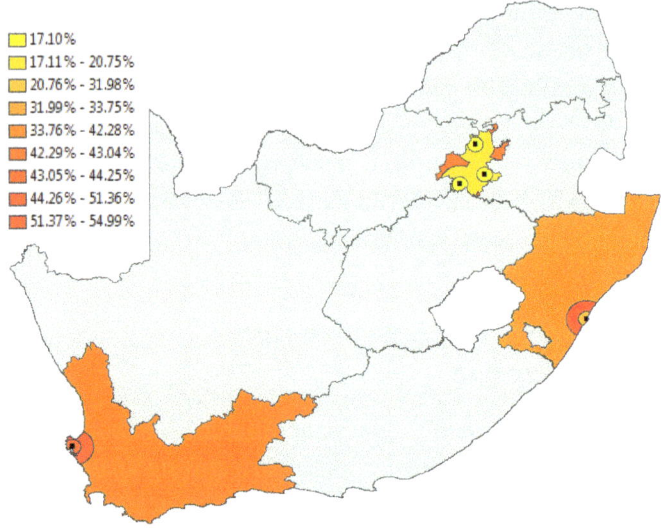

Fig. 3.   Spatial model of steel prices across three provinces — KwaZulu-Natal, Guateng and Western Cape.

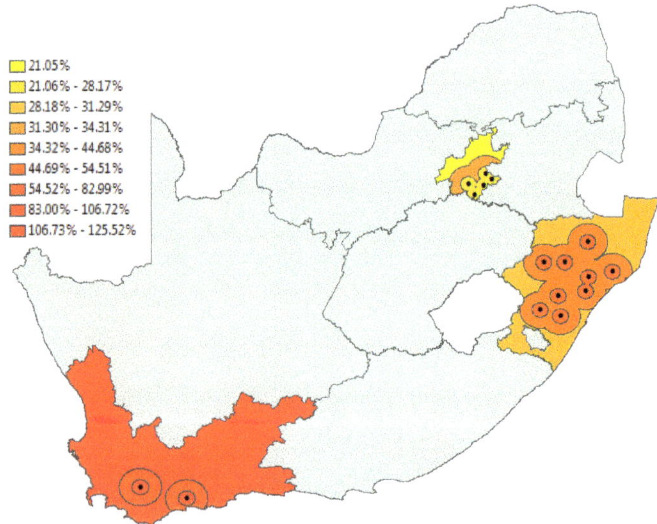

Fig. 4.   Spatial model of timber prices across three provinces — KwaZulu-Natal, Guateng and Western Cape.

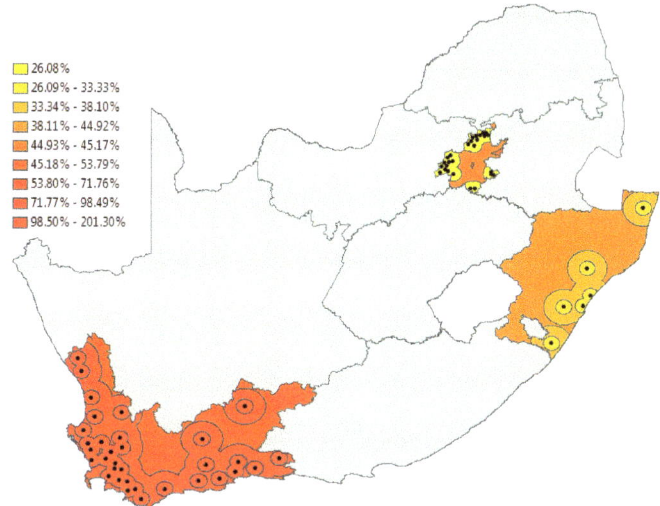

Fig. 5.   Spatial model of clay bricks prices across three provinces — KwaZulu-Natal, Guateng and Western Cape.

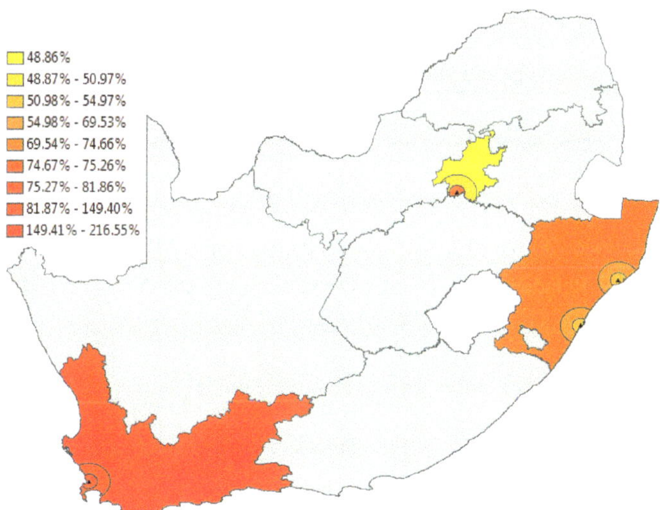

Fig. 6.   Spatial model of bitumen prices across three provinces.

Table 3. Number of retailers and average price differential of cement within the three buffers across three provinces in South Africa.

	KwaZulu-Natal	Gauteng	Western Cape
1st Buffer (20 km radius)			
Number of Stores	4	14	5
Differential Price (%)	40.98	17.33	60.66
Distance–Price Relationship	Inverse	Direct	Inverse
2nd Buffer (50 km radius)			
Number of Stores	8	9	13
Differential Price (%)	25.80	28.83	54.51
Distance–Price Relationship	Inverse	Direct	Inverse
3rd Buffer (>50 km radius)			
Number of Stores	10	2	10
Differential Price (%)	39.26	21.74	47.26
Distance–Price Relationship	Inverse	Direct	Inverse

Table 4. Number of retailers and average price differential of steel within the three buffers across three provinces in South Africa.

	KwaZulu-Natal	Gauteng	Western Cape
1st Buffer (20 km radius)			
Number of Stores	11	8	10
Differential Price (%)	31.98	17.10	44.25
Distance–Price Relationship	Direct	Direct	Direct
2nd Buffer (50 km radius)			
Number of Stores	1	12	10
Differential Price (%)	51.36	20.75	54.99
Distance–Price Relationship	Direct	Direct	Direct
3rd Buffer (>50 km radius)			
Number of Stores	10	5	8
Differential Price (%)	33.75	43.04	42.28
Distance–Price Relationship	Inverse	Direct	Inverse

Table 5. Number of retailers and average price differential of timber within the three buffers across three provinces in South Africa.

	KwaZulu-Natal	Gauteng	Western Cape
1st Buffer (20 km radius)			
Number of Stores	3	9	1
Differential Price (%)	54.51	28.17	106.72
Distance–Price Relationship	Inverse	Direct	Direct
2nd Buffer (50 km radius)			
Number of Stores	11	10	2
Differential Price (%)	44.68	34.31	82.99
Distance–Price Relationship	Inverse	Direct	Inverse
3rd Buffer (>50 km radius)			
Number of Stores	6	6	25
Differential Price (%)	31.29	21.05	125.52
Distance–Price Relationship	Inverse	Inverse	Direct

Table 6. Number of retailers and average price differential of clay bricks within the three buffers across three provinces in South Africa.

	KwaZulu-Natal	Gauteng	Western Cape
1st Buffer (20 km radius)			
Number of Stores	4	8	7
Differential Price (%)	33.33	26.08	201.30
Distance–Price Relationship	Direct	Direct	Inverse
2nd Buffer (50 km radius)			
Number of Stores	7	8	16
Differential Price (%)	38.10	45.17	98.49
Distance–Price Relationship	Direct	Direct	Inverse
3rd Buffer (>50 km radius)			
Number of Stores	9	9	5
Differential Price (%)	44.92	53.79	71.76
Distance–Price Relationship	Direct	Direct	Inverse

Table 7. Number of retailers and average price differential of bitumen within the three buffers across three provinces in South Africa.

	KwaZulu-Natal	**Gauteng**	**Western Cape**
1st Buffer (20 km radius)			
Number of Stores	7	13	21
Differential Price (%)	54.97	75.26	149.40
Distance–Price Relationship	Direct	Inverse	Inverse
2nd Buffer (50 km radius)			
Number of Stores	8	8	5
Differential Price (%)	69.53	50.97	81.86
Distance–Price Relationship	Direct	Inverse	Inverse
3rd Buffer (>50 km radius)			
Number of Stores	6	4	2
Differential Price (%)	74.66	48.86	216.55
Distance–Price Relationship	Direct	Inverse	Direct

Table 3 shows the highest price differential (60.66%) observed within 20 km radius from cement manufacturers in the Western Cape province while the lowest price differential (17.33%) is found within 20 km radius from cement manufacturers in Gauteng.

Figure 2 illustrates the inverse relationship between the distance from the cement manufacturers and cement price across Western Cape and KwaZulu-Natal provinces within a graduated scale. While across Gauteng the price of cement has a direct relationship to the distance from cement manufacturers. Based on these findings, it can be concluded that the farther the retailer, the cheaper is the cement price in the Western Cape and KwaZulu-Natal provinces, while the reverse is the case in Guateng.

Table 4 shows similar results obtained for differences in cement prices based on distance, for steel. The highest steel price differential (54.99%) was observed within 20–50 km radius from steel manufacturers in Western Cape province and the lowest price differential (17.10%) was the found within 20 km radius from steel manufacturers in Gauteng.

Figure 3 illustrates the direct relationship between the distance and steel price differential across Gauteng. In Kwazulu-Natal, the price of steel in the 2nd buffer significantly increases with distance from the manufacturer however in the 3rd buffer this direct relationship is very slight. The steel prices in Western Cape vary regardless of the increase in the distance.

Table 5 shows the significant differential in the price of timber across the Western Cape province particularly in the 1st and 3rd buffers, respectively, at 106.72% and 125.52%. Unexpectedly, the lowest price differential (17.10%) is observed farther than 50 km from timber suppliers in Gauteng.

Figure 4 illustrates the significant inverse relationship of timber prices to distance in Kwazulu-Natal although the timber prices in Western Cape and Gauteng vary regardless of the increase in the distance.

Table 6 shows the significant differential in the prices of clay bricks (201.30%) across Western Cape within 20 km radius from clay bricks suppliers.

Figure 5 illustrates the systematic direct relationship between clay bricks price and the distance from the suppliers across Gauteng and Kwazulu-Natal provinces. While the spatial model proved the strong inverse relationship in the Western Cape province. It can be inferred from the spatial model presented in Fig. 5 that the farther a retailer is from the manufacturer, the higher are the prices of clay bricks in Guateng and KwaZulu-Natal provinces of South Africa, while the opposite is the case in the Western Cape province.

Table 7 shows the significant price differential across the Western Cape province for Bitumen. It is the highest price differential among the five key building materials, and is determined in the 3rd buffer (216.55%).

Figure 6 illustrates the consistent direct relationship between the distance from the refinery and bitumen price across the KwaZulu-Natal province as well as the inverse relationship across Gauteng. However, the bitumen prices in Western Cape province vary regardless of the increase in the distance from the refinery.

7. Discussion of Findings

The study found that there are significant differences in the average prices of the five key building materials, namely cement, steel, timber, clay bricks and bitumen obtained from the five selected retailers across three major provinces of South Africa (Western Cape, Gauteng and KwaZulu-Natal). This variation is aligned to the results of earlier studies by Vrijhoef and Koskela (2000), Reddy and Jagadish (2003), and Thisse (2009), who found that the prices of building materials vary across different geographic locations in a country. Western Cape was found to have the highest average cement, steel and clay brick prices and the lowest average bitumen prices amongst the three provinces surveyed, as shown in Table 2 and summarized in Fig. 1, while KwaZulu-Natal had the highest average timber price and Gauteng had the highest average bitumen price and the lowest average cement, steel, timber and clay brick prices.

The distribution of cement manufacturing plants across South Africa displayed in Table 1 illustrates that Gauteng has three cement plants, Western Cape has four cement plants and KwaZulu-Natal has one cement plant. Since there is only one cement plant in KwaZulu-Natal, the average cement prices in KwaZulu-Natal were expected to be the highest among the provinces since the retailers would have to source the cement from a farther distance but the data gathered were contrary to expectations (see Fig. 1). Also, Fig. 2 illustrates the inverse relationship between distribution distance and price of cement in Western Cape and KwaZulu-Natal. This inverse relationship and lower price of cement in KwaZulu-Natal shows that the transport cost is not the cause of price differentials in the province. Gauteng province had the lowest average steel price amongst the three regions, as shown in Fig. 1. This is probably because there are more steel suppliers in a small Gauteng province than in the other two provinces that have only one steel supplier. The presence of three steel suppliers may result in less transportation cost on steel prices due to the shorter distribution distance and the absence or shortage of steel suppliers in Western Cape and KwaZulu-Natal will lead to higher steel prices because of higher transportation cost (see Fig. 3).

Also, the study found that KwaZulu-Natal had the highest average timber price compared to Western Cape and Gauteng. The average price of a cubic metre of timber in Western Cape was 2.41% greater than Gauteng while the mean price of a cubic metre of timber in KwaZulu-Natal was 26.31% higher than in Gauteng. The distribution of timber sawmills across three provinces displayed in Fig. 4 illustrates the significant inverse relationship between distribution distance from sawmills and timber price in KwaZulu-Natal. KwaZulu-Natal had the highest number of sawmills and paradoxically they had the highest average price as well. This is not logical or aligned to the basic principles of economics.

Furthermore, Fig. 1 shows that the clay brick prices were highest in Western Cape and lowest in Gauteng province. The average prices of clay bricks in Western Cape and KwaZulu-Natal were 138.53% and 87.74% more than what was available in Gauteng, respectively, this shows the extent of the price variation. Although Western Cape had the highest number of clay brick suppliers in South Africa (36 suppliers), as shown in Table 1 and Fig. 4, clay brick prices were highest in the Western Cape, and lowest in Gauteng due to higher density of suppliers. Also, Fig. 4 illustrated the strong inverse relationship between distribution distance and price of clay bricks in Western Cape which is not aligned to earlier findings in a different context by Porter (1998) and Skitmore *et al.* (2006).

Higher bitumen prices are recorded in Gauteng compared to Western Cape or KwaZulu-Natal, respectively (see Fig. 1). Gauteng and Western Cape provinces have only one refinery while KwaZulu-Natal has two refineries. The price differentiation may, therefore, be as a result of other economic factors such as monopoly of supplier in Gauteng and Western Cape provinces, and not transportation cost due to the inverse relationship observed between the distance from the refinery to retailers and the bitumen price across retailers in Gauteng (see Fig. 6).

The five spatial models confirm that distance is not the key cause of variation in the price of the building materials studied in the three provinces. The spatial models seem to suggest that the farther the retailers are from the manufacturer, the higher the transport costs and prices for materials like cement and steel in Gauteng, clay bricks in Gauteng and Kwazulu-Natal, and bitumen in KwaZulu-Natal. Conversely, the spatial models seem to suggest an inverse or no relationship in the prices of all five building materials in Western Cape, cement, steel and timber in KwaZulu-Natal province. Based on the previous empirical studies in the area of price differentiation and location theory, retailers located farthest away from manufacturers should have more expensive products. The results suggest that the manufacturers of building materials in the Western Cape have optimally located their plants in areas of greatest demand and point to the prevalence of other factors apart from transport costs that influence the retail price of building materials.

In general, these findings suggest that retail prices of building materials within provinces up to 500 km from manufacturers do not differ as a result of distance, however, location is a key element in building material price differentiation in South Africa because the prices do not only vary by location but also differ within the same buffer zones based on location. This study extends the location theory to the retail space and highlights the need for more knowledge on the drivers of building material prices across South Africa and how the location of a project will contribute to its cost based on the source of building materials.

8. Conclusions

This study examines the role of distance in the variation in building material prices and whether there is a relationship between the distance from manufacturers to retailers and the price of building materials based on location in South Africa. The study employed a cross-sectional survey of retailers located in three provinces — KwaZulu Natal, Guateng and Western Cape — in data collection. The study found that the prices of building materials differ significantly within South Africa, and between locations. It was also discovered that that the distance of the retailer from the manufacturer is not a predictor of price in the Western Cape province of South Africa, while the most uniform relationship between the price of building material and the distance of retailers from manufacturers is found in the KwaZulu-Natal province. Based on these findings, it can be concluded that location and not the distance of the retailer is a key factor in the price variation of building materials in the Western Cape and for selected materials such as bitumen across South Africa.

This difference in building material prices implies that there will be large differences in tender prices among contractors bidding on similar national projects located across different provinces in South Africa. This will make it difficult for clients to choose a contractor for projects since there will be significant variations in the tender prices from contractors across the country based on location. It also makes it difficult for the government to have standard housing plans and prices across the country when budgeting for the construction of Reconstruction and Development Program (RDP) houses. In the long run, significant differences in building material prices will have an adverse effect on the performance of the construction industry.

The study recommends that the South African government and the construction sector stakeholders should pay particular attention to other location factors which have a major effect on the prices of building materials, such as monopoly, competition, supply and demand, income levels, since the transport cost is not a predictor of material prices across the South Africa, and there is lack of knowledge of the prices of building materials within locations when proposing national projects. It is recommended that further studies be undertaken to explain the other extraneous factors responsible for the irreconcilable building material prices in the Western Cape province of South Africa.

References

AECOM, T. C. (2014). *Property and Construction Handbook International*, London: AECOM.

Akanni, P. (2006). "Small Scale Building Material Production in the Context of the Informal Economy", *The Professional Builders*, 13–18.

Akanni, P., Oke, A. and Omotilewa, O. (2014). "Implications of Rising Cost of Building Materials in Lagos State Nigeria", *Sage Open*, 4(4), 2158244014561213.

Arayela, O. (2005). *Laterite Bricks: Before, Now and Hereafter*, Akure: Publication Committee, FUTA.

Balcilar, M., Beyene, A., Gupta, R. and Seleteng, M. (2013). "Ripple' Effects in South African House Prices". *Urban Studies*, 50(5), 876–894.

Beckman, M. (1968). *Location Theory*. New York: Random House.

Brandt, L. and Holz, C. A. (2006). "Spatial Price Differences in China: Estimates and Implications", *Economic Development and Cultural Change*, 55(1), 43–86.

Canel, C., Khumawala, M. B., Law, J. and Loh, A. (2001). "An Algorithm for the Capacitated, Multi-Commodity Multi-Period Facility Location Problem", *Computers & Operations Research*, 28(5), 411–427.

CIDB (2007). *The Building and Construction Materials Sector: Challenges and Opportunities*, Kuala Lumpur: CIDB, 11–25.

Compass (2016). *Global Construction Cost Yearbook*, 16th Annual Edition, Morrisville, Pennsylvania, USA: Compass International Consultants Inc.

Ertug, G. A. and Jacob, P. K. (2000). "Using GIS in Emergency Management Operations", *Journal of Urban Planning and Development*, 126(3), 136–149.

ESRI (2014). What is GIS? [Online]. Available at: http://www.esri.com/what-is-gis (Accessed on April 10, 2016).

Felix, O. A., Moses, O. A. and Sodiq, B. O (2014). "Comparative Study of Price Variations of Basic Civil Engineering Construction Materials", *Energy and Environment Research*, 4(3), 50–57.

Greenhut, M. (1960). "Size of Market versus Transport", *The Journal of Industrial Economics*, 18(2), 72–184.

Horvath, A. (2004). "Construction Materials and the Environment", *Annual Review of Environment and Resources*, 29, 181–200.

LaFarge Gypsum, S. (2013). Price List 2013 [Online]. Johannesburg. Available at: http://www.lafarge.co.za/Lafarge_Feb_2013_Price_List.pdf (Accessed on March 21, 2016).

Laing, P., Marcus, G. and Dhansay, A. (2011). *Factors Determining the Volatility of Building Material Prices in the Construction Industry*, University of Cape Town: Department of Construction Economics and Management.

Lalnunmawia, H. (2010). Weber's Theory of Industrial Location. [Online]. Available at: Tromana college: http://www.trcollege.net/study-materials/102-webers-theory-of-industrial-location (Accessed on August 28, 2011).

ManuPrefab (2013). Main Factors that Influence the Total Construction Cost in Costa Rica [Online]. Available at: http://manuprefab.com/main-factors-influence-total-construction-cost-costa-rica (Accessed on March 15, 2016).

Oladipo, F. and Oni, O. (2012). "Review of Selected Macroeconomic Factors Impacting Building Material Prices in Developing Countries — A Case of Nigeria", *Ethiopian Journal of Environmental Studies and Management*, 5(2), 131–137.

Porter, M. E. (1998). "Location, Clusters, and the New Microeconomics of Competition", *Business Economics*, 76, 76–90.

Rangroo, S. (2014). Building Materials Price [Online]. Available: http://constructionworld.in/buildingmaterialprices/building-material-prices-october14.pdf (Accessed on May 16, 2016).

Reddy, B. V. and Jagadish, K. (2003). "Embodied Energy of Common and Alternative Building Materials and Technologies", *Energy and Buildings*, 35(2), 129–137.

Skitmore, M., Runeson, G. and Chang, X. (2006). "Construction Price Formation: Full-Cost Pricing or Neoclassical Microeconomic Theory?", *Construction Management and Economics*, 24(7), 773–783.

Thisse, J. F. (2009). "How Transport Costs Shape The Spatial Pattern Of Economic Activity", Paris: OECD Publishing.

Udosen, J. and Akanni, P. (2010). "A Factorial Analysis of Building Material Wastage Associated With Construction Projects", *Journal of Civil and Environmental Systems Engineering*, 11(2), 81–90.

Van Wyk, L. (2003). *A Review of the South African Construction Industry. Part 1: Economic, Regulatory and Public-Sector Capacity Influences on the Construction Industry*, Pretoria: Boutek, CSIR (Council for Scientific and Industrial Research).

Vrijhoef, R. and Koskela, L. (2000). "The Four Roles of Supply Chain Management In Construction", *European Journal of Purchasing & Supply Management*, 6(3), 169–178.

Windapo, A. O. and Cattell, K. (2010). Perceptions of Key Construction and Development Challenges Facing the Construction Industry in South Africa. *Proceedings*, Durban, South Africa.

Windapo, A. O. and Cattell, K. (2013). "The South African Construction Industry: Perceptions of Key Challenges Facing Its Performance, Development and Growth", *Journal of Construction in Developing Countries*, 18(2), 65–79.

Windapo, A. O. and Cattell, K. (2014). Evaluation of Location Factors Influencing Building Material Price Variation in South Africa. In ICEC 2014 Conference, October 20–22, Milan.

Windapo, A. O. and Moghayedi, A. (2017). Evaluation of Location Factors Influencing Building Material Price Variation in South Africa. In IRC 2017 Conference, September 11–12, Manchester, UK.

Chapter 15

Assessment of Open BIM Standards for Facilities Management

S. A. Azzran*, K. F. Ibrahim,
Joseph H. M. Tah and F. Henry Abanda

*School of Built Environment, Oxford Brookes University,
Headington Campus, Headington Road,
Oxford OX3 0BP, UK*

sheikh.ali.sh.said-2015@brookes.ac.uk

Abstract

Non-digital information handover for facilities management (FM) could be quite a challenge when evaluating the quality of information in terms of its usefulness and relevance for building operations. Manually searching for errors in large datasets is inefficient, which leads to difficulties in the course of maintenance and operation tasks. Traditional FM systems, computer-aided FM (CAFM), and computerized maintenance management system (CMMS) are all capable of digitizing information. However, the handover remains in paper format, which requires the mundane and inefficient practice of re-creating information for use in FM systems. Using the emerging open building information modelling (BIM) standards during the operational stage to resolve this problem is currently

---

*Corresponding author.

being considered, and in this context, the extent of how open BIM standards can support facility management systems is being studied. As part of an ongoing effort, a desk-based research is used as the main method of investigation to achieve the aim of this study. Findings suggest that open BIM standards, such as COBie and IFC, should be utilized to streamline information and be integrated with FM systems.

Keywords: BIM; Open BIM Standards; COBie; IFC; Facilities Management System.

1. Introduction

Traditional computer-aided facilities management (CAFM)/computerized maintenance management system (CMMS) are tools that support the functions of facilities management (FM) activities. However, it has been reported that during the handover stage, most are incapable of handling digital information (Becerik-Gerber *et al.*, 2012; Patacas *et al.*, 2015), which means that during the maintenance and operation phases, building-related information had to be re-created, culminating in information that were often fragmented and difficult to be retrieved or accessed (Jylhä and Suvanto, 2014; Bosch *et al.*, 2014). Solutions were offered by utilizing the emerging open building information modelling (BIM) standards such as Construction Operation Building information exchange (COBie) and Industry Foundation Classes (IFC). These methods provide an integrated approach for project teams to transfer design and construction information to facilities managers during the handover period. However, the adoption of COBie and IFC continues to face challenges (NBS, 2015), such as the lack of methodologies in using BIM in FM, lack of knowledge and experience in open BIM standards (Kassem *et al.*, 2015; Becerik-Gerber *et al.*, 2012; Williams *et al.*, 2014), and the lack of support from software vendors (Becerik-Gerber *et al.*, 2012; Williams *et al.*, 2014). As per literature, the use of open BIM standards (i.e. COBie and IFC) has yet to be addressed. Therefore, this study intends to investigate the extent to which COBie and IFC can support CAFM/CMMS systems. The objective of this chapter is to review and assess the steps of capturing building information using several approaches to generate COBie and IFC.

Facility and asset information are invaluable in managing building operations (Becerik-Gerber *et al.*, 2012; Patacas *et al.*, 2015). During the handover of a facility, the key information regarding building operation is critical for long-term use. More often than not, the key information relevant and suitable for FM has yet to be identified (Alfarez *et al.*, 2013; Patacas *et al.*, 2015; Parsanezhad and Dimyadi, 2014). Hence, it is important for the facilities manager to understand the different approaches of capturing information to/from COBie. The information contained in COBie spreadsheet can be delivered either manually, IFC approach, COBie extension approach, or middleware approach. According to Fig. 1, there are two steps involved in capturing building information using COBie. The first step illustrates the use of different BIM utility tools (i.e. Solibri, COBie plug-in and Ecodumus) to populate information in COBie, while the second stage illustrates capturing COBie information into the CAFM/CMMS systems.

2. An Overview of Using CAFM/CMMS and Open BIM Standards

2.1. *CAFM and CMMS*

CAFM and CMMS are both information technology tools that are of great help in operating and maintaining building facilities. A CAFM system is a facility manager's administrative tool that supports planning, tracking and managing tasks, as well as report building operations (Watson and Watson, 2016; Madritsch and May, 2009). Its main function is to coordinate space management information, such as space data, space planning, assigning room numbers and areas and some cases in 3D Computer-aided Design (CAD) (Codinhoto *et al.*, 2013; Lewis, 2013a; Sapp, 2016), while, CMMS is a platform that focuses on building assets and equipment activities (Sapp, 2016). CMMS is specifically used for preventive/planned maintenance related to building's asset and equipment, and scheduling and recording operational tasks of a building (Sapp, 2016). CMMS supports the management of work orders, equipment schedule inspections, data storage, status reports and performance reports (Tretten and Karim, 2014). However, while the implementation of CAFM/CMMS systems is vital, they both have problems when streamlining facility information for operation and maintenance tasks (Jylhä and Suvanto, 2014; Bosch *et al.*, 2014). The resulting information is usually fragmented and stored in several different systems or in paper-based files (Bosch *et al.*, 2014). Aspurez and Lewis

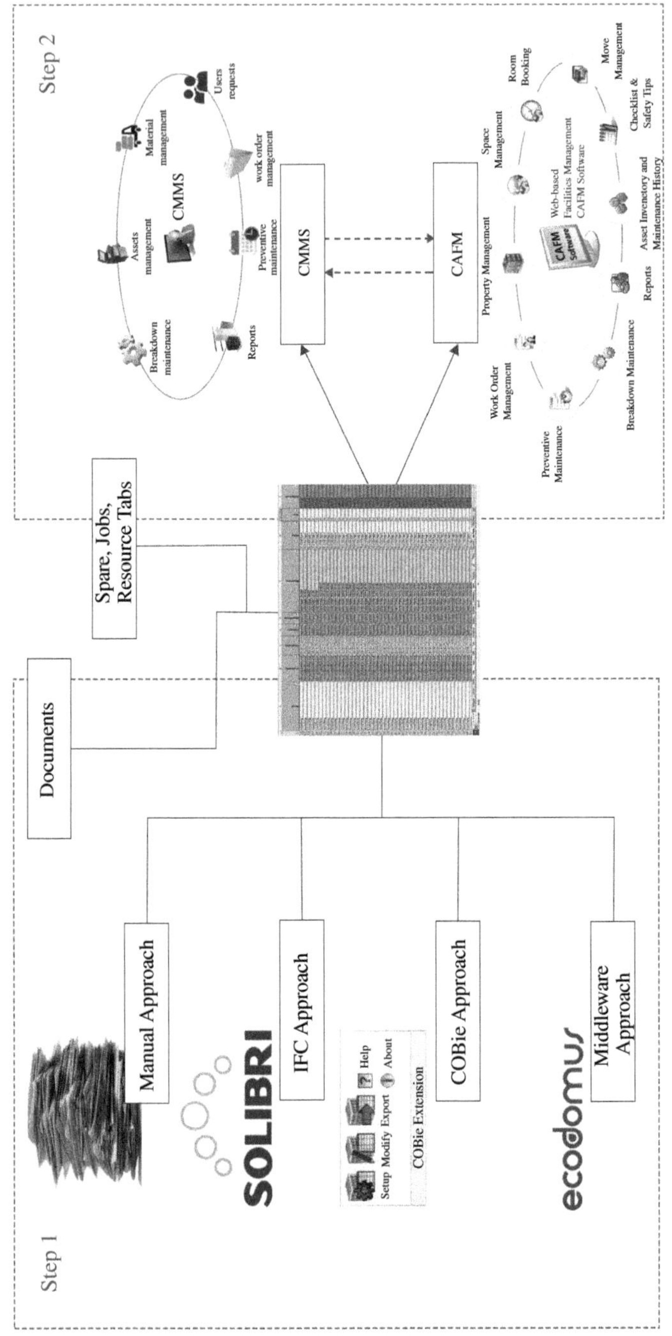

Fig. 1. Steps of capturing building information using COBie.

(2013) reported that existing CAFM/CMMS systems are underutilized and information is poorly handled by the facilities' manager that is left outdated in the process of maintaining and operating a facility. Furthermore, CAFM/CMMS systems have complex user interfaces, which makes it difficult to access and retrieve information when required for FM (Tretten and Karim, 2014; Labib, 2004). Therefore, these issues suggest that there is a need for an integrated and systematic process in managing building information.

2.2. *Open BIM standards*

Open BIM standards, specifically COBie and IFC, have been hailed as solutions for facilitating information exchange from the design and construction phases to the FM phase (Patacas *et al.*, 2015; Eastman *et al.*, 2011; East, 2013). This is due to the goal of open BIM standards being to improve the handover of facility/asset information and enhance information management in FM systems. Therefore, its adoption is important towards enabling FM to function in a BIM-driven environment.

COBie is a mechanism that presents facility information in a readable and structural format operation and maintenance phase of facility. COBie was developed by the United States Army Corps of Engineers to support the handover of operations and management information to clients (East, 2013). It resolved the lack of standardization and organization of documentation (East, 2007). While evidently useful, COBie has yet to be pervasive in FM practice. The 2015 NBS survey reported that "66% of UK construction professionals are not using COBie" (NBS, 2015). This suggests that the implementation of COBie still lags behind FM practice. One of the main reasons for this was the conflicting interest by owner/client requiring information that COBie does not provide (Parsanezhad and Dimyadi, 2014). This is further exacerbated by the perceived lack of support from software vendors and contributed to COBie's very slow adoption (Becerik-Gerber *et al.*, 2012; Williams *et al.*, 2014). Another reason was that information held within the COBie spreadsheet may contain enormous volumes of data, which increases the possible risk of missing and/or error-prone data between software applications (Venugopal *et al.*, 2012). Checking the quality of such data is labour intensive and time consuming (Williams *et al.*, 2014).

IFC is an open data format that facilitates the exchange of building information between design, construction and FM software systems (Eastman *et al.*, 2011; ISO, 2013). This allows for seamless information exchange of BIM data. Information can be reused and support the whole life cycle of the building operation. IFC provides a clearer visualization of facilities, a more effective process in handling facility and asset information, and reusable data for predicting the environmental impact and life-cycle cost (Mitchell and Schevers, 2006). However, adopting IFC remains a challenge due to uncertainties and the lack of mechanism in using the IFC data format (Williams *et al.*, 2014; Lewis, 2013a). Moreover, most facility managers are not competent in BIM authoring software (Lewis, 2013a; Lavy and Jawadekar, 2014), making it difficult for them to handle. The problem with using IFC is also exacerbated when it is not easily integrated with existing CAFM/CMMS systems, while also containing files that are enormous with irrelevant information (Lewis, 2013b). IFC data format was found to be unstable: data loss when exchanging information and deterioration of information was difficult to detect during the handover of facility information (Venugopal *et al.*, 2012; Hetherington *et al.*, 2011; Amor and Faraj, 2001). The next section will discuss the different approaches using open BIM standards for FM.

3. Assessment of Using Open BIM Standards in Facilities Management

3.1. *Open BIM standards*

This step compares the different types of linking BIM and FM to solve interoperability issues. This is divided into the following four approaches: manual approach, IFC, COBie, and proprietary middleware. A case study was collated to evaluate the different linking approaches. This comparison was made using the Solibri model checker as an IFC approach, COBie extension for BIM-authoring software (e.g. Revit) was selected as COBie approach and Ecodomus as the middleware approach. The case study was conducted on an administrative building of an electrical power plant (see Fig. 2). The project size is 2,168 m^2, and consists of two floor levels with 138 rooms.

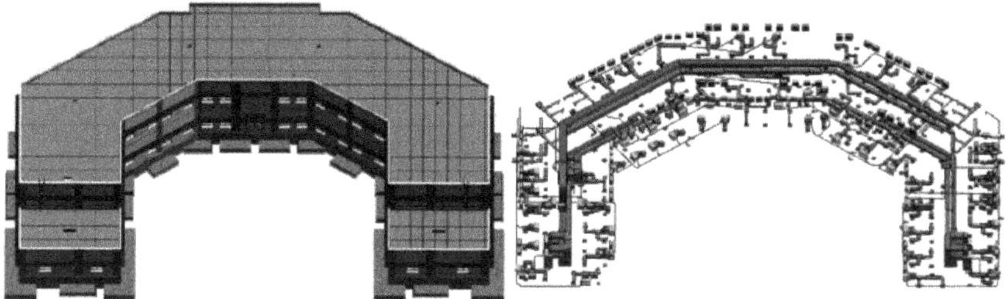

Fig. 2.   Administration building of an electrical power plant.

The BIM model was used to test the time taken to populate into COBie spreadsheet using three different approaches: Solibri, COBie Plugin and Ecodomus. The results showed that the IFC approach took ~20 h, while the COBie extension approach took ~15 h using, and Ecodomus took ~8 h (Ibrahim, 2016).

Table 1 illustrates the comparison of different BIM utility tools in capturing information onto a COBie spreadsheet. This was done to assess the capabilities of each BIM utility tool in the creation of COBie data. The first tool is Solibri — an IFC model checker — that provides quality assurance for validation and control compliances (Solibri, 2016). Solibri facilitates the visualization of the BIM model and optimizes IFC data format by removing unnecessary and redundant information. It helps users analyze and diagnose each entity and objects found in an IFC data file, and generates a summary of statistics. In the BIM collaborative environment, Solibri is incapable of working with native files or linked models. The next tool is COBie plug-in that exports a COBie spreadsheet. It is an extension of BIM-authoring software that predefines the COBie data. COBie plug-in directly facilitates the BIM model in assigning and managing parameters, such as spaces, rooms, zones and contact details. However, the function of COBie plug-in falls short in promoting a collaborative BIM environment, and data are unable to revert or be mapped to the BIM model once it is exported to a COBie spreadsheet. Finally, Ecodomus — a web-based middleware proprietary — is an intermediary approach between the BIM native model and COBie spreadsheet. Ecodomus has an integrated BIM collaboration with real-time facility information, and it is the only approach that enables a

Table 1. Methods of capturing building information using COBie and IFC.

BIM Utility Tools	Solibri	COBie Plug-in	Ecodomus
Interoperability approach	IFC	COBie	Middleware
System type	Platform	Plug-in	Web app.
Working with native file	✗	✓	✓
Online based for better collaboration	✗	✗	✓
Avoid un-needed parameters	✓	✗	✓
Automatic fill parameters	✓	✗	✓✓
Creating spaces	✓	✓✓	✗
Working with native linked models	✗	✗	✗
Working with Uniclass 2	✓	✓	✓
Working with Omniclass	✓	✓	✓
Write once/Read many	✓	✗	✓
Checking coding	✓	✗	✗
Quality check run	✓✓	✓	✓
Visualization and presentation	✓✓	✓	✓
Export to COBie .xls	✓	✓	✓
Import from COBie to native file	✗	✗	✓

bi-directional link between native files and the COBie spreadsheet. However, Ecodomus is not compatible with native linked models and is unable to assign spaces in the BIM model. The assessment suggests that none of the BIM utility tools is compatible with a native linked model. Solibri is the only tool that is able to check coding/schemas of IFC. COBie plug-in is the most useful tool in predefining COBie data and parameters. Ecodomus is better at facilitating BIM collaboration platform to import COBie and synchronise with native files.

3.2. *COBie to CAFM/CMMS software application*

In the second step, COBie spreadsheet has been tested and demonstrated in the CAFM/CMMS systems in a challenge event (East, 2014). The National Institute of Building Science (NIBS) called for a COBie Challenge for facility management software vendors to prove the ability of their products (East, 2014). Results are shown in Table 2, where it compares the capabilities FM software against each tables of COBie spreadsheet data.

A total of 15 FM software vendors presented their product capabilities in importing COBie data. It was found that most CAFM/CMMS systems lack model interface capabilities or integration. This indicates that not many CAFM/CMMS systems possess the capabilities to read IFC data formats; however, data can be captured from an Excel spreadsheet format (xlsx, csv). COBie data are related to space, type and component, and are most compatible with information to CAFM/CMMS systems. Most of this information is related to asset register and space data, such as manufacturer details, warranties, serial numbers, floor levels and room numbers, all of which could be easily manipulated in the systems. Document data are the least compatible for the majority of CAFM/CMMS systems. This type of information consists of attachment of PDF file and/or links to document files, which is difficult to process in CAFM/CMMS systems. The remaining COBie data, other than the attribute table — which represents product schedule data — are fairly compatible with most of the CAFM/CMMS systems. This information is mostly performance-related data that involves preventive maintenance schedules, safety procedures, materials, tools, training and replacement parts. Most CAFM/CMMS systems are limited when importing these data due to proprietary information mappings and configurations that differ from COBie. It was discovered that some of these types of COBie data are rather questionable and uncertain in how there support the use of CAFM/CMMS systems. In short, the capabilities of CAFM/CMMS systems vary across the different types, sizes and complexities of a building. COBie is a mechanism that represents facility information and specification to support construction handover. The information contained in COBie tables, such as space, type, component and attribute are mostly compatible with CAFM/CMMS systems, as they are important for FM use.

4. Key Literature Findings and Discussion

Traditional CAFM/CMMS is an excellent system for automating FM tasks, but is inadequate when handling digital information during the handover stage. Recreating building-related information from these

Table 2.   Capabilities of CAFM/CMMS capturing information from COBie (Adapted from NIBS, 2014).

CAFM/CMMS	Types of COBie Data										
	Model	Space	Type	Component	Job	System	Resources	Document	Spare	Attributes	Zone
Archibus	✓	✓	✓	✓	✓	✓	✓	✓	✓	✓	✓
AiM	✗	✓	✓	✓	✗	✓	✗	✓	✗	✓	✓
Bentley Facilities	✓	✓	✓	✓	✓	✓	✓	✓	✓	✓	✓
Code Book & Room Data	✗	✗	✓	✓	✗	✓	✗	✗	✗	✓	✗
Facility Online	✗	✓	✓	✓	✓	✗	✓	✗	✓	✓	✗
FaME	✗	✓	✓	✓	✗	✗	✗	✗	✗	✓	✗
FM: Interact	✗	✓	✓	✓	✗	✗	✗	✗	✗	✓	✗
Maximo	✓	✓	✓	✓	✓	✓	✓	✓	✓	✓	✓
Micromain	✗	✓	✓	✓	✓	✗	✗	✗	✓	✓	✗
Morada	✗	✓	✓	✓	✗	✗	✗	✗	✗	✗	✗
Onuma	✓	✓	✓	✓	✓	✓	✓	✓	✓	✓	✓
Planon	✗	✓	✓	✓	✓	✓	✓	✗	✓	✗	✓
Proteux MMX	✗	✓	✓	✓	✗	✗	✗	✗	✗	✗	✗
RHYTI	✗	✓	✓	✓	✗	✗	✗	✗	✗	✓	✗
WebTMA	✗	✓	✓	✓	✓	✗	✓	✓	✓	✓	✓

systems resulted in fragmented information, which is difficult to access during maintenance and operation. This could adversely affect the efficiency of FM practice within an organization. Figure 1 illustrated the methods of capturing information to COBie spreadsheet using the following four different approaches: manual approach, IFC approach, COBie approach, and middleware approach. It was discovered that using the middleware approach (i.e. Ecodumus) is most effective in generating COBie data and assuring its quality. The assessment of COBie to CAFM/CMMS showed that numerous CAFM/CMMS systems are capable of supporting a BIM model view and can read IFC data format. COBie data that involved *Space*, *Type* and *Component* were most compatible. In contrast, *Document* data were least compatible. Other than *Attributes* data, there is a limit to the ability to capture information due to technical complexities. Overall, the use of open BIM standards (COBie and IFC) in FM has demonstrated that it is possible that the information contained in COBie can be used in various FM tasks. Information can easily be automated to CAFM/CMMS systems, and this can reduce the amount of time and workload in creating data.

5. Conclusion

Poor handling of information in FM stifles conduct maintenance and operation. Traditional CAFM/CMMS has been developed to support FM activities, but often information is manually re-created into these systems. This leads to continuous inefficiency. Proper information delivery methods are required to allow seamless information exchange during the handover of a facility. The adoption of open BIM standards (COBie and IFC) can help streamline information to CAFM/CMMS systems. In this chapter, two steps of capturing COBie data have been discussed; the first is capturing information to the COBie spreadsheet using four different approaches, and the second focussed on the capabilities of FM systems in capturing information from the COBie spreadsheet. This assessment demonstrated different ways to deliver COBie and the extent to which CAFM/CMMS systems are compatible with the COBie spreadsheet. As part of a future study, the relationship of CAFM/ CMMS and open BIM standards will be investigated via a case study.

References

Amor, R. and Faraj, I. (2001). "Misconceptions about Integrated Project Databases". *Journal of Information Technology (ITcon)*, 6(5), 57–68.

Aspurez, V. and Lewis, A. (2013). "Case Study 3: USC School of Cinematic Arts", in Teicholz, P., (ed.), *BIM for Facilities Managers*, US: IFMA Foundation, pp. 185–232.

Becerik-Gerber, B., Jazizadeh, F., Li, N. and Calis, G. (2012). "Application Areas and Data Requirements for BIM-Enabled Facilities Management", *Journal of Construction Engineering & Management*, 138(3), 431–442. doi: 10.1061/(ASCE)CO.1943-7862.0000433.

Codinhoto, R., Kiviniemi, A. and Kemmer, S. (2013). BIM–FM Manchester Town Hall Complex, University of Salford in collaboration with Department of Business, Innovation & Skills and Manchester City Council.

East, E. W. (2007). "Construction Operations Building Information Exchange (COBie)", DTIC Document.

East, E. W. (2014). "buildingSMART Alliance Information Exchanges: Means and Methods". Available at: https://www.nibs.org/?page=bsa_cobiemm (Accessed on December 12, 2016).

Eastman, C., Eastman, C. M., Teicholz, P. and Sacks, R. (2011). *BIM Handbook: A Guide to Building Information Modeling For Owners, Managers, Designers, Engineers and Contractors*, Hoboken: John Wiley & Sons.

Hetherington, R., Laney, R., Peake, S. and Oldham, D. (2011). "Integrated Building Design, Information and Simulation Modelling: The Need for a New Hierarchy", in: *Building Simulation*, Sydney, Australia.

Ibrahim, K. F., Abanda, F. H., Vidalakis, C. and Woods, G. (2016). "BIM for FM: Input versus Output Data" in *Proceedings of the 33rd CIB W78 Conference*, Brisbane, Australia.

ISO (2013). "ISO/PAS 16739, Industry Foundation Classes (IFC) for Data Sharing in Construction and Facility Management Industries". Available at: https://www.iso.org/standard/51622.html (Accessed on December 8, 2016).

Kassem, M., Kelly, G., Dawood, N., Serginson, M. and Lockley, S. (2015). "BIM in Facilities Management Applications: A Case Study of a Large University Complex", *Built Environment Project and Asset Management*, 5(3), 261–277, doi:10.1108/BEPAM-02-2014-0011.

Labib, A. W. (2004). "A Decision Analysis Model for Maintenance Policy Selection Using a CMMS", *Journal of Quality in Maintenance Engineering*, 10(3), 191–202.

Lavy, S. and Jawadekar, S. (2014). "A Case Study of Using BIM and COBie for Facility Management", *International Journal of Facility Management*, 5(2).

Lewis, A. (2013a). "Case Study 5: State of Wisconsin Bureau of Facilities Management Division of State Facilities, Department of Administration", in Teicholz, P., (ed.), *BIM for Facility Managers*, Houston, US: IFMA Foundation, pp. 250–293.

Lewis, A. (2013b). "Case Study 6: University of Chicago Administration Building Renovation", in Teicholz, P., (ed.), *BIM for Facilities Managers*, Houston, USA: IFMA Foundation.

Madritsch, T. and May, M. (2009). "Successful IT Implementation in Facility Management", *Facilities*, 27(11/12), 429–444.

Mitchell, J. and Schevers, H. (2006). "Building Information Modeling for FM using IFC, CRC Construction Innovation". Available at: http://www.constructioninnovation.info/indexea25.html (Accessed on October 10, 2013).

NBS (2015). "NBS National BIM Report 2015, UK". Available at: http://www.thenbs.com/topics/bim/articles/nbs-national-bim-report-2015.asp.

Parsanezhad, P. and Dimyadi, J. (2014). Effective Facility Management and Operations via a BIM-based Integrated Information System. Joint Cib W070, W111 & W118 Conference, Technical University of Denmark, Copenhagen, May 21–23, pp. 442–453.

Sapp, D. (2016). "Computerized Maintenance Management Systems (CMMS)". Available at: http://www.myendnoteweb.com/EndNoteWeb.html?func=new& (Accessed on December 8, 2016).

Solibri (2016). "About Solibri". Available at: https://www.solibri.com/contact/solibri/ (Accessed on December 8, 2016).

Tretten, P. and Karim, R. (2014). "Enhancing the Usability of Maintenance Data Management Systems", *Journal of Quality in Maintenance Engineering*, 20(3), 290–303.

Venugopal, M., Eastman, C. and Teizer, J. (2012). Formal Specification of the IFC Concept Structure for Precast Model Exchanges, *Proceedings of Computing in Civil Engineering*, Reston, Florida, USA, pp. 213–220.

Watson, J. R. and Watson, R. (2016). "Computer-aided Facilities Management (CAFM)". Available at: https://www.wbdg.org/facilities-operations-maintenance/computer-aided-facilities-management-cafm (Accessed on December 8, 2016).

Williams, R., Shayesteh, H. and Marjanovic-Halburd, L. (2014). "Utilising Building Information Modeling for Facilities Management", *International Journal of Facility Management*, 5(1).

Chapter 16

An Investigation of the Benefits and Challenges of Adopting Alternative Building Materials (ABM) in the Construction Industry

Oluwaseun Dosumu* and Clinton Aigbavboa

Sustainable Human Settlement and Construction Research Centre, Faculty of Engineering and the Built Environment, University of Johannesburg, South Africa

**oluwaseundosumu97@gmail.com*

Abstract

The current methods of sourcing materials deplete environmental natural resources, causes climate change, carbon emission and poses health risks to human. The objective of this chapter is to investigate the benefits and challenges of alternative building materials (ABM) in the construction industry. The survey research design was adopted for the study which is quantitative in nature. The study was conducted in Gauteng province of South Africa due to its peculiarity among other provinces. Fifty (50) questionnaires were retrieved out the 98 that were distributed through the convenience sampling technique. The findings of the study

---

*Corresponding author.

revealed that professionals are neutral about the benefits of ABM to the construction industry. Analysis of variance (ANOVA) confirmed that there is no significant difference in the responses of professionals on the benefits of ABM to the construction industry. Moreover, the study indicates that all the challenges investigated in the study were barriers to the adoption of ABM except low profit margin and poor environmental friendliness. The result of ANOVA indicates that there is no significant difference in the responses of professionals on the challenges hindering the adoption of ABM. The chapter concludes that stakeholders are not convinced of the benefits of adopting ABM and this may contribute to its reluctant adoption for construction projects. The chapter also concludes that there are many challenges posing threat to the adoption of ABM in the South African construction industry. The chapter recommends that adequate awareness should be done by professional organizations on the benefits of adopting ABM. Government should also enact strict policies towards adopting ABM for construction purposes. Furthermore, government and professional organizations need to devise ways of overcoming the challenges of ABM identified in this study so that housing backlogs can be substantially improved and the living standards of citizens can be raised.

Keywords: Alternative/Sustainable Materials; Climate Change; Construction Professionals; Environmental Resources; Natural Resources.

1. Introduction

The study of construction materials is of great significance in the construction industry as the current method of sourcing materials depletes environmental natural resources and poses health risks to human. Despite the contribution of the construction industry to the growth of nations, the activities carried out have undesirable effects on the environment (Khodeir and Mohamed, 2015). Furthermore, Jayasinghe (2011) noted that construction all over the world now needs alternative building materials because the conventional materials that are being used for construction are in short supply and also cause degradation of the environment.

Generally, the conventional materials used for construction are energy intensive and their continuous use can drain the energy resources thus adversely affecting the environment (Sharma and Awchat, 2016).

Also, it is difficult to meet the growing demand for housing with the current traditional (conventional) materials and construction methods (Lin, 2015). Hence, there is a need for optimum utilization of available resources and production of sustainable (alternative) building materials and techniques to satisfy the ever-increasing demand for buildings. The concept of sustainable building construction is currently embraced in the construction industries of many countries and it has enhanced many of the construction deficiencies of the previous concepts by being more environmental friendly and improving indoor environmental quality (AlSanad, 2015).

Dulaimi *et al.* (2002) noted that alternative methods of construction have evolved over time. However, it has not been adequately accepted as planned by most construction organizations and this has inhibited the provision of adequate housing (for the masses) that fit the green building concept in many developing countries. This is particularly so for many African countries including South Africa. It is against this background that this study investigates the benefits and challenges of alternative building materials (ABM) in the South African construction industry.

2. Literature Review

Previously, studies (Hwang and Tan, 2012; Subramanian, 2007; Makenya and Nguluma, 2007) have evaluated the benefits of ABM and technologies such as prefabrication and modularization processes in housing projects. These technologies have impressively enhanced the advancement of the construction industry in relation to projects and product performance, on-site safety, customization and environmental issues. Lu (2007) stated that speed is a core benefit of alternative building technology as it reduces the construction time which invariably leads to cost savings.

According to Wyk (2013), the use of alternative building technology is predicated on the desire to reduce the time (up to 35%), cost (up to 41%) and improve the quality of construction. Ala-Juusela *et al.* (2006) claimed that buildings constructed with alternative materials can offer substantial cost savings when in use, but this advantage has not been adequately communicated to a wide audience. Rohracher (2001) argued that alternative building technologies are regarded as a socio-technical system that accounts for functional dependencies and requirements, interests, perspectives and the interaction of actors.

However, the main barrier to alternative building material is not developing the needed technology for it but distributing it. This is largely due to the absence of available services, lack of collaboration among professionals and construction companies, non-existence of expressive demands and inappropriate regulations. An investigation conducted by Priemus (2002) in the Netherlands shows that key obstacles to ABM are uneven distribution of costs and benefits between various stakeholders and the absence of collaboration and teamwork between them.

The lack of information regarding environmental performance and behaviours of ABM is a substantial barrier to construction design process (Giesekam *et al.*, 2016). Accessibility to reliable information regarding alternative building material is crucial to the improvement of ABM production (Jonsson, 2000). Seo (2002) stated that information regarding construction materials is frequently inadequate or difficult to understand, especially in situations when building project teams are assembling materials from various building industries.

In most countries, the common problems that affects the selection of innovated (alternative) materials and technologies are availability/ suitability of raw materials, accessibility of skilled labour, complexity of construction, cost difference with conventional materials, availability of sufficient power for production of components, typology founded on geo-climatic conditions, disaster-resistant requirements, environmental aspects and acceptability by people (Kebede, 2013).

Adoption of ABM may be hindered when clients are concerned with the risks that may be developed in the process (Nelms *et al.*, 2005). These risks may be caused by the use of unfamiliar techniques, inferior knowledge, further analysis and examination in construction, inadequate manufacturer and supplier backing and deficiency of performance data.

A costing analysis, using real cost data for a broad range of sustainability technologies and design solutions contradicts the assumption of high costs of ABM and demonstrates that significant improvements in environmental performance can be achieved with very little additional cost (Ala-Juusela *et al.*, 2006). Also, Bon and Hutchinson (2000), Hydes and Creech (2000) and Zhou and Lowe (2003) stated that the primary barriers to the implementation of system buildings are the misperceptions of incurring higher capital costs and inadequate market value. Zhou and Lowe (2003) noted that investors and developers hold the misconception that capital costs will raise when sustainable construction methods are adopted; meanwhile, they lack the understanding of the economic benefits of sustainable construction.

According to Bartlett and Howard (2000), consultants have been overestimating the capital costs of energy efficient measures and underestimating the potential cost savings. Higher costs may also come from increase in consultants' fees and unfamiliarity of the design team and contractors with alternative building methods (Hydes and Creech, 2000). Sodagar and Fieldson (2008), Sayce *et al.* (2007) opined that one obstacle for the wide uptake of environmental building design is the fear of extra construction cost and to overcome it, financial incentives and innovative fiscal arrangements need to be available.

Sodagar and Fielsdon (2008) further noted that ABM are hindered when there is lack of knowledge regarding the development of project brief which indicates the targets and mitigation strategies for sustainable impacts. Adetunji *et al.* (2008) mentioned that emphasis on the price of procurement processes and the low risk culture are key barriers to the adoption of ABM.

ABM are hindered because some stakeholders get involved too late in the process (Williams and Dair, 2007). Riyel *et al.* (2003) emphasized that construction organizations need to be part of the design process of construction projects as they will be responsible for giving estimates of services to be rendered. The absence of common framework for integrating the activities of ABM with construction practice at the functional level is a barrier to adopting ABM for sustainable construction.

Rydin (2008) stated that there is an opportunity to learn about ABM through partnership projects. Rohracher (2001) argued that ABM cannot be correctly utilized without a closer interaction of system providers, experts and users. Even though stakeholders in the built environment claim to subscribe to sustainability, it does not mean it is practiced at any noticeable level (Rohracher, 2001). Williams and Dair (2007) found that the absence of information is the key barrier to ABM. Rydin *et al.* (2006) discovered that, even though designers show confidence in their capabilities to access and use information in general, this confidence falls when application of ABM technologies are required. Lyons (2009) attested that the selection of building materials has constantly been a concern for construction industry professionals.

3. Research Method

In this study, close-ended structured questionnaire was used as the primary source of data to assess the benefits and challenges of ABM in the South African construction industry.

The study was conducted in the Gauteng province of South Africa. The choice of Gauteng is based on the fact that it is located where most construction works, designers/consultants and contractors' offices are located in South Africa among others. Gauteng is also the economic hub of South Africa.

The study used the convenient sampling technique because of the unwilling nature of respondents to supply useful information. Therefore, those that were willing to participate in the study were used. The respondents were construction managers/project managers, quantity surveyors, architects, building inspectors and engineers from both the public and private sector within the Gauteng province. Fifty (50) questionnaires were returned out of the 98 that were sent out and that gives a response rate of 51%. The questionnaires were analyzed basically with frequencies, percentages, mean scores and analysis of variance (ANOVA).

4. Results and Discussion

Table 1 shows the result of the biographical analysis of respondents' characteristics and their organizations. The result obtained from the analysis of data shows that 44% of females and 56% of males responded to the study. The result also shows respondents' highest educational qualifications to be 6% for grade 12 (Standard 10), 48% for Diploma/Certificate, 32% for Bachelor's degree and 14% for Post-graduate Degrees. Furthermore, the study indicates that the respondents' years of experience in the construction industry are 10% for 1–2 years of work experience,

Table 1. Background information of respondents and their organizations.

Biographic Data	Percentage (%)
Gender	
Male	44
Female	66
Education	
Standard 10	6
Diploma/Certificate	48
Degree	32
Post-graduate degree	14
Experience in construction industry	
1–2	10
2–5	24
5–10	32
11–15	16
16–20	4
Above 20	14
Type of organization	
Client (government)	23
Client (private)	21
Consultant quantity surveyor	27
Contractor	29
Profession of respondents	
Quantity surveyor	12
Architect	14
Construction manager	22
Engineer	16
Contractor	12
Building inspector	10
Project managers	14

24% for 2–5 years of experience, 32% for 5–10 years of work experience, 16% for 10–15 years of work experience and 4% for over 20 years of work experience. In addition, the study indicates that 29% of the respondents are contractors, 27% are consultant quantity surveyors, 21% are private clients and 23% are government clients. Also, 22% of the respondents were construction managers, 16% were engineers, 14% were architects, 14% were project managers, 12% were contractors, 12% were quantity surveyors and 10% were building inspectors. This demographic data indicate that the genders of the respondents for this study are evenly spread and they are educated enough to supply useful information for the study. Also, the organizations and professions of the respondents are evenly spread.

Table 2 shows the benefits of ABM to the South African construction industry based on the perspective of professionals. Generally, the professionals ranked energy efficiency (3.42) as the highest benefit of alternative materials to the construction industry. This is followed by time conservation (3.38), minimum site delays (3.32), pollution prevention (3.24), cost effectiveness (3.24), less material wastage (3.20), low maintenance (3.18), biodegradability (3.18), design certainty (3.14) and non-toxicity of materials among others. However, the mean values of the general rating indicate that all the benefits considered in this study are below the agreement level (4.00) and this shows that the respondents are not really sure or they are neutral about the benefits of ABM they rated. This invariably boils down to the level of awareness of these professionals on the available ABM and their benefits to the construction industry.

Among individual professionals, it could be seen that the quantity surveyors agree to energy efficiency (3.83) and time conservation (3.50) as the benefits of ABM. The architects are only certain of the renewability of materials as the benefits of ABM to the construction industry. The construction managers basically stayed neutral on all the benefits considered in this study. The engineers agree to energy efficiency (3.50), minimum site delays (3.50), cost-effectiveness (3.50), biodegradability (3.75) and non-toxicity of materials (3.50) as benefits of ABM. According to the contractors, only low maintenance (3.50) was agreed to be the benefit of ABM to the construction industry; the building inspectors agreed that energy efficiency (3.80), time conservation (3.80), cost-effectiveness

Table 2.   Benefits of ABM in the construction industry.

Benefits of ABM	QS	ARC	CM	EN	CON	BI	PM	Mean	Rank	*p*-Value
Energy efficiency	**3.83**	3.43	3.09	**3.50**	3.00	**3.80**	**3.57**	3.42	1	0.297
Time conservation	**3.50**	3.43	3.00	3.13	3.33	**3.80**	**3.86**	3.38	2	0.215
Minimum site delays	3.17	3.43	3.18	**3.50**	3.00	3.40	**3.57**	3.32	3	0.884
Pollution prevention	3.33	3.14	3.36	3.63	3.00	3.40	2.71	3.24	4	0.377
Cost-effectiveness	3.00	3.00	3.00	**3.50**	3.00	**3.80**	**3.57**	3.24	4	0.446
Less material wastages	2.50	3.29	3.36	3.13	3.00	3.40	**3.57**	3.20	6	0.345
Low maintenance	3.33	3.00	3.00	3.00	**3.50**	3.20	3.43	3.18	7	0.914
Biodegradability	3.00	3.14	3.09	**3.75**	3.00	2.80	3.29	3.18	7	0.587
Design certainty	3.33	3.29	3.00	3.25	3.00	3.00	3.14	3.14	9	0.968
Non-toxicity of material	2.67	3.29	3.09	**3.50**	3.00	3.20	2.71	3.08	10	0.409

(*Continued*)

Table 2. (*Continued*)

Benefits of ABM	QS	ARC	CM	EN	CON	BI	PM	Mean	Rank	*p*-Value
Cost savings	2.83	2.86	3.00	3.38	3.17	3.40	3.00	3.08	10	0.802
Competitiveness of materials	3.00	2.86	2.73	3.25	3.33	3.00	3.29	3.04	12	0.838
Renewability of materials	2.83	**3.71**	3.00	2.88	2.67	3.20	2.86	3.02	13	0.179
Durability of materials	2.83	3.14	2.73	3.00	2.83	**3.60**	3.14	3.00	14	0.610
Structural safety	2.67	3.14	2.82	3.00	3.00	3.20	3.14	2.98	15	0.331
Less training required for use	2.83	2.86	3.00	3.38	3.17	3.40	3.00	2.78	16	0.916
Local availability	2.83	3.00	2.73	3.38	2.83	2.80	1.86	2.72	17	0.021

Notes: 1 = Very low (VL); 2 = Low (L); 3 = Neutral (N); 4 = High (H); 5 = Very High (VH). Q/S = Quantity surveyors, ARC = Architects, CM = Construction managers, EN = Engineers, CON = Contactors, BI = Building inspector, PM = Project manager.

(3.80) and durability of materials (3.60) are benefits of ABM to the construction industry.

The project managers also agreed that energy efficiency (3.57), time conservation (3.86), minimum site delays (3.57), cost-effectiveness (3.57) and low material wastages (3.57) are the benefits of ABM to the construction industry. The difference in the responses of the professionals was tested with ANOVA and it was discovered that there is no significant difference (that is, p-values exceed 0.05) in the responses of all the professionals on the benefits of ABM to the South African construction industry except on the local availability of the materials ($p = 0.021$). These statistics shows that the professionals in the construction industry are likely unaware of the ABM available for construction purposes; hence, they are unable to ascertain the benefit that could accrue from adopting ABM.

Table 3 indicates the challenges of alternative building material in the construction industry. The general response of the professionals indicates that all the challenges considered in this study are agreed to be valid (as mean scores are either 4.00 or approximately 4.00) except for low profit margin and poor environmental friendliness. That is, the professionals are collectively aware that adopting ABM for construction would not reduce their profit margins and they are environmental friendly. Individual responses of professionals also indicate that all the challenges considered in this study were agreed to except in few cases.

For instance, the quantity surveyors declined that marginal cost advantage (3.33), incompatibility with other materials (3.17) and poor environmental friendliness (3.00) are challenges of ABM to the construction industry. For the architects, poor attitude of stakeholders towards these materials (3.29), poor acceptability by the public (3.29), low aesthetic value (3.43), public perception, incompatibility with other materials (3.29), marginal cost advantage (3.29) and public perception (3.43) are not challenges of ABM to the construction industry. For the construction managers, only public perception and low competency of labour are not challenges of ABM. The challenges that fall below approximation to 4.0 (Agree) are boldened in the table to show that professionals did not agree to them as challenges. Table 3 also indicates test of difference in the responses of professional using ANOVA and the result shows that there is

Table 3.   Challenges of ABM in the construction industry.

Benefits of ABM	QS	ARC	CM	EN	CON	BI	PM	Mean	Rank	p-Value
Uncertainty of the source of information on alternative building methods	4.17	3.86	4.36	4.25	3.67	3.60	3.71	4.00	1	0.325
Non-commercial status of the alternative materials	3.67	3.57	4.36	4.38	3.50	4.00	4.14	4.00	1	0.274
Doubtful durability and lifespan	4.50	3.86	4.00	4.38	3.67	3.80	3.71	4.00	1	0.430
Poor clients' interest	3.67	3.86	3.82	4.13	4.17	4.20	4.14	3.98	4	0.798
Inadequate knowledge of the material	4.50	3.86	3.64	4.50	3.67	3.80	3.86	3.96	5	0.152
Difficulty in obtaining finance from banks for sustainable projects	3.50	4.14	4.27	4.38	**3.30**	3.80	3.71	3.94	6	0.160
Poor attitude of stakeholders towards these materials	4.17	**3.29**	4.00	4.13	3.83	4.00	4.14	3.94	6	0.195
Lack of government support	4.17	3.86	3.55	4.25	3.50	3.80	4.14	3.88	8	0.625

Unavailability of the ABM in the market	3.83	3.71	3.91	4.13	3.50	3.60	4.00	3.84	9	0.338
Poor acceptability by the public	4.00	**3.29**	4.00	3.75	4.00	3.60	4.00	3.82	10	0.284
Lack of standards and specifications	4.33	4.14	3.64	4.25	3.50	**3.20**	3.57	3.82	10	0.147
Low aesthetic value	4.00	**3.43**	4.00	3.88	4.00	4.20	**3.29**	3.82	10	0.179
Public perception	3.50	**3.29**	**3.36**	3.88	4.17	4.40	4.00	3.74	13	0.227
Low level of competency of labour	3.83	4.14	**3.45**	4.00	**3.17**	**3.40**	3.57	3.66	14	0.437
Inadequate supply of the materials	3.50	3.86	3.55	3.75	**3.33**	**3.40**	4.00	3.64	15	0.862
Marginal cost advantage	**3.33**	**3.29**	3.73	3.88	**3.33**	**3.00**	4.14	3.58	16	0.688
Non-compatibility with other materials	**3.17**	**3.43**	4.00	3.88	**3.17**	**3.00**	**3.43**	3.52	17	0.342
Low profit margin	3.50	3.57	3.55	**3.25**	3.50	**3.00**	3.57	**3.44**	18	0.882
Poor environmental friendliness	**3.00**	3.71	3.73	**2.75**	3.67	**3.40**	**3.29**	**3.38**	19	0.500

Notes: 1 = Very low (VL); 2 = Low (L); 3 = Neutral (N); 4 = High (H); 5 = Very High (VH). Q/S = Quantity surveyors, ARC = Architects, CM = Construction managers, EN = Engineers, CON = Contactors, BI = Building inspector, PM = Project manager.

no significant difference ($p > 0.05$) in the responses of professionals in the construction industry on the challenges of ABM to the construction industry.

5. Conclusion

The study investigates the benefits of ABM to the construction industry from the perspectives of professionals in client (government and private), consulting and contracting organizations. Based on the finding of the study, it was concluded that professionals in the construction industry are not sure (they are neutral) of the benefits of ABM to the construction industry. This may be due to their level of awareness of the available ABM and their benefits to the construction industry. The study also concludes that there is no significant difference in the knowledge of professionals in the construction industry on the benefits of ABM to the construction industry. It was also concluded that all the challenges considered in this study are true and valid among construction professionals except for the reduction of profit margin and having poor environmental friendliness. Also, there is no significant difference in the responses of professionals in the construction industry on the challenges of ABM to the construction industry.

Based on these conclusions, it was recommended that professional organizations and the government should team up to harmonize the issue of awareness and adoption of ABM in the construction industry. Government is expected to institute a strict policy towards the adoption of ABM for construction purposes. If these steps are not taken, the housing shortage that is currently being experienced because of expensive construction cost may continue to linger in the country. Because of the inherent advantage of reduction in the overall life cycle cost of construction projects, ABM has the potential for reducing the backlog of housing shortages in the country. Besides, if the issue of adoption of alternative building material for construction purposes is not quickly addressed, the natural resources of the country will continue to be depleted, environmental hazards will continue to rise and emission of carbon dioxide will be on the increase thus posing dangers to the health and well-being of citizens.

In addition, the lack of government support is one of the major challenges of adopting ABM. The aspect of poor acceptability by the public needs to be particularly addressed by the government by way of creating awareness of the advantages of using the ABM. Government should particularly sponsor certain developments with ABM in order to convince prospective clients of its inherent advantages.

References

Adetunji, I., Price, A. D. and Fleming, P. (2008). "Achieving Sustainability in the Construction Supply Chain", *Engineering Sustainability*, 161(3), 161–172.

Ala-Juusela, M., Huovila, P., Jahn, J., Nystedt, A. and Vesanen, T. (2006). "Energy Use and Greenhouse Gas Emissions from Construction and Buildings". Final report provided by VTT for UNEP. Parts of the text published in: UNEP (2007) Buildings and Climate Change Status, Challenges and Opportunities, Paris, UNEP.

AlSanad, S. (2015). "Awareness, Drivers, Actions, and Barriers of Sustainable Construction in Kuwait", *Procedia Engineering*, 118, 969–983.

Barlett, E. and Howard, N. (2000). "Informing the Decision Makers on the Cost and Value of Green Building", *Building Research and Information*, 28(5–6), 315–324.

Bon R. and Hutchinson K. (2007). "Sustainable Construction: Some Economic Challenges", *Building Research and Information*, 28(5–6), 310–314.

Dulaimi, M., Ling, F., Ofori, G. and Silva, N. (2002). "Enhancing Integration and Innovation in Construction", *Building Research and Information*, 30(4), 237–275.

Giesekam, J., Barrett, J. R. and Taylor, P. (2016). "Construction Sector Views on Low Carbon Building Materials", *Building Research & Information*, 44(4), 423–444.

Hwang, B. G. and Tan, J. S. (2012). "Sustainable Project Management for Green Construction: Challenges, Impact and Solutions", in *World Construction Conference 2012 — Global Challenges in Construction Industry*, June 28–30, Colombo, Sri Lanka, pp. 171–179.

Hydes, K. and Creech, L. (2000). "Reducing Mechanical Equipment Cost: The Economics of Green Design". *Building Research and Information*, 28(5–6), 403–407.

Jayasinghe, C. (2013). "Embodied Energy of Alternative Building Materials and Their Impact on Life Cycle Cost Parameters". Available at: www.civil.mrt.ac.lk/conference/ICSECM_2011/SEC-11-166.pdf (Accessed on April 7, 2017).

Johnson, A. (2000). "Tools and Methods for Environmental Assessment of Building Products-Methodological Analysis of Six Selected Approaches", *Building and Environment*, 35, 223–238.

Kebede, S. (2013). "The Impact of Alternative Construction Technology on Condominium Housing Project: The Case of Addis Ababa". Master's degree thesis Submitted to the Urban Management Master's Program, Ethiopian Civil Service University, Ethiopia.

Khodeir, L. M. and Mohamed, A. H. (2015). "Identifying the Latest Risk Probabilities Affecting Construction Projects in Egypt According to Political and Economic Variables from January 2011 to January 2013", *Housing and Building National Research Center*, 11, 129–135.

Lin, T. (2015). "More Houses Needed to Meet Auckland Demand, Says Westpac Report". Available at: www.stuff.co.nz (Accessed on April 7, 2017).

Lu, N. (2007). "Investigation of the Designers and General Contractors' Perception of Offsite Construction Techniques in the United States Construction Industry". A Ph.D thesis in Education Career and Technology Education, Clemson University, Clemson.

Lyons, M. (2009). "Comparative Analysis between Steel, Asonry and Timber Frame Construction in Residential Housing". Available at: www.repository.up.ac.za (Accessed on April 2, 2017).

Makenya, A. R. and Nguluma, H. M. (2007). "Optimization of Building Materials and Design towards Sustainable Building Construction in Urban Tanzania", in *CIB World Building Congress*, pp. 2083–2093.

Nelms, C., Russell, A. D. and Lence, B. J. (2005). "Assessing the performance of sustainable technologies for building projects", *Canadian Journal of Civil Engineering*, 32(1), 114–128.

NHBRC (2016). "Annual Performance Plan 2016/2017". Available at: www.nhbrc.org.za (Accessed on April 2, 2017).

Priemus, H. (2002). "Public Private Partnerships for Spatio-economic Investments: A Changing Spatial Planning Approach in the Netherlands", *Planning Practice and Research*, 17(2), 197–203.

Rohracher, H. (2001). "Managing the Technological Transition to Sustainable Construction of Buildings: A Socio-Technical Perspective", *Technology Analysis and Strategic Management*, 13(1), 137–150.

Rydin, Y. (2008). "Reassessing the Role of Planning in Delivering Sustainable Construction Development". Available at: http://www.isurv.com/site/script/download" on April 2, 2017.

Sayce S., Ellison, L. and Parnell P. (2007). "Understanding Investment Drivers for UK Sustainable Property", *Building Research and Information*, 35(6), 629–643.

Seo, S. (2002). "International Review of Environmental Assessment Tools and Databases". Report 2001-006-B-02, CRC for Construction Innovation, Brisbane. Available at http://eprints.qut.edu.au/26860.

Sharma, V. D. and Awchat, G. D. (2016). "Excavated Soil from Construction Site as a Green Building Material in Structural Brick Masonry", *International Journal of Scientific Development and Research*, 1(8), 380–383.

Sodagar, B., Behzad, D., Fieldson, G. and Rosemary, D. (2008). "Towards a Sustainable Construction Practice", *Construction Information Quarterly*, 10(3), 101–108.

Subramanian, N. (2007). "Sustainability: Challenges and Solutions", *The Indian Concrete Journal*, 81(12), 39–50.

Williams, K. and Dair, C. (2007). "What is Stopping Sustainable Building in England? Barriers Experienced by Stakeholder in Delivering Sustainable Developments", *Sustainable Development*, 15, 135–147.

Wyk, L.V. (2013). "Innovative Building Technology: The Value Proposition". Available at: www.csir.co.za (Accessed on April 7, 2017).

Zhou, L. and Lowe, D. J. (2003). "Economic Challenges of Sustainable Construction". Available at: www.Researchgate.net/publication/269517027 (Accessed on April 2, 2017).

Chapter 17

Exploitation of BIM in Planning and Controlling the Construction Phase On-site Carbon Emissions: A 6D BIM Case Study

Ramy Al Sehrawy* and Bimal Kumar

*Engineering & Built Environment, Glasgow Caledonian University,
70 Cowcaddens Road, Glasgow G4 0BA, UK
relseh200@caledonian.ac.uk

Omar Amoudi

*Built & Natural Environment, Caledonian College of Engineering,
Seeb 111, Muscat, Oman*

Abstract

In the midst of rising concerns regarding climate change and global warming, the current construction management practices are not sufficiently considering the on-site greenhouse-gases (GHGs) including carbon emissions produced during the construction phase, and are limited in utilizing the building information modelling (BIM) techniques

---

*Corresponding author.

and keeping up with the latter's fast growing implementation in the construction industry. This chapter proposes a BIM-based framework capable of developing and exploiting a 6D BIM Model for pre-construction management, planning and control of the construction phase on-site carbon emissions to support decision makers identify the construction phase's necessary corrective actions and best practices before the phase commences. A case study is conducted for a hypothetical BIM-based construction project to validate the proposed framework. The results demonstrated that BIM can be utilized as a powerful means to assist in early holistic planning and strategic decision-making involving the construction phase key dimensions including time, cost and on-site carbon emissions. The core contribution made by this research is the facilitation for sustainability champions and others concerned with the environmental impacts of the project's construction phase to be involved within the project's BIM environment and processes, while on the other hand, enable as well those already involved with BIM to include the environmental targets and benchmarks within the project's BIM processes and collaborative common data environment, while keeping an eye and maintaining a holistic view over the other major two dimensions of the project: time and cost.

Keywords: nD BIM; Sustainability; Construction Phase; Carbon Emissions; Energy Consumption.

1. Introduction

"The world has witnessed a dramatic increase in environmental concerns and issues related to global climate change over the past decade" (Kim *et al.*, 2012, p. 982). The evolving environmental concerns stimulated by the current issues and anticipated dramatic impacts of the global climate change have provoked the development of this research. The construction industry is ranked third among all industrial sectors with 6% of the total GHG emissions (US Environmental Protection Agency (USEPA), 2008) and is responsible for more than 40% of the total energy consumption (Casals, 2006), being fairly one of the main culprits responsible for the global greenhouse-gases (GHGs) emissions and proved worthy of any research efforts attempting to help in controlling and reducing the GHG including its key component which is the carbon emissions. Thus, the construction industry has got some crucial and necessary radical moves to take in the direction towards reducing the carbon emissions (Stadel *et al.*, 2011).

The global warming and climate change are most probably related to the GHG emissions (Intergovernmental Panel on Climate Change (IPCC), 2006). Lim *et al.* (2016) stated that only carbon is responsible for 88.6% of the total GHGs emissions, and as the total energy-related CO_2 emissions worldwide reached a peak in 2010 with 30.6 Gt of CO_2 emissions released to the global atmosphere (Wong *et al.*, 2013), reducing and controlling the carbon emissions has lately become an area of interest among the main environmental topics being researched and discussed (Hong *et al.*, 2010).

When splitting the construction project life cycle into two phases, operational and non-operational phases, obviously, most of the previous research efforts are focused on the operational phase and the selection of materials; however, the GHG and carbon emission performance throughout the non-operational phase (i.e. construction phase) is rarely studied (Janssen, 2014; Ahn *et al.*, 2013; Ren, 2012; Sharrard *et al.*, 2007; Wong *et al.*, 2013) despite being a rising challenge (Environmental Protection Agency (EPA), 2009).

Several policies and initiatives were introduced to help reduce the energy consumption and emissions during the operational phase (Newton and Tucker, 2011; Peña Mora *et al.*, 2009). While for the construction phase, the key well-known initiatives and sustainability assessment methods, such as BREEAM (2016), require sole monitoring of the construction phase emissions without necessarily following a specific benchmark, which may not result in a significant progress in reducing the emissions. In light of these currently most used environmental schemes and assessment tools (e.g. BREEAM, LEED) which lack the ability to support the pre-construction planning or analyzing the effects of planning and controlling decisions about the construction phase emissions (Heydarian and Golparvar-Fard, 2011), this challenge cannot be encountered without clear and consistent management and control practices (Ahn *et al.*, 2013). In addition, there is an urgent need to handle this challenge in the context of the fast growing usage of building information modelling (BIM) in construction projects.

In the UK, the construction processes were estimated to account for about 1% of the total carbon footprint of the construction industry in 2008 (Innovation & Growth Team (IGT), 2010), considering the site operations, including the site activities, offsite assembly and offsite offices, but excluding refurbishment and demolition (Business Innovation & Skills (BIS), 2010). Apparently, such a numerical figure (i.e. 1%) had a strong influence over many scholars and researchers in believing that the carbon emitted during the construction phase is way less significant compared to emissions and energy consumed during the operational phase. Nonetheless, Peng (2016, p. 453) pointed out that "the average carbon emissions per working area per year of non-operational stages are far greater than those of the operational phase."

The impact of the construction activities was highlighted through the US statistics, stating that 40% of the non-transportation mobile carbon emissions are produced from the construction activities (EPA, 2009). In England, the energy and fuel consumed in construction sites account for almost 33% of the total construction industry emissions (Strategic Forum for Construction (SFfC), 2008). What could be even more dramatic is that the estimated fuel consumption by the construction plant in the US is nearly twice the amount suggested officially in the US reports (Sharrard *et al.*, 2007), and the emerging results indicate that the construction operations produce more carbon than estimated (Ren, 2012).

On the other hand, the rapid growth and expansion of BIM over the globe and its mandatory implementation in some countries motivate the involvement and utilization of BIM in any of the processes carried out within construction projects, including the process of planning and controlling the project's construction phase carbon emissions. However, integrating the construction phase carbon planning and management with BIM is far from being well explored compared to other phases like the operation phase.

Even though Arnold (2008) stated that no research has fully integrated BIM with energy consumption during the project construction phase, recently, scholars have started to propose BIM-based methods and techniques that incorporate and visualize carbon emissions during the construction phase (Li *et al.*, 2012; Wong *et al.*, 2013, 2014; Yung and Wang, 2014). Yet, no researcher has fully embraced the construction phase carbon emissions with the other fundamental dimensions of the project (i.e. time and cost) in a holistic way that is capable of handling the pre-construction planning and decision-making on a macro scale rather than just focusing on specific activities.

Therefore, this research is intended to exploit BIM in providing a pre-construction holistic BIM-based method for planning and controlling the construction phase on-site carbon emissions before the commencement of the construction works at site within the project's common data environment and with a full consideration to the other dimensions, such as time and cost; hence, this ensures in keeping up with the fast growing usage and implementation of BIM.

2. Methodology and Framework

Achieving the aim of this research and filling the gaps identified provoked the creation of a BIM-based framework that embraces and integrates the key project dimensions (i.e. time, cost and carbon emissions) during the planning of the construction phase through the generation of a 6D BIM model that is capable of simulating the construction process while considering the construction costs and production of on-site carbon emissions.

In the Architecture, Engineering and Construction (AEC) industry, the sixth dimension of BIM commonly refers to the post-construction phase, including facility management (FM), and the seventh dimension refers to sustainability; many have considered the seventh dimension as the most important of all because of its direct influence over the design (3D), time (4D), cost (5D) and FM (6D) (Kapogiannis *et al.*, 2015). However for many, 6D mostly represents sustainability (O'Keeffe, 2012; Parsanezhad, 2015; Yung and Wang, 2014). Others consider both definitions as correct, where the FM targets the operational efficiency which overlaps with the building's life cycle performance, in other words its sustainability (Nicał, and Wodyński, 2016). This study is concerned with the construction phase and the sixth dimension proposed herewith is the on-site carbon emissions.

The framework (Fig. 1) is initiated by acquiring the project's 3D BIM model (Fig. 1: A). The 3D model is utilized to develop a set of all the necessary construction activities needed to complete the project (Fig. 1: A.2), and automatically calculate the quantities of project's elements and materials. The list of activities, the quantified list of materials along with database of estimated activities productivity rates are all used to set the required number of crews, and the formation of each crew

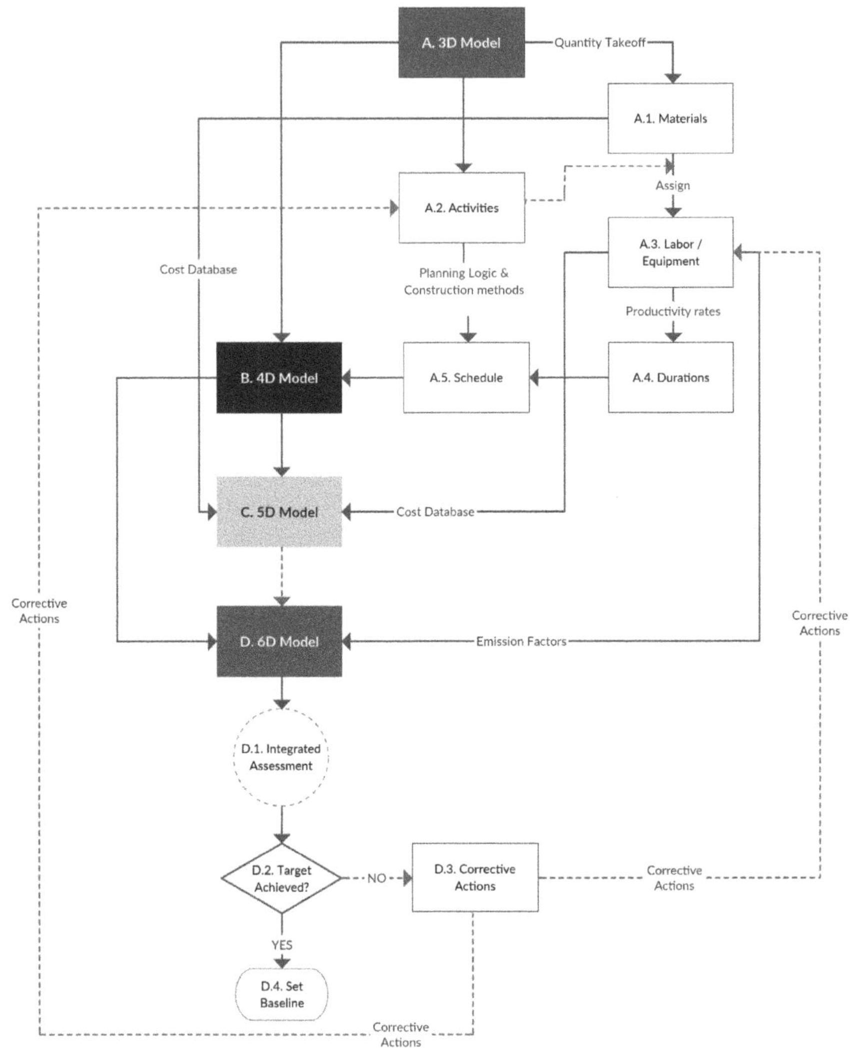

Fig. 1.   Proposed 6D BIM framework.

(Fig. 1: A.3) which defines the combination of labour and equipment types of resources needed to complete a certain activity in a specific duration (Fig. 1: A.4).

Consequently, the project time schedule is developed (Fig. 1: A.5), then linked with the 3D model, and upgraded to a 4D model (Fig. 1: B).

The construction time schedule is then enhanced by assigning the predefined resources needed for each activity (i.e. materials, labour and equipment) to that specific activity in the schedule. Each type of these resources includes cost-related data. The schedule is then re-synchronized to include the information related to cost, hence, upgrading the model to a 5D model (Fig. 1: C).

To further upgrade the 5D model and generate the 6D BIM model proposed in this research, the data related to the carbon emissions produced by the labour and equipment resources are incorporated within the construction time schedule, and subsequently the 5D model, in order to develop the 6D model (Fig. 1: D). The amount of carbon emission data for the equipment and labour types of resources assigned to any of the construction activities is calculated using Eqs. (3) and (4), respectively.

In case of equipment type, initially with reference to Engine Type (ET_i) whether Diesel or Gasoline; the Ton of Oil Equivalent (TOE_i) and Carbon Emission Factor (CEF_i) for the ith type of equipment are determined as shown in (1) and (2), respectively (IPCC, 2006).

$$(TOE_i) = 0.000845 \ (ET_i: \text{diesel}) \ / \ 0.000740 \ (ET_i: \text{gasoline}), \quad (1)$$

$$(CEF_i) = 0.837 \ (ET_i: \text{diesel}) \ / \ 0.783 \ (ET_i: \text{gasoline}). \quad (2)$$

Hence, the total amount of carbon emissions produced by ith type of equipment assigned to jth construction activity $(KgCO_{2i}^j)$ is calculated as shown in (3) by multiplying the TOE_i by CEF_i by the total equipment hours spent by the ith type of equipment during the construction activity (T_i^j); by the average fuel hourly consumption rate for ith type of equipment (HC_i). Then to convert to carbon dioxide, all is multiplied by the ratio of the molecular weight of carbon dioxide to carbon and by 1000 to calculate the emissions in kilograms.

$$(KgCO_{2i}^j) = (TOE_i) \times (CEF_i) \times (T_i^j) \times 44/12 \times 1000. \quad (3)$$

On the other hand, in case of labour type of resources, by referring to Sherwood (2006), it is assumed that a non-construction person inhales 0.5 l/breath comprising 3.97% of carbon dioxide, whereas for every

22.4 l of air inhaled by a construction labourer, 44 g of carbon dioxide is produced. Moreover, as the level of activity differs from one labourer to another based on the nature of work being handled, the respiratory rate is assumed to be as shown in Table 1 (Sherwood, 2006; Int Panis *et al.*, 2010). However, considering an idle level of activity as a minimum during a man's normal life, 12 breaths per minute are deducted from all levels of activity in order to calculate the net amount of carbon emissions produced by the labourer due to working in the project activity being assigned to.

Thus, the total amount of carbon emissions produced by *i*th type of labourer assigned to *j*th construction activity $(KgCO_{2i}^{j})$ is calculated as shown in (4) by multiplying the *i*th labourer number of breaths per hour (HC_i) based on the labourer activity rate; by 0.5 l of air per hour; by 0.0.397 representing the percentage of carbon dioxide; by 44/22.4 to measure the weight of carbon dioxide produced; by 0.001 to unify the resulting units to kilograms.

$$(KgCO_{2i}^{j}) = (HC_i) \times (CEF_i) \times (T_i^{j}) \times 44/12 \times 1000. \tag{4}$$

As the processes of the proposed framework proceed, the output of the 6D model is used in variable ways for the purpose of conducting an integrated assessment of the overall performance of the project construction plan which involves the three variables of time, cost and carbon emissions (Fig. 1: D.1). This performance is then evaluated or compared with the targets set for the project if any (e.g. benchmarks, client or local authorities' requirements, project constraints, etc.) (Fig. 1: D.2). Based on the results of the integrated assessment, the suggested plan could be set as a baseline (Fig. 1: D.4) for the project's construction phase in case targets

Table 1. Labourers' levels of activity and corresponding respiratory rates.

Level of Activity	Respiratory Rate (Breaths Per Minute)	Net Respiratory Rate (Breaths Per Minute)
Idle	12	0
Low	18	6
Medium	28	16
High	34	22
Intensive	40	28

were met, otherwise possible corrective actions shall be taken in an attempt to meet the aforementioned targets (Fig. 1: D.3).

3. Case Study, Results and Discussion

3.1. *Case study*

The case study undertaken in this research involves a hypothetical project that is used to demonstrate and test the applicability, usability and effectiveness of the proposed framework and its output mainly represented in the created 6D BIM Model in an attempt to answer the research questions. This hypothetical project is an educational institution in UK comprising the main building and surrounding landscape. The project is built on a plot of a total area equals 8,000 m^2, and consists of three floors with a total built-up area of 5,018 m^2, distributed over the three floors (Fig. 2).

The case study introduced is utilized to apply the framework demonstrated and explained above using three sets of data. The first set comprises the 3D BIM Model, while the second set includes the information related to the cost of the different types of resources used and assigned to the project (i.e. material, labour and equipment). The third set of data holds the information about the equipment type, power source and fuel consumption as shown in Table 2, along with the information related to the different labour and corresponding levels of activity shown in Table 1, which are all critical to estimate the project carbon emissions. The resulting 6D model represented a basic scenario model (Fig. 3).

Fig. 2.   Project Front perspective. (Revit®, 2015)*.

Table 2.   List of major project's equipment.

Type of Equipment	Power Source (Fuel/Electric) — Engine Type (Diesel/Gasoline)	Fuel Consumption (L/h)	Cost Per Hour ($/h)
Wheel dozer	Fuel — Diesel	48.2	60
Roller soil compactor	Fuel — Diesel	13.3	55
Shovel	Fuel — Diesel	216	110
Articulated dump truck	Fuel — Diesel	20.7	60
Mobile crane	Fuel — Diesel	10	150
Concrete vibrator	Fuel — Diesel	1.1	13
Materials hoist	Fuel — Diesel	2	25

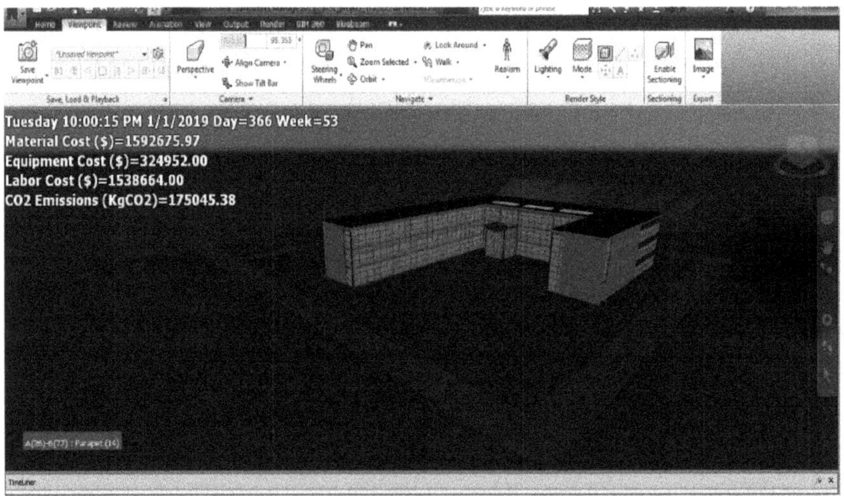

Fig. 3.   6D BIM — A screenshot of the basic scenario model at project completion.

3.2. *Carbon emissions*

The amount of carbon emissions produced every week as tabulated and plotted to illustrate the distribution of the total carbon emissions produced from the construction phase over the latter's duration (Fig. 4), and hence

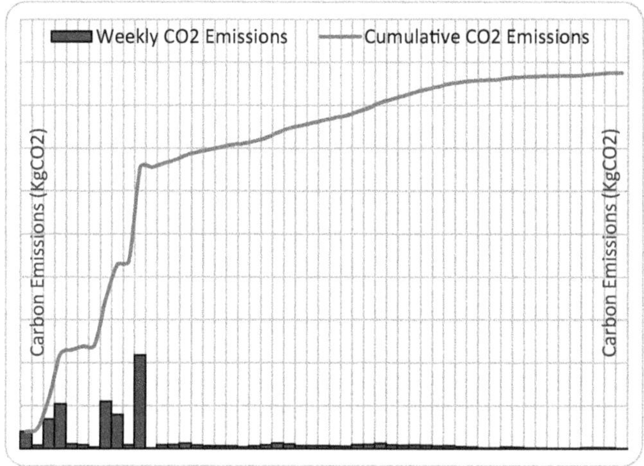

Fig. 4. Construction phase weekly/cumulative carbon emissions against time.

identifying the peaks and critical weeks that will need to be thoroughly considered through the decision-making process in order to bring about the most effective and efficient corrective actions.

The cumulative curve in Fig. 4 explicitly proves that the majority of the emissions are approximately produced in the first quarter of the project, whereas almost 75% are already emitted after the first 11 weeks from a total project duration of 52 weeks with the peak of carbon emissions produced per week detected in week 11 with a value of 43,462 $kgCO_2$ in one week (Fig. 5). This is apparently due to the relatively intensive use of equipment at the first quarter of the project in performing substructure works specifically in earthwork activities. Obviously excessive care should be taken during this period of the project especially at the 1st, 3rd, 4th, 8th, 9th and 11th weeks; whereas these weeks combined are responsible for a total of 123,556.4 $kgCO_2$ which is about 70% of the total construction phase of on-site carbon emissions.

3.2.1. *Carbon emissions of work packages*

However, in order to identify the intensive types of work packages with respect to carbon emissions, the emissions were allocated to the related

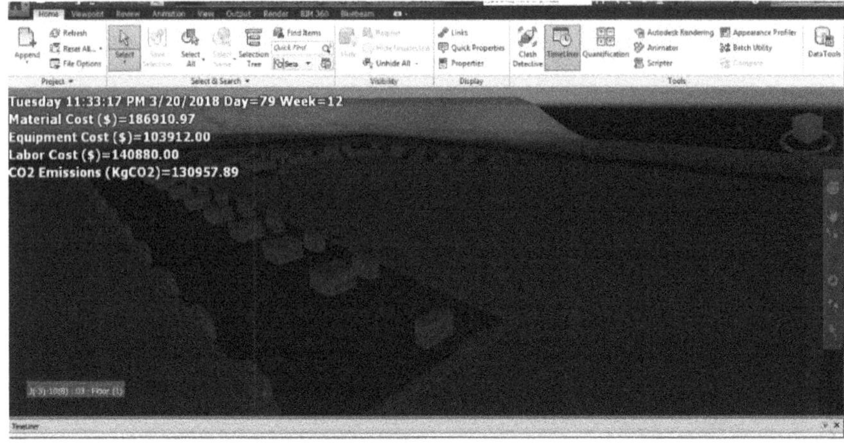

Fig. 5.   6D BIM — Basic scenario model at the end of week 11.

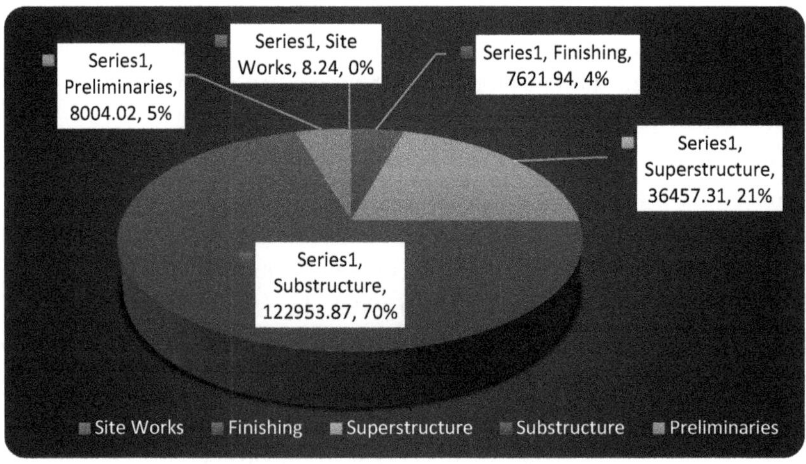

Fig. 6.   Percentages of contribution of project's work packages to the construction phase on-site carbon emissions.

work packages and the percentages of contribution of each work package are demonstrated in Fig. 6. Obviously, the substructure is the prime contributor with 70% of the total emissions.

3.2.2. *Equipment vs. labour carbon emissions*

On the other hand, the carbon emissions produced by labour type of resources were compared to the total amounts produced by the equipment, and the results proved, without any doubts, that the equipment type of resources is the leading culprit of the construction phase with 97% of on-site carbon emissions, compared to labour type of resources which was responsible for only 3% (Fig. 7(a)).

With a deeper look into the carbon emissions produced by equipment, it is illustrated in Fig. 7(b) that shovel, mobile crane and dump truck are the highest contributors to the construction phase on-site carbon emissions, resulting in 40%, 23% and 21%, respectively, of the total emissions produced by project's equipment. It is worth to mention that shovel and dump truck which are responsible together for producing 61% of the project's equipment carbon emissions were both used only in the earthwork activities (i.e. excavation, backfilling and compaction), whereas this significantly high percentage justifies the strong interest and focus of many of the previous researches in studying the earthwork activities more specifically than any other activities.

However, although the building consists of only three floors, the mobile crane showed a relatively considerable contribution that is worthy of attention. The 21% contribution by the mobile crane to the total emissions produced by equipment proves the criticality of the activity of lifting materials in construction projects in terms of carbon emission production, which has not gained enough attention within the previous micro-scale type of research and studies which were meant to focus on studying the carbon emissions of specific construction activities.

Obviously, the labour with the least levels of activity had the least contribution, while those with higher levels of activity and consequently higher respiratory rates due to the intensive nature of work were the top contributors, especially if they got involved in many of the project activities like the helpers, who were responsible for 46% of the carbon emissions generated by the labour force at site during the construction phase as illustrated in Fig. 7(c).

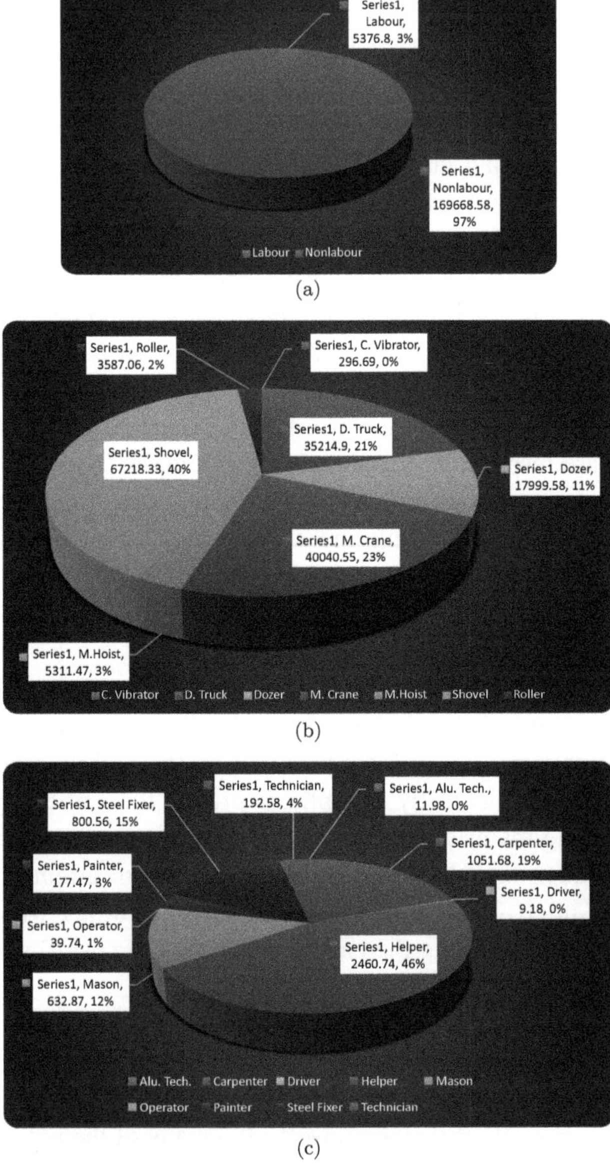

Fig. 7. (a) Contribution of labor and equipment to the construction phase on-site carbon emissions. (b) Contribution of different types of equipment to the equipment's construction phase on-site carbon emissions. (c) Contribution of different types of labour to the labour's construction phase on-site carbon emissions.

3.3. *Cost*

3.3.1. *Total costs of resources*

When comparing the overall cost for each of the three types of resources, the results highlight to a great extent the labour intensive nature of the construction industry, where the labour cost comes in the second place with 45% as shown in Fig. 8(b), close enough to the first place taken by the material costs with 46%, while the cost of hiring and operating the major equipment required for the project activities weighs only 9% of the total project costs.

On the other side, the total costs of all types of resources (i.e. materials, labour and equipment) utilized every week were calculated and plotted to illustrate the distribution of the total direct costs paid throughout the construction phase (Fig. 8(a)), and hence identify the peaks and critical weeks in terms of expenditure, which will need to be more considered through the decision-making process in order to take the most cost-effective corrective actions. While the 6D model showed a high concentration of the total carbon emissions in the first quarter of the project, specifically in the first 11 weeks, the project direct costs were more evenly distributed over the total duration, resulting in a smoother cumulative S-curve. The visually identified peaks are in 20th, 28th and 37th weeks. These three critical weeks are responsible for approximately 15% of the total construction direct costs, whereas costly activities overlap leading to the illustrated rigorous sudden increases in total costs as shown in Figs. 9(a)–9(c).

3.3.2. *Cost of labour and equipment*

However, as the 6D BIM model here was mainly developed to be used by construction managers during the planning of the construction phase; materials and subsequently material costs are assumed to be fixed with no possible room for changes or modifications despite the considerable material costs. It is obvious that the equipment cost curve is more front loaded, reaching its peak in week 11 with AU\$ 25,360 per week, which is somehow a similar trend that shares the same peak with the one shown in Fig. 4 concerning the overall project's carbon emissions, where both

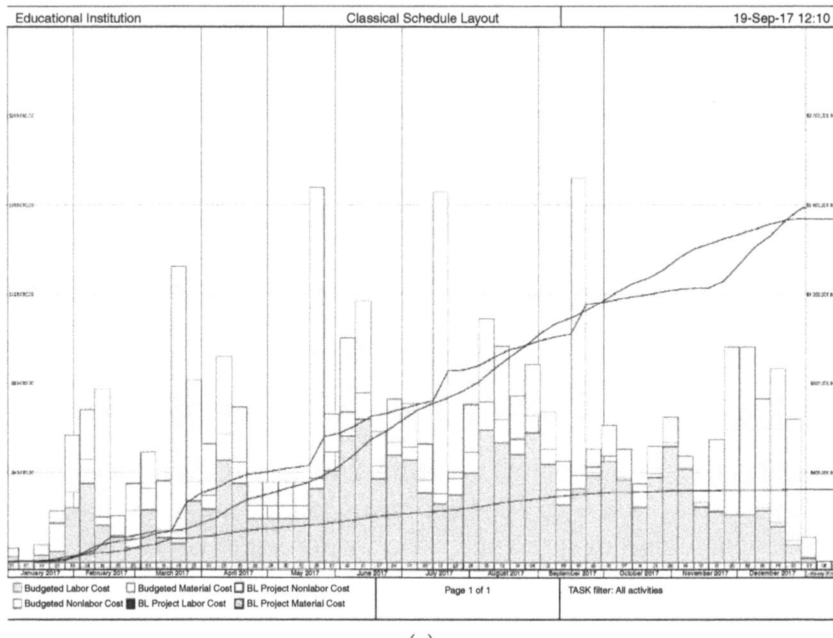

(a)

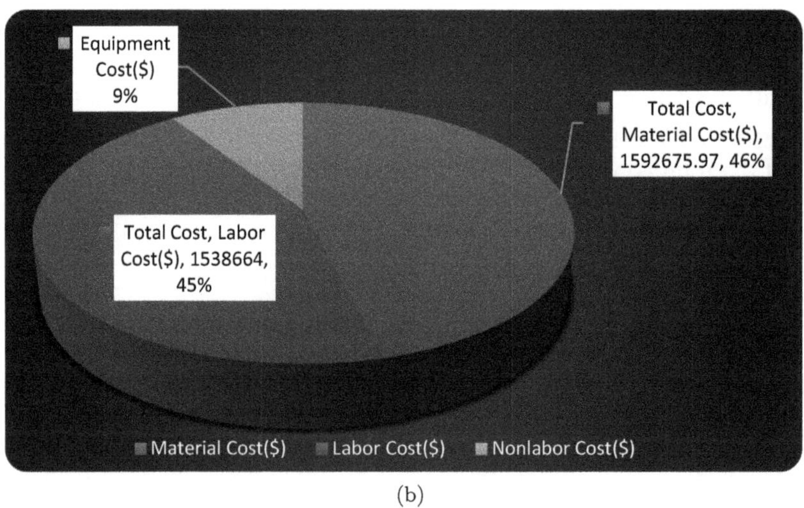

(b)

Fig. 8. (a) Construction phase weekly/cumulative total cost (material, labour and equipment) against time and (b) Percentages of contribution of different types of resources to the construction phase total cost.

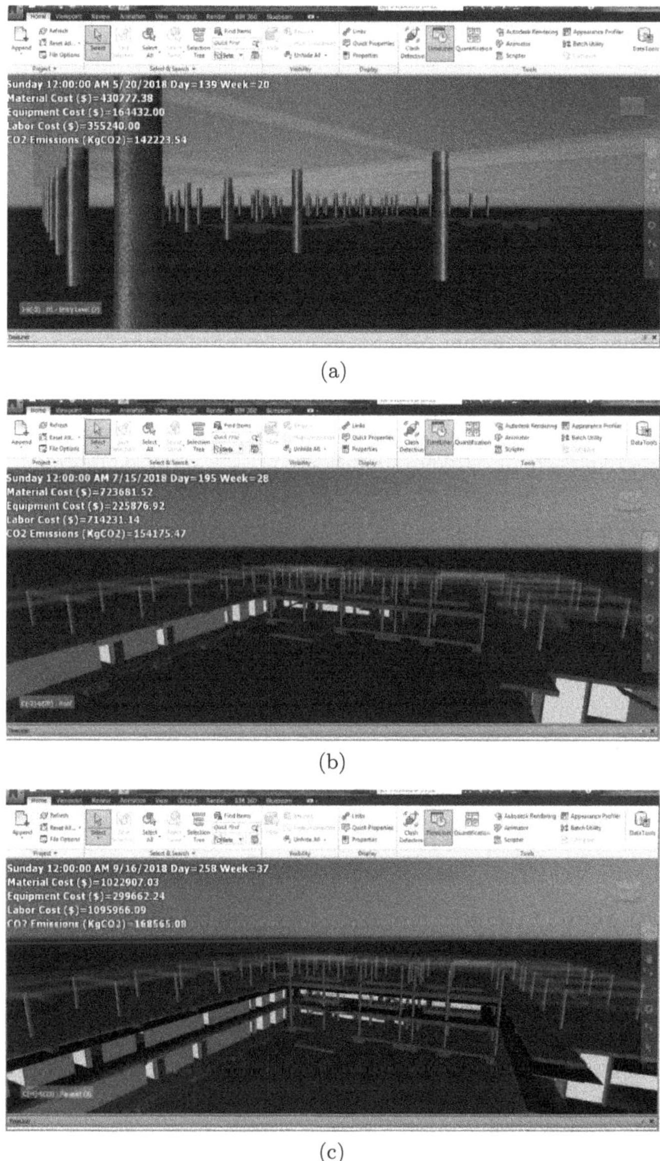

Fig. 9. (a) 6D BIM model showing ongoing activities in Week 20. (b) 6D BIM model showing ongoing activities in Week 28. (c) 6D BIM model showing ongoing activities in Week 37.

figures reflect the intensive usage of equipment during the first quarter of the project.

On the contrary, the labour costs had the most evenly distributed curve and are significantly higher when compared to the equipment costs. Both labour and equipment costs together showed a peak in week 23, including AUD 64,040 and AUD 12,000 for labour and equipment, respectively, that also witnessed multiple activities including shuttering works, fixing rebar, and block works in the ground floor (Fig. 10).

3.4. *Alternative scenarios and corrective actions*

The preceding detailed analysis of the developed 6D BIM model had sharpened the view in terms of identifying the weak points within any of the project dimensions, and those areas that needed to be addressed the most in order to develop a more effective 6D BIM model and efficient enough were set as the project's baseline construction plan.

For instance, three main superstructure types of activities (i.e. formwork installing, formwork uninstalling and rebar fixing) proved to have a leading role in controlling the project's total cost and time. With reference

Fig. 10.   6D BIM model showing ongoing activities in Week 23 (maximum labour and equipment costs).

to the project time schedule, these three types of activities were the longest with zero total float, acting as the core activities of the critical path. While the aforementioned activities are labour intensive by nature, they have more power and influence over the project's total cost, as the three types of labours, including carpenters, steel fixers and helpers who were mainly utilized in these activities, were responsible for 72% of the labour-related costs, and the mobile crane mainly used during the these activities was on the top of all equipment with 71% of the equipment-related costs.

Subsequently, a suggestion could be working two shifts per day (i.e. night shift) whenever the above-mentioned three labour intensive types of activities are being performed. However, in terms of carbon emissions, working at night means consuming power for lighting up the site and hence producing more carbon emissions. Working at night during the above-mentioned activities resulted in 77 night shifts. Apart from the changes in time, the carbon emission data were also adjusted to consider and include the carbon emissions produced as a result of the electricity consumed to light the site up during the 8 h night shift for 77 working nights. Using Eq. (1), the lux of 80 Lux (Health and Safety Executive (HSE), 1997) is multiplied by the total site area (approximately 8,000 m^2) and divided by the luminous efficacy of LED lights (e.g. 75 Lm/W) suggested to maintain sufficient lighting at site to calculate the electric consumption, then divided by 1,000 to obtain the result in kWh, which is multiplied by 8 h per night shift; by 77 nights and by 0.46254, which is the carbon emission factor per kWh in UK (Table 3).

$$\mathrm{KgCO_2} = (\mathrm{Lux} \times \mathrm{Area}) / (\text{luminous efficacy} \times 1000) \times 8 \times 77 \times 0.46254. \tag{1}$$

Table 3.   Change in time, cost and carbon emissions due to scheduled night shifts.

Dimension	Basic Scenario	Alternative Scenario with Night Shifts	% of Change
Time (days)	262	187	−28.6259542
Cost ($)	3,456,291.99	3,456,291.99	0
Emissions (KgCO$_2$)	175,045.38	177,478.58	1.390039543

4. Conclusion

4.1. *Assessment of research findings*

This research highlights the absence of a holistic view that considers and embraces all the project's major dimensions (i.e. time, cost and carbon emissions) crucial for the early strategic planning and decision-making on a macro scale. Moreover, the utilization of BIM in the process of pre-construction management, planning and controlling of the construction phase on-site carbon emissions is still far away from being mature, and there is an urgent need to bring the dimension of carbon emissions into the BIM environment for the benefit of both BIM implementation process as well as achieving the environmental long-term goals and ambitions.

Hence, the introduced BIM-based method — including the developing of a 6D BIM model — has opened doors for the process of planning and controlling the carbon emissions to be carried out at earlier stages and within the project's BIM environment. Involving the dimension of carbon emissions within the BIM environment *per se* has offered a genuine opportunity for better coordination between all major project dimensions.

The proposed 6D BIM model showed a great ability to fully embrace all the major project dimensions (i.e. time, cost and carbon emissions) on a macro scale for all project activities rather than a specific activity or a work package. Thus, we present a holistic view that is essential for early planning and strategic corrective decisions with sufficient and balanced realization of all the project's main variables and dimensions.

4.2. *Future research*

Further research may consider developing a software tool or a plug-in to incorporate the whole framework within a common platform, which will significantly ease the process, enhance the interoperability, reduce time and efforts, and help decision makers to create more scenarios and alternative 6D BIM plans in a less time with less effort.

Another avenue for further studies would be enriching the model by including the embodied carbon in construction materials along with the carbon generated due to the transportation of materials with the integration of spatial data and related supporting applications (e.g. Google Maps) thus supporting decision-making while choosing the best materials suppliers, by including the environmental impacts along with traditional criteria.

Future studies may include more application for the framework on real construction projects with further real-time monitoring at site using primary data instead of secondary data, by recording actual progress, cost and actual carbon emissions at site and updating the 6D BIM model to be compared with the baseline set at the pre-construction and planning phase that is developed based on actual measurements of carbon emissions and productivity rates. This will help in better visualizing, understanding and controlling the six dimensions, as well as supporting the process of benchmarking; thus, a built 6D model could be delivered by the end of construction.

References

Ahn, C., Lewis, P., Golparvar-Fard, M. and Lee, S. (2013). "Integrated Framework for Estimating, Benchmarking, and Monitoring Pollutant Emissions of Construction Operations", *Journal of Construction Engineering and Management*, 139(12).

Arnold, A. (2008). Development of a Method for Recording Energy Costs and Uses during the Construction Process. PhD, Texas A&M University.

BREEAM (2016). "BREEAM International New Construction 2016", BRE Global Ltd. Available at: http://www.breeam.com/new-construction (Accessed on May 30, 2017).

Casals, X. G. (2006). "Analysis of Building Energy Regulation and Certification in Europe: Their Role, Limitations And Differences", *Energy & Buildings*, 38(5), 381–392.

Environmental Protection Agency (2009). "Potential for Reducing Greenhouse Gas Emissions in the Construction Sector", EPA. Available at: http://www.epa.gov (Accessed on May 30, 2017).

Health and Safety Executive (1997). *Lighting at Work*, Norwich: HSE Books.

Heydarian, A. Golparvar-Fard, M. (2011). "A Visual Monitoring Framework for Integrated Productivity and Carbon Footprint Control of Construction Operations", in *Proceedings of the 2011 ASCE International Workshop on Computing in Civil Engineering*, Technical Council on Computing and Information Technology of ASCE, Miami, Florida, pp. 504–511.

Hong, W., Kim, H., Kim, J., Kim, S., Kim, J. T., Park, S., Lee, S. and Yoon, K. (2010). "A New Apartment Construction Technology With Effective CO_2 Emission Reduction Capabilities", *Energy*, 35(6), 2639–2646.

Innovation & Growth Team (2010). *Low Carbon Construction: Final Report*, London: Department for Business, Innovation & Skills, IGT.

Int Panis, L., De Geus, B., Vandenbulcke, G., Willems, H., Degraeuwe, B., Bleux, N., Mishra, V., Thomas, I. and Meeusen, R. (2010). "Exposure to Particulate Matter in Traffic: A Comparison of Cyclists and Car Passengers", *Atmospheric Environment*, 44(19), 2263–2270.

Intergovernmental Panel on Climate Change (2006). Climate Change 2007: Synthesis Report. Contribution of Working Groups I, II And III to the Fourth Assessment Report of the Intergovernmental Panel on Climate Change. *The IPCC 4th Assessment Rep.*, C. W. Team, R. K. Pauchauri, and A. Reisinger, eds., IPCC, Geneva.

Janssen, R. M. J. (2014). Assessing Onsite Energy Usage: An Explorative Study, Masters, University of Twente, Netherlands.

Kapogiannis, G., Gaterell, M. and Oulasoglou, E. (2015). "Identifying Uncertainties Toward Sustainable Projects", *Procedia Engineering*, 118, 1077–1085.

Kim, B., Park, H., Kim, H. and Lee, H. (2012). "Greenhouse Gas Emissions from Onsite Equipment Usage in Road Construction", *Journal of Construction Engineering and Management*, 138(8), 982–990.

Lim, T., Gwak, H., Kim, B. and Lee, D. (2016). "Integrated Carbon Emission Estimation Method for Construction Operation and Project Scheduling", *KSCE Journal of Civil Engineering*, 20(4), 1211–1220.

Newton, P. W. and Tucker, S. N. (2011). "Pathways to Decarbonizing the Housing Sector: A Scenario Analysis", *Building Research and Information*, 39(1), 34–50.

Nicał, A. K. and Wodyński, W. (2016). "Enhancing Facility Management through BIM 6D", *Procedia Engineering*, 164, 299–306.

O'Keeffe, S. E. (2012). Developing 6D BIM Energy Informatics for GDL LEED IFC Model Elements. *Proceedings of the 2012 International Conference on Industrial Engineering and Operations Management*, 2328–2333.

Parsanezhad, P. (2015). "An Overview of Information Logistics for FM&O Business Processes", in *Ework and Ebusiness in Architecture, Engineering and Construction — Proceedings of the 10th European Conference on Product and Process Modelling*, ECPPM 2014, pp. 719–725.

Peña-Mora, F., Ahn, C., Golparvar-Fard, M., Hajibabai, L., Shiftehfar S., An, S. and Aziz, Z. (2009). "A Framework for Managing Emissions from Construction Processes", in *Proceedings International Conference & Workshop on Sustainable Green Bldg. Design & Construction, National Science Foundation.*

Peng, C. (2016). "Calculation of a Building's Life Cycle Carbon Emissions Based on Ecotect and Building Information Modeling", *Journal of Cleaner Production*, 112, 453–465.

Ren, Z., Chrysostomou, V. and Price, T. (2012). "The Measurement of Carbon Performance of Construction Activities", *Smart and Sustainable Built Environment*, 1(2), 153.

Revit®. (2015). "*Advanced Sample Project*". Autodesk®.

Sharrard, A. L., Matthews, H. S. and Roth, M. (2007). "Environmental Implications of Construction Site Energy Use and Electricity Generation", *Journal of Construction Engineering and Management*, 133(11), 846–854.

Sherwood, L. (2006). *Fundamentals of Physiology With Infotrac: A Human Perspective.* London: Cengage Learning.

Stadel, A., Eboli, J., Ryberg, A., Mitchell, J. and Spatari, S. (2011). "Intelligent Sustainable Design: Integration of Carbon Accounting and Building Information Modeling", *Journal of Professional Issues in Engineering Education and Practice*, 137(2), 51–54.

Strategic Forum for Construction (2008). "Carbon Assessment Report NIST, National Building Information Modelling Standard, Version 1.0 — Part 1.0. Overview, Principles and Methodologies", National Institute of Standards Technology, USA.

U.S. Environmental Protection Agency (2008). Quantifying Greenhouse Gas Emissions In Key Industrial Sectors. EPA 100-R-08-002, Sector Strategies Division, Washington, DC. Available at: http://www.epa.gov (Accessed on May 30, 2017).

Wong, J. K. W., Li, H., Wang, H., Huang, T., Luo, E. and Li, V. (2013). "Toward Low-Carbon Construction Processes: The Visualisation Of Predicted Emission Via Virtual Prototyping Technology", *Automation in Construction*, 33, 72–78.

Yung, P. and Wang, X. (2014). "A 6D CAD Model for the Automatic Assessment of Building Sustainability", *International Journal of Advanced Robotic Systems*, 11(8), 131.

Chapter 18

Energy Efficiency in Australia: Occupant Behaviour

Vivian W. Y. Tam*, Laura M. M. C. E. Almeida,
and Khoa N. Le

*School of Computing, Engineering and Mathematics,
Western Sydney University, Locked Bag 1797, Penrith,
NSW 2751, Australia*
**College of Civil Engineering, Shenzhen University, Shenzhen, China*

**vivianwytam@gmail.com*

Abstract

One of the main aspects that has to be carefully accounted and controlled in sustainable buildings is energy use, which is directly related to greenhouse-gas (GHG) emissions. During a building life cycle, the operation is the most energy intensive stage due to the presence of technical systems, equipment, occupants, etc. In green buildings (GB), design teams predict the expected energy performance of a building at the design stage, by means of a building energy simulation model, and study the most sustainable solutions in order to minimize the use of energy before the construction process occurs, so as to reduce the impacts that the

*Corresponding author.

operation stage will have in the environment. Meanwhile, researchers by comparing real energy performance results with the predicted ones have identified significant differences between them. These differences can be related to low maintenance, inefficiencies and occupants. Nevertheless, the way occupants behave in terms of energy use is the most impacting factor to the performance of a building (Norford *et al.*, 1994). Therefore, this chapter presents the influence occupants have in the energy use of a building using a simulation model that analyzes occupant interaction with different building systems, having as baseline the Australian National Construction Building Code (ABCB, 2016) requirements for office buildings. The model has been developed using the simulation tools DesignBuilder and EnergyPlus in order to determine the annual energy performance of a typical Australian office building in Sydney and consequent GHG emissions. By varying certain parameters it is possible to estimate the impact occupants will have in the energy use of a building as a whole and which system will be most affected by their action.

Keywords: Energy Use; Building Simulation; Environment; Green Building; Greenhouse-gas Emissions; Office Building.

1. Introduction

Green buildings (GB) incorporate a constant improvement, in which "today's best practices become tomorrow's standard practices" (USGBC, n.d.). The concept of GB includes planning design, construction, operations and end of life of buildings with the aim of creating environmentally, socially and economically communities. This requires critical thinking and a life cycle approach is essential, where several systems such as materials, resources, energy, people and information are taken into account.

In Australia, buildings represent 58% and 34% of all the electrical and natural gas final uses, respectively (IEA, 2014). Therefore, buildings are relevant energy intensive users that need careful analysis and attention throughout its whole life cycle. Despite the fact that a 1.8% decrease of the energy intensity in buildings was achieved in 2015, due to energy efficiency measures and policies implemented by national governments, in order to drive the world towards decarbonized energy systems, countries need to reach a 2.6% reduction per annum in the intensity of energy (IEA, 2016).

The way occupants use energy is one of the main contributions to high rates of energy intensity in buildings. Past studies showed that the difference between predicted energy and real energy use is mainly due to the way occupants behave in terms of energy use (Norford *et al.*, 1994, Branco *et al.*, 2004). This behaviour has several direct and indirect factors that may have influence in the way occupants consume energy. These factors are designated as "driving forces", according to the International Energy Agency (IEA), and represent parameters that have influence on the way occupants interact with buildings and its control systems. Figure 1 is a representation of the driving forces that have influence in the way occupant behave when using energy in cooling, heating, ventilation (including window operation), lighting, domestic hot water (DHW), appliances and cooking (Polinder *et al.*, 2013). Moreover, "driving forces" are divided into Internal Driving Forces (IDF) that are intrinsic to the person and may be related with the perception of comfort, social activities, psychological and biological characteristics of the individuals (Al-Mumin *et al.*, 2003, Steemers and Yun, 2009), and External Driving Forces (EDF) which are physical and environmental aspects such as the climate, buildings location and orientation, timeframe, etc. (Mahdavi *et al.*, 2008, Bluyssen *et al.*, 1996, Polinder *et al.*, 2013).

This chapter focuses on the operational stage of an existing building that served as a baseline for a case study, and aims, by means of a building simulation, to mimic the behaviour of occupants when interacting with the lighting, cooling and heating systems. Therefore, the present research is directly focused on External Driving Forces affecting occupant behaviour, and in an indirect way touches the Internal Driving Forces more specifically in terms of the perception of comfort.

2. Description of the Building

For the purpose of this case study, the building chosen for research was the Western Sydney University (WSU) Building Y, in Kingswood campus. The reason for choosing this building was mainly related with the easy access to information. The type of construction should be traditional from where it is located and data related to its thermal and physical characteristics should be available and easily accessed.

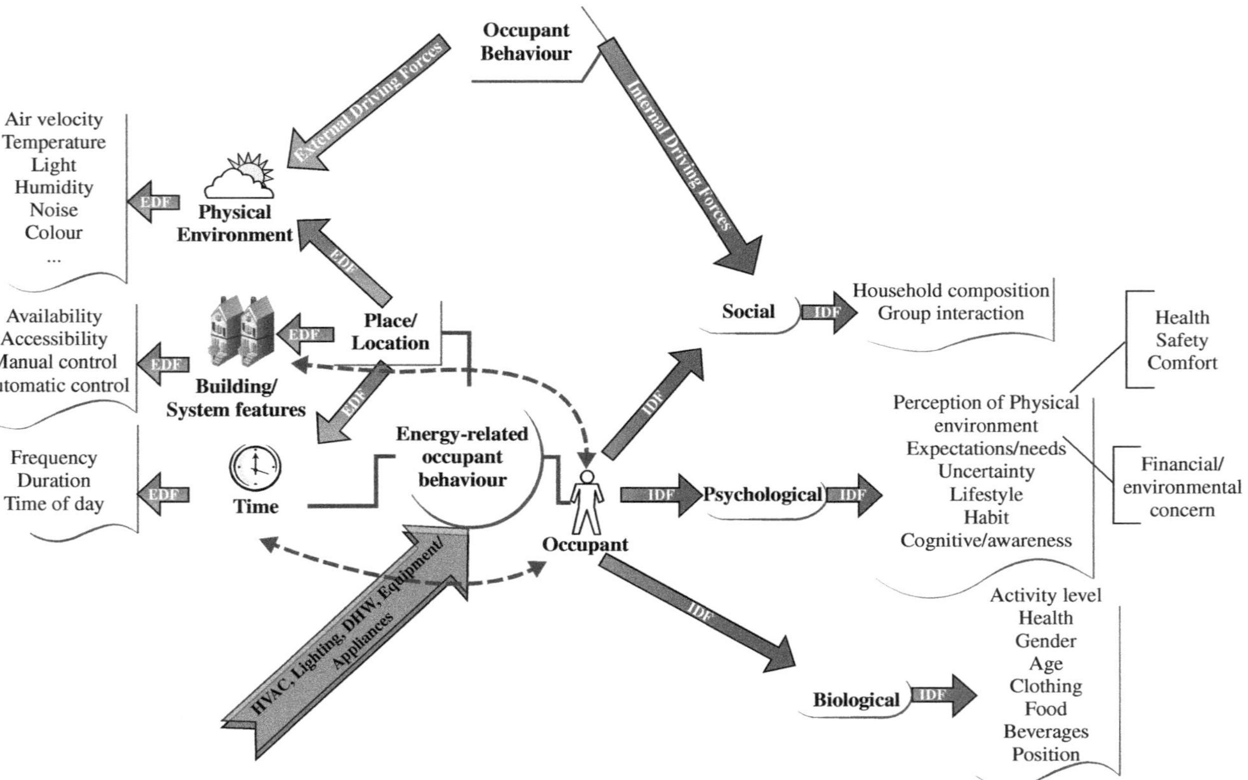

Fig. 1.   Driving forces influencing occupant behavior when interacting with energy, according to IEA.

WSU Building Y serves the School of Computing, Engineering & Mathematics complex and, being a university building, is composed by several different occupant activity types, such as offices, laboratories, computer rooms, technical areas, data centres, corridors and stairs. The present study is mainly focused on the "offices" activity, and its designation according to the Building Code of Australia (GBCA), in terms of building usage, is that of a Class 5 (ABCB, 2016) activity type that corresponds to an "office building used for professional or commercial purposes". Figure 2 represents the exterior view of the building.

The building with 5156 m^2 has four levels above the ground and the last one is basically the technical area for the air conditioning system. It is orientated from west to east and has facades in all four main orientations. There are some neighbourhood buildings that shade WSU Building Y and architectural features that provide shading to the internal rooms.

In terms of occupancy, equipment/appliances and lighting system, an energy audit was performed at the building in order to collect the real existing data for each one of the main activities addressed previously.

As an existing building, but in 1989, some data were not available at the time of this present case study and, therefore, some assumptions were made having as reliable support the American Society of Heating, Refrigerating and Air-Conditioning Engineers (ASHRAE) standards (ASHRAE, 2007, 2009), European Directives (MAOTESESS, 2013) and Australian standards and regulations (ABCB, 2006, 2016; GBCA, 2015).

As a result, for the present case study, cooling season is assumed as the heating is provided by an electrical system with an energy efficiency

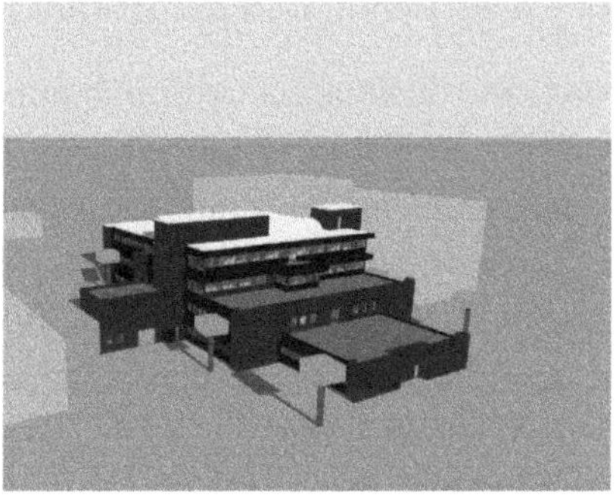

Fig. 2.   WSU Building Y, Kingswood Campus.

rating (EER) of 4.25 and a Coefficient of Performance (COP) of 3.85. The set point temperature for the cooling season is 23°C and for the heating season is 22°C. These data are real and were provided by the Capital Works Department of the WSU.

3. Building Simulation

3.1. *Simulation software*

DesignBuilder is used as a 3D Model interface to simulate in EnergyPlus the dynamic load of the building and evaluate the interaction that occupants have in its energy performance.

EnergyPlus is a software that enables the calculation of energy use in a model of a building and associated energy systems when exposed to diverse operating and environmental conditions. In order to do so, EnergyPlus is based on fundamental heat balance principles simulated in transient state (DoE, 2016) in a nodal system. Heat balance includes all the main heat transfer phenomena, such as radiation, convection and conduction, performed in a 3D way.

3.2. *3D model*

The 3D Model of WSU Building Y is based on the real architectural plans provided by the Capital Works Department of WSU, as well as in real data collected and measured during the energy audit to the building.

(a) Construction and infiltration rates

Due to the fact that the present building is originally from 1989, some data were not available at the time of the data collection, more specifically related with the building construction characteristics. Therefore, some assumptions were made having as support data available in the National Construction Code of Australia (NCC) (ABCB, 2016). Table 1 represents the heat transfer coefficients assumed for each different type of envelope in the WSU Building Y.

Windows, in general, are assumed to have medium coloured drapes as internal shading. It is assumed that the minimum fresh air admission per

Table 1. Heat transfer coefficients of WSU Building Y envelope.

Envelope	Position	U (W/(m²·°C))
Pitched roof	Exterior	0.160
Flat roof	Exterior	0.248
Wall Type1	Exterior	0.350
Wall Type2	Exterior	0.516
Wall Type3	Interior	1.770
Floor	Interior	1.230
Floor	Ground	2.009
Glazing	Exterior	3.835

occupant is 10 L/s with an infiltration rate of 0.7 air changes per hour, due to the fact that windows are not either well insulated or permanently closed.

(b) Thermal zones

The 3D Model is divided into 52 different thermal zones according to the occupants' activity, internal loads and orientation (DoE, 2016; ABCB, 2016; GBCA, 2015). The main activities present in WSU Building Y are laboratories, offices, circulations, computer labs, server room, workshop areas, technical and storage areas. The thermal zones for the ground and third floors are represented in Fig. 3, as an example.

(c) Internal loads

Occupancy, light intensity and equipment are defined according to real internal loads of WSU Building Y collected for all rooms during the energy audit. The real intensities are presented in Table 2.

In terms of the Air Conditioning System, as referred previously, and due to the fact that since this building is from 1989, some data were missing or inaccessible, it was assumed at this stage that the cooling and heating systems are provided by a centralized electrical heat pump system with an EER of 4.25 and a COP of 3.85 (MAOTESESS, 2013).

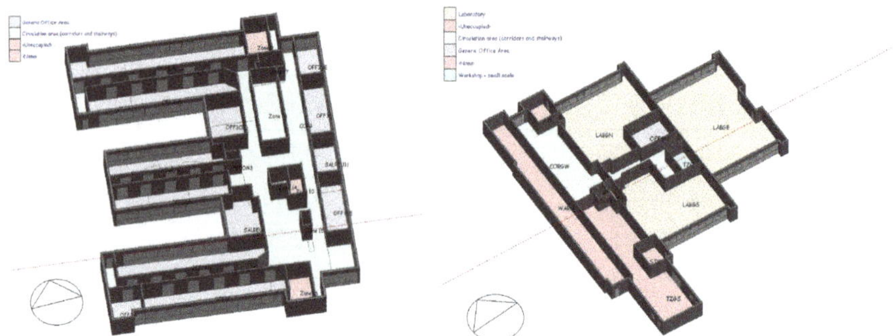

Fig. 3.   Thermal zones for the ground and third floors.

Table 2.   Internal loads from WSU Building Y.

Floor	Area (m^2)	Light Intensity (W/m^2)	Occupancy (Occ)	Equipment (W/m^2)
Ground	1449.4	8.57	81	35.02
1st	1617.2	9.74	168	15.23
2nd	1107.8	11.47	174	42.69
3rd	894.6	11.21	82	27.76
4th	91.4	1.2	—	—
Exterior Light			4375 W	

For the present case study, technical data pertaining to the hot water for consumption purposes in spite of being residual are assumed real and that all the hot water is being produced by means of an electrical resistance.

(d) Schedules

The occupancy schedules used in the present case study were the ones provided by the NCC (ABCB, 2016).

3.3. *Simulations*

In order to mimic energy-related occupant behaviour and draw conclusions how it will impact the overall performance of WSU Building Y, five dynamic simulations are performed. The first one serves as a baseline for the others and has as main assumptions the real data collected during the energy audit, which includes all the aspects referred previously and schedules according to the NCC in order to align with the same assumptions that are done during the design stage of a building.

Meanwhile, the sequential four simulations are designed by changing several parameters related to occupants and/or their interactions with the building systems, such as the air conditioning, equipment/appliances and lighting. Table 3 illustrates the different scenarios created in each simulation in order to evaluate the energy-related occupant impacts in the performance of WSU Building Y as a whole. The assumptions made in all scenarios are based on average data from previous existing research.

Table 3.   Studied scenarios.

Simulations	Scenario	Aim
Baseline	— Schedules according to the NCC — Real data collected in the energy audit	This simulation serves as baseline
LU	— Reduction of 50% in light use	Simulates occupants more aware in terms of lighting systems
L&E	— Reduction of 50% either in light use as well as in equipment	Mimics occupants more aware in terms of lighting systems and equipment, by reducing their use when not needed
LE&AC	— Reduction of 50% either in light use as well as in equipment — Variation of a 1.5°C in the air conditioning set-point, for both seasons	Intends to show occupants awareness when interacting with the lighting and air conditioning systems, as well as with equipment/appliances, reducing their use when not in need
LEAC&AI	— Reduction of light during daytime — Equipment and lighting turned off whenever there is no occupancy — Variation of a 1.5°C in the Air Conditioning set-point, for both seasons — Air infiltrations decreased to 0.5 ac/h	Intends to show occupants awareness when interacting with the lighting and air conditioning systems, as well as with equipment/appliances, reducing their use when not in need; and more awareness in windows and/or doors opening

The aim is to simulate occupant behaviour by making them more aware of their active implications in the energy consumption of a building.

4. Results

The outputs from the baseline simulation show that the most high energy intensity use is for cooling, followed by interior lighting and equipment. Figure 4 is a representation of the energy distribution per end use in the baseline simulation.

4.1. *Simulation outputs*

Figures 5–7 represent the outputs from the simulations. Due to the amount of results produced, only the baseline simulation and the LEAC&AI that correspond to the worst case scenario and the best case scenario, respectively, will be presented.

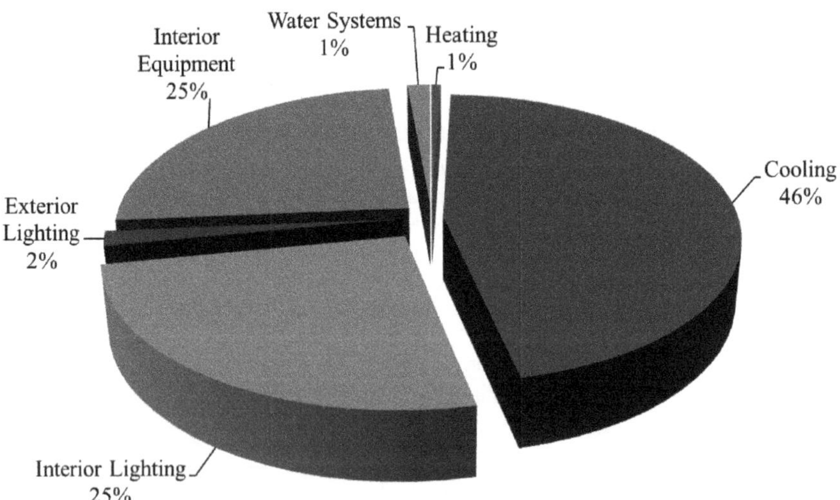

Fig. 4.   Energy distribution per end use from the baseline simulation.

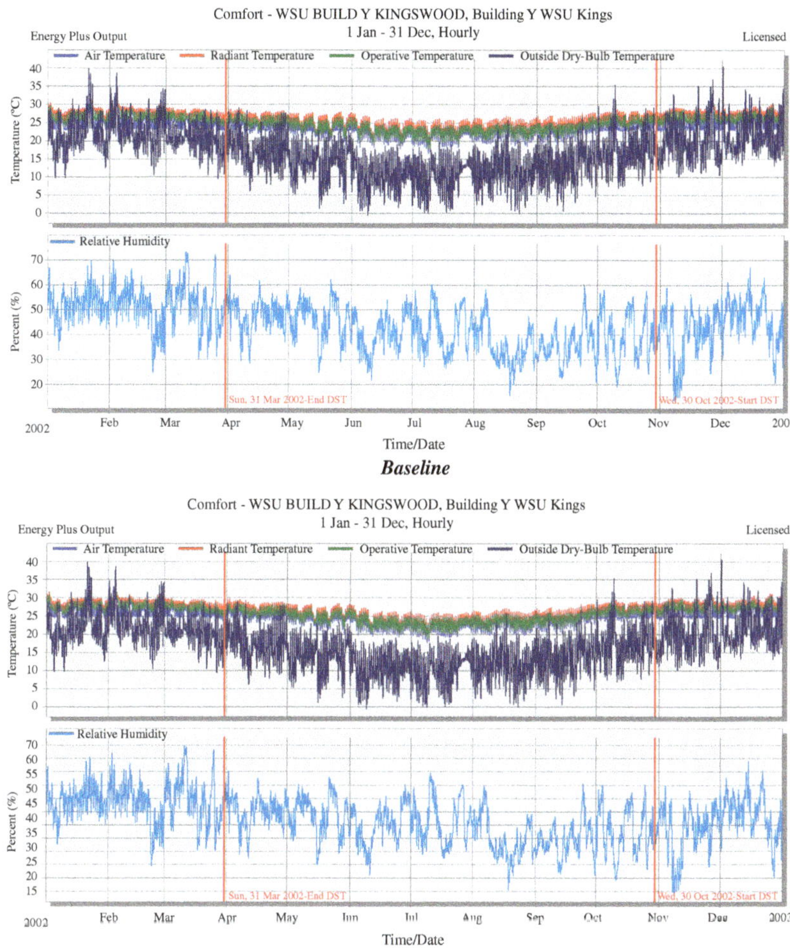

Baseline

LEAC&AI

Fig. 5. Hourly temperature and relative humidity outputs for the baseline and the LEAC&AI simulations.

4.2. *GHG emissions calculation and associated costs*

The GHG emissions, due to electricity consumption, are determined using the following equation and the conversion factors provided by the Green Building Council of Australia (GBCA, 2015):

$$\text{GHG}_{\text{emissions}} = f_{\text{Ele}} \times E_{\text{Ele}}, \tag{1}$$

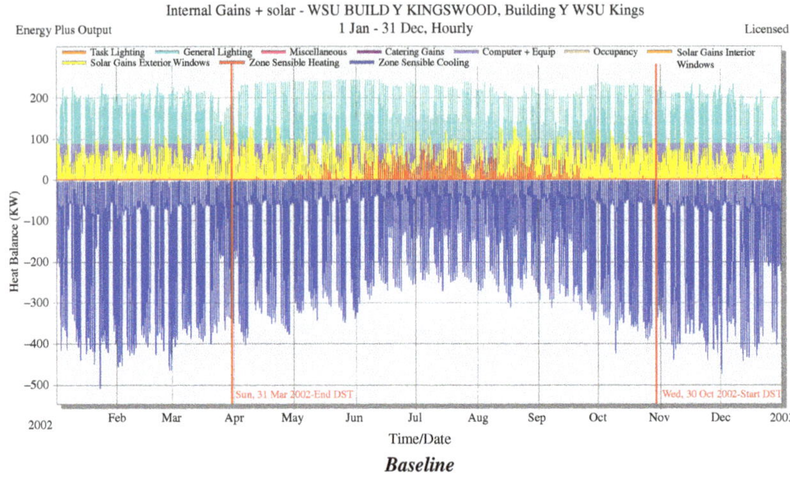

Baseline

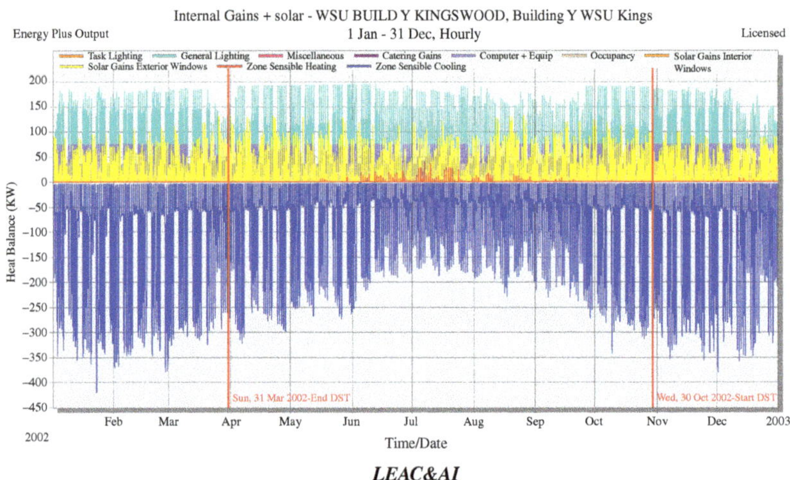

LEAC&AI

Fig. 6.   Hourly internal gains for the baseline and the LEAC&AI simulations.

where f_{Ele} is a conversion factor for GHG emissions related to the production of electricity ($kgCO_{2-eq}/kWh$), $GHG_{emissions}$ are the emissions due to electricity ($kgCO_{2-e}$), E_{Ele} is the electricity use (kWh).

The costs associated with the electricity consumption are determined with Eq. (2) and the current electricity costs (AUSGRID, 2016–2017):

$$\text{Cost}_{electricity} = \frac{C_{Ele} \times E_{Ele}}{100}, \tag{2}$$

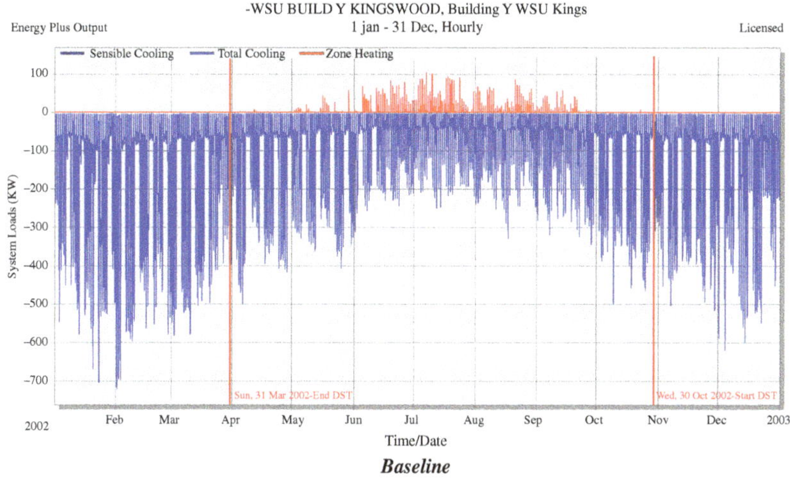

Baseline

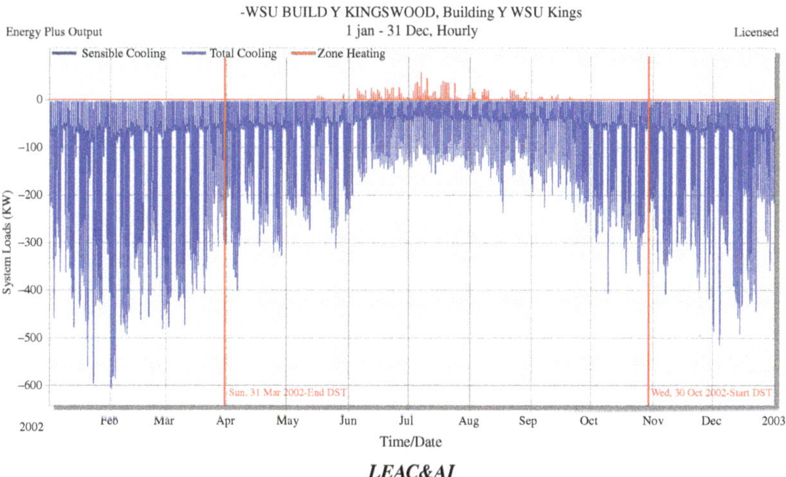

LEAC&AI

Fig. 7. Hourly air conditioning load gains for the baseline and the LEAC&AI simulations.

where C_{Ele} is the cost of elect electricity (cent AUS Dollar/kWh), $Costs_{electricity}$ are the costs due to electricity (AUS Dollar), E_{Ele} is the electricity use (kWh).

4.3. *Final results*

Figure 8 represents the percentage of reduction in energy per end use, after the implementation of the assumptions from Table 3, and comparing

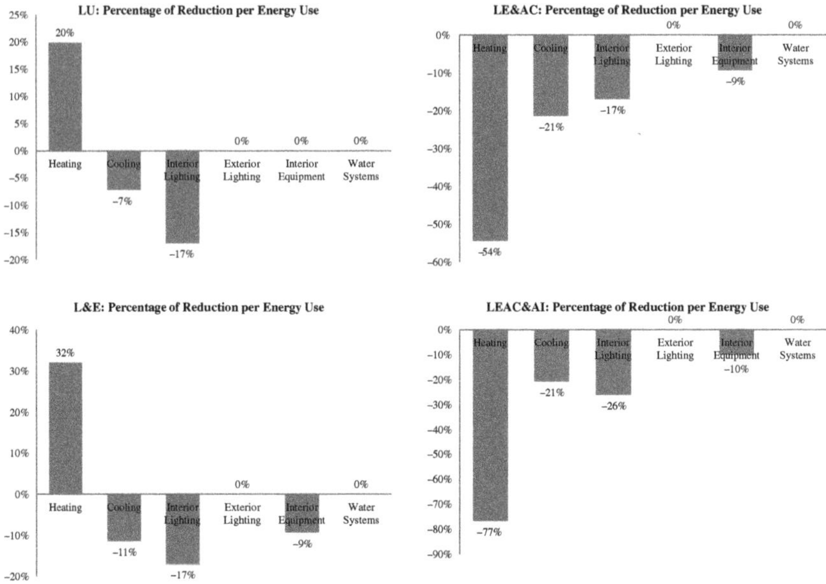

Fig. 8.   Percentage of reduction in energy, per end use, for each individual simulation.

each individual simulation to the baseline. It is important to underline that the heating function in WSU Building Y is residual and therefore the graphics from Fig. 8 are merely indicators.

When comparing the worst case scenario with the best case scenario, it is possible to achieve nearly 20% reduction in the energy consumed. This reduction is due to an increase in occupants' awareness of their actions when interacting with lighting and air conditioning systems as well as with equipment and windows/doors opening. Therefore, the total annual energy consumption for the baseline simulation and for the best case scenario is 2,328,422 kWh and 1,882,249 kWh, respectively, representing a total reduction of 446,173 kWh and 473 ton $CO_{2\text{-eq}}$ per annum. Finally, the energy intensity reduction per unit area is 87 kWh/(m^2 p.a), which implies a reduction of 92 kgCO$_{2\text{-eq}}$/(m^2 p.a) and annual savings around AU$25,120.

The representation and compilation of all the results from the five different simulations performed to WSU Building Y are addressed in Fig. 9 and Table 4.

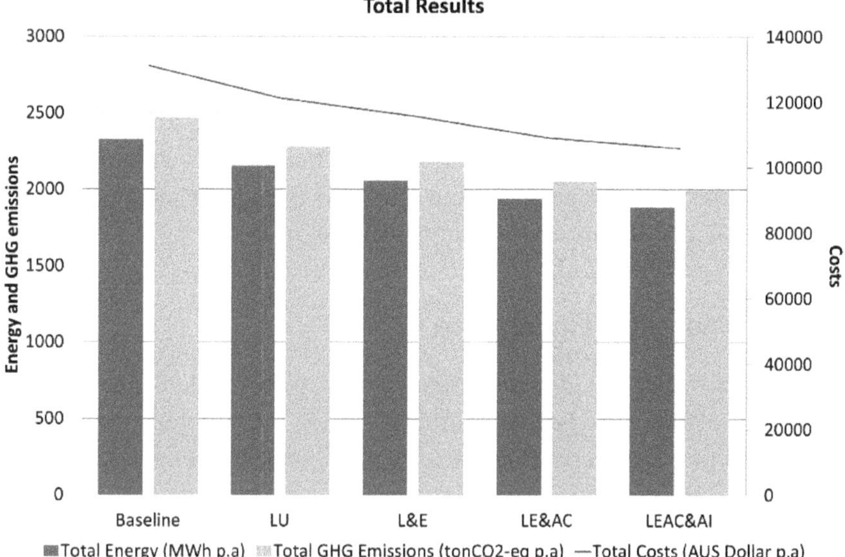

Fig. 9.   Total annual results of energy, GHG emissions and costs for all simulations.

Table 4.   Results.

Simulation	Energy (kWh/p.a)	GHG Emissions (ton $CO_{2\text{-eq}}$/p.a)	Costs (AU\$/p.a)	Total Reduction
Baseline	2,328,422	2,468	131,090	—
LU	2,153,390	2,283	121,236	−34 kWh/m² p.a −36 kgCO$_{2\text{-eq}}$/m² p.a −9854 AUS Dollar/p.a
L&E	2,056,308	2,180	115,770	−53 kWh/m² p.a −56 kgCO$_{2\text{-eq}}$/m² p.a −15320 AUS Dollar/p.a
LE&AC	1,938,199	2,054	109,121	−76 kWh/m² p.a −80 kgCO$_{2\text{-eq}}$/m² p.a −21970 AUS Dollar/p.a
LEAC&AI	1,882,249	1,995	105,971	−87 kWh/m² p.a −92 kgCO$_{2\text{-eq}}$/m² p.a −25120 AUS Dollar/p.a

Table 4 is expressed in terms of energy consumption, GHG emissions and costs.

As a final note, it is important to note that the current case study of the WSU Building Y is only focused in accounting for the occupant variation related with a Class 5 "offices" activity. Future work will take into account all the activities performed in the WSU Building Y, and therefore a higher reduction in the annual energy use is expected.

5. Discussion and Conclusion

The outputs from the dynamic simulations performed at the WSU Building Y model are conclusive in determining the difference between the best and the worst case scenarios. After analyzing the results, it is possible to conclude that if occupants become more aware of the impact they have when interacting with windows, equipment, light and air conditioning systems, significant improvements can be expected in the energy performance of buildings due to a reduction and more efficient use of energy.

In the present case study, it was possible to achieve around 20% reduction in the annual energy use for the studied building when comparing the best case scenario with the worst one. This annual reduction will imply the use of less than 446,173 kWh (with an energy intensity reduction of 87 kWh/m^2 p.a) in electricity, less 473 ton CO_{2-eq} and annual savings around AU$ 25,120.

This study takes into account only the "offices" activity, which corresponds approximately to one-third of all the activities performed in the WSU Building Y. Therefore, it is expected that the percentage of reduction in annual energy use will increase when all the activities performed in WSU Building Y are accounted within the next future work.

It is also interesting to highlight that all variables/parameters in a building are interconnected, as it can be confirmed when comparing the LU simulation with the baseline, where a reduction of approximately 17% in lighting will lead to an increase of 20% in the heating function and a reduction of 7% in the cooling function. Therefore, as conclusion, just by changing one parameter this could impact other factors that at a first glance were not accounted for. Occupants are one of the most impacting variables in a building due to the fact that they are permanently interacting with all its systems and features (Norford *et al.*, 1994). Finally, there is a need for more detailed and reliable studies based on real models that take into account the interaction of all the variables in buildings.

As our final conclusion, the results from this study are a good indication of the implications that occupant behaviour has in terms of energy consumption. Moreover, this case study validates and supports the thesis presented in the literature where the predicted energy use, during the design stage, is always different from the operational stage, due to the fact that occupant behaviour is being ignored, leading to an overestimation of the energy use during the design stage (REA, 1984). As it may be confirmed, the worst case scenario, in the present study, is the baseline and this scenario is performed with standard figures that are used by professionals to predict the operational energy use in buildings during the design stage.

Acknowledgement

The authors wish to acknowledge the financial support from the Australian Research Council (ARC) Discovery Project under grant number DP150101015.

References

ABCB (2006). "Protocol for Building Energy Analysis Software, for Class 3, 5, 6, 7, 8 and 9 Buildings". Australian Building Codes Board.

ABCB (2016). *National Construction Code, Building Code of Australia, Class 2 to Class 9 Buildings*. Volume One.

Al-Mumin, A., Khattab, O. and Sridhar, G. (2003). "Occupants' Behavior and Activity Patterns Influencing the Energy Consumption in the Kuwaiti Residences", *Energy and Buildings*, 35, 549–559.

ASHRAE (2007). "ASHRAE 90-1: Energy Standard for Buildings Except Low-Rise Residential Buildings", *American Society of Heating, Refrigerating and Air-Conditioning Engineers,* Inc, I-P Edition.

ASHRAE (2009). "Non-residential Cooling and Heating Load Calculations", *ASHRAE Handbook — Fundamentals,* Chapter 9, 18.

AUSGRID (2016–2017). Network Price List 2016–2017. *AUSGRID.* Available at: https://www.ausgrid.com.au/~/media/Files/Industry/Regulation/Network%20 prices/AUSGRID%20NETWORK%20PRICE%20LIST%20FY201617.pdf (Accessed on May 4, 2017).

Bluyssen, M., de Oliveira Fernandes, E., Groes, L., Clausen, G., Fanger, P. O., Valbyorn, O., Bernhard, C. A. and Roulet, C. A. (1996). "European Indoor Air Quality Audit Project in 56 Office Buildings", *Indoor Air-International Journal of Indoor Air Quality and ClimateIndoor Air-International Journal of Indoor Air Quality and Climate*, 6, 221–238.

Branco, G., Lachal, B., Gallinelli, P. and Weber, W. (2004). "Predicted Versus Observed Heat Consumption of a Low Energy Multifamily Complex in Switzerland Based on Long-Term Experimental Data", *Energy and Buildings*, 36, 543–555.

DoE (2016). "Engineering Reference". *EnergyPlus™ Documentation*, Version 8.7.

GBCA (2015). "Energy Consumption and Greenhouse Gas Emissions Calculator Guide". *Green Star — Design & As Built and in Green Star — Interiors*, 4.

IEA (2014). "Australia: Balances for 2014", *International Energy Agency*. Available at: https://www.iea.org/statistics/statisticssearch/report/?country= AUSTRALI=&product=balances (Accessed on December 6, 2016).

IEA, I. E. A. (2016). "Energy Efficiency Market Report 2016", *IEA, International Energy Agency*, 17.

Mahdavi, A., Mohammadi, A., Kabir, E. and Lambeva, L. (2008). "Occupants' Operation of Lighting and Shading Systems in Office Buildings", *Journal of Building Performance Simulation*, 1, 57–65.

MAOTESESS (2013). "Portaria no. 349-D/2013. *Diário da República, 1.ª série, N.º 233*.

Norford, L. K., Socolow, R. H., Hsieh, E. S. and Spadaro, G. V. (1994). "Two-to-one Discrepancy Between Measured and Predicted Performance of a 'Low-Energy' Office Building: Insights from Reconciliation Based on the DOE-2model", *Energy and Buildings*, 21, 121–131.

Polinder, H., Schweiker, M., der Aa, A. V., Schakib-Ekbatan, K., Fabi, V., Andersen, R., Morishita, N., Wang, C., Corgnati, S., Heiselberg, P., Yan, D., Olesen, B., Bednar, T. and Wagner, A. (2013). "Occupant behavior and modeling", *International Energy Agency, Energy in Buildings and Communities*, 2.

Rea, M. S. (1984). "Window Blind Occlusion: A Pilot Study", *Building and Environment*, 19, 133–137.

Steemers, K. and Yun, G. Y. (2009). "Household Energy Consumption: A Study of the Role of Occupants", *Building Research & Information*, 37, 625–637.

USGBC (n.d.). "Green Building and LEED Core Concepts", *Green Building and LEED Core Concepts Guide,* Second Edition.

Chapter 19

An Economical Approach to Geo-Referencing 3D Model for Integration of BIM and GIS

Junxiang Zhu*, Jun Wang and Xiangyu Wang

Australasian Joint Research Centre for Building Information Modelling, Curtin University, Perth, Australia
**junxiang.zhu@postgrad.curtin.edu.au*

Yi Tan

Department of Civil and Environmental Engineering, The Hong Kong University of Science and Technology, Hong Kong

Abstract

The integration of building information modelling (BIM) and geographic information system (GIS) has been studied for a long period of time by researchers from both domains. The foundation of the integration is the interoperability of the data, which means the BIM data has to have a right reference system, allowing it to be read correctly by GIS software applications. 3D models from the BIM world may not be correctly geo-referenced, which impairs the data interoperability. Instead of rebuilding

---

* Corresponding author.

models from scratch, which is time- and labour-consuming, this paper proposes a more efficient, economical, two-step alternative mainly based on Affine transformation. In the first step, the modification is made against the x- and y-coordinates. A number of control points would be selected to form displacement links. Based on these, the transformation parameters would be calculated, and the 2D footprint of the model would be rectified by Affine transformation. Then, the z-value (height information) of each vertex in the model would be adjusted using the scaling factor f. This method could obtain a 3D model without a geographical coordinate system correctly geo-referenced, and thus, it could be further read by GIS applications, consuming the vast spatial analysis tools supported by the GIS world, achieving more than just visualization. The key to the success of this study is creating an accurate right footprint of the model and selecting appropriate control points to guarantee the accuracy of transformation. By far, this approach has only been tested with a bridge model, its performance on other building models needs to be further studied.

Keywords: GIS; BIM; 3D Geo-referencing.

1. Introduction

Research on the integration of building information modelling (BIM) and geographic information system (GIS) has emerged in these years because of its significant benefits to both worlds (Ebrahim *et al.*, 2016; Li *et al.*, 2017). The GIS world could use the detailed 3D models created by BIM to overcome its drawback in the building of 3D models, while the BIM world might use the vast spatial analysis tools provided by GIS to expand its capacity (Chong *et al.*, 2017; Shou *et al.*, 2015).

Approaches to combining BIM and GIS could be categorized into three levels: application level, process level and data level (Amirebrahimi *et al.*, 2015). Among them, the data level is the most fundamental, as the other two levels are more or less dependent on it. In the data level, the data format is transformed from one to another. For example, the most common way is to transform from Industrial Foundation Class (IFC) to City Geography Markup Language (CityGML) (Deng *et al.*, 2016). IFC is an open file format for facilitating interoperability in the architecture, engineering and construction (AEC) industry, while CityGML is an open standardized data model and exchange format to store 3D models of cities and landscapes based on Geography Markup Language (GML) defined by Open Geospatial Consortium (OGC) in Extensible Markup Language (XML) format (Kang and Hong, 2017). Both of them are designed to store and exchange data in their own domain. Another alternative is to translate IFC file into shapefile, a data format developed by Environmental Systems Research Institute, Inc. (ESRI) and widely accepted open format in the geospatial domain for data storage and analysis (Amirebrahimi *et al.*, 2016). The difference between the two approaches is that CityGML is more often used for data exchange, while the shapefile is for spatial analysis.

The barrier that cannot be neglected during the data transformation is the incompatibility of spatial reference systems adopted by the two domains. GIS uses global reference system such as geographic coordinate system in the form of latitude, longitude and altitude, whereas BIM uses a local coordinate system in the form of x, y and z in a 3D Cartesian Coordinate System (Karan and Irizarry, 2015; Wang *et al.*, 2015). Without a geographic coordinate system, the BIM model would introduce errors when imported into a GIS platform, such as incorrect size or location. All models need to be geo-referenced before they can be properly processed by GIS applications. To achieve the integration of BIM and GIS, the BIM world introduced geographic coordinate system, and the GIS world starts to support local coordinate system defined in CityGML (Gröger *et al.*, 2014).

Nevertheless, obstacles still exist. The geographic coordinate system may still not be defined at the beginning of a BIM project due to designer's preference to local coordinate system, or some old BIM models may also not include a proper geographic coordinate system. In this situation, if those models are to be used in GIS, they have to be either rebuilt or geo-referenced. Compared with the first option, the second is much more economical.

The GIS world has a mature experience in geo-referencing 2D data, such as remote sensing imagery, contour map, and other data with 2D geometry (Turner *et al.*, 2014). However, as a beginner in 3D domain, it does not have a well-accepted approach to geo-reference 3D data from BIM, considering its complexity.

This paper proposes a two-step approach to geo-reference 3D models from the BIM world without a geographical coordinate system. No new complex mathematical algorithm is going to be developed, instead, the traditional 2D geo-referencing algorithm will be utilized and extended. To demonstrate the feasibility of the proposed method, a bridge model is selected, processed by the approach, and the final geo-referenced model is displayed.

2. Methodology

2.1. *Data*

A 3D model of a bridge with a width of 8.5 m and a length of 223 m, from a project in design stage in Guangdong Province, China, is used, as shown in Fig. 1. This model is built with Tekla in a local coordinate system, and exported in IFC format, making it able to be exchanged to other applications. A footprint of the model was also created in ArcGIS with a WGS 1984 World Mercator coordinate system, indicating the real location of the bridge, as presented in Fig. 2.

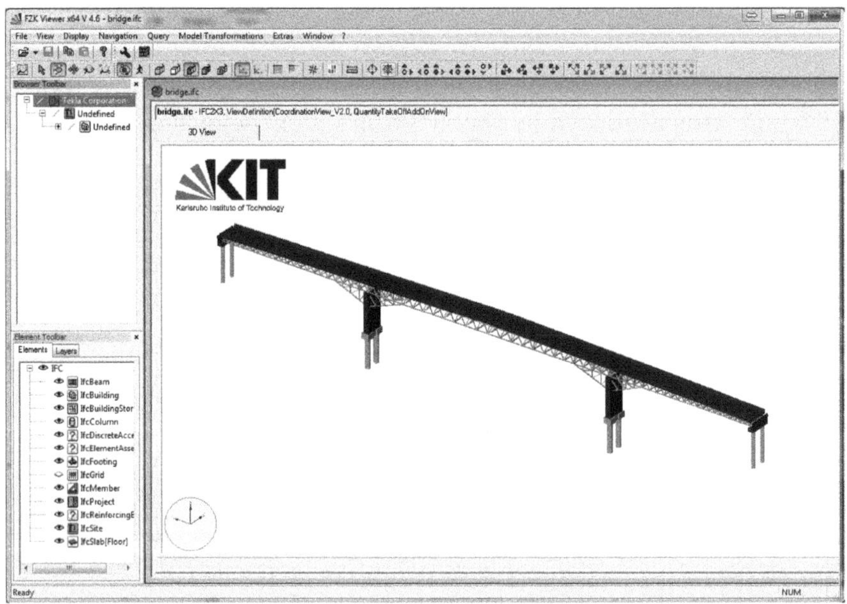

Fig. 1.   Bridge model in IFC format.

Fig. 2.   The planned location of the bridge in real world.

2.2. *Geo-referencing*

The purpose of geo-referencing is to assign geographic coordinates to an image or misplaced vector data to display them correctly.

The available algorithms for geo-referencing include Affine transformation, similarity transformation and projective transformation (INC., 2017a). In this study, Affine is adopted because only a simple linear translation as well as a shift is required, and it is the mostly used method in the 2D coordinate system transformation.

An affine transformation can differentially scale the data, skew it, rotate it and translate it. It could be represented by

$$\begin{pmatrix} x \\ y \end{pmatrix} = \begin{pmatrix} A & D \\ D & E \end{pmatrix} \begin{pmatrix} \hat{x} \\ \hat{y} \end{pmatrix} + \begin{pmatrix} C \\ E \end{pmatrix}, \tag{1}$$

where x and y are the adjusted coordinates (or target coordinates), and $\hat{x}$ and $\hat{y}$ indicate the original coordinates (or source coordinates) (Song *et al.*, 2014). A, B, C, D, E, and F are transformation parameters, which could be calculated from control points.

2.3. *Workflow*

The workflow in Fig. 3 shows the steps from the beginning until it is finished. It spans two areas, BIM and GIS, and connects many platforms (Tekla, ArcMap, ArcScene and ArcGIS Pro) to geo-reference the data.

The data are first transformed from IFC to geodatabase by ArcGIS Data Interoperability tool, supported by Feature Manipulation Engine (FME) (INC., 2017b). After that, a two-step process is conducted to geo-reference the model. The first step, completed by ArcMap, is to rectify its x- and y-coordinates (2D footprint). The second is to adjust the z-value of each point in the model, and this part is finished by ArcScene.

2.3.1. *Rectification of x- and y-coordinates*

In this section, the x- and y-coordinates would be rectified. A "right" footprint (RF) has to be created first using the right size and location information of the bridge. Meanwhile, the original footprint of the model is referred to as "wrong" footprint (WF).

Then, control points are selected from both RF and WF, and the mathematic relationship between them is established. A pair of control points, one from RF and another from WF, constitute a displacement link. In a 2D case, at least three displacement links are required. With those control points, the transformation parameters in Eq. (1) could be derived and the transformation could be completed.

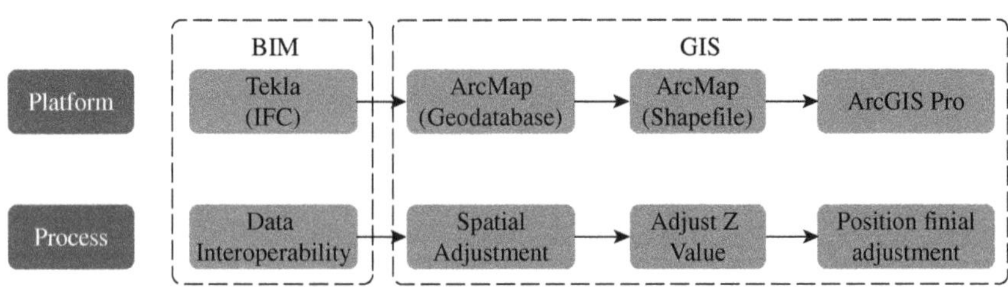

Fig. 3.   Workflow.

2.3.2. *Height information adjustment*

After x- and y-coordinates are rectified, the model is distorted, as it is not scaled proportionally. The height information or z-value must also be adjusted to maintain its structure correctness. The z-value could be adjusted by

$$Z_a = f \times Z_b, \tag{2}$$

where Z_a is the adjusted value, Z_b is the original value, and f stands for a scaling factor, indicating whether the value is increasing ($f > 1$) or decreasing ($f < 1$), and it could be calculated by

$$f = L_r / L_w, \tag{3}$$

where L_r is the length of the RF, whereas L_w stands for the length of the WF, both of them could be acquired by measuring using internal applications provided by ArcMap. Note that the f could also be calculated using the width of the bridge in this case, but length is preferred, as it could reduce the measuring error. The f is the key to guarantee the model's structure correctness.

3. Results

3.1. *WF and RF*

Figure 4 shows the WF of the model as well as the RF, with a background of a satellite image covering part of Guangdong Province. The WF is represented as a red box because of its wrong referencing system, its size presented in the map is not real and considerably larger than the RF, which is too small to be displayed in the map, and an exaggeration has been made, as shown in the green box. It could be learnt that if a 3D model imported into GIS platform is not correctly geo-referenced, it could not be displayed and analyzed appropriately. As the project is still in the design stage, the bridge has not been built yet, nevertheless, its location has been required by consulting the designer.

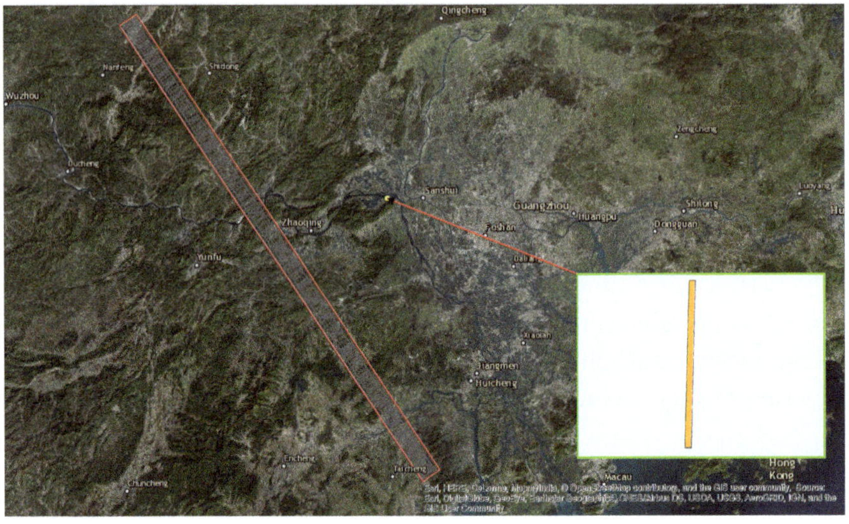

Fig. 4.　The WF (red box) and the RF (green box).

3.2. *Footprint rectification*

In order to perform the transformation, the parameters, namely A, B, C, D, E, and F in Eq. (1), are required. They could be derived by choosing a number of control points. A point in WF and its corresponding point in RF make up a displacement link. All the displacement links are shown in Fig. 5, each arrow stands for a displacement link.

Table 1 presents the control points selected. X Source and Y Source are coordinates from the WF, while the X Destination and Y Destination are from the RF. Each row of the table denotes a displacement link. The residual error is a measure of the fit between the true locations and the transformed locations of the output control points and is generated for each displacement link. It is calculated by

$$\mathrm{RE} = \sqrt{(x_a - x_d)^2 + (y_a - y_d)^2},\qquad (4)$$

where RE is residual error, x_a and y_a are adjusted coordinates of X Source and Y Source calculated using the affine transformation, while x_d and y_d

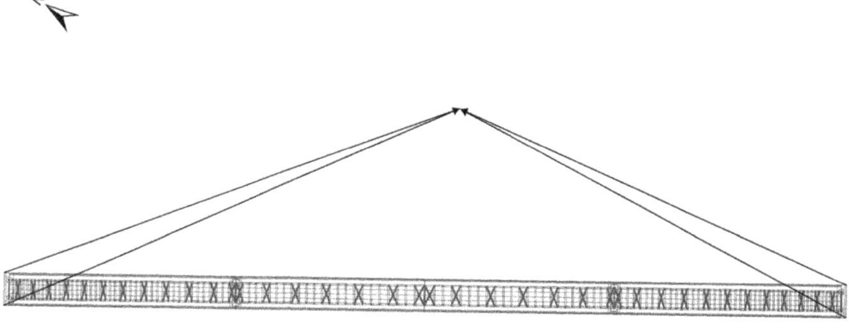

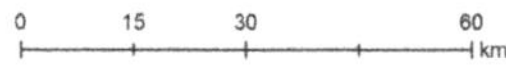

Fig. 5.   Displacement link.

Table 1.   Control points.

ID	X Source	Y Source	X Destination	Y Destination	Residual Error
1	12446488.628050	2709125.359343	12550256.642045	2634645.825519	0.000000
2	12439460.110040	2704345.198607	12550248.151620	2634646.084748	0.000000
3	12571785.074786	2524895.501584	12550250.684949	2634424.219664	0.000000
4	12564756.556776	2520115.340848	12550242.194525	2634424.478929	0.000000

are values of X Destination and Y Destination, respectively. All the residual errors here are zero, indicating a high transformation accuracy. Figure 6 illustrates the rectified footprint of the bridge.

3.3. *Z-value adjustment*

The Z-value is adjusted by the scaling factor f, which is the ratio of the true footprint's size to the false footprint's size. In this study, the width of the true footprint is 8.5 m (map units), while that of the false footprint is 8500 m (map units), calculated by measuring tool in ArcMap, thus the value of scaling factor f is 0.001.

Fig. 6.   Rectified footprint.

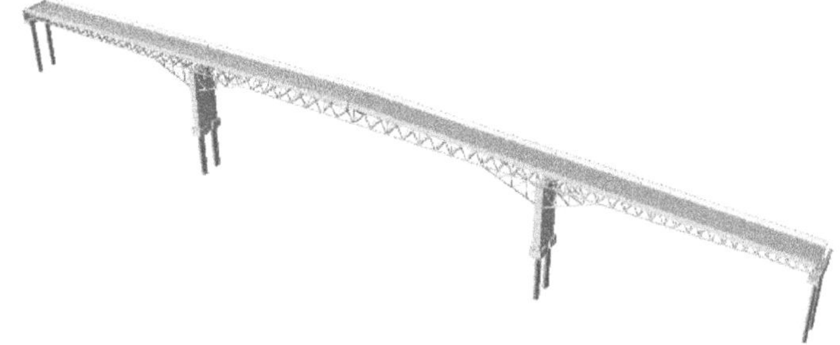

Fig. 7.   Rectified bridge model.

The fully rectified model was shown in Fig. 7, which is just the same as the original one shown in Fig. 1. The advantage of combining BIM and GIS is to utilize the excellent 3D visualization of BIM model and the vast spatial analysis functions provided by GIS to make the most of the model. Figure 8 puts the bridge model in the "real" scene created by GIS. The GIS links the isolated bridge model to its surroundings, which introduces richer environmental information to it.

Fig. 8.   The rectified model in a real scene.

4. Discussion

4.1. *Semantic information loss*

As a well-recognized issue in the integration of BIM and GIS, the semantic information loss was also noted in this study. During the transformation of data format from IFC to shapefile, the file was separated into six thematic files. This process created a file for each type of element, namely slab, member, footing, discrete accessory, column and beam defined in BIM. As a result, the hierarchy structure of those elements was destroyed, causing semantic information loss. However, by checking each file's attribution table, we found they contain a field, ifc_parent, pointing to each component's parent, which means the hierarchy structure was not totally lost and could be restored in some way.

4.2. *Selection of control points*

The accuracy of the geo-referencing process could be affected by many aspects, one of the most significant aspects roots in the 2D footprint rectification, introduced by the selection of control points. The effect that control points have on the adjustment depends on the number and location of those points. In this study, the selection of control points was relatively simple, as the outline of the bridge footprint is a rectangle. Four vertices from the four corners of the footprint were selected, and high transformation accuracy was achieved. However, in other similar applications, if the footprint of an object was not rectangular or a square, the authors would suggest to select those points at the corner, as they are easier to identify than those in the middle of a line.

4.3. *Enrichment of model property*

Apart from the default information, such as global ID, name, description, tag and parent, imported from the IFC file, other customized attributes could also be created and added to the model later on the GIS side to enrich the property of the model for various purposes. For example, the material, manufacturer, manufacture date and installation date information could be incorporated for the maintenance of the bridge. With the assistance of an additional sensor network, a bridge health monitoring system could be established. Once a component is found to be a risk to the bridge, replacement and manufacturer information could be quickly obtained, and even the order process could be automated, as all the needed information are within the model, thus improving the efficiency of bridge maintenance.

5. Conclusion

This paper presents an easy-to-conduct but efficient approach to geo-referencing 3D models without a geographical coordinate system. A 3D bridge model was used to demonstrate the performance of the proposed approach, and good result was achieved. This approach is economical, as the non-georeferenced 3D model does not have to be rebuilt from scratch. If the bridge used in this study is to be rebilt, it will take about 20 h as estimated by the builder himself, while using this method, it will take only about 2 h. The correctly geo-referenced bridge model could utilize the vast functions provided by spatial analysis and can provide more than just 3D visualization.

However, since manual intervention is required, this method only suits situations where merely a small number of models are to be geo-referenced. In a scenario where there are a large number of models, authors would suggest developing an automated way. By far, this method has only been applied to a bridge model. Theoretically, any model with a wrong referencing system could adopt this approach to rectify the model, however, its real performance on those models needs to be further validated.

Acknowledgement

This research was undertaken with the benefit of a grant from the Australian Research Council Linkage Program (Grant No. LP130100451).

References

Amirebrahimi, S., Rajabifard, A., Mendis, P. and Ngo, T. (2015). "A Framework for a Microscale Flood Damage Assessment and Visualization for a Building using BIM–GIS Integration", *International Journal of Digital Earth*, 9, 363–386.

Amirebrahimi, S., Rajabifard, A., Sabri, S. and Mendis, P. (2016). "Spatial Information in Support of 3D Flood Damage Assessment of Buildings at Micro Level: A Review", *ISPRS Annals of Photogrammetry, Remote Sensing & Spatial Information Sciences*, 4.

Chong, H.-Y., Lee, C.-Y. and Wang, X. (2017). "A Mixed Review of the Adoption of Building Information Modelling (BIM) for Sustainability", *Journal of Cleaner Production*, 142, 4114–4126.

Deng, Y., Cheng, J. C. and Anumba, C. (2016). "Mapping between BIM and 3D GIS in Different Levels of Detail Using Schema Mediation and Instance Comparison", *Automation in Construction*, 67, 1–21.

Ebrahim, M. A.-B., Mosly, I. and Abed-Elhafez, I. Y. (2016). "Building Construction Information System Using GIS", *Arabian Journal for Science and Engineering*, 41, 3827–3840.

Gröger, G., Kolbe, T., Nagel, C. and Hafele, K. (2014). "OpenGIS City Geography Markup Language (CityGML) Encoding Standard (OGC 12-019)". Version 2.0. 0. OGC 12-019. Open Geospatial Consortium.

INC. E. (2017a). "About Spatial Adjustment Transformations" [Online]. Available at: http://desktop.arcgis.com/en/arcmap/latest/manage-data/editing-existing-features/about-spatial-adjustment-transformations.htm (Accessed on 2017).

INC. S. S. (2017b). "Integrate Esri ArcGIS Shapefile (SHP) Using FME" [Online]. Available at: https://www.safe.com/integrate/arcgis-shp/ (Accessed on 2017).

Kang, T. W. and Hong, C. H. (2017). "IFC-CityGML LOD Mapping Automation Using Multiprocessing-Based Screen-Buffer Scanning Including Mapping Rule", *KSCE Journal of Civil Engineering*, 22, 373–383.

Karan, E. P. and Irizarry, J. (2015). "Extending BIM Interoperability to Preconstruction Operations Using Geospatial Analyses and Semantic Web Services", *Automation in Construction*, 53, 1–12.

Li, X., Wu, P., Shen, G. Q., Wang, X. and Teng Y. (2017). "Mapping the Knowledge Domains of Building Information Modelling: A Bibliometric Approach", *Automation in Construction*, 84, 195–206.

Shou, W., Wang, J., Wang, X. and Chong, H. Y. (2015). "A Comparative Review of Building Information Modelling Implementation in Building and Infrastructure Industries", *Archives of Computational Methods in Engineering*, 22, 291–308.

Song, Z., Zhou, S. and Guan, J. (2014). "A Novel Image Registration Algorithm for Remote Sensing Under Affine Transformation", *IEEE Transactions on Geoscience and Remote Sensing*, 52, 4895–4912.

Turner, D., Lucieer, A. and Wallace, L. (2014). "Direct Georeferencing of Ultrahigh-Resolution UAV Imagery", *IEEE Transactions on Geoscience and Remote Sensing*, 52, 2738–2745.

Wang, J., Zhang, X., Shou, W., Wang, X., Xu, B., Kim, M. J. and Wu, P. (2015). "A BIM-based Approach for Automated Tower Crane Layout Planning", *Automation in Construction*, 59, 168–178.

Chapter 20

Biomimicry Approaches for Innovative Sustainable Solutions in the Construction Industry

Olusegun A. Oguntona* and Clinton O. Aigbavboa[†]

Department of Construction Management and Quantity Surveying, Faculty of Engineering and the Built Environment, University of Johannesburg, Doornfontein Campus, Johannesburg, South Africa

**201572103@student.uj.ac.za*
[†]architectoguntona12@gmail.com

Abstract

Attaining the goal of sustainability in the construction industry is a demanding task due to the complexities and challenges that come with it. A concise and environmental-friendly approach is imperative to mitigate these challenges. Consulting and learning from the forms, processes and strategies of life's genius (nature) offer a rich source of inspirational learning strategies that lead to sustainable outcomes. The purpose of the study was to explore biomimetic approaches to solving

---

*Corresponding author.

sustainability issues in the construction industry. The study is conducted with reference to the existing literature, published and unpublished research studies. An extant review of the literature on biomimicry and its strategies for innovative sustainable solutions were adopted. The findings from the survey revealed problem-based and solution-based biomimicry approach for proffering sustainable solutions to human challenges. This calls for a holistic adoption and multi-disciplinary collaboration in maximizing the sustainable solutions and innovative ideas offered by the adoption and application of biomimicry.

Keywords: Biomimicry; Sustainable Construction; South Africa Construction Industry (SACI); Nature; Sustainability.

1. Introduction

The construction industry plays a pivotal role in South Africa's economy and social development by providing the physical infrastructure which is fundamental to the country's development (Marx, 2014). It is a large-scale provider of employment opportunities (especially for the least-skilled members of society). The industry plays a cogent role in the development and transfer of technology, creates many opportunities for enterprises, and contributes directly to enhancing the quality of life of the users of its products. (Construction Industry Development Board [CIDB], 2016; Windapo and Cattell, 2013). The sole biggest client of the construction industry is the South African government and its affiliated arms, resulting in between 40% and 50% of the total domestic construction expenditure (Dlungwana *et al.*, 2002). This is because it is believed that, to ensure and achieve economic growth of the country, the government needs to give further attention to the development of the construction industry (Ntuli and Allopi, 2013).

However, Aigbavboa and Thwala (2014) indicated that several fundamental and more complex challenges have been identified as confronting and influencing the performance, development and growth of the South African construction industry (SACI). Some of the identified challenges have existed for some time; however, there is little evidence to show that the issues raised in the past are no longer current owing to the paucity of relevant and reliable information on the subject (Windapo and Cattell, 2013).

These difficulties and challenges sit alongside the general situation of socio-economic and political issues which were underpinned by the previous apartheid regime of segregation (Aigbavboa and Thwala, 2014). It is also worth noting that the construction industry in South Africa is not exempted from the environmental challenges facing its counterparts in other parts of the world. High consumption of energy and natural resources resulting in environmental degradation and pollution and coupled with greenhouse gas emissions are but a few of the numerous characteristics of the industry (Windapo, 2014).

A report by the United Nations Environment Programme — Sustainable Buildings & Climate Initiative (UNEP-SBCI) — suggested prioritizing and a national focus on the building sector as the top two high-level recommendations for reducing the environmental impact of the SACI (Milford, 2009). The role of the government and stakeholders is of utmost importance in the process of enhancing the performance of the industry towards delivering on its benefits. Sustainable construction (SC) is now the medium through which the industry can live up to its potential as the major driver of economic growth and environmental and social development in South Africa. The first giant step to redirecting the industry from the conventional way to the sustainable paradigm is to develop indicators through effective government and stakeholders' participation. These indicators are best transmitted via different plans, agendas and frameworks of both government and construction stakeholders towards the adoption and implementation of SC.

In the quest for solutions to the challenges of sustainability, scientists, engineers, architects, designers and innovators around the world are now consulting life's genius (nature) to create new products, processes and policies. As posited by Benyus (2011), nature has managed to survive for 3.8 billion years with organisms as models that manufacture without heat, beat and treat; ecosystems that run on sunlight and create opportunities rather than waste. Their resulting designs are found to be functional, sustainable and aesthetically pleasing as well. Hence, a new field of discipline has evolved, biomimicry, which studies nature's models and then emulates their forms, processes, systems and strategies to solve human problems sustainably (Rao, 2014). This chapter presents how biomimicry approaches can influence the birth of innovations that have sustainable attributes in the SACI. It begins with the discussion of environmental issues attributed to the construction industry, followed by an overview of sustainability and SC, and finally, biomimicry potential in the drive towards sustainability in the SACI.

1.1. *Why sustainability in the construction industry*

The significant impact of the construction industry on the environment and society makes it a major sector involved in achieving sustainability (Shi *et al.*, 2012). Sustainability issues in the construction industry are therefore very important, as the industry often adversely impacts on the environment through its use of large amounts of unrecoverable natural resources and generates large amounts of waste, to mention but a few of the impacts. Shen *et al.* (2000) and (Wang, 2014) identified fossil fuel, generic resources and mineral consumption, waste production requiring land disposal and pollution of the living environment as forming part of the environmental impacts of construction. The products of the construction industry are known to continuously emit large amounts of pollution. These are pollutant concentrations within buildings emanating from finishes, paints, backing materials and other components. Indoor pollutant level is often higher than those outside, based on a study by the United States Environmental Protection Agency (USEPA), indicating that indoor air levels of many pollutants may be 2.5 times and occasionally more than 100 times higher than outdoor levels (Pearce *et al.*, 2012). The release of greenhouse gas (GHG) emissions by the construction industry remains a significant contributor to the depletion of the ozone layer and its associated global warming and climate change. Most notably, cement production is a major source of GHG emissions. As recorded by Akadiri and Fadiya (2013), the construction industry is accountable for 50% of the UK carbon emissions while in China, carbon emissions associated with building energy use reached 1260 million tonnes in 2008 (Diamond *et al.*, 2013).

Kibert (2013) identified three forces propelling the shift to sustainability as the pressures experienced by the increasing demand for natural resources resulted in shortages and higher prices for materials; secondly, the massive and accelerated destruction of planetary ecosystems and biodiversity, alteration of biogeochemical cycles, significant increase in population and consumption, all resulting in the threat of global warming, depletion of marine life, deforestation, and desertification, among many others; and thirdly, the transformation movement coupled with the various approaches adopted in agriculture, tourism, manufacturing, medicine and the public sector, among others, towards greening their activities. These have, however, led to the growing recognition and importance of adopting the principles of sustainability in the construction industry (Plank, 2005).

2. Sustainable Construction

The shift of the construction industry from the traditional paradigm towards sustainable development has become a global movement in the form of "SC". This is vitally important considering the damage done to our environment and bearing in mind that natural resources are finite. The statement is further accentuated by the rapid economic growth in several highly populated areas of the world, thereby significantly increasing the potential environmental impact globally (Plank, 2008), as SC outlines the tenet that the construction industry can reach sustainability (Shi *et al.*, 2014). The terms "green", "high-performance", "environmentally friendly", and "SC" are often used interchangeably in the construction industry (Du Plessis, 2007; Azis *et al.*, 2012; Zabihi *et al.*, 2012; Kibert, 2013).

Pietrosemoli and Monroy (2013) describe SC as the result of the common efforts of investors, construction leaders, service representatives, industry suppliers, communities and other stakeholders directed to develop new buildings considering the environmental, energy, socioeconomic and cultural conditions needed to bring integrated solutions in the society. SC is also defined as the creation and responsible management of a healthy environment based on resource efficient and ecological principles (Tan *et al.*, 2011; Marhani *et al.*, 2013). The aim of SC is to reduce or eliminate environmental problems and issues associated with built facilities and construction activities while maximizing the potential benefits to society and the economy (Pearce *et al.*, 2012). Suliman and Omran (2009) also state that SC aims to enhance the quality of life and offer customer/client satisfaction, offer flexibility and the potential to cater for user changes in the future, provide and support desirable natural and social environments, and maximize the efficient use of resources. As indicated by Cotgrave and Riley (2013), re-use of materials as much as possible, reduced resource consumption, accommodation/incorporation of recyclable and reusable materials, non-use/exclusion of toxic elements, selection and use of only materials with low embodied energy, protection and nurturing of the ecology of the area, adoption of the whole life costing techniques, efficient use of energy, water and resources, and protection of the occupants' health are other aims of SC.

3. Overview of Biomimicry

Biomimicry is taken from a combination of the Greek words *bios* (life) and *mĭmēsis* (imitation), which literally means life imitation (Arnarson, 2011; Gamage and Hyde, 2012; Murr, 2015). The term "biomimicry" (also known as biomimetic) became popularized and widely circulated in 1997 through a book titled *Biomimicry: Innovation Inspired by Nature*. The book was written by Janine M. Benyus, a biologist, author, and co-founder of the Biomimicry Guild and who is widely recognized as the founder of this novel field of study (Goss, 2009). Benyus describes biomimicry as the quest of men and women, exploring nature's masterpieces (photosynthesis, self-assembly, natural selection, self-sustaining ecosystems, eyes and ears and skin and shells, talking neurons and natural medicines) and then copying these designs and manufacturing processes to solve their own problems (Benyus, 1997). The idea is that nature has developed highly efficient systems and processes, which have the potential to propel solutions to the waste, management and other challenges confronting humanity today (Hargroves and Smith, 2006).

Biomimicry proponents believed that the industry needs to study the highly successful Research and Development (R&D)) lab that has been operational on earth for over 3.8 billion years in which 10–30 million species have learned to do everything humans want to do, without polluting the environment or mortgaging the common future of generations to come (Strategic Direction, 2008). It is believed that innovations, ideas and solutions inspired by nature are now the panacea to the challenges facing humanity. However, biomimicry encourages the teaming up of biologists with designers and professionals from various fields (industrial design, medical science, materials science, architecture and interior design) to study processes and materials in nature, and then apply those natural models to the man-made or built environment. Rather than utilizing nature for human advantage and use, biomimicry focuses on the identification and integration of ideas that are predominantly sustainable and amenable to earth's capacity (Goss, 2009).

4. Biomimicry Approaches/Design Strategies

These are approaches that help to redefine the levels of biomimicry and its potential as a tool for sustainability. They offer unique focal points and step-by-step paths under the larger umbrella of biomimicry (Niewiarowski and Paige, 2011) to provide methodologies through which biomimicry can be incorporated into various disciplines in order to arrive at a sustainable, effective and efficient solution. They are also referred to as the biomimicry thinking process which is a multiplex procedure combining design and systemic thinking with life principles (sustainability). They furnish the context to what, why, how and why biomimicry application fits into any discipline or design scale (Planet, 2016). Biomimicry proponents believe that strict adherence to the steps constituting the approaches will result in sustainable outcomes, either in tackling human challenges or in innovative design solutions.

4.1. *Biomimicry problem-based approach*

The problem-based approach is also known as design looking to biology approach (Zari, 2007), problem-driven biologically inspired approach (Helms *et al.*, 2009), top-down approach (Knippers, 2009; El-Zeiny, 2012), design-to-biology approach (Pandremenos *et al.*, 2012), problem-to-solution approach (Buck, 2015). This approach entails defining a challenge by understanding and conceptualizing the processes and structures exhibited by natural organisms or ecosystems in resolving similar problems (Gamage and Hyde, 2012). Here, nature is looked at and turned to for solutions by first identifying the problems and then pairing such problems with organisms in nature that have resolved similar problems. In this approach, natural models are emulated to proffer solutions to a specific challenge (Pandremenos *et al.*, 2012).

This approach is effectively spearheaded by designers, innovators and engineers identifying preliminary design objectives and frameworks (El Ahmar, 2011). An example of innovations birthed by this approach is Eco-cement, a natural and strength-enhancing carbon sequestering cement inspired by the sea snail. Here, they look to the living world of nature for solutions by first identifying the challenges which biologists need to correspond to organisms that have resolved similar cases (Zari, 2007). The pattern of this approach follows a progression of different steps. Practically, this is nonlinear and dynamic in the sense that the output from later stages customarily influences previous stages, providing iterative feedback and refinement loops (Helms *et al.*, 2009). This approach is also believed to be a step towards transiting the built environment from an unsustainable state to an effective and resilient paradigm (McDonough and Braungart, 2010; El-Zeiny, 2012), as it helps proffer sustainable solutions to identified existing challenges.

4.2. *Biomimicry solution-based approach*

The solution-based approach is also known as biology influencing design approach (Zari, 2007), solution-driven biologically inspired approach (Helms *et al.*, 2009), bottom-top approach (Knippers, 2009; El-Zeiny, 2012), solution-to-problem approach (Pandremenos *et al.*, 2012; Buck, 2015). This approach involves identifying a definite attribute, behaviour or function in an organism or ecosystem then translating it into human designs, solutions or innovations (Zari, 2007). This describes a situation whereby biological knowledge influences the human design, solution or innovation. The collaborative design process is dependent on people having knowledge of relevant biological or ecological research rather than on determined human design problems (El-Zeiny, 2012). This approach has informed novel and amazing innovations which are inspired by the results of nature's constant and endless experimenting and reinvesting in the face of new challenges. An example of innovations birthed by this approach is the Mercedes-Benz Bionic car inspired by the boxfish.

5. Levels of Biomimicry

Biomimicry delivers on triadic levels (forms, processes and ecosystems) of increasing requirements in terms of sustainability (Gamage and Hyde, 2012) to achieve solutions, designs and innovations that awe in terms of sustainable performance (Kennedy *et al.*, 2015). These levels are foundational and fundamental in order to maximise the sustainable benefits biomimicry offers. They serve as inspirational eyes through which biomimicry is perceived and subsequently applied to proffer sustainable solutions to human challenges. As posited by Benyus (2011), this triad of natural forms, natural processes and natural ecosystems are levels through which biomimicry delivers its supreme objective of achieving sustainability wherever it is applied. This objective of holistic sustainability is achievable and guaranteed when biomimicry is applied up to the third level (natural ecosystems).

5.1. *Natural forms level of biomimicry*

This entails emulating part or the whole structure of the organisms (Zari and Storey, 2007), making it the most straightforward level of biomimicry (Kenny *et al.*, 2012). It could be the emulation of the pattern, shape or microstructure of a surface, such as a lotus leaf or a larger physical trait that can be observed with the naked eye, such as the kingfisher's beak. Although it is believed that this will solve a specific challenge, there is no assurance that it will culminate in an environmentally sustainable solution (Volstad and Boks, 2008). An example of this is imitating the hooks and barbules of an owl's feather to create a fabric that opens anywhere along its surface or imitating the frayed edges that grant the owl its silent flight (Benyus, 2011). Another example is the innovation of a lightweight but structurally strong building panel which can be inspired by the shape and support ribs of the giant leaves of the Amazon water lily (Attenborough and Graham, 1995; Kennedy *et al.*, 2015). This innovation may be sustainable depending on the constituent materials. In a situation where the costs of such innovations exceed the benefits and the constituting materials are toxic and harmful to the environment, such innovations or solutions become unsustainable and definitely are not biomimetic.

5.2. *Natural processes level of biomimicry*

This is a deeper level which relates to the emulation of the processes (manufacturing, building and recycling) operational in the identified natural organism (Biomimicry Europa, 2016). It entails the study and emulation of a series of operations or behaviours in organisms. These processes are understood not to be wasteful or polluting and are less energy intensive. For example, nature is known to assemble structures at ambient temperature and pressure using non-toxic chemistry compared to the human method of bending, melting, casting or manipulating huge blocks of raw material at high temperatures and pressures (Kennedy *et al.*, 2015). At this level, identified and relevant processes in natural organisms are studied and emulated in order to solve a particular challenge. An example of this is the attempt by the unfurling field of green chemistry to emulate the benign process of how owl feather self-assembles at body temperature without toxins or high pressure but by way of nature's chemistry (Volstad and Boks, 2008; Benyus, 2011). Another example of biomimicry application at this level is the creation of algorithms (step-by-step procedures for calculations) by computer scientists based on the way swarming bees or flocking birds coordinate their movements as a group.

5.3. *Natural forms level of biomimicry*

This is the deepest and most holistic level which relates to the behaviour and performance of an identified natural organism with respect to the ecosystem. It involves creating an integrated system that efficiently optimizes materials or energy, or both, in an ongoing cycle such as that obtained found in the natural ecosystem (Biomimicry Institute, 2014). It is noteworthy that nothing exists or operates in isolation in the natural world (Kenny *et al.*, 2012), as every organism is part of a biome that is also part of a larger biosphere (Kennedy *et al.*, 2015). The survival and sustenance of natural organisms lie in their ability to operate and perform at this level. For example, the owl feather is gracefully nested; it is part of an owl that is part of a forest that is part of a biome that is part of a sustaining biosphere (Benyus, 2011). This literally means that every organism's continued prosperity is dependent on the health of the bio-sphere (McDonough and Braungart, 2010; Kennedy *et al.*, 2015).

It is important to note that biomimicry reaches its full potential of sustainability only when it is applied in the context and at the level of the ecosystem (Volstad and Boks, 2008). The natural ecosystem level, how-ever, assures of a sustainable solution to human challenges as it entails the application of biomimicry in a holistic manner (combination of natural forms, natural processes and natural ecosystems). In the same way, our owl-inspired fabric must be part of a larger economy that works to restore rather than deplete the earth and its inhabitants (Benyus, 2011). Biomimics and the proponents of biomimicry, therefore, advocate for the application and practice of biomimicry on the three levels to achieve sustainable solu-tions to human challenges. It is believed that this will ensure that the human environment begins to do what all well-adapted organisms have learned to do, which is to create conditions conducive to life and for gen-erations to come.

6. Biomimetic Innovative Solutions in the Construction Industry

Biomimicry has become a leading edge among emerging sustainable tech-nologies and process improvement systems. Since the integration of bio-mimicry into different sectors and fields, there have been records of novel innovative and sustainable breakthroughs made. Bio-inspired adhesives and coatings, self-cleaning materials, bio-inspired tough composite mate-rials, biotechconcrete, and energy conversion and conservation are few of the numerous successes recorded with the application of biomimicry in the construction industry. Examples of these innovations are high-strength ceramics inspired by abalone shells, waterproof adhesives inspired by the mussel's ability to attach to the ocean floor and high-strength fabric inspired by spider silk (Benyus, 1997). Table 1 shows few biomimetic products and technologies applicable in the construction industry with their source of inspiration and function.

Table 1.   Biomimetic products and technologies applicable in the construction industry (AskNature, 2017).

Product	Bio-inspiration	Function
Lotusan® Paint	Morpho butterfly and Sacred lotus	Automatic self-cleaning coat after the mere rinse of a rain shower
Eco-Cement	Sea snail	Neutral and strength-enhancing carbon sequestering cement
Self-repairing Concrete	Pipevine, vertebrates bone, cells, human skin	Self-repairing concrete that increases durability of structures and reduces life cycle cost
I2™ Modular Carpet	Forest floor	Individually replaceable and recyclable carpet tiles
Purebond	Blue mussels	Formaldehyde-free wood glue
Lotus Clay Roofing Tiles	Morpho butterfly and Sacred lotus	Self-cleaning clay roofs
Dye-Sensitised Solar Cells and Panels	Cooke's Koki'o (photosynthesis)	Low-cost and efficiently produced electricity by artificial photosynthesis
ORNILUX insulated glass	Orb-web spider	Insulated architectural glass that prevents bird collisions
HotZone™ radiant heater, using Irlens™	Lobster	Spot heater that focuses radiant heat onto users rather than heating an entire space
bioWAVE™	Bull kelp	Harnessing wave energy for power generation
Biolytix® System	Soil ecosystem	Waste treatment and water filtering system
COMOLEVI Forest Canopy	Shade trees	Leaf colour and shape enhance cooling effects
Eco-Machine	Forests	Waste water treatment system that purifies water without chemicals
Chaac-ha	Spiders and Bromeliads	Water system collector for rainwater and dew
Sharklet™	Shark	Surfaces that prevent biofouling without causing resistance
Aquaporin Membrane Technology™	Human and cells	Water filtration and purification system
Sage Glass, Quantum Glass (Europe)	Bobtail squid and Hummingbird	Electrochromic smart windows that provide energy savings by reducing cooling/heating costs

7. Findings

The environmental impacts of the construction industry are severely felt globally with the rising population, which comes with the need for infrastructural developments. This has encouraged and aided the importance of embracing sustainability in the construction industry. To successfully transfer the industry from the present traditional paradigm to the sustainable approach, the onus is now on the professionals and stakeholders to spearhead the movement. However, biomimicry offers amazing potentials to herald an era of novel and innovative sustainable solutions in the construction industry. The study of nature's architecture (forms, processes, strategies and systems) proffers sustainable solutions to human challenges, designers, engineers, planners, and innovators among many others who are now embracing biomimicry as guide and source of inspiration. Findings from the study revealed that lack of well-defined approach is a major barrier hindering the employability of biomimicry. It is also discovered that there are two major approaches to the application of biomimicry, namely problem-based approach and solution-based approach. It is believed that these approaches when strictly adhered to and followed will culminate in the birth of sustainable solutions to the challenges of sustainability in the construction industry. As related to the construction industry, examples of innovations borne out of the application of biomimicry include Lotusan paint, Eco-cement, Purebond, I2 modular carpet, Lotus clay roofing, Biolytix systems, Chaah-ha and bioWave among others. These bio-inspired materials and technologies, known to possess outstanding and environmental-friendly attributes, are results of biomimicry application using either the problem-based approach or the solution-based approach.

8. Conclusion

It can be concluded from the findings that the construction industry through its activities is majorly responsible for the degradation of the human environment. Biomimicry has therefore presented both problem-based and solution-based paths to provide solutions to these challenges in a sustainable way. The use of biomimetic materials borne out of the application of biomimicry through the two approaches offers sustainable options to the traditional ones in use in the construction industry. This alone is believed to lower the atmospheric concentration of Greenhouse gases (GHGs) and prevent the release of additional pollutants (indoor and outdoor) emanating from the traditional construction materials in use. Biomimicry also has the potential of heralding an era where innovative and sustainable technologies and strategies will be birthed to upscale the transition of the industry to a scale with less or no environmental footprint. However, institutional and governmental funding and support for research should be encouraged to ensure the birth of these innovative technologies a reality. Multi-disciplinary collaboration among professionals and stakeholders should also be encouraged and supported with biologists serving as guides while these ideas are incubated and perfected. This deliberate focus on nature as *mentor, model* and *measure* as suggested by biomimicry proponents will change human orientation towards the natural environment thereby supporting, preserving and protecting them. This will ensure their optimum performance in improving human well-being and environment by limiting overheating, sequestering carbon to offset CO_2 emissions, and reducing the risk of flooding amongst many others.

References

Aigbavboa, C. and Thwala, W. (2014). "An Assessment of Critical Success Factors for the Reduction of the Cost of Poor Quality from Construction Projects in South Africa", in Raiden, A. B. and Aboagye-Nimo, E. (eds.), in *Proceedings of the 30th Annual ARCOM Conference*, September 1–3, 2014, Portsmouth, UK, Association of Researchers in Construction Management, pp. 773–782.

Akadiri Oluwole, P. and Fadiya Olaniran, O. (2013). "Empirical Analysis Of The Determinants Of Environmentally Sustainable Practices In The Uk Construction Industry", *Construction Innovation*, 13(4), 352–373.

Arnarson, P. O. (2011). *Biomimicry: New Technology*. Reykjavík University. Retrieved 16 November, *2016* from: http://olafurandri.com/nyti/papers2011/ Biomimicry%20-%20P%C3%A9tur%20%C3%96rn%20Arnarson.pdf.

Ask Nature (2017). *Inspired Ideas*. The Biomimicry Institute. Retrieved 27 January 2017 from: https://asknature.org/?s=&page=0&hFR%5Bpost_type_label%5D%5B0%5D=Inspired%20Ideas&is_v=1#.Wd8kXGiCzIU.

Attenborough, D. and Graham, N. (1995). *The Private Life Of Plants: A Natural History Of Plant Behaviour*, Princeton, NJ.: Princeton University Press.

Azis, A. A. A., Memon A. H., Rahman I. A., Nagapan S. and Latif Q. B. A. I. (2012). "Challenges Faced By Construction Industry In Accomplishing Sustainability Goals", in *2012 IEEE Symposium of Business, Engineering and Industrial Applications (ISBEIA)*, pp. 630–634.

Benyus, J. M. (2011). *A Biomimicry Primer*. The Biomimicry Institute and the Biomimicry Guild.

Benyus, J. M. (1997). *Biomimicry: Innovation Inspired by Nature*. New York, USA: William Morrow & Company.

Biomimicry Europa. (2016). *Biomimicry?* Brussels, Belgium: Biomimicry Europe. Retrieved 6 March 2016, from: http://www.biomimicry.eu/en/ biomimicry/.

Biomimicry Institute. (2014). *Biomimicry in Youth Education: A Resource Toolkit for K-12 Educators.*

Construction Industry Development Board (cidb). (2016). *About the cidb.* Retrieved on 16 April 2016 from: http://www.cidb.org.za/AboutUs/Pages/ default.aspx.

Cotgrave, A. and Riley, M. eds., (2012). *Total Sustainability in the Built Environment.* Palgrave Macmillan.

Diamond, R. C., Ye, Q., Feng, W., Yan, T., Mao, H., Li, Y., Guo, Y. and Wang, J. (2013). "Sustainable Building in China — A Green Leap Forward?" *Buildings*, 3(3), 639–658.

Dlungwana, W. S., Nxumalo, X. H., Van Huyssteen, S., Rwelamila, P. D. and Noyana, C. (2002). "Development And Implementation Of The South African Construction Excellence Model (SACEM)", in *International Conference on Construction in the 21st Century (CITC2002)*, April 25–26, Miami, Florida, USA.

Du Plessis, C. (2007). "A Strategic Framework For Sustainable Construction In Developing Countries", *Construction Management and Economics*, 25(1), 67–76.

El Ahmar, S. (2011). Biomimicry As A Tool For Sustainable Architectural Design: Towards Morphogenetic Architecture. Unpublished Master's thesis, Alexandria University.

El-Zeiny, R. M. A. (2012). "Biomimicry As A Problem-Solving Methodology In Interior Architecture", *Procedia — Social and Behavioral Sciences*, 50502–50512.

Gamage, A. and Hyde, R. (2012). "A Model Based On Biomimicry To Enhance Ecologically Sustainable Design", *Architectural Science Review*, 55(3), 224–235.

Goss, J. (2009). *Biomimicry: Looking to Nature for Design Solutions.* Corcoran College of Art and Design, ProQuest Dissertations Publishing.

Hargroves, K. and Smith, M. (2006). "Innovation Inspired by Nature: Biomimicry", *Ecos*, (129), 27–29.

Helms, M., Vattam, S. S. and Goel, A. K. (2009). "Biologically Inspired Design: Process and Products", *Design Studies*, 30(5), 606–622.

Kennedy, E., Fecheyr-Lippens, D., Hsiung, B., Niewiarowski, P. H. and Kolodziej, M. (2015). "Biomimicry: A Path to Sustainable Innovation", *Design Issues*, 31(3), 66–73.

Kenny, J., Desha, C., Kumar, A. and Hargroves, C., (2012). "Using Biomimicry To Inform Urban Infrastructure Design That Addresses 21st Century Needs", in *1st International Conference on Urban Sustainability and Resilience: Conference Proceedings.* UCL London.

Kibert, C. J. (2002). "Policy Instruments for a Sustainable Built Environment", *Journal of Land Use & Environmental Law*, 17(2), 379–394.

Kibert, C. J. (2013). *Sustainable Construction: Green Building Design and Delivery*, Third Edition, Hoboken, N.J.: John Wiley & Sons.

Knippers, J. (2009). "Building and Construction as a Potential Field for the Application of Biomimetic Principles", in *International Biona Symposium*, November 27, Stuttgart, Germany.

Marhani, M. A., Adnan, H. and Ismail, F. (2013). "OHSAS 18001: A Pilot Study of Towards Sustainable Construction in Malaysia", in *Procedia — Social and Behavioral Sciences*, pp. 8551–8560.

Marx, H. (2014). Results of the 2014 Survey of the CIDB Construction Industry Indicators. University of the Free State, pp. 1–28.

McDonough, W. and Braungart, M. (2010). *Cradle to Cradle: Remaking the Way We Make Things.* MacMillan.

Milford, R. (2009). *Greenhouse Gas Emission Baselines and Reduction Potentials From Buildings in South Africa*, Paris: United Nations Environment Programme.

Murr, L. E. (2015). "Biomimetics and Biologically Inspired Materials", in *Handbook of Materials Structures, Properties, Processing and Performance*, Springer International Publishing, pp. 521–552.

Niewiarowski, P. H. and Paige, D. (2011). *Proceedings of the First Annual Biomimicry in Higher Education Webinar. The Biomimicry Institute.*

Available at: https://biomimicry.org/shop/proceedings-higher-ed-webinar-2012/ (Retrieved February 27, 2016).

Ntuli, B. and Allopi, D. (2013). "Impact of Inadequate Experience and Skill on the Construction Sector in KwaZulu-natal, South Africa". *Engineering, Technology & Applied Science Research*, 4(1), 570–575.

Pandremenos, J., Vasiliadis, E. and Chryssolouris, G. (2012). "Design Architectures in Biology", in *Procedia CIRP*, pp. 3448–3452.

Pearce, A. and Ahn, Y. H. (2012). *Sustainable Buildings and Infrastructure: Paths to the Future*, Routledge.

Pietrosemoli, L. and Monroy, C. R. (2013). "The Impact of Sustainable Construction and Knowledge Management on Sustainability Goals. A Review of the Venezuelan Renewable Energy Sector", *Renewable and Sustainable Energy Reviews*, pp. 27683–27691.

Planet, (2016). "Biomimicry. Biomimicry Thinking Approach". Available at: http://planet.wemimic.it/biomimicry.html (Retrieved November 16, 2016).

Plank, R. J. (2005). "Sustainable Construction — A UK Perspective", in *Structures Congress 2005*, April 20–24, New York, United States, pp. 1–7.

Plank, R. (2008). "The Principles of Sustainable Construction", *The IES Journal Part A: Civil & Structural Engineering*, 1(4), 301–307.

Rao, R. (2014). Biomimicry in architecture. *International Journal of Advanced Research in Civil, Structural, Environmental and Infrastructure Engineering and Developing*, 1(3):101–107.

Shen, L. Y., Bao, Q. and Ip, S. L. (2000). "Implementing innovative Functions In Construction Project Management Towards The Mission Of Sustainable Development", in *Proceedings of the Millennium Conference on Construction Project Management*, Hong Kong, October 24, 2000, pp. 77–85.

Shi, L., Ye, K., Lu, W. and Hu, X. (2014). "Improving the Competence of Construction Management Consultants to Underpin Sustainable Construction in China". *Habitat International*, 41236–41242.

Shi, Q., Zuo, J. and Zillante, G. (2012). "Exploring the Management of Sustainable Construction At The Programme Level: A Chinese Case Study", *Construction Management and Economics*, 30(6), 425–440.

Strategic Direction. (2008). "Nature's Inspiration: Solving Sustainability Challenges", *Strategic Direction*, 24(9), 33–35.

Suliman, L. K. M. and Omran, A. (2009). "Sustainable Development and Construction Industry in Malaysia", *Manager (University of Bucharest, Faculty of Business & Administration)*, (10), 76–85.

Tan, Y., Shen, L. and Yao, H. (2011). "Sustainable Construction Practice and Contractors' Competitiveness: A Preliminary Study", *Habitat International*, 35(2), 225–230.

Taylor Buck, N. (2015). "The Art of Imitating Life: The Potential Contribution Of Biomimicry in Shaping The Future of Our Cities", *Environment and Planning B: Planning and Design*.

Volstad, N. L. and Boks, C. (2008). "Biomimicry — A Useful Tool For The Industrial Designer?" in *DS 50: Proceedings of NordDesign 2008 Conference*, August 20–22, 2008, Tallinn, Estonia, pp. 275–284.

Wang, N. (2014). "The Role of the Construction Industry in China's Sustainable Urban Development", *Habitat International*, 44442–44450.

Windapo, A. O. (2014). "Examination of Green Building Drivers in the South African Construction Industry: Economics Versus Ecology", *Sustainability*, 6(9), 6088–6106.

Windapo, A. O. and Cattell, K. (2013). "The South African Construction Industry: Perceptions Of Key Challenges Facing Its Performance, Development and Growth", *Journal of Construction in Developing Countries*, 18(2), 65.

Zabihi, H. and Habib, F. (2012). "Sustainability in Building and Construction: Revising Definitions and Concepts", *International Journal of Emerging Sciences*, 2(4), 570.

Zari, M. P. and Storey, J. B. (2007). "An Ecosystem Based Biomimetic Theory for a Regenerative Built Environment", in *Sustainable Building Conference 07*, Lisbon, Portugal.

Zari, M. P. (2007). *Biomimetic Approaches to Architectural Design for Increased Sustainabilit*, SB07 Auckland, New Zealand.

Chapter 21

Antecedents of BIM Training Effectiveness Amongst AEC Firms in New Zealand

James Rotimi*

*Built Environment Engineering,
Auckland University of Technology, New Zealand
jrotimi@aut.ac.nz*

Chris Arasanmi

*Business, Information Technology & Creative Arts,
Toi-Ohomai Institute of Technology, New Zealand*

Funmilayo Rotimi

Construction Programme, NorthTec, New Zealand

Chamila Ramanayaka

*Department of Construction Management,
Curtin University, Australia*

Abimbola Komolafe

Department of Building, Kaduna Polytechnic, Nigeria

---

*Corresponding author.

Abstract

Building information modelling (BIM) is radically changing the face of construction project management. BIM adoption is significant in alleviating the attendant problems of poor detailing, communication and coordination because of its inherent integrative technology. Several BIM training initiatives have occurred within the construction industry in New Zealand, but the knowledge base on the success of these training initiatives is at its infancy. Furthermore, there is a dearth of research on how acquired knowledge and skills in BIM implementation have been transferred to other end users post training. This chapter seeks to understand BIM knowledge transfer mechanisms among AEC members. The chapter proposes a model of the enablers of BIM knowledge transfer and effective utilization of skills and knowledge in a BIM environment, and presents the psychometric analysis of the research instrument in this exploratory phase of the study.

Keywords: BIM; Knowledge Transfer; Training and New Zealand.

1. Introduction

Building information modelling (BIM) is changing practices within the construction project management. The construction industry, although slow to change, is steadily tapping into the benefits of BIM adoption and implementation, which include but not limited to increased efficiency, productivity and quality of whole-of-industry performance (Eastman *et al.*, 2008). BIM uptake in countries such as New Zealand and Australia is on the increase. Support generally comes through major clients, especially government agencies and BIM specialist solution providers who market BIM knowledge aggressively. Furthermore, the implementation of national BIM mandates (Wong *et al.*, 2010) ensures that architectural, engineering and construction (AEC) firms see BIM implementation as the forward-thinking and best practice.

The adoption of BIM in New Zealand is anticipated to alleviate the attendant problems of poor detailing, communication and stakeholder coordination, which characterize the New Zealand construction industry (NZ BIM Industry Training Group, 2016). However, information on how acquired knowledge and skills from diverse training interventions have been applied within AEC firms is in its infancy. Training is only meaningful when end users take the knowledge and skills learned in training episodes and apply them to situations in their workplaces (Arasanmi *et al.*, 2016). Training transfer, as the goal of training interventions, enhances the assessment of the impact of training for individuals and organizational performance. It is important for end users to maximize their knowledge of different information system software associated with BIM; however, this can only be realized by users with requisite skills. This remains to be a challenge in information system studies (Jasperson *et al.*, 2005), wherein skills acquired on information system training are effectively transferred to work environments for whole-systems benefit. Therefore, the key question addressed in this chapter is: What predicts post-training knowledge transfer among BIM adopters in their task environments?

To answer this question, this chapter presents a conceptual model and the pilot result of the psychometric properties of some operationalized constructs in this ongoing study. The aim of this study is to assess the antecedents of knowledge transfer in the BIM environment as a measure of BIM training effectiveness. Therefore, the objective of this chapter is to find the relationship between proposed exogenous variables and post-training utilization of BIM. This chapter covers the review of the relevant literature, statements of the proposed research model and hypotheses, study methods, contributions and conclusions emanating from this exploratory study.

2. Literature Review

2.1. *BIM adoption and training*

"BIM is a digital representation of the physical and functional characteristics of a building. As such, it serves as a shared knowledge resource for information about a building, forming a reliable basis for decisions during its life cycle from inception onward" (NZ BIM Handbook, 2004). Sacks *et al.* (2010) usefully summarized the main functions of BIM as: visualization and rapid generation of design, building performance prediction, automation of model integrity and reports, providing a communication platform and promoting collaboration between design and construction.

Miettinen and Paavola (2014) acknowledge the propensity for inter-organizational collaboration and increased productivity with the adoption of BIM. Furthermore, Ghaffarian Hoseini *et al.* (2016) documented a plethora of benefits associated with BIM, which may include, but not limited to, enhanced technical knowledge, standardization, integration, economy, planning and decision support potentials of the suit of software that make up BIM within the construction industry. While these studies have highlighted the benefits of BIM to the construction industry, there is a huge gap in knowledge on how well these benefits cut across intra-organization units. Specifically, research has neglected issues around BIM post-training knowledge transfer among AEC members. Though adoption issues feature prominently in the previous studies, BIM acceptance and adoption issues are centred on predictors of BIM implementation. This narrow focus of BIM adoption studies fail to account for end-user knowledge transfer as a measure of training effectiveness. Information systems studies show that there is an increased resistant behaviour among end users in using information system software after deployment. This may be due to lack of the required skill set, which hampers routinized and deep use of associated information system software. Remarkably, prior adoption studies in the BIM environment did not show the adequacy of how BIM knowledge adoption represents a measure of training transfer (Blume *et al.*, 2010). Though training provision is a critical success factor of BIM implementation (Park *et al.*, 2011), researchers should therefore account for the effectiveness of BIM training by assessing the post-training usage of acquired knowledge and skills in a BIM-enabled practice environment. The state of research in this area is still shrouded concerning end users' transfer of BIM knowledge acquired from training as an appropriate measure of a training intervention or impact in AEC organizations. BIM training would be considered effective if knowledge, skills and attributes acquired during a BIM training programme are effectively applied by learners within their respective organizations and consequently whole-of-industry upskill. Such practice connotes the capability to use acquired learning and knowledge in the workplace (Baldwin and Ford, 1988). This chapter posits that the transfer phase is where learners can reproduce acquired learning in the workplace.

2.2. *Transfer of training and BIM training effectiveness*

Research attention has been skewed particularly towards the adoption issues and the benefits of BIM. Discussions on BIM as an outstanding innovation and the associated challenges and risk factors are unprecedented (Ghaffarian Hoseini *et al.*, 2016; Ahuja *et al.*, 2016). For instance, Ahuja *et al.* (2016) developed a TOE-based model to analyze the impact of technology, organization and environmental influences on BIM adoption and highlighted the importance of top management support for BIM adoption among construction professionals. Several studies have recognized skills as a core reason for BIM adoption among AEC firms (Li *et al.*, 2010; Zakaria *et al.*, 2013), while different theoretical models have been applied to explain BIM implementers' behavioural intentions (Son *et al.*, 2015). There is a consensus, however, that the complexities surrounding the deployment and implementation of BIM knowledge base makes training provision a critical success factor (Park *et al.*, 2011).

Training is a prevalent method for updating individual knowledge and skills in the workplace (Arthur *et al.*, 2003; Rowold, 2007). Furthermore, the amplification of knowledge created by individuals within an organization helps to connect it to an organization's knowledge system (Nonaka and Krogh, 2009). Training is the systematic acquisition of useful knowledge and skills required in the execution of work tasks. Training is equally important during BIM implementation because BIM exploration requires associated skills, knowledge, experience and capabilities from end users. Research suggests that end user's intention to adopt BIM is linked to the capability of a project team (Ding *et al.*, 2015). This made BIM training to be largely focused on updating the capability and knowledge base of AEC members. While it is challenging to determine end user's skills and knowledge during training sessions, it is agreed that end-user post-training performance and use of learning need to be assessed in the work environment. End users' capability to use acquired skills and knowledge in the work environment is termed as "transfer of training". Transfer of training therefore is a measure of the effectiveness of organized training interventions by specifically assessing how acquired knowledge and skills are transferred and applied in the job environment.

2.3. *Research model and hypotheses*

Transfer of training represents a change in work knowledge, attitude and skills in achieving meaningful changes in work performance. The use of skills improves individuals and overall outcomes. On the other hand, knowledge sharing is the behaviour when an individual disseminates his acquired knowledge to other members within an organization (Ryu *et al.*, 2003). Simply, it is the transmission of knowledge between people or entities. Information-sharing mechanisms comprise practices, such as transfer of personnel, training, technology transfer, routine replications, discussion and interactions with suppliers and customers, and other forms of inter- and intra-organizational dealings. While the acquisition and management of important experience-based knowledge are vital in a BIM environment, the ability to transfer BIM training in this environment could lead to intra-organizational knowledge creation among work teams. Information sharing is proposed as a central construct in the proposed model. It is assumed that end users' effective post-implementation adoption of BIM knowledge will be influenced by information-sharing behaviours.

Noe (1986) suggests that an individual's desire to learn affects his motivation to transfer such learning in a task environment. Transfer motivation captures the essence of training as a means of achieving future benefits (Chiaburu and Tekleab, 2005) because motivated end users will seek out the opportunity to transfer skills on the job. Prior research shows that transfer motivation influence end users' post-training outcomes (Garavan *et al.*, 2010; Klein *et al.*, 2006).

Information system studies aver that social cognitive variables affect learning and post-learning performance. Importantly, computer self-efficacy which represents an individual's perception of his/her abilities to use a computer in the accomplishment of work-related tasks has been acknowledged as an important antecedent of post-training usage of software. Self-efficacy has been found to positively relate to system usage (Garavan *et al.*, 2010).

Likewise, end users' perception concerning ease-of-use (PEOU) of an information system (IS) is a significant predictor of IS usage. PEOU is the perception that learning and using a software will be free of effort (Davis 1989). The cognitive effort required in learning an information system affects end user's motivation to learn and post-learning usage of the software (Gefen, 2003).

Based on the above, end users with increased computer self-efficacy (self-confidence) and positive perception concerning BIM knowledge will be more willing to learn and apply the acquired skills and knowledge in a BIM task environment effectively. The information systems and training transfer perspectives underpin this conceptual model. The model proposed that two exogenous variables, i.e. computer self-efficacy and perceived ease-of-use are antecedents of knowledge transfer and transfer performance, which is represented by BIM implementation. The model also positioned knowledge transfer as a mediating construct in this model. The model is depicted in Fig. 1. The operationalization of the constructs is presented in Table 1, followed by a listing of the hypotheses that have been developed within the study framework.

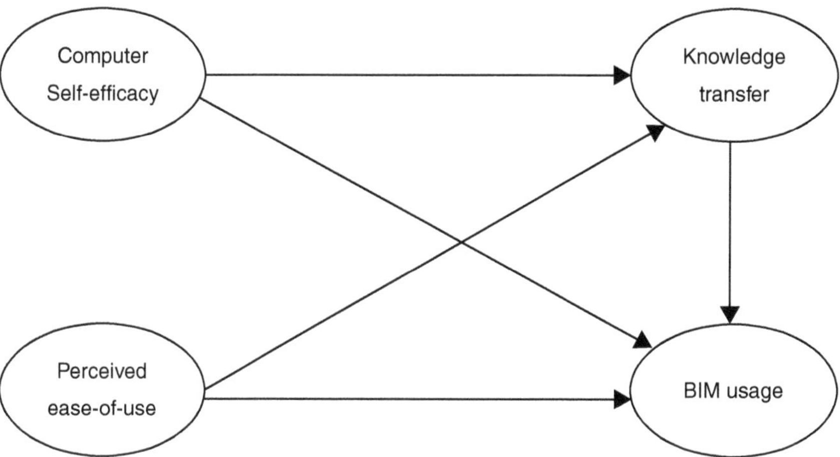

Fig. 1.   Conceptual model.

Table 1.   Operationalization of constructs.

Constructs	Operationalization of Constructs	References
Perceived ease-of-use	Perceived ease-of-use (PEOU) as the degree of belief that using a computer-based system will be free of efforts	Davis (1989)
Computer self-efficacy	Computer self-efficacy (CSE) represents an individual's perceived ability to accomplish tasks in a computerised environment	Compeau and Higgins (1995)
Knowledge transfer	Knowledge sharing is the behaviour when an individual disseminates his acquired knowledge to other members within an organization	Ryu *et al.* (2003)
Transfer of training (BIM usage)	Transfer of training comprises the application and use of skills in achieving meaningful changes in the workplace	Baldwin and Ford (1988)

Consequently, the following hypotheses are developed for testing the relationships between the constructs in this study:

Hypothesis 1: Knowledge sharing will influence BIM implementation among AEC work teams.

Hypothesis 2: Knowledge sharing will mediate the relationship between computer self-efficacy and perceived ease-of-use and BIM implementation.

Hypothesis 3: Computer self-efficacy will influence knowledge sharing and BIM implementation among AEC work teams.

Hypothesis 4: PEOU will influence knowledge sharing and BIM usage knowledge transfer among AEC work teams.

3. Research Methods

This is a cross-sectional quantitative study wherein an online survey method was employed to collect information from previous BIM training participants within New Zealand AEC firms. Data collection through the survey is ongoing. The research participants are AEC members who have previously attended BIM training for BIM users within the last 2 years. This sample is suitable for the study because these AEC members are expected to be deep and routine adopters of the BIM knowledge. The measurement items are drawn from the previous studies in this area. To ensure consistency, all items in the study were measured using a five-point Likert-scale ranging from "strongly disagree" to "strongly agree".

3.1. *Analysis strategy for the pilot study*

Pilot testing is used in the study to clear confusion and to test the psychometric properties of the research that has been based on self-completed questionnaires (Bryman and Bell, 2007). In this study, a sample frame similar to that for actual data collection was used. The respondents' classifications are provided in terms of gender and professional affiliations. The pilot sample comprised 19 respondents drawn from construction (17.8%), architecture (17.8%) and civil engineering (44.8%) organizations.

Table 2.   Psychometric properties of research instrument.

Construct	Items	Loadings	Cronbach Alpha
BIM Usage	BIM Usage1	0.847	0.959
	BIM Usage2	0.902	
	BIM Usage3	0.931	
Perceived ease-of-use	PEOU1	0.506	0.825
	PEOU3	0.778	
	PEOU4	0.788	
Computer self-efficacy	CSE1	0.862	0.958
	CSE2	0.844	
	CSE3	0.856	
Knowledge transfer	KS1	0.854	0.899
	KS2	0.678	
	KS3	0.798	
	KS4	0.771	
	KS6	0.740	

The validity and the psychometric properties of the survey items were tested using the Principal Component Method. The Principal Component Method is a form of exploratory factor analysis in the SPSS 24.0 software version. Table 2 shows that the survey items and the Cronbach alpha of all the items met the benchmark of 0.7.

The descriptive analysis and the measurement validity assessment are provided in the following section. Table 3 gives the descriptive statistics of the key features of the data in the study. The analysis shows that 83.3% of the respondents confirmed that the BIM implementation enhances their task performance when compared with routine task performance. About 61.1% of the respondents equally established that BIM implementation improved their respective job performance. Notably, 94.4% of the respondent (overwhelming majority) affirmed that BIM was revolutionary in the construction industry.

With respect to the usefulness of BIM training in the construction industry, 94.4% of the participants agreed that the training they had received on BIM exposed them to wide-ranging knowledge in

Table 3.   Descriptive analysis outcomes.

Descriptive Outcomes	Percentage
End user's perception of the impact of BIM on task performance	83.3
End user's perception of the impact of BIM on task improvement	61.1
End user's perception concerning BIM as a revolutionary organizational tool	94.4
End user's perception of the usefulness of BIM training in a work environment	94.4
Impact of BIM training on users' self-efficacy in a work environment	66.7
Post-training sharing of BIM knowledge among users	61.1
Perceptions of colleagues' confidence of the impact of BIM in the workplace	88.9

construction automation and project management. A substantial proportion of the research participants (66.7%) aver that the BIM training they received increased their self-efficacy and they were now able to transfer the knowledge acquired to their jobs. On the issue of knowledge sharing among colleagues in the work environment, 61.1% agree that it is important to share BIM knowledge in a task environment. Furthermore, 88.9% declared that their work colleagues are now more confident to implement BIM because of the training that they had passed on within their respective workplaces.

3.2. *Ramifications of the study*

The model proposed a knowledge transfer mechanism in view of the challenges associated with ascertaining BIM training effectiveness within AECs. Specifically, it highlights the importance of training transfer in understanding BIM training effectiveness, which hitherto was not adequately covered in the BIM adoption research domain. The propositional assumptions of this model suggest some practical and managerial implications.

Firstly, the model contributes to the understanding of BIM training by introducing the concept of transfer as a way of understanding BIM training effectiveness. The model posits that computer self-efficacy and perceived ease-of-use are critical antecedents of post-training performance in the BIM environment. End users with increased perceptual confidence in their abilities are more inclined to use associated BIM software in their task environments.

Secondly, end users' positive perception concerning the ease-of-learning acquired from BIM training will affect their motivation to learn as a proximal predictor of BIM implementation.

Thirdly, knowledge and information sharing encourages the dissemination of acquired knowledge and skills among AEC work teams. Post-training information sharing enhances the success of some of the train-the-trainer workshops organized by practitioners in this area among adopters of BIM and associated software. It is affordable for trained or experienced trainees to train others about BIM. Information sharing among end users will lead to the spread of BIM knowledge and processes and consequently improve BIM practice environments.

The pilot study findings give some indication that some of the propositions in the developed hypotheses may be accepted, when larger data are analyzed for the study. BIM training was largely useful and the respondents had indicated that the training improved their task performance within their respective organizations. There exists a greater need for collaboration and knowledge sharing for whole-of-industry benefits on BIM implementation. It is illuminating to observe that a significant number of the respondents are transferring their acquired BIM knowledge to colleagues at their workplaces. It seems apparent that there is an increase in confidence on the implementation of BIM in New Zealand.

4. Conclusion

The aim of this chapter is to provide insights on the antecedents of training transfer in BIM environments, especially within AECs. BIM training would be relevant in the task environment when an AEC community of practitioners can apply their learned skills on the job, since the goal of such training is to facilitate routine practices and the enhancement of BIM knowledge base among AEC members.

The chapter presents a pilot data which validate the research instrument designed to test the conceptual model at the second phase of this study. This model, though at a crude state, provides some insights on the importance of skill and knowledge transfer as a measure of training effectiveness in this changing environment and the theoretical relevance of computer self-efficacy, perceived ease-of-use and knowledge transfer in assessing post implementation of BIM in New Zealand.

The model trialled in this chapter has the potential to contribute to the theoretical development and understanding of adoption issues from the train-the-trainer perspectives. Other classical models on organizational learning are useful in a holistic approach to training and knowledge transfer. For example, Nonaka's knowledge-creation model (Nonaka and Nihihara, 2018) explains the collaborative working required of organizational members, which will enable an internalization process that supports learning opportunities. Engeström's expansive learning model builds on other models to suggest exploration and exploitation as learning dimensions that yield different learning typologies (Engeström and Sannino, 2010). Research has shown that learning and/or knowledge-building activities within organizations is multi-dimensional. However, it is apparent that less attention has been given to knowledge transfer and its influence on team dynamics, knowledge sharing and BIM performance. This study is therefore limited to BIM knowledge transfer effectiveness within AEC firms and would seem to be a pioneering effort in this regard.

References

Ahuja, R., Jain, M., Sawhney, A. and Arif, M. (2016). "Adoption of BIM by Architectural Firms in India: Technology-organization-environment Perspective", *Architectural Engineering and Design Management*, 12(4), 311–330.

Arasanmi, C. N., Wang, W. Y. C. and Singh, H. (2016). "Examining the Motivators of Training Transfer in an Enterprise Systems Context: Enterprise Information Systems". Available at: http://dx.doi.org/10.1080/17517575.2016.1177206.

Arthur, W. J., Bennett, W. J., Edens, P. S. and Bell, S. T. (2003). "Effectiveness of Training in Organizations: A Meta-analysis of Design and Evaluation Features", *Journal of Applied Psychology*, 88(2), 234–245.

Baldwin, T. T. and Ford, J. K. (1988). "Transfer of Training: A Review and Directions for Future Research", *Personnel Psychology*, 41, 63–105.

Blume, B. D., Ford, J. K., Baldwin, T. T. and Huang, J. L. (2010). "Transfer of Training: A Meta-analytic Review", *Journal of Management Information Systems*, 39, 1065–1105.

Bryman, A. and Bell, E., eds. (2007). *Business Research Methods*, Third Edition, New York: Oxford University Press.

Chiaburu, D. S. and Lindsay, D. R. (2008). "Can Do or Will Do? The Importance of Self-efficacy and Instrumentality for Training Transfer", *Human Resources Development International*, 11(2), 199–206.

Chiaburu, D. S. and Marinova, S. V. (2005). "What Predicts Skills Transfer? An Exploratory Study of Goal Orientation, Training Self-Efficacy and Organizational Supports", *International Journal of Training and Development*, 9(2), 110–123.

Chiaburu, D. S. and Tekleab, A. G. (2005). "Individual and Contextual Influences on Multiple Dimensions of Training Effectiveness", *Journal of European Industrial Training*, 29(8), 604–626.

Compeau, D. and Higgins, C. (1995). "Application of Social Cognitive Theory to Training For Computer Skills", *Information Systems Research*, 6(2), 118–143. doi:10.1287/isre.6.2.118.

Davis, F. D. (1989). "Perceived Usefulness, Perceived Ease of Use, and User Acceptance of Information Technology", *MIS Quarterly*, 13(3), 319–339. doi:10.2307/249008.

Eastman, C., Teicholz, P., Sacks, R. and Liston, K. (2011). *BIM Handbook: A Guide to Building Information Modeling for Owners, Managers, Designers, Engineers and Contractors*, Hoboken, N.J.: John Wiley & Sons.

Engeström, Y. and Sannino, A. (2010). "Studies of Expansive Learning: Foundations, Findings and Future Challenges", *Educational Research Review*, 5(1), 1–24.

Garavan, T. N., Carbery, R., O'Malley, G. and O'Donnell, D. (2010). "Understanding Participation in e-learning in Organizations: A Large Scale Empirical Study of Employees", *International Journal of Training and Development*, 14(3), 155–168.

Ghaffarianhoseini, A., Tookey, J., Ghaffarianhoseini, A., Naismith, N., Azhard, S., Efimovaa, O. and Raahemifar, K. (2016). "Building Information Modelling (BIM) Uptake: Clear Benefits, Understanding its Implementation, Risks and Challenges", *Renewable and Sustainable Energy Reviews*, 75, 1046–1053.

Klein, H. J., Noe, R. A. and Wang, C. (2006). "Motivation to Learn and Course Outcomes: The Impact of Delivery Mode, Learning Goal Orientation and Perceived Barriers and Enablers", *Personnel Psychology*, 59, 665–702.

Miettinen, R. and Paavola, S. (2014). "Beyond the BIM Utopia: Approaches to the Development and Implementation of Building Information Modelling", *Automation in Construction*, 43, 84–91.

Noe, R. (1986). "Trainees' Attributes and Attitudes: Neglected Influences on Training Effectiveness", *Academy of Management Review*, 11, 736–749.

Nonaka, I. and Krogh, G. V. (2009). "Tacit knowledge and Knowledge Conversion: Controversy and Advancement in Organizational Knowledge Creation Theory", *Organisation Science*, 20(3), 635–652. https://doi.org/10.1287/orsc.1080.0412.

Nonaka, I. and Nishihara, A. H. (2018). Introduction to the Concepts and Frameworks of Knowledge-Creating Theory. In *Knowledge Creation in Community Development*, London: Palgrave Macmillan, pp. 1–15.

Park, Y., Son, H. and Kim, C. (2012). "Investigating the Determinants of Construction Professionals' Acceptance of Web-Based Training: An Extension of the Technology Acceptance Model", *Automation in Construction*, 22, 377–386.

Rowold, J. (2007). "Individual Influences on Knowledge Acquisition in a Call Centre Training Context in Germany", *International Journal of Training and Development*, 11(1), 21–34.

Sacks, R., Koskela, L., Dave, B. A. and Owen, R. (2010). "Interaction of Lean and Building Information Modeling in Construction", *Journal of Construction Engineering and Management*, 136(9), 968–980. doi:10.1061/(ASCE)CO.1943-7862.0000203.

Son, H., Lee, S. and Kim, C. (2015). "What Drives the Adoption of Building Information Modeling in Design Organizations? An Empirical Investigation of the Antecedents Affecting Architects' Behavioural Intentions", *Automation in Construction*, 49, 92–99.

Wong, A. K. D., Wong, F. K. W. and Nadeem, A. (2010). "Attributes of Building Information Modelling Implementation in Various Countries", *Architectural Engineering and Design Management*, 6(4), 288–302.

Zakaria, Z., Ali, M. N., Haron, A. T., Ponting, A. M. and Hamid, Z. A. (2013). "Exploring the Adoption of Building Information Modelling (BIM) in the Malaysian Construction Industry: A Qualitative Approach", *International Journal of Research in Engineering and Technology*, 2(8), 384–395.

Chapter 22

Corporate Social Responsibility (CSR) Reporting Practices in the Australian Construction Industry

Bo Xia*, Nur Zainal and Robin Drogemuller

*Queensland University of Technology,
Brisbane City, QLD 4000, Australia
paul.xia@qut.edu.au

Linlin Xie

*South China University of Technology,
Guangdong Sheng, China*

Yong Liu

*Zhejiang Sci-Tech University, Hangzhou,
Zhejiang, China*

Xiaoyan Jiang

*Hefei University of Technology, Hefei Shi,
Anhui Sheng, China*

---

*Corresponding author.

Abstract

Corporate social responsibility (CSR) has gained increasing attention among academics and practitioners in the construction industry. To evaluate CSR practices in the Australian context, 20 CSR reports provided by leading construction companies on their public websites are selected for content analysis. The findings reveal that CSR reporting is normally incorporated in published documents such as annual reports, sustainability reports and corporate responsibility and sustainability (CR&S) reports, with annual reports being the most frequently used ones. The analysis shows that CSR practices of these leading construction companies cover the themes of corporate governance, environment management and healthy working environment (HWE). A total of 28 CSR practices relating to each category were identified, among which the ethics and moral values within company structure, community development and employment opportunities and OHS training and engagement are the most frequently adopted by the sample companies. The findings of content analysis provide insight as to how Australian top construction companies conducted CSR reporting and provide recommendations for the improvement.

Keywords: Sustainability; Corporate Social Responsibility (CSR); CSR Reporting; Construction Industry; Australia.

1. Introduction

Corporate social responsibility (CSR) is regarded as the voluntary initiative carried out by an entity to express its ethical contributions in enhancing the quality of life of the employees and the community (Asongu, 2007; Tan *et al.*, 2011). It requires the company to do the right thing, be a good citizen, build up sound reputation among the employees, unions, suppliers, community representatives, and provide an accountable and transparent CSR reporting (Yadong, 2007).

In the construction industry, construction activities not only influence the national Gross Domestic Product, provide job opportunities to the residents and allow for diversification in buildings and facilities to satisfy the demand of consumers, but also lead to serious environmental issues such as dust and gas emission, noise pollution, waste generation, energy consumption and misuse of water and land. Therefore, construction companies are obliged to fulfil the requirements of CSR and contribute to the implementation of sustainable construction (Brammer and Pavelin, 2006; Jose and Lee, 2007; Petrovic-Lazarevic, 2010; Glass, 2012; Wu *et al.*, 2014). Furthermore, it is revealed that companies performing well in engaging CSR will be able to boost company image and push the profit margin to its above-average level (Jose and Lee, 2007; Zhao *et al.*, 2012).

In order to demonstrate the company's transparency and integrity on its CSR strategies, actions and achievements, CSR reporting is normally adopted as the major means. CSR reporting is a tool to communicate company achievements and best practices (Myers, 2005; Bouten *et al.*, 2011), a means to reach out to the interest groups showing that the company has fulfilled their expectations (Andrews, 2011), and a media to exemplify that the company is capable of performing its duty and obligations to the society at large (Reynolds and Yuthas, 2008). Spence (2009) and Sutantoputra *et al.* (2012) believe that some investors even depend on CSR reporting to gather market information for the assessment of their investment decisions.

However, even though CSR reporting is becoming increasingly favourable, there are still no legislations or specific international consensus/standard outlining the ideal reporting strategy for CSR (Davidson, 2011), with Australia being no exception. The existing CSR reports are largely voluntarily produced based on international guidelines and customized for company's own need and preference (Petrovic-Lazarevic, 2008; Zhao *et al.*, 2012). As a result, there are significant variations in the content style and quality of reporting (Jose and Lee, 2007; Glass, 2012), making it difficult to evaluate the CSR commitment or achievements of construction companies (Zhao *et al.*, 2016).

Therefore, this chapter aims to investigate the CSR reporting practices of Australian construction companies and reveal what CSR practices/aspects are covered in their CSR reports. The findings of this study will help to understand how the Australian construction industry perceives the CSR and paves the way for developing a national standard of CSR reporting practice.

2. Literature Review

Although there are various notions contributing to the development of the definition for CSR, a universal meaning is yet to be established. World Business Council for Sustainable Development (WBCSD) suggests that CSR is an organization's or entity's ethical contribution in providing better life for workforces, local community and society (Asongu, 2007). Meanwhile, Commission for European Communities describes CSR as the organization's voluntary initiative to integrate the environmental and social aspects within its business operation (Tan *et al.*, 2011).

In the construction industry, the engagement of CSR practice shows the voluntary recognition of the company by taking into account the externalities from the construction players' behaviour, construction works and transaction activities (Gond *et al.*, 2011). CSR practices in construction industry should comprise a variety of aspects such as occupational health and safety, sustainable construction, human resources, customers and communities, human rights and ethical conduct, and corporate governance (e.g. Jones and Comfort, 2006; Petrovic-Lazarevic, 2010). In recent years, large and leading construction companies around the world, normally multinational ones, increasingly recognize the impact of their construction business on environmental, economical and social development, and begin to commit CSR practices among various stakeholders, such as their customers, employees and the community residents (e.g. Jones *et al.*, 2006; Change *et al.*, 2016).

With CSR getting more attention in construction activities, it creates opportunities in promoting competitive advantages among construction companies through productive human resources and efficient management in the supply chain (Asongu, 2007). Huang and Lien (2012) point out that CSR creates a "reputation insurance" that acts as a protection upon any impacts due to the risks on profitability and the values of company organizations. Zhao *et al.* (2012) agree that CSR has the potential to enhance construction company's reputation and financial position. Such CSR reputation can help construction companies improve relations with bankers and investors and eventually facilitate their access to capital (Orlitzky *et al.*, 2003). Corporate value and care for the environment are seen as compelling factors to attract more investors and customers due to the company's ability in meeting the expectations (Azim *et al.*, 2009; Galbreath, 2010). The recent research by Wang *et al.* (2016) confirms a positive and linear impact of CSR practices on return of assets (ROA) and earnings per share (EPS) of construction companies.

In order to demonstrate a company's commitment to CSR, CSR reporting has been widely used in various industries. CSR reporting is defined as "the process of communicating the social and environmental effects of the economic actions of organizations to particular interest groups within society and to society at large" (Gray *et al.*, 1996). It is regarded as one of the most effective instruments for companies to not only convey external communication of CSR but also the internal CSR management processes (Paul *et al.*, 2006; Lock and Seele, 2015). As a result, recent years have witnessed an increasing number of companies resorting to CSR reporting practices globally (Cooper and Owen, 2007).

However, CSR reporting is unregulated in most countries, and there is no mandatory requirement for CSR reporting. According to Nielsen and Thomsen (2007), there is no established framework, but only guidelines on how to communicate consistently about CSR. As a result, companies choose different tools for reporting and different content styles in CSR reports, resulting in different quality of reporting (Jose and Lee, 2007; Glass, 2012). Furthermore, many organizations find it difficult to develop consistent strategies for CSR reporting and communication (Nielsen and Thomsen, 2007).

Although there is no common approach for CSR reporting, a growing number of global schemes and standard frameworks have been developed to help manage the uniformity of CSR reporting practices, such as Global Reporting Initiative (GRI), UN Global Compact, and the global Sullivan principles. CSR reports produced by construction companies in Australia are normally based on the Global Reporting Initiative (GRI) guideline or in accordance with industry standards, such as AS/NZS ISO 14001, AS/NZS 4801, OHSAS 18001 and AS/NZS ISO 9001. In particular, the Global Reporting Initiative (GRI) framework covers the triple bottom-line of social, environmental and economic aspects, and has been widely accepted by many industrial sectors because it forms a basis of standardized and comparable reporting for organizations on sustainability and CSR (Reynolds and Yuthas, 2008; Azim *et al.*, 2009; Andrews, 2011; Zuo *et al.*, 2012).

3. Research Methods

This study focused on evaluating CSR reporting practices of construction companies in Australia. Given that CSR reporting is not mandatory in Australia and normally only the large companies (usually stock-listed ones) provide CSR reports, this study adopted a purposive sampling to identify the sample group. The purposive sampling technique is a type of non-probability sampling, where researchers decide what needs to be known and set out to find appropriate samples providing the information by virtue of knowledge or experience (Tongco, 2007). Purposive sampling was carried out based on the following two criteria:

(1) The construction companies were listed as "Australia's Top 500 Private Companies in 2013" in Business Review Weekly Magazine (Ruthven, 2013) and the "Top 50 Construction Firms in Australia" in WhiteCard Australia website (Cutforth, 2012) and
(2) The companies have company websites publicly available.

From the purposive sampling process, 70 construction companies across Australia were chosen for this study. These companies were further classified according to the CSR reporting information they provided. Only those companies that managed to produce CSR reports, either as a separate document or incorporated in other reports (e.g. annual reports, sustainability reports, etc.), and made them available on their company websites, are assessed to form the basis of analysis of this study. Finally, 20 companies were selected for the research purpose.

After that, a comprehensive content analysis of CSR reports was conducted. Content analysis is an observational research method that is used to systematically evaluate symbolic content of all forms of recorded communications (Kolbe and Burnett, 1991). By means of counting the frequency of events or a topic depicted, content analysis is a useful approach to capture the viewpoints of various participants which consequently are coded into key constructs or emerging themes (Nayak and Taylor 2009; Xia *et al.*, 2012, 2013). The contents of CSR reports of each company were analyzed. First, all CSR practices provided in the reports were recorded. Second, practices with similar meanings were assembled and subsequently coded into different main themes or categories. Finally, the overall frequency of each practice throughout the entire collection of CSR reports was examined to show the popularity of these practices.

4. Results and Analysis

4.1. *CSR reporting categories*

Of the 70 companies selected as the sample for this study, they were further classified into three categories based on availability levels of CSR practices from their public websites:

(1) Category A — online CSR reports were provided,
(2) Category B — no online reports but some CSR information was mentioned,
(3) Category C — no CSR information provided in the website.

As shown in Fig. 1, 20 companies (28.6%) provided CSR reports on their websites, 30 companies mentioned CSR information without CSR reports, and the remaining 20 companies, which might have adopted CSR practices within their business operations and construction activities, did not provide any CSR information on their websites. Such differences can be attributed to the additional cost incurred to form special management team/committee to administer the implementation and reporting of CSR (Brammer and Pavelin, 2006), voluntary nature of reporting (Sutantoputra *et al.*, 2012), and the lack of legal enforcement in reporting CSR practices (Myers, 2005).

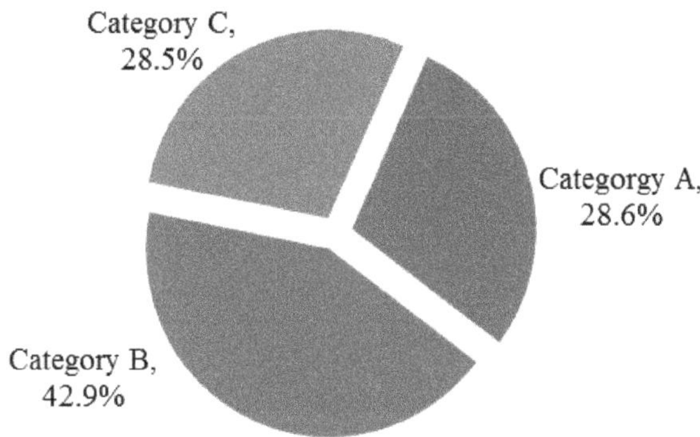

Fig.1.   Classification of construction companies based on CSR information.

It is noticed that in all 20 categories, A companies are tier 1 and tier 2 construction organizations. In addition, 75% (15 out of 20) of them are the top 50 contractors in Australia. The remaining ones are also large and leading companies in construction, consulting and specific urban development sectors (e.g. transport infrastructure, resources and energy sectors, etc.). This is understandable given that normally large and well-established construction companies have a preference and are more likely to produce a standalone CSR report or allocate a specific section of it in the annual financial report due to relative cost advantages (Schreck and Raithel, 2015; Chang *et al.*, 2016). For those companies that did not engage with CSR, they only produced traditional reports with the financial aspect of the company's performance.

Of these 20 companies providing online CSR reports, 55% provided CSR reporting through Annual Reports, 35% communicated it through Sustainability Reports, and the remaining 10% disclosed their CSR practices through corporate responsibility and sustainability (CR&S) Reports. All these reports were the latest versions that can be found on their public websites, ranging from 2009 until 2014 (when this research was conducted).

Compared with the annual report, both Sustainability Report and CR&S Report are standalone/independent non-financial reports, which provide the information on corporate responsibility — environmental and social performance (Parliament of Australia, 2010). Different companies use different names for these reports, however, both reports are aimed at communicating the achievements of the company in sustainability and corporate responsibility to the stakeholders and public at large (Global Reporting Initiative, 2014).

Furthermore, eight companies (40%) mentioned that they provided CSR reporting according to the GRI Sustainable Reporting Guidelines. The remaining 12 (60%) of companies adopted the company-specific framework and benchmarked it against the industry standards such as AS/NZS ISO 14001, AS/NZS 4801, OHSAS 18001 and ISO 9001.

4.2. *Analysis of CSR practices*

Twenty-eight (28) CSR practices were identified and recorded through the content analysis of CSR reports (see Table 1). Clusters of themes of these CSR practices were summarized and organized based on their meanings.

Table 1.   Coded categories of CSR practices.

No.	CSR Practices	Frequency	Percent
1	Corporate governance		
1.1	Ethics and moral values within company structure	20	100
1.2	Community development and employment opportunities	19	95
1.3	Charity foundation and financial support	18	90
1.4	Introduction of Social Responsibility Committee or equivalent	18	90
1.5	Risk management strategy	18	90
1.6	Leadership development strategy/ opportunity	18	90
1.7	Stakeholders management strategy	17	85
1.8	Internal and/or external audit system	16	80
1.9	Training on business conduct	14	70
1.10	Supplier management strategy	10	50
2	Environmental Management		
2.1	Management on waste disposal, water and energy consumption, and carbon emission	17	85
2.2	Training of environmental awareness improvement	17	85
2.3	Engagement of environment management system (EMS)	15	75
2.4	Engagement on Green Star and/or ESD projects	12	60
2.5	Annual assessment of environment management system	12	60
2.6	Systematic documentation on environmental issues	11	55
2.7	Environmental risks management	11	55
2.8	Supply chain management	9	45

(Continued)

Table 1.   (*Continued*)

No.	CSR Practices	Frequency	Percent
3	Healthy working environment (HWE)		
3.1	OHS training and engagement	19	95
3.2	Safety performance indicator and reporting	17	85
3.3	Diversity and equal opportunity	17	85
3.4	Graduate or apprenticeship program	17	85
3.5	Employee support service	16	80
3.6	Employee performance assessment	15	75
3.7	Employee engagement survey	12	60
3.8	Development of mobile application and online programs	12	60
3.9	Flexible working hour and adequate sick/ parental leave	9	45
3.10	Fair remuneration structure	7	35

Finally, three main categories of these CSR practices were coded, comprising (1) corporate government, (2) environmental management and (3) health working environment.

4.2.1. *CSR practices of corporate governance*

Good corporate governance can be characterized as ethical and moral obligations in doing the right thing and good relationship with external entities such as suppliers and community (Yadong, 2007). Petrovic-Lazarevic (2010) points out that good corporate governance provides a good platform for construction companies to express their awareness in sustainability and then benefit the society. Corporate governance is assumed to be company's intangible assets which can be directly related to the development of company reputation among stakeholders and community (e.g. Petrovic-Lazarevic, 2010; Zuo *et al.*, 2012a).

As shown in Table 1, all companies managed to demonstrate their commitment in the engagement of ethics and moral values within the company's corporate structure (item 1.1). Similarly, majority of them established internal and/or external audit systems (item 1.8) and conducted training on business conduct (item 1.9). All clearly show the fundamental importance of doing business ethically and legally. In addition, most companies valued community development and employment opportunities (item 1.2), provided charity foundation and financial support (item 1.3), and adopted stakeholder management strategies (item 1.7). These CSR practices show the contribution of construction companies to their community. According to Galbreath (2010), societal contribution leads to an enhanced reputation and stakeholders' satisfaction.

4.2.2. *CSR practices of environmental management*

In Australia, the construction and demolition sector accounts for 38% of total waste. Therefore, management on waste disposal, water and energy consumption, and carbon emission (item 2.1) is the most frequently adopted CSR practice (85%) by these construction companies.

Engaging Environmental Management System (items 2.2, 2.3, 2.5 and 2.6) is also another common CSR practice. An Environment Management System (EMS) is a tool for managing the impacts of an organization's activities on the environment (Department of Environment, 2015). The implementation of environmental management system within business operation of construction companies may develop a good relationship with the stakeholders and improve company image (Davidson, 2011; Freeman and Hasnaoui, 2011). The CSR practice of engaging Green Star and/or Environmental Sustainable Design projects (item 2.4) is in accordance with the increasing interest from the investors and public in sustainable development and green buildings. Many companies adopted rating tools such as Green Star rating and National Australian Built Environment Rating System (NABERS) for commercial buildings to ensure that the projects are carried out to meet the environmental performance target. However, given the significant cost associated with the delivery of sustainable development, it is understandable that the adoption rate of this practice is relatively low (60%).

4.2.3. *CSR practice of healthy working environment*

HWE is the commitment of the construction company to engage in the development of employee productivity and the community improvement through welfare support (Zhao *et al.*, 2012). Many construction jobs are considered to be risky with high possibility for on-site accidents and injuries. Therefore, occupational health and safety (OHS) has been given the most attention and priority by construction companies (items 3.1 and 3.2).

The diversification and equal opportunity in the workplace (item 3.3) is an important aspect of HWE in the Australian construction industry. It covers the aspects of gender, racial and age diversity as well as equal employment opportunity. Construction companies in Australia recognize and provide equal employment opportunities to women in non-traditional roles, site-based roles and leadership roles, in compliance with the requirements outlined in Workplace Gender Equality Act (WGEA) 2012. Equal employment opportunity has also been given to the indigenous people to promote cultural diversification for community development. Some companies also encourage the participation of the disabled to eradicate the issue on employment-related discrimination.

In order to improve the working productivity and employee career development, CSR practices of graduate or apprenticeship program, employee support service, employee performance assessment, and employee engagement survey are commonly adopted by Australian construction companies. The content analysis results also indicate that flexible working hours and paid sick leave/parental leave policy (item 3.9) help employees to have a balanced life and healthy well-being. In terms of remuneration structure (item 3.10), the adoption of equitable remuneration approach will provide justice in workplace and encourage career development. Finally, with regard to the use of technological advancement (item 3.8), 60% sample companies have mobile applications and e-learning programmes to help their employees get access to company's intranet and other online systems.

5. Discussion and Conclusions

The content analysis aims to provide insight into to how construction companies conducted CSR reporting. The findings reveal that CSR reporting is normally incorporated in the published documents such as Annual Report, Sustainability Report and CR&S Report. Although Sustainability Report provides the best coverage for CSR reporting, the Annual Report is the most widely used to incorporate practices of CSR reporting as producing a stand-alone sustainability report will impose additional capital expenses for reporting and quality monitoring.

The analysis results show that CSR practices mainly cover the categories of corporate governance, environment management and HWE. A total of 28 CSR practices relating to each category were identified, among which the ethics and moral values within company structure (100%), community development and employment opportunities (95%), and OHS training and engagement (95%) are the most frequently adopted ones. However, CSR practices relating to economic and social dimensions of sustainability were rarely covered.

The findings suggest that unlike financial reporting, the information on corporate responsibility practices is subjective in nature and no acceptable auditing standards exist in evaluating the environmental and social data. Thus, construction companies are recommended to conduct the CSR reporting in accordance with GRI reporting standards and industrial standards such as AS/NZS ISO 14001, AS 4801, OHSAS 18001, AS/NZS 9001. Additionally, given the increasing demand for CSR disclosure to internal managers and external communities, it is of increasing importance to improve communication with stakeholders, especially with the external ones.

Furthermore, given that the CSR practices adopted by those leading contractors are focused predominantly on health and safe working environment and environmental dimension of sustainability, it implies that these two categories of CSR are currently perceived as the most important issues in the current Australian construction industry.

As the GRI Guidelines for sustainability reports is the most prominent and influential international initiative, a comparison between the categories of CSR practices from this study and those recommended in GRI framework was conducted to reveal sample companies' unique perception on sustainability.

As shown in Table 2, the economic category of sustainability practices has been rarely covered in sample construction companies. This is not surprising as the primary goal of most construction companies is to

Table 2. Comparison of CSR practices between GRI and current study.

Global Reporting Initiatives (GRI)	Findings of Current Study
Economic dimensions	Rarely covered
Environmental dimensions	Mostly covered
Social dimensions	
Labour practice and decent work	Mostly covered
Human rights	Rarely covered
Society	Rarely covered
Product responsibility	Rarely covered

maximize profit (Myer, 2005). Meanwhile, the environmental sustainability category of CSR practices has been mostly covered in the sample CSR reports. It is mainly due to the wide recognition of the society on negative impacts of the construction industry on the eco-environment system. In the social sustainability category, the dimension of labour practice and decent work has been widely covered in the sample CSR reports, however, the dimensions of human rights, society and product responsibility were rarely, if not none, covered in the sample construction companies. This is confirmed with current studies not only in Australia but also globally (e.g. Myers, 2005; Petrovic-Lazarevic, 2008; Zhao *et al.*, 2016).

Even though this empirical study of content analysis revealed the current state of the art of CSR reporting and real CSR practices adopted in the Australian construction industry, this study suffered a number of limitations. First, considering that all the sample companies are large and leading contactors in the industry, the findings may not be applicable to the whole industry, where most of construction companies are small-to-medium ones. In addition, as the companies without public available CSR report can still practice CSR in their internal processes, as such, the sample may be biased.

However, given the unique features of the sample companies (i.e. all top construction companies in a specific geographical region) and the nature of content analysis (i.e. studying the "mute evidence" of texts and artifacts), this study contributed to a better understanding of

Australian leading contractors' perception on social corporate responsibility (e.g. what CSR practices to commit, how to convey the message, etc.). In addition, after an in-depth comparison between the CSR practices/categories adopted by those companies and those recommend by Global Reporting Initiatives (GRI), this chapter is able to reveal the sector specific findings in the CSR area, which is another major contribution to the current body of knowledge.

References

Andrews, O. (2011). "Getting Started on Sustainability Reporting," *Environmental Quality Management*, 11(3), 3–11.

Asongu, J. J. (2007). "Innovation as an Argument for Corporate Social Responsibility", *Journal of Business and Public Policy*, 1(3), 1–21.

Azim, M., Ahmed, S. and Islam, M. (2009). "Corporate Social Reporting Practice: Evidence from Listed Companies in Bangladesh", *Journal of Asia-Pacific Business*, 10(2), 130–145.

Bouten, L., Everaert, P., Van Liedekerke, L., De Moor, L. and Christiaens, J. (2011). "Corporate Social Responsibility Reporting: A Comprehensive Picture?" *Accounting Forum*, 35(3), 187–204.

Brammer, S. and Pavelin, S. (2006). "Voluntary Environmental Disclosures by Large UK Companies", *Journal of Business Finance & Accounting*, 33(7–8), 1168–1188.

Chang, R., Zuo, J., Soebarto, V., Zhao, Z., Zillante, G. and Gan, X. (2016). "Sustainability Transition of the Chinese Construction Industry: Practices and Behaviors of the Leading Construction Firms", *Journal of Management in Engineering*, 32(4), 05016009.

Cooper, S. and Owen, D. (2007). "Corporate Social Reporting and Stakeholder Accountability: The Missing Link", *Accounting, Organizations and Society*, 32: 649–667.

Cutforth, P. (2012). "Top 50 Construction Firms in Australia". Available at: http://www.whitecardaustralia.com.au/white-card-news/top-50-construction-firms-in-australia/ (Accessed on July 9 2015).

Davidson, K. (2011). "Reporting Systems for Sustainability: What Are They Measuring?", *Social Indicators Research*, 100(2), 351–365.

Freeman, I. and Hasnaoui, A. (2011). "The Meaning of Corporate Social Responsibility: The Vision of Four Nations", *Journal of Business Ethics*, 100(3), 419–443.

Galbreath, J. (2010). "How does Corporate Social Responsibility Benefit Firms? Evidence from Australia", *European Business Review*, 22(4), 411–431.

Glass, J. (2012). "The State of Sustainability Reporting in the Construction Sector", *Smart and Sustainable Built Environment*, 1(1), 87–104.

Global Reporting Initiative (2014). "About Sustainability Reporting". Available at: https://www.globalreporting.org/information/sustainability-reporting/Pages/default.aspx (Accessed on July 9, 2015).

Gond, J., Kang, N. and Moon, J. (2011). "The Government of Self-regulation: On the Comparative Dynamics of Corporate Social Responsibility", *Economy and Society*, 40(4), 640–671.

Gray, R., Owens, D. and Adams, C. (1996). *Accounting and Accountability: Changes and Challenges in Corporate Social and Environmental Reporting*, London: Prentice-Hall.

Huang, C. F. and Lien, H. C. (2012). "An Empirical Analysis of Influences of Corporate Social Responsibility on Organizational Performance of Taiwan's Construction Industry: Using Corporate Image as a Mediator", *Construction Management and Economics*, 30(4), 263–275.

Jones, P., Comfort, D. and Hillier, D. (2006). "Corporate Social Responsibility and the UK Construction Industry", *Journal of Corporate Real Estate*, 8(3), 134–150.

Jose, A. and Lee, S. M. (2007). "Environmental Reporting of Global Corporations: A Content Analysis Based on Website Disclosures", *Journal of Business Ethics*, 72(4), 307–321.

Kolbe, R. H. and Burnett, M. S. (1991). "Content Analysis Research: An Examination of Applications with Directives for Improving Research Reliability and Objectivity", *Journal of Consumer Research*, 18, 243–250.

Lock, I. and Seele, P. (2015). "Analyzing Sector-Specific CSR Reporting: Social and Environmental Disclosure to Investors in the Chemicals and Banking and Insurance Industry", *Corporate Social Responsibility and Environmental Management*, 22, 113–128.

Myers, D. (2005). "A Review of Construction Companies' Attitudes to Sustainability", *Construction Management and Economics*, 23(8), 781–785.

Nayak, N. V. and Taylor, J. E. (2009). "Offshore Outsourcing in Global Design Networks", *Journal of Management in Engineering*, 25(4), 177–184.

Nielsen, A. E. and Thomsen, C. (2007). "Reporting CSR — What and How to Say It?", *Corporate Communications: An International Journal*, 12(1), 25–40.

Orlitzky, M., Schmidt, F. L. and Rynes, S. L. (2003). "Corporate Social and Financial Performance: A Meta-analysis", *Organization Studies*, 24(3) 403–441.

Parliament of Australia (2010). "Sustainability Reporting". Available at: http://www.aph.gov.au/About_Parliament/Parliamentary_Departments/Parliamentary_Library/Browse_by_Topic/ClimateChange/responses/economic/Sustainability (Accessed on July 9, 2015).

Paul, K., Cobas, E., Ceron, R., Frithiof, M., Maass, A., Navarro I. and Palmer, L. (2006). "Corporate Social Reporting in Mexico", *The Journal of Corporate Citizenship*, 22 (Summer 2006), 67–80.

Petrovic-Lazarevic, S. (2008). "The Development of Corporate Social Responsibility in the Australian Construction Industry", *Construction Management and Economics*, 26(2), 93–101.

Petrovic-Lazarevic, S. (2010). "Good Corporate Citizenship in the Australian Construction Industry", *Corporate Governance*, 10(2), 115–128.

Porter, M. and Kramer, M. (2006). "Strategy and Society: The Link between Competitive Advantage and Corporate Responsibility", *Harvard Business Review*, 84(12), 78–92.

Reynolds, M. and Yuthas, K. (2008). "Moral Discourse and Corporate Social Reporting", *Journal of Business Ethics*, 78(1–2), 47–64.

Ruthven, P. (2013). "Revealed: Australia's 500 top private companies in 2013". Available at: http://www.brw.com.au/p/business/revealed_australia_top_private_companies_KoGuU8Lc6xBkz6FH0v28BN (Accessed on July 9, 2015).

Schreck, P. and Raithel, S. (2015). "Corporate Social Performance, Firm Size, and Organizational Visibility Distinct and Joint Effects on Voluntary Sustainability Reporting", *Business & Society*, doi: 0007650315613120.

Spence, C. (2009). "Social and Environmental Reporting and the Corporate Ego", *Business Strategy and the Environment*, 18, 254–265.

Sutantoputra, A., Lindorff, M. and Prior Johnson, E. (2012). "The Relationship between Environmental Performance and Environmental Disclosure", *Australasian Journal of Environmental Management*, 19(1), 51–65.

Tan, Y., Shen, L. and Yao, H. (2011). "Sustainable Construction Practice and Contractors' Competitiveness: A Preliminary Study", *Habitat International*, 35(2), 225–230.

Tongco, M. D. C. (2007). "Purposive Sampling as a Tool for Informant Selection", *Ethnobotany Research & Applications*, 5, 147–158.

Wu, P., Xia, B., Pienaar, J. and Zhao, X. (2014). "The Past, Present and Future of Carbon Labelling for Construction Materials — A Review", *Building and Environment*, 77, 160–168.

Xia, B., Skitmore, M. and Zuo, J. (2012). "Evaluation of Design-Builder Qualifications Through the Analysis of Requests for Qualifications", *Journal of Management in Engineering*, 28(3), 348–351.

Xia, B., Zuo, J., Skitmore, M., Pullen, S. and Chen, Q. (2013). "Green Star Points Obtained by Australian Building Projects", *Journal of Architectural Engineering*, 19(4), 302–308.

Yadong, L. (2007). *Global Activities of Corporate Governance*, Oxford: Blackwell Publishing.

Zhao, Z., Zhao, X., Davidson, K. and Zuo, J. (2012). "A Corporate Social Responsibility Indicator System for Construction Enterprises", *Journal of Cleaner Production*, 29(30), 277–289.

Zhao, Z. Y., Zhao, X. J., Zuo, J. and Zillante, G. (2016). "Corporate Social Responsibility for Construction Contractors: A China Study", *Journal of Engineering, Design and Technology*, 14(3), 614–640.

Zuo, J., Zillante, G., Xia, B., Chan, A. and Zhao, Z. (2015). "How Australian Construction Contractors Responded to the Economic Downturn", *International Journal of Strategic Property Management*, 19(3), 245–259.

Zuo, J., Zillante, G., Wilson, L., Davidson, K. and Pullen, S. (2012a). "Sustainability Policy of Construction Contractors: A Review", *Renewable and Sustainability Energy Reviews*, 16, 3910–3916.

Zuo, J., Jin, X. H. and Flynn, L. (2012b). "Social Sustainability in Construction — An Explorative Study", *International Journal of Construction Management*, 12(2), 51–63.

Chapter 23

Application of DMAIC Approach to Improve the Centralized Procurement Process in Construction Logistics Enterprises

Zhongya Mei and Maozeng Xu

*School of Economics and Management,
Chongqing Jiaotong University, Chongqing, China*

Yi Tan*

*Department of Civil and Environmental Engineering,
The Hong Kong University of Science and Technology,
Hong Kong, China*
ytanai@connect.ust.hk

Abstract

Due to the advantages of the centralized procurement of materials in the AEC industry, the number of emerging logistics enterprises has been on a rise in recent years in China. According to the manner of payments in construction projects, such logistics enterprises play the role of pur-

---

*Corresponding author.

chasing the construction materials for projects under construction by advancing funds. However, redundant processes in construction logistics enterprises result in massive funds being piled up and lower efficiency, which usually weakens the advantages of the centralized procurement. Therefore, this chapter presents a case study on the process improvement to identify the appropriate method to eliminate the valueless process in centralized procurement and improve the productivity in the construction logistics enterprise. Six Sigma is an advanced self-control process that aims to produce or deliver near-perfect products and services, which is implemented in this chapter by integrating DMAIC approach and flow chart. Taking account of the distinct properties between the AEC industry and the manufacturing industry, this chapter first constructs the configuration of construction logistics under centralized procurement, which facilitates the process differently from that existing in most previous researches. The current process in a construction logistics enterprise is represented in the flow chart format. Then according to the flow chart in practice, the process improvement is achieved by using DMAIC approach containing five basic steps. Finally, a case study is used to verify the performance of the improved process by comparing it with the existing process. The entire process time and the actual work time have been reduced by 52% and 53%, respectively, in this case.

Keywords: Construction Productivity Development; Construction Logistics Enterprises; Process Improvement; DMAIC; Flow Chart.

1. Introduction

In an effort to reduce waste in construction projects, hence promoting value creation, the focus on the construction logistics becomes increasingly important, especially with increasing complexity of the project (Vitra, 2012). Due to the manner of payments in construction projects, the emerging construction logistics enterprises can purchase materials by advancing funds for projects to facilitate the fund turnover on sites. Accordingly, integrating and purchasing the quantities of materials required on sites can benefit the stakeholders by economic scale advantages. Therefore, considering the economical efficiency and productivity, the construction logistics is urged to move from the traditional decentralized procurement pattern towards the cost economy and time effective method of the centralized procurement pattern.

However, challenges exist in the implementation of the centralized procurement process. The first challenge is that a construction project normally consists of several suppliers, sub-contractors (Sub) and other stakeholders, which increases the complexity of material procurement on sites as shown in Fig. 1. The complex relationships among suppliers and Sub make the material procurement not cost-effective and inefficient. Secondly, as an emerging enterprise in construction material management, there is no reference standard for the implementation of the centralized procurement process, which leads to difficulties in evaluating and improving the performance of the process. In order to overcome these barriers for better utilizing the centralized procurement process in construction logistics enterprises, a systematic approach is needed.

The application of Six Sigma in construction has been widely studied in the past few years (Marhani *et al.*, 2013; Youssouf *et al.*, 2014). In fact, there is a trend to adopt the Six Sigma approach to deal with the various process problems in construction. Heravi and Firoozi (2016) used value stream mapping (VSM) as a lean technique in the production phase of prefabricated steel frames in a case study and the results showed that lean principles can be successfully applied to prefabricated construction. In the project delivery, Purnapandhega and Adhiutama (2016) utilized the Lean Six Sigma to improve the process of delivering the bridge project with efficient cost and meet the quality requirement.

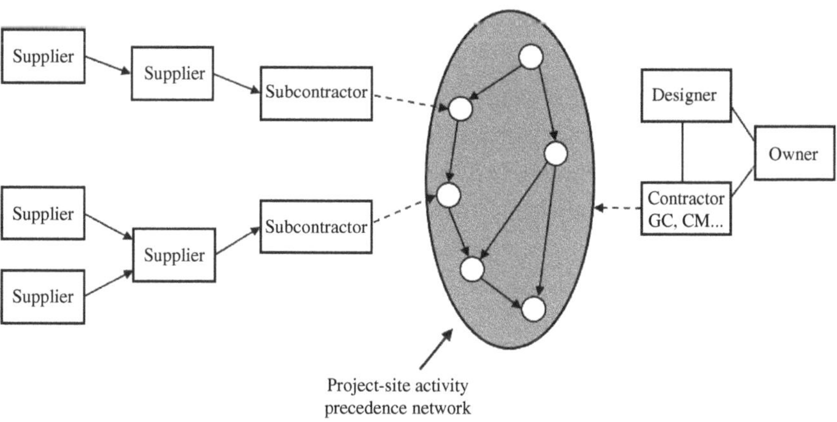

Fig. 1.   Conceptual view of the project SC (O'Brien *et al.*, 2009).

The Six Sigma principle was also applied to control slab cracks in the framework in construction stage (Eom *et al.*, 2015). Due to the large influence on the overall performance of the purchasing function, the maturity of the purchasing function in construction firms was measured and explained (Lith *et al.*, 2015); however, the purchasing processes have not been discussed. By far, the previous studies have addressed some process issues in construction enterprises by implementing the Six Sigma approach, but the researches focussing on the material procurement process are limited, especially the centralized procurement process in construction logistics enterprises. Hence, there is a need to identify the appropriate approach to improve the process in practice.

The structure of this chapter is organized as follows: In Section 2, the configuration of construction logistics in centralized procurement is constructed. Section 3 presents the detailed implementation of the DMAIC approach. Section 4 concludes the chapter and suggests future work.

2. Configuration of Construction Logistics

As stated in Section 1, the conventional construction project supply chain, in which the site managers have to purchase materials from several suppliers simultaneously, causes time delays and low productivity. In order to improve the performance of the construction and reduce large waste, many researchers have emphasized the benefits of supply chain management to the construction industry (Irizarry *et al.*, 2013). However, distinct structures of supply chains impact the methods that could be applied in the process improvement. In order to identify the properties of the supply chain in centralized procurement, this chapter first constructs the configuration of construction logistics according to the practice (see Fig. 2).

In the configuration, the construction sites referring to the customers in the manufacturing supply chain first submit the material plans to the construction logistics enterprise, which plays the role of the third party logistics (TPL) enterprise in the manufacturing industry. After receiving the plans from several construction sites, the construction logistics enterprise should generate the whole purchasing plan and invite potential suppliers to tender for material supply. Selecting the

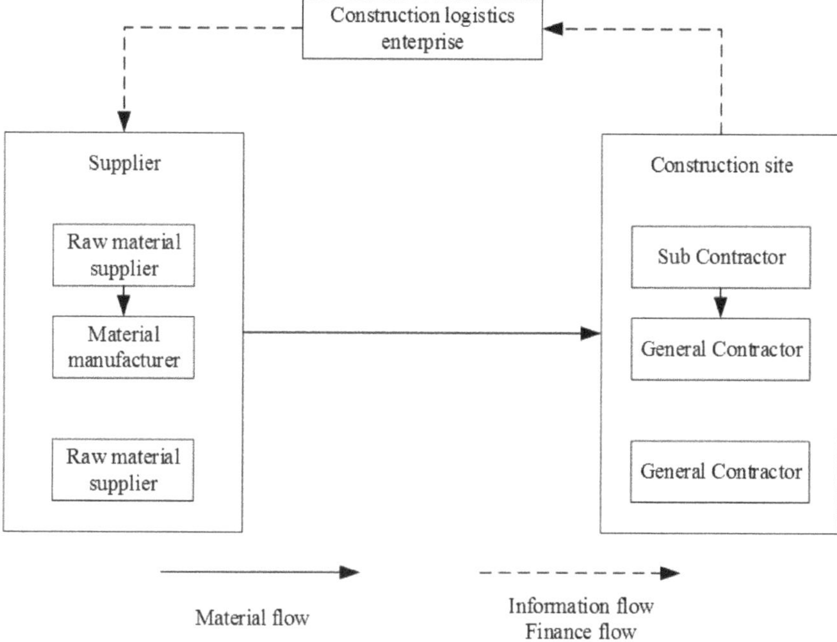

Fig. 2.   Configuration of construction logistics in centralized procurement.

material suppliers, the supply process is implemented between suppliers and sites. Then, the construction logistics enterprise should advance the material funds to suppliers for each construction site. Finally, the construction sites will return the funds and interests to the logistics enterprise after the settlement. According to O'Brien (2009), construction projects are composed of multiple supply chains, each with specific behaviours, which is strongly determined by the type of product that is delivered to a project site. Therefore, in this configuration, materials could be delivered by the raw material suppliers or the material manufacturers (Said and El-Rayes, 2014), and the customers could be divided into general contractors and Sub (Fig. 1) that need to complete some sub-projects for the general contractor.

In spite of the advantages in centralized procurement in the construction supply chain, the redundant processes in emerging construction logistics enterprises weaken the performance. After constructing the configuration of construction logistics in centralized procurement, the DMAIC approach is implemented to improve the procurement process in construction logistics enterprises in the next section.

3. Approach Implementation

The Six Sigma concept was conceived in the 1980s at Motorola and several models such as DMAIC, DMADV, DCCDI and IDOV can be used to implement Six Sigma (EOM, 2015). In particular, the flow of the DMAIC model processes with respect to defining, measuring, analyzing, improving and controlling has proven itself in multitude environments (Nagi and Altarazi, 2017) including manufacturing and service sectors, private and public sectors, governmental and non-profit organizations, and distinct organizations and supply chains. In this case, the JQ construction logistics enterprise founded in 2008 is a wholly owned subsidiary of a construction engineering group corporation in China and its primary businesses focus on purchasing construction materials for projects under construction by advancing funds, such as steel products, cements, installation materials, construction equipment, etc. Before improving the process quality of centralized procurement of materials in construction logistics, the overall operation of construction logistics enterprises is developed in a flow chart. As shown in Fig. 3, the overall process of implementing DMAIC is represented and the details at each phase are also presented as follows.

3.1. *Define*

As depicted in their research, Nagi and Altarazi (2017) explained "Define" as the first phase of the DMAIC approach where a project's team clearly defines and quantifies problems by developing project's charter, estimating costs of poor quality for existing processes, determining more precisely the criteria that are critical to the customer, etc. In this study, the purchasing service provided by construction logistics enterprises is regarded as a product, and the SIPOC and VOC are simultaneously utilized in this phase.

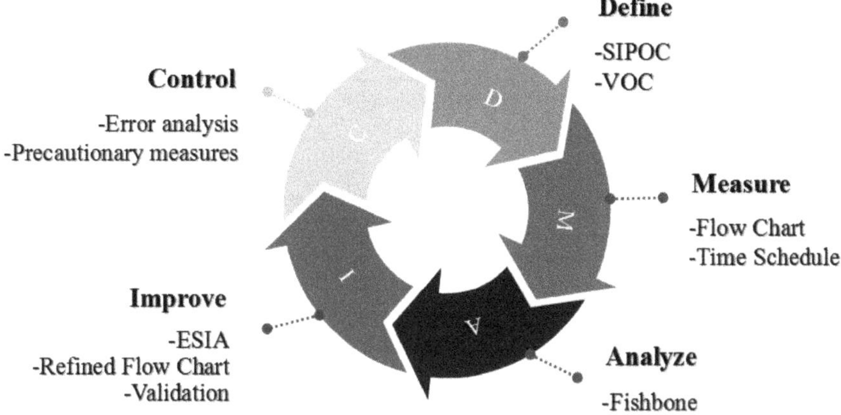

Fig. 3.   Overall process of implementing DMAIC in this research.

3.1.1. *SIPOC analysis*

SIPOC analysis refers to the technique for analyzing a process relative to these parameters to fully understand their impacts (Ghosh, 2017). The common parameters are summarized in the acronym (SIPOC): Supplier, Input, Process, Output and Customer. For the process in centralized procurement, the material requirement plans are presented by construction sites, the centralized material purchasing plan is generated by construction logistics enterprises and the materials are ultimately delivered to sites. The parameters are explained as follows (see Table 1):

- Supplier denotes the organizations or enterprises that provide the purchasing information for the process, such as the category, amount, specification of materials.
- Input represents the purchasing information, for example, the material requirement plans, provided by the supplier.
- Process expresses the workflow that certainly converts input into value in practice. After receiving the input information, the construction logistics enterprise examines and approves the plans, generates the entire purchasing plan and invites potential material suppliers to join in the bidding.

Table 1.   SIPOC analysis for the process in centralized procurement.

Supplier	Input	Process	Output	Customer
Project site A	Category, amount, specification, delivery time and site	Examination and approval, centralized procurement	Material supply plan, ordering cycle, delivery cost	Project site A
Project site B				Project site B
Project site C				Project site C
...				...

- Output represents the product or service generated by process. For the situation that the materials are ultimately delivered to construction sites, the output should be the indicator representing customers' satisfaction regarding material supply plan, ordering cycle and delivery cost.
- Customer denotes the cross-functional departments or enterprises who accept the product or service in the centralized procurement process. Therefore, the project sites accepting the material delivery should be the customers of the process and the performance of the process is closely related to the satisfaction of the project sites.

3.1.2. *Voice of the customer (VOC)*

Due to the core purpose of DMAIC, the customer's requirement is the objective that the process is optimized to satisfy (Malik *et al.*, 2016) and VOC is to understand customer needs, which is the essence of Six Sigma (Uluskan, 2016). Therefore, the critical-to-quality (CTQ) of the process is obtained by the VOC which was conducted in a real construction logistics enterprise by collecting the complaints from the construction sites. The CTQ converts VOC into measurable metrics to settle the final objective of the process improvement.

The complaints from construction sites were collected in the form of post-service evaluation by the JQ construction logistics enterprise for two years and the evaluation complaints focused on the tedious processes, approval time and long-ordering cycle. Meanwhile, the investigated customers raised a claim that the duration of the process implementation should not exceed 15 days. Therefore, the process improvement objective is set as decreasing the actual work time in the procurement process to improve the satisfaction of customers.

3.2. *Measure*

During this phase, the measurement system is verified and data are collected to establish a baseline measurement for the current process (Anderson and Kovach, 2014). Taking account of the features in construction logistics in centralized procurement, the flow chart depicts the current state of the process. Accordingly, the relative work hours in practice are obtained to generate the time schedule by the author after working and tracking in three construction sites with the whole procurement process.

3.3. *Flow chart*

Achieving the objective mentioned in the Define phase, the applied method needs to identify the problems in the process. However, there is no baseline for the process in centralized procurement, as the construction logistics enterprise is an emerging industry in China. Therefore, by combining the practice on sites, the flow chart for the current centralized materials procurement process in JQ construction logistics enterprise is presented as follows.

The whole process has been presented in the proposed configuration (in Section 2) and the details implemented in the actual construction sites are shown in Fig. 4. There are five departments in the process, namely project sites, plan approval department, approval manager, plan-supervision department and material supplier, which are in three distinct enterprises, respectively. The activities in the process are shown in the flow chart by 12 rectangles and all activities in the same pool are taken by the corresponding process department. During the whole process, three documents are generated consisting of the MRP submitted by project sites, the entire purchasing plan integrated by the JQ enterprise and the

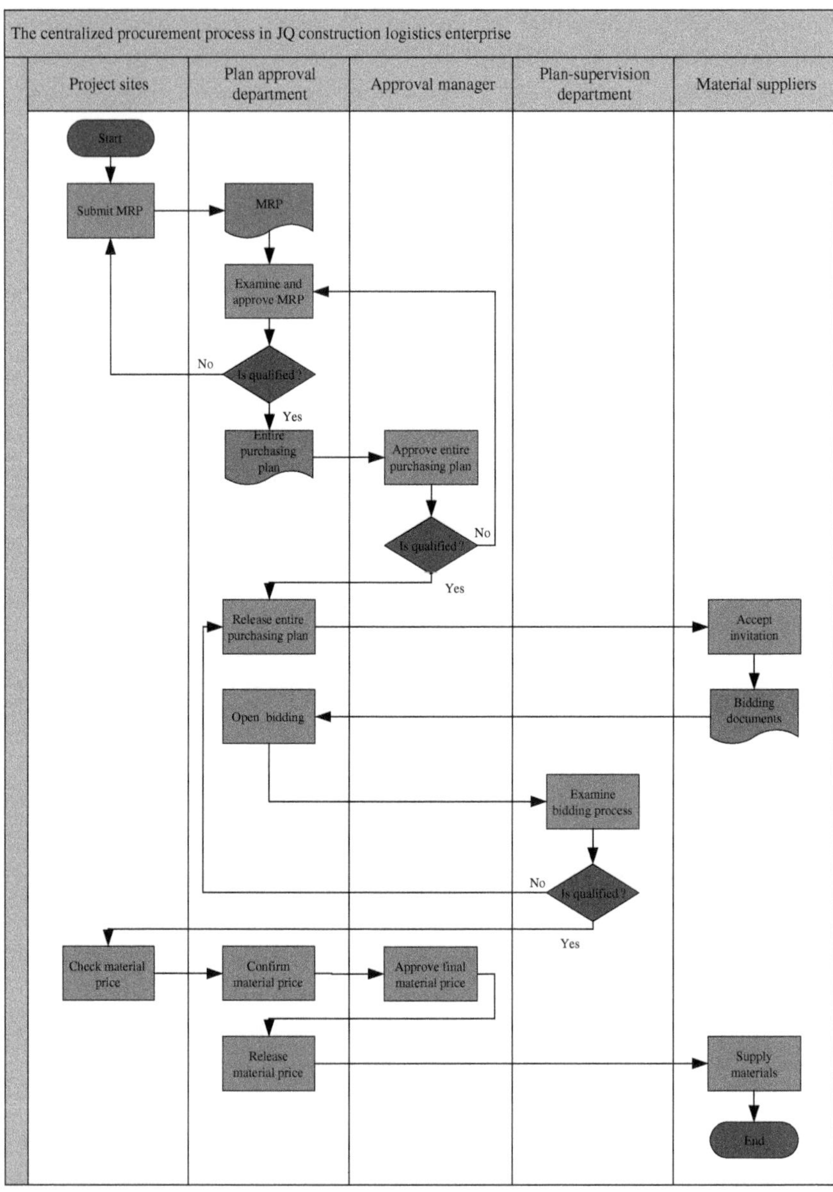

Fig. 4.   Flow chart of the current centralized procurement process.

bidding documents tendered by material suppliers. With the detailed centralized procurement process, the process improvement baseline is confirmed and the next step is to measure the consumed time in each activity.

3.3.1. *Time schedule*

With respect to the 12 activities in the process, the consumed time for each one impacts the actual work time, which is set as the improvement objective. Therefore, the time schedule is used to collect the data for process improvement. Through working and tracking in three construction projects under supply by JQ enterprise, the work time for activities is obtained. In order to illustrate the approach, the data about the activity time in centralized procurement process in July 2014 are presented in Fig. 5.

As shown in Fig. 5, the activities and their work time are presented in the Gantt Chart in the sequence of the process implementation. The days in the light yellow lines are the weekends when the staff in JQ enterprise would not dispose tasks. According to the schedule, the duration for implementing the process is 25 days and the actual work time is 19 days. It is worth noting that although the activity, supply materials, is at the end of the procurement process, it actually belongs to the supply process. Therefore, the work time for this activity is not presented in the Gantt Chart.

Above all, integrating the information in VOC, flow chart and time schedule, the quantitative improvement objective in the centralized procurement process is proposed (see Table 2). As discussed in section 3.1.2, the purpose to improve the procurement process is to decrease the actual operation time, so the procurement process duration is set as the CTQ in the improvement. According to the actual working and tracking on construction sites, the current process duration has been obtained as 25 days. Considering the procurement interval and lead time suggested by site managers, the objective value set as 15 days for procurement process improvement is reasonable. Therefore, this paper selects these two indexes to quantify the CTQ before and after improving to evaluate the improvement efficiency.

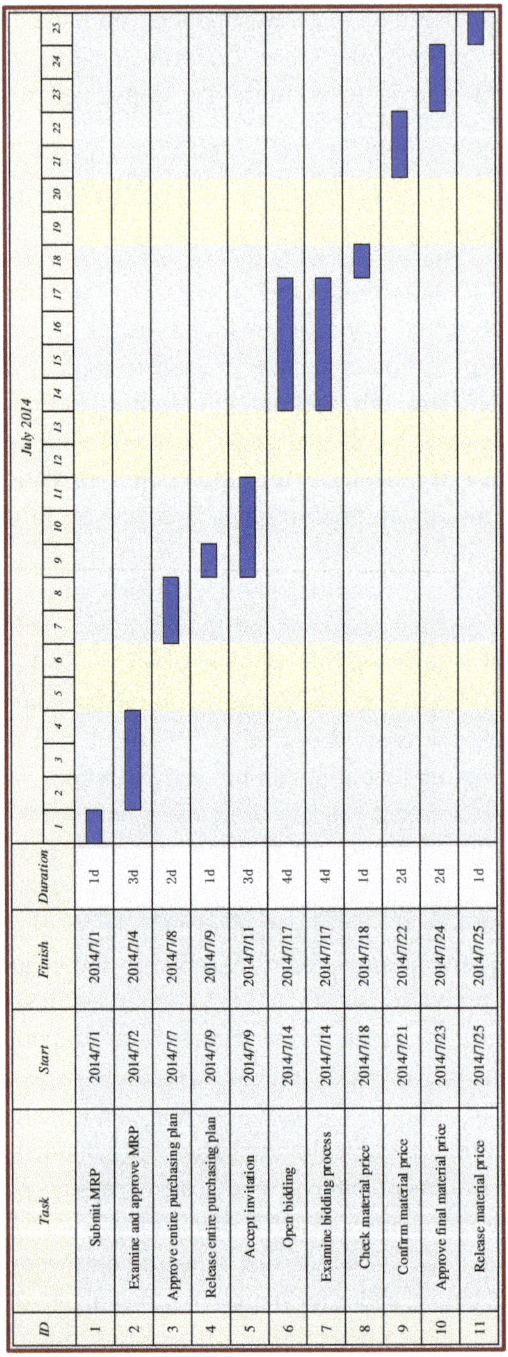

ID	Task	Start	Finish	Duration
1	Submit MRP	2014/7/1	2014/7/1	1d
2	Examine and approve MRP	2014/7/2	2014/7/4	3d
3	Approve entire purchasing plan	2014/7/7	2014/7/8	2d
4	Release entire purchasing plan	2014/7/9	2014/7/9	1d
5	Accept invitation	2014/7/9	2014/7/11	3d
6	Open bidding	2014/7/14	2014/7/17	4d
7	Examine bidding process	2014/7/14	2014/7/17	4d
8	Check material price	2014/7/18	2014/7/18	1d
9	Confirm material price	2014/7/21	2014/7/22	2d
10	Approve final material price	2014/7/23	2014/7/24	2d
11	Release material price	2014/7/25	2014/7/25	1d

Fig. 5.   Gantt Chart for the time schedule in the current process.

Table 2.   Quantitative CTQ of before and after improving.

CTQ	Time for Implementing Process (Day)	
	Current	Objective
Procurement process	25	15

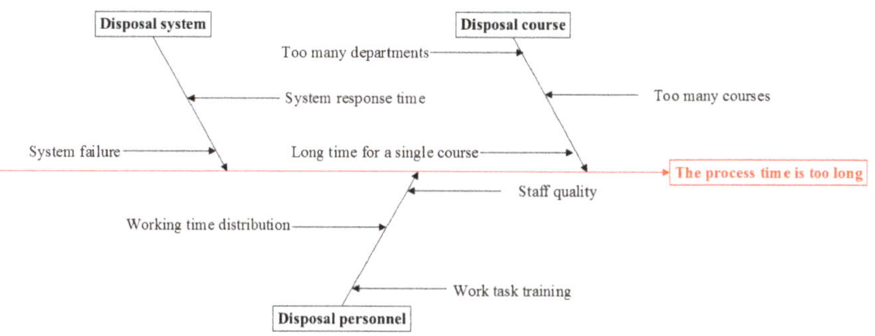

Fig. 6.   Fishbone Chart for cause analysis in process.

3.4. *Analyze*

In the Analyze phase, information about the problem and underlying process are analyzed to identify potential causes of the problem (Kumar *et al.*, 2007). Normally, factors affecting customer satisfaction include personnel, machines, materials, environment, etc. In order to facilitate the objective to decrease the actual work time, the Fishbone Chart is created for the causes that may lead to the redundant process.

As shown in Fig. 6, the identified problem in the centralized procurement process is the length of the entire process. The primary causes consisting of disposal course, disposal personnel and disposal system are sorted by the influence on the problem from right to left. Accordingly, several subordinate causes are included in each primary cause to illustrate the influence in detail. In the primary cause, for example in the disposal course, three subordinate causes are enumerated as long time for a single course, too many courses, and too many departments. The distance between the subordinate causes and the problem line in red represents the degree of influence which is the same as for the primary causes.

Understanding the customer's requirements and the causes for the process problem are accessed in the Improve phase to improve the process.

3.5. *Improve*

In this phase, the solutions to the root causes detected in Analyze phase have been developed by eliminate, simply, integrate and automate (ESIA) approach. In addition, the refined flow chart for improved process is generated. Finally, in order to measure the performance of the approach, the data of a practical case implemented by the improved process are collected to validate the results.

3.5.1. *ESIA approach*

After analyzing the VOC data and brainstorming with practitioners in JQ enterprise, three primary causes are obtained, which have significant influence on the problem in the centralized procurement process. Furthermore, the disposal course is regarded as the most significant factor causing the problem. Therefore, the ESIA approach is used to improve the process first.

- *Long time for a single course*: According to the interviewed practitioners, the examine-and approve-MRP activity is conducted by the staff to collect, approve, sort and integrate the material requirement plans from sites. It is proved to be time and manpower consuming and human errors frequently happen in the process. Therefore, the office automatic (OA) system is developed to automate the process to approve the MRP and generate the entire purchasing plan by the given document forms. In addition, the bidding course consisting of open-bidding and examine-bidding processes also needs to be integrated because it takes up too much process time with valueless activities and thus the bidding process needs to be approved by several departments.
- *Too many courses*: Analyzing the courses implemented in the process, the activities, for example the approval of the entire purchasing plan, and examining the bidding process are part of the internal staff performance assessment in JQ enterprise, which are not the value-added

process for the centralized procurement. In addition to the examination and approval, the activity confirms the material price, which is only used to aggregate the price data, that could be conducted by the OA system automatically. These valueless courses are eliminated to refine the process.

- *Too many departments*: As shown in Fig. 4, the departments, plan approval department, approval manager and plan-supervision department are the cross-functional departments in the JQ enterprise, some of which aim to assess the internal performance. Therefore, these departments are integrated into a comprehensive department named integrated plan disposal department.

3.5.2. *Refined flow chart*

Based on the above ESIA approach, the refined centralized procurement process is presented by the flow chart (see Fig. 7). In the improved process flow chart, only three departments are included. In addition, the number of the disposal curses is reduced from 12 to 8 and the redundant processes have been eliminated.

3.5.3. *Validation in practice*

In order to illustrate the usability of the improved process, the data of an implemented case were collected to obtain the entire work time in the process. Considering the feasibility of the improved process implementation, the data were collected in November 2015 when the improved process had been operated for half a year. The time schedule for the improved process in JQ enterprise is presented in Fig. 8.

After eliminating, simplifying, integrating and automating the centralized procurement process as depicted in Section 3.4.1, the duration for implementing the improved process is 12 days and the actual work time is 10 days. In comparison with the time consumed in Fig. 5, the entire process time and the actual work time have been decreased by 52% and 53%, respectively. Moreover, the entire work time is less than the CTQ objective in Table 2.

With the ESIA approach, the centralized procurement process is improved and the customer requirements are also satisfied. Accordingly,

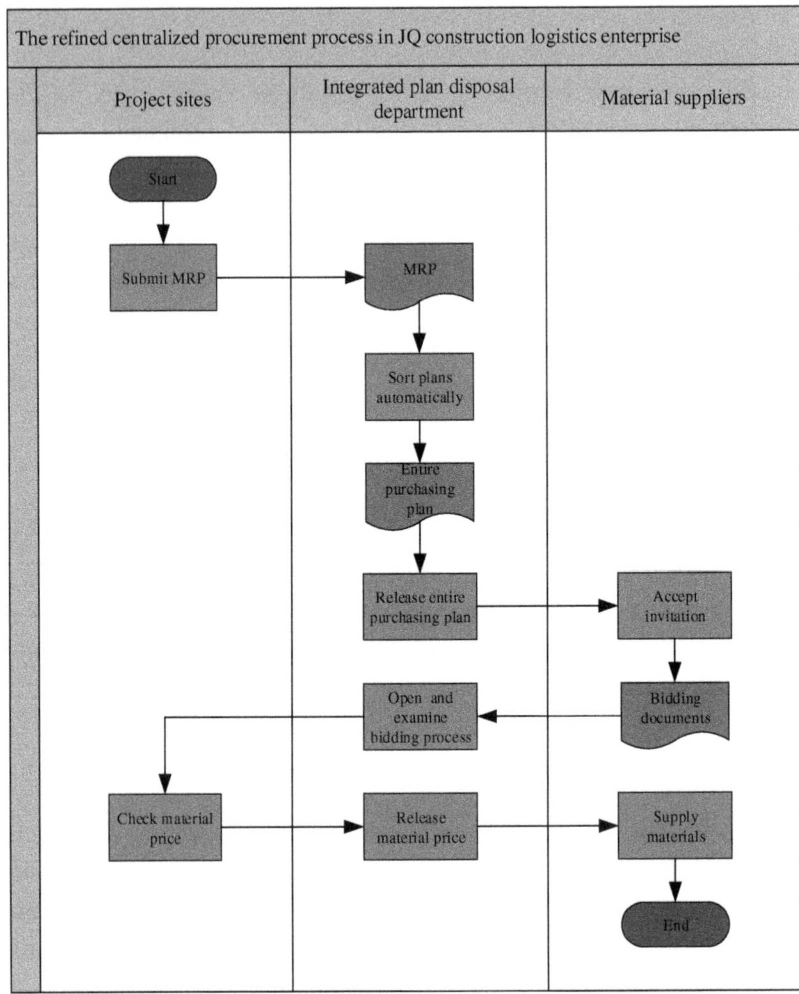

Fig. 7. Flow chart of the refined centralized procurement process.

to facilitate the achievements, some suggestions and measures are developed against the disposal personnel and disposal system as follows:

- *Disposal personnel*: The staff quality is the first subordinate cause and it significantly influences the efficiency of the staff in the process. Therefore, the enterprise should develop some appropriate

ID	Task	Start	Finish	Duration	November 2015														
					1	2	3	4	5	6	7	8	9	10	11	12	13	14	15
1	Submit MRP	2015/11/2	2015/11/2	1d															
2	Sort plans automatically	2015/11/3	2015/11/3	1d															
3	Release entire purchasing plan	2015/11/4	2015/11/4	1d															
4	Accept invitation	2015/11/4	2015/11/6	3d															
5	Open and examine bidding process	2015/11/9	2015/11/11	3d															
6	Check material price	2015/11/12	2015/11/12	1d															
7	Release material price	2015/11/13	2015/11/13	1d															

Fig. 8.   Gantt Chart for the time schedule in the improved process.

performance indicators to employ and evaluate the staff. Training the staff with the station technologies over the entire process can solve the problem caused by the working time distribution and work task training.

- *Disposal system*: According to the ESIA approach, the OA system in the JQ enterprise can provide some necessary actions to simplify the process. Hence, based on the disposal process requirements, developing some functions in the OA system and doing the regular maintenance on the system could improve the stability of the disposal system.

3.6. *Control*

Normally, this phase is used to monitor the stability of the improved system and related to the Define phase since these two phases represent the process before and after implementing the corrective actions (Nagi and Altarazi, 2017). In the current improvement work, brainstorming with practitioners in JQ enterprise, error analysis and precautionary measures are implemented in this phase.

3.6.1. *Error analysis*

Identifying the errors that might happen in the process in advance could facilitate the performance of the centralized procurement process. Talking with the practitioners, two types of errors in the process are summarized in the entire purchasing plan generation and staff opera-tions. The entire purchasing plan includes the information of the types, quantities and delivery time of the materials and is also the basis for the material suppliers to tender. The errors in plan generation could result in a dispute among suppliers, the construction logistics enterprise and site customers. On the other hand, the errors by the operation staff could cause inaccuracy and lower efficiency in the process, which results from the misunderstanding of the station tasks or the low level of proficiency in the disposal process.

3.6.2. *Precautionary measures*

For the former errors, the JQ enterprise set two full-time staff to make a double check, meaning that the staff should check the plan independently and then contrast the independent results to ensure the accuracy of the plan. For the later errors, a systematic training should be provided to the staff to improve the handling ability.

4. Discussions and Conclusions

This chapter has explored the ability of the DMAIC approach to improve the centralized procurement process in a construction logistics enterprise. The major contribution of the implemented case study is to provide a research approach for process improvement in an emerging construction logistics. Unlike the conventional construction supply chain management, the centralized procurement in construction logistics has its advantages. Therefore, this chapter constructs the configuration of the construction logistics in centralized procurement. Then, considering the properties in the configuration, the DMAIC approach was used to improve the process. The SIPOC and VOC are implemented to obtain the CTQ objective for the improvement. Taking account of the lack of baseline for the process, the flow chart is used to depict the process in JQ enterprise. Accordingly, the data about the work time are collected to quantify the improved objec-tive. In addition to the objective, the potential causes of the problem are presented by the Fishbone Chart. Understanding the improved objective and the potential causes, the ESIA approach is employed to refine the current process. With the improvement approach, the entire process time and actual work time have been reduced by 52% and 53%, respectively. Finally, approaches to facilitate the achievements are also proposed in the control phase.

The implemented case study has provided a research approach for both researchers and practitioners in construction logistics, however, limitations still exist in this study. For example, this case study worked on the process improvement in JQ construction logistics enterprise and ignored the situations in distinct enterprises. Therefore, one potential extension of the current work is to construct a baseline to improve the centralized procurement process in construction logistics enterprises in the future.

Acknowledgements

The support of the Graduate Research Innovation Project of Chongqing Jiaotong University (No. 2017B0101) is gratefully acknowledged.

References

Anderson, N. C. and Kovach, J. V. (2014). "Reducing Welding Defects in Turnaround Projects: A Lean Six Sigma Case Study", *Quality Engineering*, 26(2), 168–181.

Eom, S. J., Jang, W.-S. and Kim, S.-C. (2015). "Managing Concrete Crack Information Through Correction of the Slab Rebar Arrangement Based on Six Sigma", *KSCE Journal of Civil Engineering*, 19(7), 1973–1981.

Ghosh, S. S. (2017). "Financial Incentives — A Potent Weapon for Higher Productivity", *International Journal of Productivity and Performance Management*, 66(4), 554–571.

Heravi, G. and Firoozi, M. (2016). "Production Process Improvement of Buildings' Prefabricated Steel Frames Using Value Stream Mapping", *The International Journal of Advanced Manufacturing Technology*, 89(9–12), 3307–3321.

Irizarry, J., Karan, E. P. and Jalaei, F. (2013). "Integrating BIM and GIS to Improve the Visual Monitoring of Construction Supply Chain Management", *Automation in Construction*, 31, 241–254.

Kumar, M., Antony, J., Antony, F. J. and Madu, C. N. (2007). "Winning Customer Loyalty in an Automotive Company through Six Sigma: A Case Study", *Quality and Reliability Engineering International*, 23(7), 849–866.

Malik, A. L., Akbar, M. and Irianto, D. (2016). "Improving Quality of Seal Leak Test Product using Six Sigma", *IOP Conference Series: Materials Science and Engineering*, 114, 012054.

Marhani, M. A., Jaapar, A., Bari, N. A. A. and Zawawi, M. (2013). "Sustainability Through Lean Construction Approach: A Literature Review", *Procedia — Social and Behavioral Sciences*, 101, 90–99.

Nagi, A. and Altarazi, S. (2017). "Integration of Value Stream Map and Strategic Layout Planning into DMAIC Approach to Improve Carpeting Process", *Journal of Industrial Engineering and Management-Jiem*, 10(1), 74–97.

O'Brien, W. J., Formoso, C. T., Vrijhoef, R. and London, K. A. (2009). *Construction Supply Chain Management Handbook*, Boca Raton: CRC Press.

Purnapandhega, P. and Adhiutama, A. (2016). Lean Sigma Application for Bridge Improvement Project. *Proceedings of the 2016 Global Conference on Business, Management and Entrepreneurship*, (eds.), A. G. Abdullah, R. Hurriyati, A. B. D. Nandiyanto *et al.*, 15, 296–300.

Said, H. and El-Rayes K. (2014). "Automated Multi-Objective Construction Logistics Optimization System", *Automation in Construction*, 43, 110–122.

Uluskan, M. (2016). "A Comprehensive Insight into the Six Sigma DMAIC Toolbox", *International Journal of Lean Six Sigma*, 7(4), 406–429.

van Lith, J., Voordijk, H. Matos Castano, J. and Vos, B. (2015). "Assessing Maturity Development of Purchasing Management in Construction", *Benchmarking: An International Journal*, 22(6), 1033–1057.

Vitra, D. (2012). *Procurement of a Third Party Logistics Company a Possible Benchmarking Method*, Stockholm: Stockholm University.

Youssouf, A., Rachid, C. and Ion, V. (2014). "Contribution to the Optimization of Strategy of Maintenance by Lean Six Sigma", *Physics Procedia*, 55, 512–518.

Chapter 24

Deepwater Oil and Gas Project Performance: A Study of Cultural Intelligence

Quek Tiang Beng* and Shavarebi Kamran[†]

Faculty of Science, Technology, Engineering and Mathematics, International University of Malaya-Wales, Kuala Lumpur, Malaysia
**ngptrck@yahoo.com*
[†]kamran56@iumw.edu.my

Abdul-Rahman Hamzah

Faculty of Built Environment, University of Malaya, Kuala Lumpur, Malaysia
arhamzah@gmail.com

Abstract

Mammoth-scale Greenfield deepwater projects with specialized and novel technologies demand cross-border subject matter experts and leaders to deliver the Triple Constraint. Cross-border project leaders must be able to explicitly accommodate the environments at different

---

*Corresponding author.

geographical locations. Factually, cultural barriers in the context of language, religion, values and so on cause cultural split and schizophrenic divide among individuals, which escalates even greater when two or more cultures have a slight acquaintance with each other and had proven to significantly expose respective organization to business and reputation risk. Cultural intelligence (CQ) at the organizational level constitutes the organization's capacity to reconfigure its capability to function and manage effectively in culturally diverse environments and to sustain and gain its competitive advantages. So far, studies on CQ are limited to non-deepwater oil and gas megaprojects, non-Malaysians and cover studies from different clusters of personnel in organizations. Explicit study on CQ with respect to deepwater oil and gas project performances in Malaysia is needed and will be inculcated in this chapter. This study embraces four underpinning leadership theories with aims to establish the importance of CQ as cross-border leadership enablers to enhance the performance of deepwater oil and gas megaprojects in Malaysia. This qualitative case study exploratory research will facilitate an inductive approach and may lead to a unique, contextual attestation of CQ as key enablers in cross-border project leadership to enhance the performance of future deepwater oil and gas megaprojects. Data will be thematically analyzed from interviews conducted with selected project leaders and team leaders with a minimum of five years distinct deepwater megaprojects experience. An initial conceptual framework was developed with the objective of progressive refinement along the study phases. The final framework will be useful as a strategic tool to tackle the soft-skills challenges that frequently appear in megaprojects and is envisaged to continuously improve the future cross-border project manning strategy in international oil and gas organization, thus potentially mitigating the prospective destruction of shareholder wealth and enterprise reputation.

Keywords: Cultural Intelligence; Cross-border Leadership; Exploration and Production; Oil and Gas Industry; Megaprojects; Project Performance.

1. Introduction

This chapter presents the summary of a doctoral research work in progress and provides an outline of the research idea from the main author for a doctoral research proposal. It describes the set-up of the study and conceptualization of the gap, based on in-depth literature review, pilot interviews and the vast industrial experience of the author. The purpose is to pinpoint the importance of cultural intelligence (CQ) in cross-border project leaders to enhance future oil and gas megaproject performance by studying the case of two deepwater projects in Malaysia.

1.1. *Background of the study: Preliminary investigation and problem recognition*

Oil and gas projects are getting significantly larger and more complex over the past decade or so. Its globalized business nature gave rise to unavoidable exposure to cross-cultural project environments. Leaders who work in cross-border contexts possess complex social problems of leadership due to cultural background influences, prototypes and schemas with respect to appropriate leadership behaviors, for example, expectations about preferred leadership styles, managerial behaviors, and the nature of relationships are all influenced by culture (Littrell and Nicolae Valentin, 2005). CQ skills are increasingly becoming beneficial for cross-border project leaders, as those who have high CQ can quickly adapt their management style in order to handle issues raised by internal and external stakeholders from a multi-cultural background. Project leaders with high CQ are able to speedily adapt their management style in order to handle issues raised by internal and external stakeholders from diverse cultural backgrounds. When certain projects get more complex and increase in size, managing them can become very difficult (Mišić and Radujković, 2015). Some of these common issues are serious slips in completion schedules, operability problems and cost overruns (Rui *et al.*, 2017; Olaniran *et al.*, 2015). A large number of these massive projects end up as the biggest disappointment for their investors, a fair number of these projects destroy almost all the shareholder wealth, and even few of these projects turn out to be horrendous for everyone involved in them, such as the environment, investors and local population.

In this regard, if the megaproject disasters become public, all their investors' reputation can be damaged and their existence can be in jeopardy. As an instance, studies have shown that in the industrial megaprojects, oil and gas industry has fared very poorly (Edward W Merrow, 2011). Upstream oil and gas megaprojects have been more fragile than their non-oil and gas industry cousins. This was linked to the poor functional integration, which characterizes upstream project organizations. The poor functional integration can make these projects highly sensitive to poor preparation, can result in loss of project leadership continuity and may also lead to schedule aggressiveness (Edward W Merrow, 2011). For instance, the US$10-plus billion overrun of Shell Sakhalin-2 Project has ruined the reputation of the Shell and made a good excuse for Kremlin to nationalize a massive portion of the project (Lustgarten, 2007).

Even though there are a large number of techniques and standards for project management and many expert project leaders are available, the whole coordination and management of megaprojects have shown not to be a straightforward exercise in the industry, which has been fragmented with individuals working in silos and diverse cultural teams (Winter *et al.*, 2006). It has been found that silo working environment will be further compounded if either the contractor or consultant on a construction megaproject is a JV or merger of companies and worsened in a case where that megaproject is in a foreign international environment including JV partners from different countries. Despite forming an international joint venture (IJV), executing a megaproject has been a risky global business decision and is susceptible to failure, these IJVs have been established commonly to spread risks, gain entry and maintain the projects and corporate interests of foreign territories (Ozorhon *et al.*, 2007). However, many companies that enter themselves in IJV agreements have no knowledge and essential understanding about the foreign environment, where the project is located; they also have no knowledge of the required communication skills and management abilities to achieve success in the project (Lu and Hébert, 2005). The lack of understanding has been the main reason for disasters that need to be avoided. It has been shown that one-third of the cross-border project leaders or international managers have been underperforming in their assignments, which subsequently showed that the effort of multinational companies in order to develop a global manager has fallen far beyond the optimum (Manning, 2003). According to the studies, the leaders with higher CQ can be more effective at management of the culturally diverse expectation and they can perform better in minimization of the exclusionary reactions, which may happen in cross-border contexts (Punj, 2016).

By nature, it is essential for the project managers to have fundamental technical project management hard skills, however soft skills, in particular CQ, have become extremely important for the cross-border project leaders in order to enhance project performance. International project managers need to cultivate soft skills and in an overseas environment and also need to be culturally astute from the get-go, with a keen awareness of organizational politics and the ability to clearly communicate, listen and instill a sense of trust and one-teaming in all communications.

The leadership style in Asia has been unique and for many appeared to be autocratic, unstructured and unsystematized. The varying strains of Asians based on their beliefs and ethics in certain businesses such as trust, confrontation avoidance and networking have many influences, which cannot be discounted easily. For instance, Malaysia, the land that has been well known as "truly Asia", is a multiracial society, which points to an ambiguity in the concept that there is only one culture or one Malaysian leadership style in the country (Kennedy and Mansor, 2000). While the ethnic groups in Malaysia might share some common values and beliefs, each of them originates from a distinct cultural and religious heritage, which contributes to different behaviors and beliefs in Malaysian managers and creates diversity in leadership styles in the country as well as other regions of the world (Kennedy and Mansor, 2000). For instance, being forthright or direct is considered as a rude and insensitive behavior, it has been very important to preserve the "face", and the leaders have been inhibiting openness in order to maintain the social harmony (Kennedy and Mansor, 2000). On the other hand, even though there has never been an urgent need for global business leaders, serious deficiencies have existed in the preparation of corporate managers, since they usually deal with the interpersonal realities of global business (Alon and Higgins, 2005). According to the studies on project management practices in the construction industry, it has often been assumed by academics that the best practice is the Western Way. However, in order to ensure the efficiency of the "Project Way", there has been a need for cross-border project leaders who participate in megaproject assignments outside of their home country to be aware of the situation and environment of the location they operate. As an instance, without an appropriate awareness about the totality of the environment, these leaders will definitely not be able to achieve the one-teaming principles of megaprojects in Malaysia.

Megaproject challenges have been studied enormously with respect to soft skills including competencies and leadership styles, behavioral and cultural issues; however, the CQ, explicitly as a cross-border leadership enabler in deepwater oil and gas megaproject performance, need further studies and researchers have to be more focused on this topic. Previous studies in CQ literature have focused on a set of abilities or capabilities at an individual or group level, but few have examined a set of capabilities at an organizational level. Yitmen (2013) conducted studies on CQ at the organizational level as the main cross-cultural competence, however these were limited to establishing and increasing the performance of international strategic alliances. A number of studies have also been conducted on the role of CQ in the effectiveness of the cross-border leadership, however all these studies have been limited to the military context in Australia with male Caucasian participants (Rockstuhl *et al.*, 2011). An integrated model of three types of intelligences, known as CQ, general intelligence (IQ) and emotional intelligence (EQ), has been tested by researchers; it has been realized that the model is necessarily incomplete. Dulewicz and Higgs (2003) studied three types of competences, which explain most managerial performances, such as emotional (EQ), intellectual (IQ) and managerial (MQ) skills, however this study was limited to the focus on evidence from the Board members in the United Kingdom (UK) organizations. According to their result, EQ competencies have been a very important factor for managerial performance, but their results were limited to the domestic samples that were taken from the UK directors (Dulewicz *et al.*, 2003). Researchers have included that in multi-cultural teams, those cross-border managers need more than three competence types for addressing the challenges in deepwater megaproject execution.

Nevertheless, none of the studies has examined the relationship between organizational CQ and deepwater megaproject performance, particularly in Malaysia. Therefore, studies on CQ competency on cross-border leaders are very important and are expected to be a significant contribution to the Malaysian and global oil and gas industry and project management body of knowledge in general. In this research, the researcher emphasizes the importance of this enabler to yield better understanding of its interconnection and interdependency with the project performance through in-depth case studies of two deepwater mega-projects in Malaysia.

2. Country and Industry Contexts

Light, sweet crude oil resources are in short supply, while the less expensive heavy, sour crudes are more plentiful worldwide. Oil and gas companies were left with limited options but to develop their capital assets, primarily via projects with increased risks and challenges. Unplanned and unrecognized risks have been "project killers" (Heldman, 2010). Examples of complex, high-risk projects are the upstream oil and gas projects, happening globally in diverse geographical and socioeconomic environments (Salazar-Aramayo *et al.*, 2013). On risky international exploration and production (EP) projects, the joint collaboration between large producers has been common (Kolios and Luengo, 2016). For two reasons, the first decade of the 21st century has seen very large and complex projects executed by the oil and gas industry. The first reason was that the national resource holders were able to more easily control the developed crude hydrocarbons, remote geographical locations, harsher environmental conditions, forcing oil and gas companies to explore and operate in increasingly deeper waters (Merrow, 2011). The next reason was that the easily accessed hydrocarbon resources close to markets had massively depleted (Merrow, 2011). These projects were tending to be characterized by large budgets and problems associated with complex systems, including uncertainty, nonlinearity and irregularity (Rezvani *et al.*, 2016). Furthermore, complex projects like these normally attract the attention of political quarters and the public because of their substantial national, international, environmental and social implications, which have been in association with the failure as well as the success of such enterprises (Whitty and Maylor, 2009). The performances of these complex and large projects have been mostly disappointing. A large number of these complex projects usually experience considerable cost delays in completion and overruns, and they even fail to achieve their objectives (Eden *et al.*, 2005; Williams and Samset, 2010; Chang *et al.*, 2013). Deepwater is an oil and gas field with the water depth from 305 m (1000 feet) and above as defined by the United States Mineral Management Service (MMS), known as the Bureau of Ocean Energy Management (BOEM) from the year 2010. The usage of extensive infrastructure as well as unconventional extraction processes in deepwater oil and gas fields have brought a large number of investors, and the amounts of their investment have exceeded the numerical threshold of US$1 billion, which has been defined as megaprojects (Flyvbjerg, 2014). These complex projects have satisfied both domestic and worldwide demands for energy and other finished oil and gas products. Megaprojects are programs, which integrated strategically aligned projects into a single large project and have been used by the oil and gas industry for delivering key strategic assets (Pinto and Morris, 2004; Miller *et al.*, 2001). Traditional modes of project delivery have been based on the megaproject manager's basic experience and training, therefore they have usually adopted it. The energy of the manager in this mode has been devoted to low level duties, such as contract management, focusing mainly on cost and time, which has been a project efficiency with insufficient attention on how to get the best overall results, thereby negating the longer-term strategic views. (Halman and Braks, 1999; Asrilhant *et al.*, 2007; Shenhar and Dvir, 2007; Turner *et al.*, 2009; Turner *et al.*, 2010). As program managers, their focus must be more on business benefits and strategic objectives, which require much of a leadership role other than management role (Shao, 2010; Shao and Müller, 2011; Murray-Webster and Thiry, 2000). Megaprojects can take 5–12 years from inception to the start of revenue generation, while the operational life spans between 7 and 30 years (Eweje *et al.*, 2012). This long-term nature made the megaprojects vulnerable to ambiguity and uncertainty, which has accentuated a need for strategic decision-making instead of tactical short-term efficiency management of traditional project execution in their management (Office of Government Commerce, 2007; Project Management Institute, 2008).

For an instance with regard to this case, Malaysia has the second largest oil and gas reserves in South East Asia with the average production of domestic crude oil and having condensate decreases from 667,000 barrels per day (bpd) in 2006 to 576,000 bpd in 2013 (*Source*: Chapter 7, 11th Malaysia Plan 2016–2020; MIDA, Meet Malaysia Oil & Gas, 2013). This reduction in average production has pushed the national oil company (NOC) and its other production-sharing contract (PSC) partners to seek into the deepwater zones, which has posed high operating costs and required considerable technical expertise. KIKEH block was the country's first deepwater oil and natural gas discovery which began operation in January 2007 followed by GUMUSUT-KAKAP (GK) deepwater fields which began production in 2014 (planned 2012). Thereafter, a third deepwater project known as MALIKAI (MLK) began production in 2016, (planned 2014), with perceived schedule slippages, respectively. Each of these deepwater projects has been executed by separate PSCs, with different International Oil Companies (IOCs). These PSCs have been offering unique technology transfer to host country, thus leading to the need to mobilize expatriate (cross-border) subject matter experts into project leadership roles by IOCs, Contractors and Consultants. The internal policy in the respective organization jointly plays a part in the appointment of human resources in key positions and international assignments which normally led to the discrepancy between the requirement and/or balance of hard (technical) and soft (managerial) skills, dominantly the managerial skills involving soft skills.

Teams with higher levels of CQ tend to gradually exhibit the higher rates of performance improvement (Moon, 2013). Soft skills include the cluster of personal habits, friendliness, personality traits, optimism, social grace and facility with language, which marked people to different degrees (Bancino and Zevalkink, 2007). It was found by researchers that the challenges in complex projects have been primarily associated with managerial rather than technical problems (Sauser *et al.*, 2009; Dvir *et al.*, 2006). EQ as a managerial skill has been defined by Mayer *et al.* (2004) and has been explained as an ability to be aware of, to manage emotions, to utilize and to understand self and others (Caruso *et al.*, 2002). Even though both the EQ and GQ have been connected with leadership effectiveness in domestic contexts, neither of them were dealing explicitly with the ability to operate in multi-culture and cross-border contexts. Earley and Ang (2003) drew on multi-dimensional perspective on intelligence, which has been developed by Sternberg and Detterman (1986) in order to address the unique aspects of culturally diverse settings and there upon develop the conceptual model of CQ (Sternberg and Detterman, 1986; Earley and Ang, 2003). CQ has been defined by Ang and colleagues as an individual's capability to perform effectively in situations characterized by cultural diversity (Ang *et al.*, 2008). Rockstuhl *et al.* (2011) have defined CQ as the leadership capability to manage effectively in culturally diverse settings (Rockstuhl *et al.*, 2011). With regard to the impacted performance of oil and gas megaprojects, particularly the Malaysian deepwater, this research provides the extensive and in-depth study of Leadership and Culture elements in cross-border project leaders in order to enhance the respective project performances. This leads towards the discussion on interconnection an interdependency among CQ, cross-border leadership and deepwater oil and gas megaproject performance.

3. Literature Review

Megaproject stakeholders are numerous, as is the diversity of their objectives (Shao, 2010). Significant stakeholders in the oil and gas industry include local communities, employees (including the project team), Governments and NOCs, shareholders and NGOs with interest in socio-cultural aspects. In view of the stakeholder's constellation, it is unavoidable that the megaprojects attract many interests from political and socioeconomic, as well as high industrial and public fields. Therefore, it is key that both the sponsor's and stakeholder's goals and the way they link to the project are clear (Turner *et al.*, 2009). Megaprojects have been mostly placed under extreme time and cost pressure (Miller *et al.*, 2001; Edouard *et al.*, 1988). According to the project managers surveyed, it has been found that 85% were under moderate to very high cost and time pressure, which has generated systemic influences within the projects. For instance, it has been shown that the decision quality can be impaired by time pressure (Chu and Spires, 2001). Many studies have been carried out on leadership styles appropriate to multi-cultural projects (Ochieng and Price, 2010; Limsila and Ogunlana, 2008; Sunindijo *et al.*, 2007; Matveev and Milter, 2004).

Björkman and Schaap (1992) proposed that expatriate managers can adopt one of the three following styles categorized by them:

(1) **Didactical:** They sell ideas by analogy and site visits.
(2) **Organization design:** They carefully choose team members to design out potential conflict.
(3) **Culturally blind:** They do not recognize cultural differences.

Emotion, stability, intellect, openness/extroversion, agreeability and conscientiousness are some of the personality traits to cope with cultural differences. Most project managers have adopted task-oriented styles, which are inappropriate in multi-cultural situations, whereas some have adopted the other much appropriate relationship-oriented and people-oriented styles (Mäkilouko, 2004).

3.1. *Global cross-border leadership effectiveness: Positioning cultural intelligence (CQ)*

The global economy has been producing a competitive landscape, which became increasingly more dynamic, ambiguous and complex for the firms operating across borders. According to the Price Waterhouse Coopers' 14th Annual Global CEO Survey (2011), "bridging the global skills gap" was one of the most important concerns in the future, since companies have been always looking for better ways to grow and deploy staff globally. Globally competent business leaders have been defined as persons who have the ability to operate successfully in today's global environment and ameliorate the organizational performance across all geographic markets, stated to be critical for the ability of the firms to compete and succeed internationally. Due to the growing demand for the globally competent business leaders, 62% of the firms globally have reported on having a global leadership development program of some form (American Management Association, 2010). Many global leadership development programs have been including talent management and leadership succession programs, which contain a variety of developmental experiences. These organization-initiated developmental experiences include in-country training or coaching, formal instructional programs, involvement in global teams, global rotational programs, cross-national mentors, international assignments and global travel that encourages learning from colleagues in different countries (Caligiuri and Di Santo, 2001; Mezias and Scandura, 2005; Oddou *et al.*, 2000; American Management Association, 2010; Mendenhall and Stahl, 2000). Even though many activities may exist in a global leadership development program, only half of the 939 firms who have surveyed in the American Management Association study agreed that their global leadership development programs have been significantly effective and improved the skills of leadership in the participants. Similarly, a 2010 study conducted by IBM on over 700 chief human resource executives globally reveals that developing future leaders was rated as the most important business capability needed to achieve future global business objectives. However, it has also been rated as one of the least effective capabilities in their firms. In order to come up with the most proper and efficient program for global leadership development, it is essential to understand the ways that global leadership competencies can be gained, which has not yet been understood well. The explanatory mechanisms through which the experiential global leadership development opportunities can be effective for developing global competencies and there upon how these competencies can subsequently influence the ultimate goal of the programs will be addressed in this study. These two steps have been stated to be necessary for improving the ability of leaders to operate effectively in multi-cultural and cross-cultural environments. Thus, the leaders have to function effectively in both domestic contexts and cross-border situations.

Leaders have been required to cope effectively with contrasting cultural, economic and political practices in order to be able to work in cross-border contexts. Careful selection, grooming, and development of those leaders who are capable of working effectively in globalized environment constitute a pressing need for contemporary organizations. According to the findings so far, researchers have mainly focused on domestic leadership effectiveness only, therefore there is a critical need to understand how differences in contexts need different leadership capabilities (Johns, 2006; Youssef and Luthans, 2012). Similar to all leaders, global leaders have been responsible for their performances at the jobs as well as fulfilling their individual goals. Thus, general effectiveness, which has been stated as the effectiveness of observable actions that managers need to take in order to fulfil their goals, has been very important for global leaders (Rockstuhl *et al.*, 2011). Moreover, realizing the leader's unique responsibilities in their international jobs, which include cross-border responsibilities, has been crucial too (Rockstuhl *et al.*, 2011).

Following are the leadership responsibilities that a leader in cross-border contexts need to undertake (Bartlett and Ghoshal, 1991):

(1) Adoption of multi-cultural perspective other than country-specific perspective;
(2) Balancing global and local demands which can be contradictory;
(3) Working with multiple cultures simultaneously other than working with a single dominant culture.

The effectiveness of global leaders is based on their explicit recognition and emphasis on the unique challenges of heterogeneous cultural, national and institutional contexts. Effective leadership can be measured based on the ability to solve complex social and technical problems. Due to the differences in cross-border and domestic contexts, it has been unlikely that the leadership effectiveness is similar. Therefore, this study will jointly address these differences by focusing on ways that leadership competencies have been similar and different with relevance to different contexts (domestic vs. cross-border).

The social problems for leaders who work in cross border contexts have been shown to be especially complex, since the cultural heritage usually affects the prototypes and schemas regarding proper leadership behaviors. For instance, many factors such as managerial behaviors, nature of relationships and expectations on preferred leadership styles have all been influenced by culture (Littrell and Nicolae Valentin, 2005). Even though it has been shown that both EQ and GQ have been connected to leadership effectiveness in domestic contexts, none of them has been dealing explicitly with the ability to perform in cross-border contexts (Caruso *et al.*, 2002; Judge *et al.*, 2004). Those with high metacognitive CQ have been consciously aware of norms of different societies as well as cultural preferences prior to and during interactions. Whereas metacognitive CQ has been focusing on higher order cognitive processes, cognitive CQ has been the knowledge of conventions, norms and practices in diverse cultures acquired from personal and educational experiences. This includes the knowledge of cultural differences and cultural universals. Leaders with higher cognitive CQ possess sophisticated mental maps of cultural environments and know to how embed this knowledge in cultural contexts. These knowledge structures have been providing them with a starting point for anticipation and understanding of cultural systems, which could shape and influence the patterns of social interaction in a culture. Motivational CQ is the capability to direct energy and attention to learning and operating in the culturally diverse occasion. The value and expectations in association with successfully fulfilling a task have been influencing the magnitude and direction of energy channeled toward that task (Ang and Van Dyne, 2015). The leaders with high motivational CQ, based on their confidence in the intercultural effectiveness and intrinsic interest in cultures, have been directing their energy and attention toward cross-cultural situations (Bandura, 2002; Ang and Van Dyne, 2015). Finally, in interaction with people from other cultures, behavioural CQ has the capability to represent culturally proper verbal and non-verbal actions. It also involves the sensible use of speech acts using proper phrase and words in communication.

Leaders with high behavioural CQ demonstrate flexibility in their intercultural interactions, they have also been trying to adapt their behaviours to put others at ease and facilitate effective interactions. Cognitive, motivational, behavioural CQ and metacognitive, as rooted in differential biological bases, have been representing qualitatively different capabilities to function, different forms of overall CQ and different capability to manage effectively in culturally diverse settings (Earley and Ang, 2003). Many researchers have demonstrated the importance of CQ for performance effectiveness in cross-border contexts. The community in business environments has been tending to recognize only the visible parts of a culture. However, the most important aspects of culture including philosophies, convictions, values, opinions, attitudes and viewpoints remain unseen. Specific training is essential to understand some of these above-mentioned values that may drive behaviour. Many corporations have been using group facilitation and online training modules to help provide these insights. The scientists have been suggesting a mindset on developing CQ in the organization, which says that the entities have to incorporate CQ in decision-making, and they also must adopt a CQ training plan. The training must be so interesting that it should enhance the motivation and target certain job requirements for the employee. During those training session, employees can be engaged by employers to develop an individual plan to spread the CQ. The skills and language classes, as well as training and coaching opportunities can be offered by employers in order to fulfil the individual plan. The employees in the organization must be rewarded for development and practices of CQ. The knowledge includes an understanding of the basic cross-cultural problems. This understanding is the foundation for the strategy development, which raises the ability to engage in action. Positive results from action have generated motivation to initiate the cycle again. Ideally, each time the cycle has been repeated, cultural intelligence skills (CQS) became more effective for an individual and subsequently increased the CQS score. Based on all the theoretical insights into the ways of building a foundation for CQ enhancement in the organization, a process involving different steps has been defined. In this study, since no specific organization has been analyzed, the design of actions will be more universal.

4. Brief Methodology

This study was conducted based on a qualitative case study methodology. Generally, a case study is defined as a study of a certain case or set of cases and explaining or describing the events of the case(s) (Yin, 2011). The qualitative case study has been allowing the researcher to participate in an inductive investigative strategy, appointing the researcher as the primary research instrument in order to find out the meaning and understanding of the research's respondents on the research questions, which has been leading to a rich, descriptive account at the end (Merriam and Tisdell, 2015). It has been asserted by Yin (2011) there might be no fixed design for qualitative study and the researcher might customize the design based on the way that the study has been conducted.

This study follows a semi-structured interview approach enabling a deeper insight into the research problems by achieving a free flow of information from interviewees. The semi-structured interview is based on grounded/underpinning theories of leadership (Barnard, 1938; House, 1971) and cultural competencies (Bandura, 1977; Tajfel, 1981; Turner *et al.*, 1987); hence, perspectives of each coded interviewee, projects and study themes Deepwater Project Performances, Project Leadership, Culture and Project Leadership Improvements were recorded. The interview questions were validated by two senior project directors via preliminary interview. Data were coded to break them down into manageable categories by marking segments of data with symbols, descriptive words or names of categories. Explanation-building and conceptualization including both de-contextualization (taking apart into smaller pieces) and re-contextualization (emerging of a larger and consolidated picture) were carried out in line with the data reduction process (Creswell, 1994).

Interviewees were selected based on the researcher's personal involvement in Malaysia's deepwater megaprojects. They included few distinct project leadership team members and their direct reports who were involved in two of the three deepwater projects in Malaysia up to the year 2016.

A thematic framework method (Ritchie and Spencer, 2002) was adopted as the method of qualitative data analysis for this study for its simplicity and flexibility to suit the conceptual framework. Through this flexibility, thematic analysis allows for rich, detailed and complex description data obtained from the interviews. It is an analytical process, which involves a number of distinct and highly interconnected stages which involve a systematic process of sifting, charting and sorting out material according to key dimensions and themes. In the context of this study, since the focus of the research was to establish the Key Leadership Enablers (KLE) for cross-border project leaders to better manage multicultural teams to enhance deepwater megaprojects performance in Malaysia, the transcripts of each case were thoroughly examined and were coded. The themes were then viewed and arranged based on how project managers managed their projects. The development of the coding scheme was an iterative process throughout the study of all the transcripts for each of the two cases.

5. Preliminary Data Collection and Findings

The author completed Preliminary Interviews with two Senior Project Managers/Directors securing full support on the validity of this Study and Interview Questions, endorsement on the proposed Initial Conceptual Framework and interview protocols without reservation. The semi-structured interviews were completed with findings of additional cross-cultural competencies which will be updated in the Conceptual Framework. Case study and interviews were ongoing at the time of writing this chapter.

6. Significance of the Study

The outcome of this study is expected to continuously improve the decisions of major international oil and gas companies in developing and executing cross-border megaprojects especially in the selection and assignment of "the right" human resources to improve the success for deepwater oil and gas megaprojects. This study is important for international companies seeking involvement in IJVs for the following fundamental reasons:

(1) The largest construction projects in the world are executed by IJVs managed in a multi-cultural context and require considerable amounts of effort, resources and understanding;
(2) The relationships between partners on IJVs on construction megaprojects always encounter major roadblocks to project success;
(3) The style of business and management across borders is alien to many project managers and needs to be at the very least understood and appreciated;
(4) The 'host team' always has an advantage and as such the host country partner on a IJV will always have the upper hand.

This study is expected to be beneficial not only for companies both actively involved and considering involvement in IJVs, but also host country partners, expatriate project managers and individuals involved in multi-national/cultural as well as dispersed project teams.

7. Expected Research Limitations

The study is expected to be limited by the available data and resources from the execution of the two deepwater oil and gas megaprojects in Malaysia. Interviews and validations are limited to Malaysian project leadership members (from Client, Consultant and Contractors organization) and its direct reports. It excludes insights from other project stakeholders. "Restricted and Control" data by the Clients from IOCs and NOCs are not published herewith. It, however, provides the author with an in-depth knowledge to complete this thesis holistically.

Despite constraints in the sampling of certain sensitive and confidential data, the personal involvement of the author as one of the project leadership team members in the project did not retard the study and can still provide valuable guidance and insights for oil and gas companies to strategize and position their future deepwater megaproject investments for success in Malaysia and at any geographical location globally.

8. Conclusion

At the time this chapter was written, the author recognized that in the environment of increasing diversity and globalization in many countries, which is leading to a growing number of international projects, cultural differences can either be a source of creativity and enlarged perspectives or they can be a source of difficulties and miscommunication. The literature on cultural differences is steadily increasing and now provides very helpful conceptual frameworks for understanding the different points of view encountered when managing cross-cultural differences in projects. Managers of multi-cultural project teams can increase their effectiveness and their firm's competitiveness by making use of this literature.

In order to achieve project goals and avoid potential risks, cross-border project leaders should possess CQ and promote creativity and motivation through flexible leadership. Without them, it is destined to underperform.

References

Alon, I. and Higgins, J. M. (2005). "Global Leadership Success Through Emotional and Cultural Intelligences", *Business Horizons*, 48(6), 501–512.

Ang, S. and Van Dyne, L. (2015). Handbook of Cultural Intelligence: Routledge; 1 edition, January 28, 2015, New York.

Ang, S., Van Dyne, L. and Tan, M. L. (2008). Cultural Intelligence. In Sternberg, R. J. &. Kaufman, S. B. (eds.), *The Cambridge Handbook of Intelligence*, Cambridge University. New York: Cambridge Press.

Asrilhant, B., Dyson, R. G. and Meadows, M. (2007). "On the Strategic Project Management Process in the UK Upstream Oil and Gas Sector", *Omega*, 35(1), 89–103.

Bancino, R. and Zevalkink, C. (2007). "Soft Skills: The New Curriculum for Hard-Core Technical Professionals", *Techniques: Connecting Education and Careers (J1)*, 82(5), 20–22.

Bandura, A. (1977). "Self-efficacy: Toward a Unifying Theory of Behavioral Change", *Psychological Review*, 84(2), 191.

Barnard, C. I. (1938). *The Functions of the Executive*, Cambridge, Massachussetts: Harvard University.

Bartlett, C. A. and Ghoshal, S. (1991). "Global Strategic Management: Impact on the New Frontiers of Strategy Research", *Strategic Management Journal*, 12(S1), 5–16.

Björkman, I. and Schapp, A. (1994). "Outsiders in the Middle Kingdom: Expatriate Managers in Chinese-Western Joint Ventures", *European Management Journal,* 12(2), 147–153.

Caligiuri, P. and Di Santo, V. (2001). "Global Competence: What is it, and can it be Developed Through Global Assignments?" *People and Strategy,* 24(3), 27–35.

Caruso, D. R., Mayer, J. D. and Salovey, P. (2002). "Emotional intelligence and emotional leadership", *Paper presented at the Kravis-de Roulet Leadership Conference, 9th, Apr, 1999, Claremont McKenna Coll, Claremont, CA, US.*

Chang, A., Chih, Y.-Y., Chew, E. and Pisarski, A. (2013). "Reconceptualising Mega Project Success in Australian Defence: Recognising the Importance of Value Co-creation", *International Journal of Project Management,* 31(8), 1139–1153.

Chu, P.-C. and Spires, E. E. (2001). "Does Time Constraint on Users Negate the Efficacy of Decision Support Systems?" *Organizational Behavior and Human Decision Processes,* 85(2), 226–249.

Dulewicz, V., Higgs, M. and Slaski, M. (2003). "Measuring Emotional Intelligence: Content, Construct and Criterion-related Validity", *Journal of Managerial Psychology,* 18(5), 405–420.

Dvir, D., Ben-David, A., Sadeh, A. and Shenhar, A. J. (2006). "Critical Managerial Factors Affecting Defense Projects Success: A Comparison Between Neural

Network and Regression Analysis", *Engineering Applications of Artificial Intelligence,* 19(5), 535–543.

Earley, P. C. and Ang, S. (2003). "Cultural Intelligence: Individual Interactions Across Cultures". *Stanford University Press*, Washington.

Eden, C., Williams, T. and Ackermann, F. (2005). "Analysing Project Cost Overruns: Comparing the "Measured Mile" Analysis and System Dynamics Modelling", *International Journal of Project Management,* 23(2), 135–139.

Eweje, J., Turner, R. and Müller, R. (2012). "Maximizing Strategic Value from Megaprojects: The Influence of Information-feed on Decision-making by the Project Manager", *International Journal of Project Management,* 30(6), 639–651.

Flyvbjerg, B. (2014). "What You Should Know About Megaprojects and Why: An Overview", *Project Management Journal,* 45(2), 6–19.

Halman, J. and Braks, B. (1999). "Project Alliancing in the Offshore Industry", *International Journal of Project Management,* 17(2), 71–76.

Heldman, K. (2010). "Project Manager's Spotlight on Risk Management". *John Wiley & Sons*, Jossey Bass, San Francisco.

House, R. J. (1971). "A Path Goal Theory of Leader Effectiveness", *Administrative Science Quarterly,* 16(3), 321–339.

Judge, T. A., Colbert, A. E. and Ilies, R. (2004). "Intelligence and Leadership: A Quantitative Review and Test of Theoretical Propositions", *Journal of Applied Psychology,* 89(3), 542–552.

Kennedy, J. and Mansor, N. (2000). "Malaysian Culture and the Leadership of Organizations: A GLOBE Study", *Malaysian Management Review,* 35(2), 44–53.

Kolios, A. and Luengo, M. M. (2016). "Operational Management of Offshore Energy Assets", *Paper presented at the Journal of Physics: Conference Series.*

Limsila, K. and Ogunlana, S. O. (2008). "Performance and Leadership Outcome Correlates of Leadership Styles and Subordinate Commitment", *Engineering, Construction and Architectural Management,* 15(2), 164–184.

Littrell, R. F. and Nicolae Valentin, L. (2005). "Preferred Leadership Behaviours: Exploratory Results from Romania, Germany, and the UK", *Journal of Management Development,* 24(5), 421–442.

Lu, J. W. and Hébert, L. (2005). "Equity Control and the Survival of International Joint Ventures: A Contingency Approach", *Journal of Business Research,* 58(6), 736–745.

Lustgarten, Abrahm. (2007). Shell Shakedown. *Fortune Magazine, Archive. fortune.com/magazines/fortune/fortune_archive/2007/02/05/8399125/ index.htm.*

Mäkilouko, M. (2004). "Coping with Multicultural Projects: The Leadership Styles of Finnish Project Managers", *International Journal of Project Management,* 22(5), 387–396.

Manning, T. T. (2003). "Leadership Across Cultures: Attachment Style Influences", *Journal of Leadership & Organizational Studies,* 9(3), 20–30.

Matveev, A. V. and Milter, R. G. (2004). "The Value of Intercultural Competence for Performance of Multicultural Teams", *Team Performance Management: An International Journal,* 10(5–6), 104–111.

Merriam, S. B. and Tisdell, E. J. (2015). *Qualitative Research: A Guide to Design and Implementation,* Hoboken, NJ: John Wiley & Sons.

Merrow, E. W. (2011). *Industrial Megaprojects: Concepts, Strategies, and Practices for Success,* Vol. 8, Hoboken, NJ: Wiley Online Library.

Merrow, E. W., McDonwell, L. and Arguden, R. (1988). *Understanding the Outcome of Megaprojects,* Santa Monica: Rand Corporation.

Mezias, J. M. and Scandura, T. A. (2005). "A Needs-driven Approach to Expatriate Adjustment and Career Development: A Multiple Mentoring Perspective", *Journal of International Business Studies,* 36(5), 519–538.

Miller, R., Lessard, D. R., Michaud, P. and Floricel, S. (2001). *The Strategic Management of Large Engineering Projects: Shaping Institutions, Risks, and Governance*: MIT press, MA.

Mišić, S. and Radujković, M. (2015). "Critical Drivers of Megaprojects Success and Failure", *Procedia Engineering,* 122, 71–80.

Moon, T. (2013). "The Effects of Cultural Intelligence on Performance in Multicultural Teams", *Journal of Applied Social Psychology,* 43(12), 2414–2425.

Murray-Webster, R. and Thiry, M. (2000). "Managing Programmes of Projects", *Gower Handbook of Project Management,* 3, 47–64.

Ochieng, E. G. and Price, A. (2010). "Managing Cross-cultural Communication in Multicultural Construction Project Teams: The Case of Kenya and UK", *International Journal of Project Management,* 28(5), 449–460.

Oddou, G., Mendenhall, M. E. and Ritchie, J. B. (2000). "Leveraging Travel as a Tool for Global Leadership Development", *Human Resource Management,* 39(2–3), 159–172.

Olaniran, O. J., Love, P., Edwards, D., Olatunji, O. and Matthews, J. (2015). "Chaotic Dynamics of Cost Overruns in Oil and Gas Megaprojects: A Review", *World Academy of Science, Engineering and Technology, International Journal of Civil, Environmental, Structural, Construction and Architectural Engineering,* 9(7), 911–917.

Ozorhon, B., Arditi, D., Dikmen, I. and Birgonul, M. T. (2007). "Effect of Host Country and Project Conditions in International Construction Joint Ventures", *International Journal of Project Management,* 25(8), 799–806.

Pinto, J. K. and Morris, P. W. (2004). *Wiley Guide to Managing Projects,* Hoboken, NJ: John Wiley & Sons.

Punj, A. (2016). *The Relationship between Cultural Intelligence and Transformational Leadership among Law Enforcement Leaders: A Mixed Methods Study.* North Carolina Agricultural and Technical State University.

Rahim, K. A. and Liwan, A. (2012). "Oil and Gas Trends and Implications in Malaysia", *Energy Policy,* 50, 262–271.

Rezvani, A., Chang, A., Wiewiora, A., Ashkanasy, N. M., Jordan, P. J. and Zolin, R. (2016). "Manager Emotional Intelligence and Project Success: The Mediating Role of Job Satisfaction and Trust", *International Journal of Project Management,* 34(7), 1112–1122.

Rockstuhl, T., Seiler, S., Ang, S., Van Dyne, L. and Annen, H. (2011). "Beyond General Intelligence (IQ) and Emotional Intelligence (EQ): The Role of Cultural Intelligence (CQ) on Cross-border Leadership Effectiveness in a Globalized World", *Journal of Social Issues,* 67(4), 825–840.

Rui, Z., Peng, F., Ling, K., Chang, H., Chen, G. and Zhou, X. (2017). "Investigation into the Performance of Oil and Gas Projects", *Journal of Natural Gas Science and Engineering,* 38, 12–20.

Salazar-Aramayo, J. L., Rodrigues-da-Silveira, R., Rodrigues-de-Almeida, M. and de Castro-Dantas, T. N. (2013). "A Conceptual Model for Project Management of Exploration and Production in the Oil and Gas Industry: The Case of a Brazilian Company", *International Journal of Project Management,* 31(4), 589–601.

Sauser, B. J., Reilly, R. R. and Shenhar, A. J. (2009). "Why Projects Fail? How Contingency Theory can Provide New Insights — A Comparative Analysis of NASA's Mars Climate Orbiter loss", *International Journal of Project Management,* 27(7), 665–679.

Shao, J. (2010). *Impact of Program Managers' Leadership Competences on Program Success and its Moderation Through Program Context,* PhD thesis, Lille: SKEMA Business School.

Shao, J. and Müller, R. (2011). "The Development of Constructs of Program Context and Program Success: A Qualitative Study", *International Journal of Project Management,* 29(8), 947–959.

Shenhar, A. J. and Dvir, D. (2007). *Reinventing Project Management: The Diamond Approach to Successful Growth and Innovation,* Boston, Massachusetts: Harvard Business Review Press.

Sternberg, R. J. and Detterman, D. K. (1986). *What is Intelligence? Contemporary Viewpoints on its Nature and Definition*, Norwood, NJ: Praeger Pub Text.

Sunindijo, R. Y., Hadikusumo, B. H. and Ogunlana, S. (2007). "Emotional Intelligence and Leadership Styles in Construction Project Management", *Journal of Management in Engineering*, 23(4), 166–170.

Tajfel, H. (1981). *Human Groups and Social Categories: Studies in Social Psychology*, Cambridge University Press, ISBN 0 521 22839 5, New York.

Turner, R., Xue, Y. and Anbari, F. (2010). "Achieving Results from Major Infrastructure Projects in China Using a Results-based Monitoring and Evaluation System", *Proceedings of EURAM 2010*, European Academy of Management Conference.

Turner, R., Zolin, R. and Remington, K. (2009). "Monitoring the performance of complex projects from multiple perspectives over multiple time frames", *Paper presented at the Proceedings of the 9th International Research Network of Project Management Conference*.

Whitty, S. J. and Maylor, H. (2009). "And then Came Complex Project Management (revised)", *International Journal of Project Management*, 27(3), 304–310.

Williams, T. and Samset, K. (2010). "Issues in Front-End Decision Making on Projects", *Project Management Journal*, 41(2), 38–49.

Winter, M., Smith, C., Morris, P. and Cicmil, S. (2006). "Directions for Future Research in Project Management: The Main Findings of a UK Government-funded Research Network", *International Journal of Project Management*, 24(8), 638–649.

Yin, R. K. (2011). *Applications of Case Study Research*, Thousand Oaks: Sage.

Yitmen, I. (2013). "Organizational Cultural Intelligence: A Competitive Capability for Strategic Alliances in the International Construction Industry", *Project Management Journal*, 44(4), 5–25.

Youssef, C. M. and Luthans, F. (2012). "Positive global leadership", *Journal of World Business*, 47(4), 539–547.

Chapter 25

BIM- and IoT-Based Data-Driven Decision Support System for Predictive Maintenance of Building Facilities

Weiwei Chen, Jack C. P. Cheng and Yi Tan*

The Hong Kong University of Science and Technology, Clear Water Bay, Kowloon, Hong Kong

**ytanai@connect.ust.hk*

Abstract

Facility managers usually conduct reactive maintenance or preventive maintenance plans according to the condition of building components. However, reactive maintenance cannot prevent the failure of a component while preventive maintenance cannot predict the failure in advance, which leads to inefficient maintenance. Building information modelling (BIM) and internet of things (IoT) provide new opportunities to improve the efficiency of facility maintenance management. Even though significant efforts have been made on BIM and IoT applications in the architecture, engineering and construction/facility management

---

*Corresponding author.

(AEC/FM) industry, the exploration of BIM and IoT integration for FM is still at an initial stage. This chapter develops a BIM- and IoT-based data-driven predictive maintenance framework for facility management, which consists of the following four modules: condition monitoring and fault diagnosis module, condition assessment module, condition prediction module and maintenance rescheduling module. In this process, real-time data collected from IoT sensor network and historical maintenance records from FM system are used for condition prediction. Furthermore, two machine learning methods, namely support vector machine (SVM) and artificial neural network (ANN), are applied to predict the condition of critical equipment in the illustrative example to validate the feasibility of this framework.

Keywords: Building Information Modelling; Internet of Things; Facility Management; Predictive Maintenance; Artificial Neural Network; Support Vector Machine.

1. Introduction

Building maintenance is widely regarded as an activity in the larger context of facilities management (FM), since the maintenance cost accounts for more than 65% of FM cost per year (Eastman *et al.*, 2011). Actually, the maintenance cost can be reduced and the service life of building facilities can be extended by using effective maintenance strategies. In the daily maintenance process, facility condition data can be gathered to serve as an important source of maintenance decision-making. Failing to capture and use the condition data results in significant cost due to ineffective decisions. Therefore, the information technology is promisingly applied to improve the maintenance information management and enhance the decision-making process of maintenance activities.

Building information modelling (BIM) is a digital representation of the physical and functional characteristics of a facility (Eastman *et al.*, 2011). BIM has been applied to building projects and civil infrastructure projects (Cheng *et al.*, 2016; Cheng *et al.*, 2017; Tan *et al.*, 2017). BIM usually serves as a shared knowledge resource for information about a facility, forming a reliable basis for decisions during its life cycle. BIM takes into account the whole life cycle of a building, including design, construction, operation and demolition. The most obvious information in BIM is the geometric information. Semantic information, such as material property data and operations and maintenance instructions, is also included in BIM when the level of detail (LOD) reaches a certain level. In the architecture, engineering and construction (AEC)/FM industry, BIM has also been used for facilitating maintenance activities and storing maintenance information, such as failure location and maintenance documentation. Most maintenance activities have been conducted by conventional methods (e.g. MS Word and MS Excel) with limited use of 3D models for visualization and maintenance development.

Currently, there are three maintenance strategies, namely reactive maintenance, preventive maintenance and predictive maintenance. Reactive maintenance repairs broken components once they are out of functioning and cannot prevent failure in advance. Preventive maintenance can prevent some failures through regular inspection, but cannot predict the failure. Predictive maintenance can predict the failure by analyzing the condition data and historical maintenance records. However, two challenges need to be addressed when using predictive maintenance: one is the collection of condition data for facilities; and the other one is integrating BIM data, FM data and sensor data from IoT network. The first challenge can be solved by leveraging IoT technology to collect condition data of critical equipment in buildings. IoT was first coined by Kevin Ashton in 1999 in the context of supply chain management (Gubbi *et al.*, 2013). Sensors are becoming an essential part of IoT, improving the measurement accuracy and lowering the hardware cost. Besides, the sensor network has been applied to monitoring the condition of critical components in building systems. The second challenge is how to integrate BIM, IoT technology and FM system to improve the efficiency of current maintenance strategy, as well as to reduce the maintenance cost and prevent the failure of building components.

There is a lack of a systematic way for predictive maintenance system design and implementation; therefore, this research aims to develop a methodology framework of integrated BIM- and IoT-based data-driven decision support system for predictive maintenance of building facilities. The developed system captures information and knowledge from BIM and IoT sensor network, as well as considers all affected building components.

The system contains four modules: (1) condition monitoring and fault diagnosis module, (2) condition assessment module, (3) condition performance prediction module and (4) maintenance rescheduling module. The methodology framework can provide a scientific and technological information basis for decision-making.

2. Related Research

2.1. *BIM-based facility maintenance management*

One of the applications of BIM on FM is mainly related to the maintenance management. Shen *et al.* (2012) proposed a loosely coupled integration approach for decision support in facility management and maintenance. Shen *et al.* (2012) only described that the framework can perform decision-making for maintenance. However, details on how to use this framework or specific case studies were not presented. Motawa and Almarshad (2013) investigated a number of case studies and developed an integrated system to capture information and knowledge of building maintenance operations using case base reasoning (CBR), but the system could not predict the failure of building components. Motamedi *et al.* (2014) utilized BIM visualization capabilities to provide FM technicians with visualizations that allowed them to utilize their cognitive and perceptual reasoning for problem solving. However, it could not yet directly undertake automatic decision-making. Chen *et al.* (2016) proposed a framework for automatic scheduling of maintenance work orders based on BIM, which can obtain information from BIM models and FM system as well as automatically generate the maintenance schedule. However, this framework can only be used for maintenance rescheduling, and not designed for predictive maintenance.

Some studies of BIM applications to FM focus on technology integrations, such as combining BIM with geographic information system (GIS) and augmented reality (AR), to improve the maintenance strategies. Kang and Hong (2015) proposed a software architecture for the effective incorporation of BIM into a GIS-based FM system. A BG-ETL prototype was created to extract and transform data from BIM and GIS for FM to solve the extensibility problem. Lee and Akin (2011) used marker-based AR technology to detect the condition of the equipment, which can only show condition information, and visualize the maintenance and operation data, failing to conduct optimization for maintenance and operation activities. Overall, most of the previous BIM-based FM studies are only related to information visualization or information extension from IFC to a data model, analysis and optimization of facility maintenance are still limited.

2.2. *Predictive maintenance*

Researchers have studied the possibilities of applying BIM to predictive maintenance. Hallberg (2009) found that the possibilities of adopting predictive maintenance management depend on the availability of performance-over-time and service life forecasting models and methods. Hao *et al.* (2010) developed a decision support system for integrating reactive maintenance, preventative maintenance and condition-based maintenance. This study also explored the possibility of combining the condition monitoring system and BIM technology. Later on, Hallberg and Tarandi (2011) discussed how the open BIM, with the aid of IFC and Product Life Cycle Support, may facilitate the implementation of a predictive Lifecycle Management System and improve the feasibility for adopting long-term and dynamic maintenance strategy in the FM process.

Other researchers gave some suggestions about how to leverage the advantage of IoT technology for predictive maintenance. Ren and Zhao (2015) indicated that typical challenges for maintenance under environmental big data are lack of timely and accurate data of products, and lack of useful patterns and knowledge of the product life cycle. Schmidt and Wang (2016) mentioned that the cloud technology can enhance the predictive maintenance process, and there was a lack of a systematic way in predictive maintenance system design and implementation. Cheng *et al.* (2016) developed a BIM-based decision support system for predictive maintenance of building facilities, which provides a reasonable framework to enable timely information acquisition, condition assessment, remaining service life prediction and maintenance scheduling. However, there was no case study to validate the feasibility of this system.

Therefore, based on the above-mentioned research limitations, this chapter proposes a BIM- and IoT-based decision support system for predictive maintenance to minimize the research gaps and elaborates on the details about the implementation of this framework.

3. The Proposed Framework

The proposed framework aims to utilize BIM and IoT technologies to support decision-making for facility staff, as well as to provide functions of condition monitoring and condition prediction of facility performance. Figure 1 illustrates the BIM- and IoT-based decision support system for predictive maintenance process and the corresponding data flow. Four distinct stages are included in this process, from the initial data collection to the eventual maintenance action. At the beginning, the FM staff obtain attributes and parameters such as product information, location, dimensions, materials and space from the BIM model. At the condition monitoring and fault diagnosis stage, the real-time data via the RFID and sensors are gathered for the maintenance or service technicians, and then these data are visualized in the BIM models for real-time monitoring. When an abnormal signal appears, the maintenance technicians will analyze the abnormal events using analysis methods. If there is no failure event, the FM staff will assess the condition of the critical components according to inspection records and maintenance standards in the second stage. Following that, in the condition prediction stage, machine learning methods are applied to predict the condition of important components, by using historical maintenance records, sensor data, and geometric information from BIM models. After the prediction, the maintenance plan is rescheduled according to the prediction results to extend the lifetime of the components in the last stage. When conducting maintenance activities, such as repairing, a new RFID tag is attached with new information; if replacing the failure components, the components and RFID tags are replaced by new ones. Finally, the FM staff update the information to BIM models and databases for further operation.

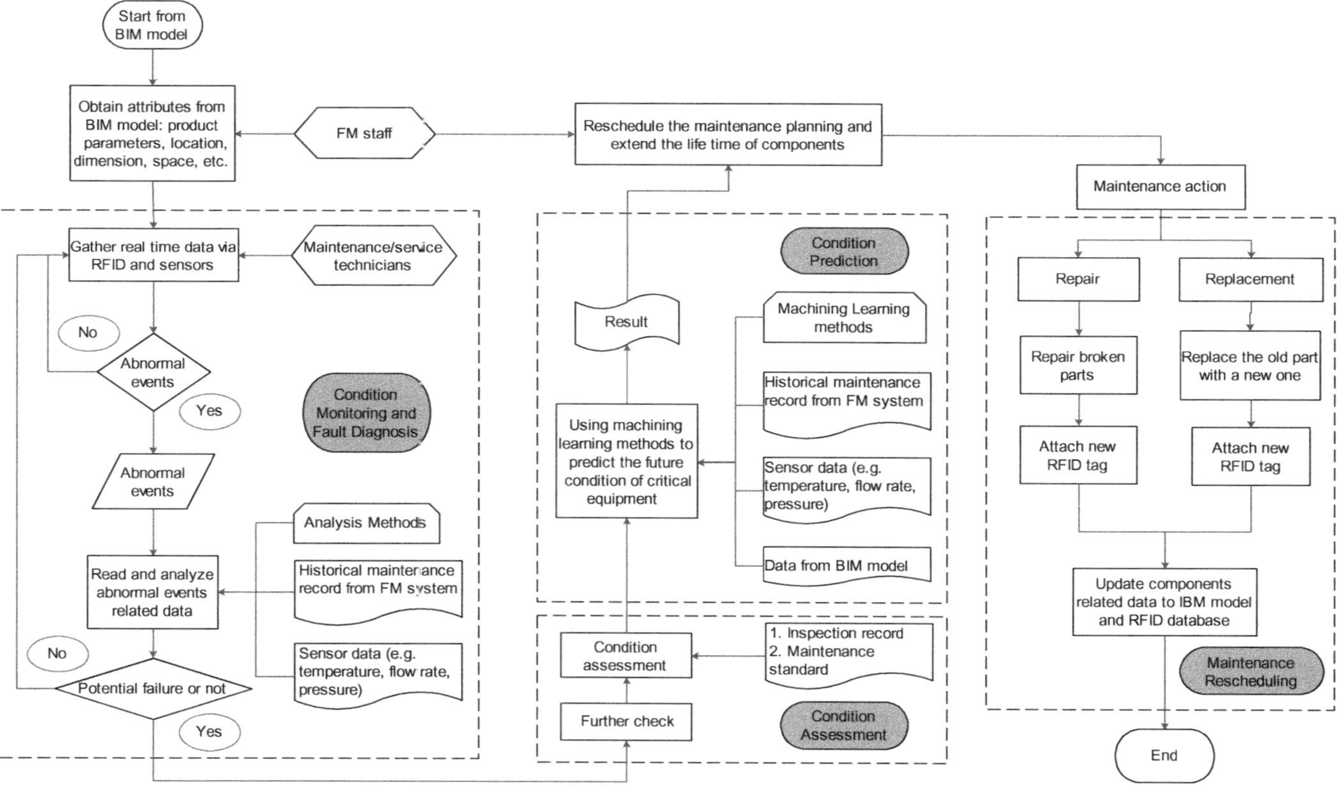

Fig. 1.   The framework of BIM- and IoT-based decision support system for predictive maintenance.

3.1. *Condition monitoring and fault diagnosis module*

In order to visualize the sensor data and monitor the condition of critical equipment through 3D models, a plug-in was developed using Autodesk Revit API. There are two functions in this plug-in, namely sensor management and condition monitoring, as shown in Fig. 2. FM staff can directly retrieve sensor data in the plug-in, and also monitor the condition of each equipment with the user interface and track the real-time data trend in the BIM models for automatic control. The properties of selected sensor are shown in the left properties panel. The sensor management panel displays the basic sensor information including sensor ID and sensor type (temperature, pressure or flow rate), and shows sensor data of current value, maximum value, minimum value and historical value. In addition, data trend of sensor data is illustrated in the condition monitoring panel so that the abnormal signal can be easily detected by facility manager.

For fault diagnosis, the decision is mainly based on abnormal events for the critical equipment, such as an abnormal temperature for

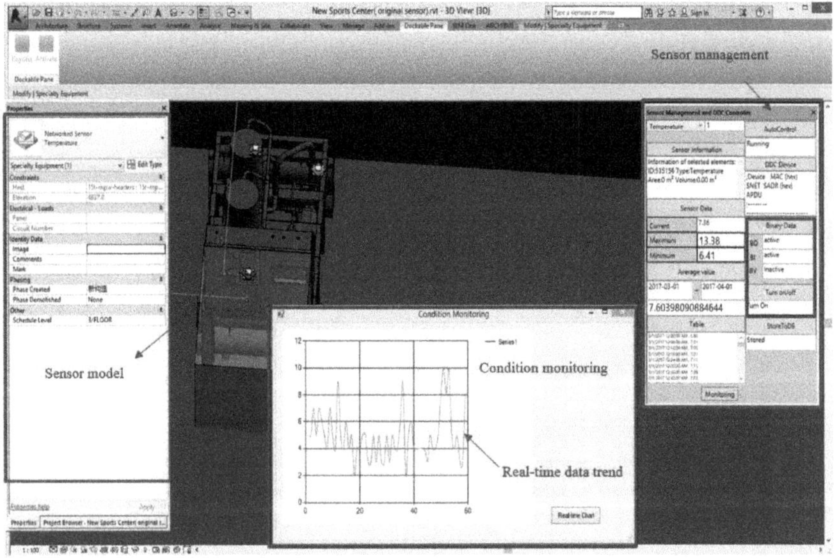

Fig. 2.   The user interface of sensor management.

the bearings of chiller or abnormal vibration of machine tool spindles. It can be diagnosed, estimated and mined periodically from field sensor data using a prognostic algorithm embedded in the decision support system. The basic rule of predictive maintenance is that if the parameter value of the equipment reaches a certain threshold, it will be diagnosed to find the cause. Data mining algorithms, such as support vector machine (SVM) and artificial neural network (ANN), can be used for prognostics and classification. Besides, the abnormal event times and reasons are recorded in the FM system and used for further assessment and prediction.

3.2. *Condition assessment*

Much research has been directed towards identifying proper evaluation criteria in order to assess the condition of building components (Ashworth, 1996; Chew and De Silva, 2003). However, regardless of the criteria used and the details of components, the results of the assessment process mainly depend on the accuracy of the subjective field inspection process. Existing systems require an experienced inspector to judge the condition of an asset during the inspection process manually. The employment of such inspectors is therefore very costly and a long time is needed for inspection. Therefore, condition assessment scale is proposed for inspection and maintenance according to the code of practice of maintenance management in Hong Kong (Building Maintenance Guidebook, 2002), maintenance handbook in US (Facility Condition Assessment Guidebook, 2015) and the experience from maintenance technicians. The HVAC system condition numeric range and its corresponding linguistic description, most possible associated features, and suggested required actions are illustrated in Table 1. The inspectors assess the condition of each system and give suggestions about maintenance action according to the condition value in Table 1. In addition, in the proposed system, some of these inspection data have been input into the BIM-based design tool in order to analyze the possibilities to store and represent inspection data directly in the 3D model. Attributes such as "date of inspection" and "condition index" have been added as additional attributes.

Table 1.   The condition assessment scale of building HVAC system.

Overall Chiller Condition	Scale	Description	Action Required
9–10	Excellent	No defects; As new condition and appearance	Regular monthly inspection. New construction, no visible defects or damage. Meets efficiency and capacity goals and maintains desired temperature and air quality throughout the facility.
7–9	Good	Minor defects; Superficial wear and tear; Some deterioration to finishes; Major maintenance not required	Minor improvement needed, may be slightly outdated and less efficient and consistent. Minor deterioration or defect with no functional impact typically addressed through routine maintenance.
5–7	Fair	Average condition; Significant defects are evident; Worn finished require maintenance; Services are functional but need attention; Deferred maintenance work exists	Repairs are needed; some deterioration exists, and maintenance needs are significant. With these, the system meets need, still within its useful life.
3–5	Poor	Badly deteriorated; Potential structural problems; Inferior appearance; Major defects; Components fail frequently	System has exceeded its useful life; fails to meet standards or needs. Components need extensive repair at a minimum. Currently does not appear to be any safety issue.
0–3	Critical	Building has failed; Not operational; Not viable; Unfit for occupancy or normal use; Environmental/contamination/ pollution issues exist	System is well past its useful life and has critical defects affecting function; its issues are beyond repair and warrant detailed review.

3.3. *Condition performance prediction module*

The aim of predictive maintenance is to provide decision support for maintenance scheduling by diagnosing the defects and predicting the future condition of building components. Based on the condition monitoring data collected from certain sensors, various data analyses and signal processing techniques can be involved to extract different features, such as statistical analysis, frequency analysis and model analysis. In the conventional approach, all the raw sensing measurements are collected and transmitted to the central server, where the analysis is performed. Several machine learning algorithms can be used to predict the condition of building components including ANN, SVM, random forest (RF), and K-nearest neighbour (K-NN). Some researchers applied ANN and SVM methods for condition prediction, such as Sousa *et al.* (2014) evaluated the performance of ANN and SVM in predicting the structural condition of sewers. In fact, ANN and SVM have different advantages in predicting the component condition and both are also dependent on the cases in the training and testing samples. The main advantage of ANN and SVM is the capability of learning from specific predefined patterns. The learning capacity may include classification, prediction and controlling of any specific task. Therefore, ANN and SVM algorithm are selected to validate the feasibility of this system.

The inspection data for selected factors are used to train the ANN and SVM models in order to obtain ANN-based condition prediction model and SVM-based condition prediction model. The inspection data were divided randomly into three sets: (1) 80% for training, (2) 10% for cross-validation and (3) 10% for testing. The training set was used to train the network, whereas the testing set was used to test the network during the model development and also to continuously correct it by adjusting the weights of the network links. The validation dataset not presented to the network during training was used to validate the model. Overall, the prediction process includes three steps: (1) training, (2) cross-validation and (3) prediction.

3.4. *Maintenance rescheduling module*

After predicting the condition of each critical equipment of a building, the maintenance action will be conducted according to the prediction results.

Fig. 3.   The general deterioration curve and corresponding maintenance methods.

Different kinds of components have different deterioration curves but the shape of these deterioration curves is similar. The general deterioration curve for typical building components is shown in Fig. 3 (Uzarski *et al.*, 2007). Combining the prediction results, the required maintenance is illustrated in Fig. 3. When the predicted condition value is above 9, no maintenance actions are required; if the value is between 7 and 9, conduct the inspection to confirm the failure condition; if the value is between 5 and 7, repair the failure component; if the value is between 3 and 5, replace components immediately; and if the value is under 3, the component is discarded.

4. Illustrative Example

4.1. *Case background*

To validate the proposed approach, the authors collected data on an educational building located at the Hong Kong University of Science and Technology (HKUST) campus. There are four chillers (WCC1-WCC4) serving for three buildings. The capacity of WCC1 to WCC 3 is 1800 kW, while the capacity of WCC 4 is 350 kW. Three types of sensor are installed in the chiller system to monitor the operational condition, including temperature sensors, pressure sensors and flow rate sensors. The signals from sensors and DDC controllers are through the sensor network, and imported into the BIM models for monitoring and controlling.

4.2. *The application process*

(1) Data collection

There are three kinds of data as follows: (i) Data from BIM models. The FM manager and maintenance technicians can obtain the properties of chillers from BIM models, such as capacity and installation year. (ii) Data from the sensor network. Real-time condition data is gathered from the IoT sensor network and visualized in BIM models for FM managers to monitor and control the HVAC system, as shown in Fig. 2. The historical sensor data in two weeks (January 1, 2017–January 14, 2017) are selected from the illustration example to show the long-term data trend, as shown in Fig. 4. The abnormal signals of the data trend indicate that the equipment may be non-functional at that moment. In addition, abnormal events are recorded into FM system for further prediction. (iii) Data from FM system. The inspection records, historical maintenance records, and historical sensor data were used for prediction. Nine factors were used to build the models, namely (a) the component ID (b) usage year (c) minor repair times/ year (d) major repair times/year, (e) total service years, (f) abnormal times (g) problem type 1, (h) problem type 2, and (i) condition. Inspection data for the chillers were collected from the facility management office of HKUST. The main characteristics and values of four chillers are shown in Table 2.

(2) Prediction methods selection and configuration

A total of 300 datasets were collected for condition prediction, including three steps, namely training, cross-validation, and prediction, according to the prediction process algorithms. ANN and SVM were chosen to predict the condition of the chillers and for evaluation. The ANN model (multilayer perceptron) was trained using 90% data sets to get the new ANN model. The configuration of the ANN model was listed as follows: $L = 0.3$, $M = 0.2$, $N = 500$, $v = 0$, $S = 0$, $E = 20$ and $H = a$. Similarly, SVM

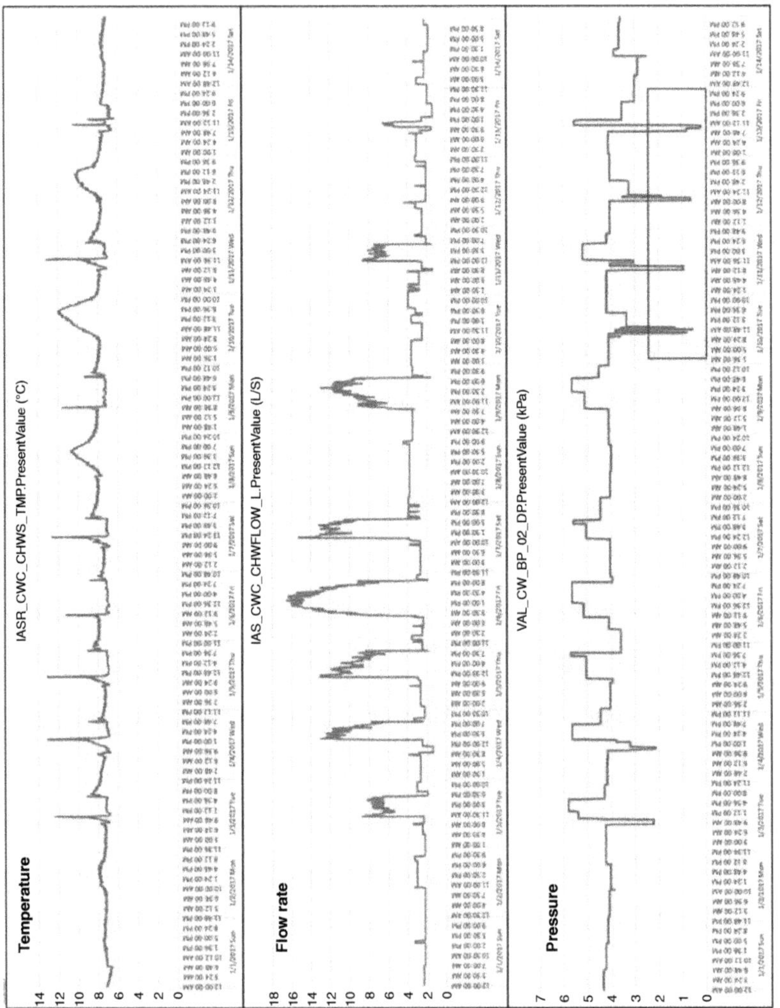

Fig. 4.   The historical sensory data of one chiller.

Table 2.   Collected chiller's main characteristics.

Characteristics	Chiller (1) WCC01	Chiller (2) WCC02	Chiller (3) WCC03	Chiller (4) WCC04
Chiller type	Water-Cooled Liquid	Water-Cooled Liquid	Water-Cooled Liquid	Water-Cooled Liquid
Capacity	1800 kw	1800 kw	1800 kw	350 kw
Location	IAS building	NABN building	NABS building	IAS building
Installation year	2015	2002	2006	2013
Usage age	2	15	11	4
Minor repair times/ year	3	3	3	2
Major repair times/ year	0	3	2	1
Total service years	26 years	26 years	25 years	25 years
Abnormal times/ year	3	3	4	2
Problem type 1	2	2	3	2
Problem type 2	1	2	1	0
Inspection times/ year	12	12	12	12
Current condition	9.2	6.7	7.8	8.3

method was trained with the following configuration: $L = 0.001$, $W = 1$, $P = 1.0\text{e-}12$, $T = 0.001$, $V = -K$. In next step, the trained models were applied to predict the condition of chillers.

(3) Comparison of predictive maintenance methods

To ensure the generality of comparative performance analysis of these algorithms, the authors conducted the condition performance prediction using the same datasets. Table 3 provides a comparison between ANN and SVM algorithms in term of prediction results. The prediction result of the correlation coefficient is around 0.993 in both methods and ANN algorithm (0.9935) is slightly more precise compared to the SVM algorithm (0.9927). However, four error indexes of ANN and SVM indicate that SVM shows better performance than ANN. On the other hand, in the processing time required for training and validation aspect, ANN costs twice the time (0.21 s and 0.14 s) than SVM (0.09 s and 0.05 s). If there are large datasets, SVM method is more efficient than ANN. Overall, the experimental results indicate that proposed SVM algorithm is slightly better when compared to the ANN algorithm.

Table 3.　The comparison of ANN and SVM.

	Training Model		Cross-validation for Model		Prediction Result	
	ANN	SVM	ANN	SVM	ANN	SVM
Correlation coefficient	0.9952	0.9949	0.9935	0.9946	0.9935	0.9927
Mean absolute error	0.1878	0.1757	0.2034	0.1813	0.1909	0.1875
Root mean squared error	0.2289	0.2154	0.2537	0.2206	0.2282	0.2267
Relative absolute error (%)	10.3296	9.663	11.1456	9.9344		
Root relative squared error (%)	10.7591	10.1263	11.8939	10.3403		
Time taken (s)	0.21	0.09	0.14	0.05		
Instance No.	240	240	240/30	240/30	30	30

(4) Maintenance rescheduling

The predicted conditions of SVM method are selected for maintenance rescheduling based on the above comparison result. The predicted conditions of these four chillers three monthly later are 9.128, 5.901, 7.523 and 8.137, respectively. According to Fig. 3, there is no maintenance action for Chiller 1; immediate inspection and replace action are required for Chiller 2; and regular inspection is required for Chillers 3 and 4.

5. Conclusions

The chapter discusses that BIM and IoT can facilitate the implementation of predictive maintenance and improve the feasibility of adopting a dynamic maintenance strategy in the FM process. The architecture of the proposed framework is illustrated in detail and data flow processes are elaborated for data integration among BIM models, IoT and FM system. Four modules are designed for predictive maintenance, including condition monitoring and fault diagnosis module, condition assessment module, condition performance prediction module and maintenance rescheduling module. The intelligent framework makes a control loop in building facility monitoring, prediction and maintenance. Furthermore, two supervised learning algorithms, ANN and SVM, are compared in order to give some suggestions for FM managers. The proposed framework has significant implications in the following aspects: achieving fault diagnosis in an early stage to improve maintenance efficiency, guaranteeing safe operation, saving in maintenance costs and improving the utilization rate of facilities in a building.

References

Ashworth, A. (1996). "Estimating the Life Expectancies of Building Components in Life-Cycle Costing Calculations", *Structural Survey*, 14(2), 4–8.

Building Department (2002). *Building Maintenance Guidebook*, Hong Kong: Building Department.

Chen, W., Chen, K. and Cheng, J. C. P. (2016). "A BIM-based Facility Management Framework for Automatic Scheduling of Maintenance Work Orders", in *Paper presented at the The 16th International Conference on Construction Applications of Virtual Reality (CONVR2016)*, Hong Kong.

Cheng, J. C., Lu, Q. and Deng, Y. (2016). "Analytical Review and Evaluation of Civil Information Modeling", *Automation in Construction,* 67, 31–47.

Cheng, J. C. P., Chen, W., Tan, Y. and Wang, M. (2016). "A BIM-based Decision Support System Framework for Predictive Maintenance Management of Building Facilities", in *Paper presented at the The 16th International Conference on Computing in Civil and Building Engineering (ICCCBE2016), Osaka International Convention Center (Grand Cube Osaka),* Osaka, Japan.

Cheng, J. C. P., Tan, Y., Song, Y., Liu, X. and Wang, X. (2017). "A Semi-Automated Approach to Generate 4D/5D BIM Models for Evaluating Different Offshore Oil And Gas Platform Decommissioning Options", *Visualization in Engineering,* 5(1), 12. doi: 10.1186/s40327-017-0053-2.

Chew, M. and De Silva, N. (2003). "Maintainability Problems of Wet Areas in High-Rise Residential Buildings", *Building Research and Information,* 31(1), 60–69.

Eastman, C. M., Eastman, C., Teicholz, P., Sacks, R. and Liston, K. (2011). *BIM Handbook: A Guide to Building Information Modeling for Owners, Managers, Designers, Engineers and Contractors,* Hoboken: John Wiley & Sons.

Facility Condition Assessment Guidebook. (2015). U.S. Department of Transportation: Federal Transit Administration.

Gubbi, J., Buyya, R., Marusic, S. and Palaniswami, M. (2013). "Internet of Things (IoT): A Vision, Architectural Elements, and Future Directions", *Future Generation Computer Systems,* 29(7), 1645–1660.

Hallberg, D. (2009). *System for Predictive Life Cycle Management of Buildings and Infrastructures,* Stockholm: KTH.

Hallberg, D. and Tarandi, V. (2011). "On the Use Of Open Bim and 4D Visualisation in a Predictive Life Cycle Management System for Construction Works", *Journal of Information Technology in Construction (ITcon),* 16(26), 445–466.

Hao, Q., Xue, Y., Shen, W., Jones, B. and Zhu, J. (2010). "A Decision Support System for Integrating Corrective Maintenance, Preventive Maintenance and Condition-Based Maintenance", in *Paper presented at the Proceedings of Construction Research Congress.*

Kang, T. W. and Hong, C. H. (2015). "A Study on Software Architecture for Effective BIM/GIS-Based Facility Management Data Integration", *Automation in Construction,* 54, 25–38.

Lee, S. and Akin, Ö. (2011). "Augmented Reality-based Computational Fieldwork Support for Equipment Operations and Maintenance", *Automation in Construction,* 20(4), 338–352.

Motamedi, A., Hammad, A. and Asen, Y. (2014). "Knowledge-assisted BIM-based Visual Analytics for Failure Root Cause Detection in Facilities Management", *Automation in Construction*, 43, 73–83. doi: https://doi.org/10.1016/j.autcon.2014.03.012.

Motawa, I. and Almarshad, A. (2013). "A Knowledge-Based BIM System for Building Maintenance", *Automation in Construction*, 29, 173–182.

Ren, S. and Zhao, X. (2015). "A Predictive Maintenance Method for Products Based on Big Data Analysis", in *Paper presented at the International Conference on Materials Engineering and Information Technology Applications*.

Schmidt, B. and Wang, L. (2016). "Cloud-enhanced Predictive Maintenance", *The International Journal of Advanced Manufacturing Technology*, 1, 1–9.

Shen, W., Hao, Q. and Xue, Y. (2012). "A Loosely Coupled System Integration Approach for Decision Support in Facility Management and Maintenance", *Automation in Construction*, 25, 41–48.

Sousa, V., Matos, J. P. and Matias, N. (2014). "Evaluation of Artificial Intelligence Tool Performance and Uncertainty for Predicting Sewer Structural Condition", *Automation in Construction*, 44, 84–91.

Tan, Y., Song, Y., Liu, X., Wang, X. and Cheng, J. C. (2017). "A BIM-based Framework for Lift Planning in Topsides Disassembly of Offshore Oil and Gas Platforms", *Automation in Construction*, 79, 19–30.

Uzarski, D. R., Grussing, M. N. and Clayton, J. B. (2007). "Knowledge-based Condition Survey Inspection Concepts", *Journal of Infrastructure Systems*, 13(1), 72–79.

Chapter 26

Feature-Based Comparison of International Green Neighbourhood Assessment Systems

H. Karimipour, Vivian W. Y. Tam* and Khoa N. Le

*School of Computing, Engineering and Mathematics,
Western Sydney University, Locked Bag 1797, Penrith,
NSW 2751, Australia*
**College of Civil Engineering, Shenzhen University, Shenzhen, China*
**vivianwytam@gmail.com*

Abstract

Together with many social and economic benefits of urbanization, there are also environmental problems. Cities comprise less than 3% of Earth's surface, but there is an extraordinary concentration of population, industry and energy use, leading to a massive local pollution and environmental degradation. Green neighbourhood has emerged to respond to these urban environmental problems. Green neighbourhood is broadly defined as being moderately dense, mixed-use, designed at a human scale, active and public transportation oriented

---

*Corresponding author.

and literally "green". A green neighbourhood assessment system is a tool that evaluates sustainability performance of a given neighbourhood against a set of criteria. This chapter evaluates 20 major international green neighbourhood assessment systems according to their particular features. Regarding regional distribution of the systems around the world, the initiative of green neighbourhood assessment system was first started in United Kingdom and European Union in 2000 and 2004, respectively and this trend reached to United States in 2005. Meanwhile, LEED (ND) with 184 certificates/projects is the main well-known system in the world followed by Envirodevelopment in Australia and SPeAR in the United Kingdom with 100 certificates/projects each. Future research can be conducted on the implementation of several green neighbourhood assessment systems on actual projects and compare their capabilities in measuring the rate of greening.

Keywords: Green Neighbourhood; Assessment Systems; Sustainability.

1. Introduction

Together with many social and economic benefits of urbanization, there are also environmental problems. Cities comprise less than 3% of the Earth's surface, but there is an extraordinary concentration of population, industry and energy use, leading to a massive local pollution and environmental degradation. In the cities, approximately 78% of carbon emissions are due to human activities. The ecological footprints of cities go (through emissions, consumption and other human activities) far beyond their urban boundaries to forests, agriculture, water and other surfaces, which supply their residents so that they have an enormous impact on the surrounding rural, regional and global ecosystem (Ksenija, 2016). Green neighbourhood has emerged to respond to these urban environmental problems. Green neighbourhood is designated as such by using various indicators well beyond traditional variables such as vegetation cover and the size of parks (Jamei *et al.*, 2016). Green neighbourhood is broadly defined as being moderately dense, mixed-use, designed at a human scale, active and public transportation oriented and literally "green". In short, these elements make neighbourhoods liveable places to live, work and play (Denis-Jacob, 2011).

Along with this new concept of green neighbourhood, a need for assessing the city neighbourhood and regions according to their rate of greening has become an issue for consideration (Karimipour *et al.*, 2015). Recent literature has discussed the importance of assessing sustainable development at the community and neighbourhood scales (Berardi, 2011; Turcu, 2012).

The important point is that a framework is required to evaluate objectives and strategies of sustainable community development (Liu *et al.*, 2015; Lee and Kim, 2016). This need has resulted in the emergence and spread of certification systems. Considering the variety of objectives, strategies, and practical approaches of sustainable development at different levels and areas, it can be stated that "certification systems" are frameworks for assessing these objectives and approaches in the city scale (Hamedani and Huber, 2012; Sharifi and Murayama, 2013).

A green neighbourhood assessment system is a tool that evaluates sustainability performance of a given neighbourhood against a set of criteria (Sharifi and Murayama, 2013). According to the classification by Sharifi and Murayama (2013), there are two types of neighbourhood sustainability assessment frameworks: decision-making tools embedded into neighbourhood scale planning (e.g. HQE2R, Ecocity, EcoDistricts, SPeAR, One Planet (Communities) Living, EcoDistricts Performance and Assessment Toolkit), and the systems created from existing third-party building assessment systems. (e.g. LEED (ND), BREEAM (Communities), CASBEE (For Urban Development), QSAS, Green Star (Communities), Green Mark for Districts, and Green Neighbourhood Index (Reith and Orova, 2015).

Previous researchers conducted comparison studies on limited assessment systems. Reith and Orova (2015) undertook a comparison of the five main international green neighbourhood assessment systems with main focus on their criteria. Sharifi and Murayama (2014) conducted a comparative case study of three main assessment systems: LEED (ND), BREEAM (Communities) and CASBEE (For Urban Development). Berardi (2011) compared BREEAM (Communities), LEED (ND) and CASBEE (For Urban Development) and their assessment criteria. Hamedani and Huber (2012) and Liu *et al.* (2010) compared six and nine main international green neighbourhood assessment tools, respectively.

Considering this new worldwide approach to the green neighbourhood and the need for assessing the communities according to their level of sustainability, the 20-main green neighbourhood systems around the world have emerged. However, there is very low or no evidence of in-detail comparison between these assessment methods in the literature. To be more precise, there have been several comparisons; however, they mainly focus on the system's criteria and their sustainability aspects. With a different look at the evaluating methods, this chapter intends to compare the particular features of these 20 assessment systems and provides a useful toolkit for the communities or sectors who desire to adapt themselves to the sustainability standards. The full terminologies of 20 green neighbourhood assessment systems are shown in Table 1.

Therefore, the aim of this comparison is as follows:

(1) To prepare a complete list of all the green neighbourhood assessment systems around the world and
(2) To compare all the green neighbourhood assessment systems in the world based on their special features along with their regions and their emergence time.

The primary aim of this comparison is to provide an insight to urban planners and stakeholders who intend to implement the green neighbourhood systems in their communities. This chapter can act as a toolkit for them to compare the different systems around the world, according to their location, living time, lifetime and the number of certificates, so that they can simply decide which system they want to adapt themselves.

2. Research Methodologies

Sustainability assessment is regarded as the latest generation of impact assessment tools, and can be defined as "any process that directs decision-making towards sustainability" (Bond *et al.*, 2012). Ambiguities in the definition of sustainable development are also reflected in the definition of its assessment (Ness *et al.*, 2007; Berardi, 2013). Numerous methodological approaches have been taken to assess sustainability.

Table 1. The main aim and type of space included in the international green neighbourhood assessment systems.

Row	System	Main Aim	Type Space Included
1	Green Infrastructure	The Green Infrastructure Project, hosted by the Botanic Gardens of South, has developed an Evidence Base for Green Infrastructure in South Australia. This Project has a vision of "South Australians living in healthy, resilient and beautiful landscapes that sustain and connect people with plants and places"	—
2	EcoCity	EcoCity Project was designed for Minimizing the use of land, energy and materials, minimizing the impairment of the natural environment, maximizing human well-being, and minimizing transport demand	All uses in an area of a city
3	Green Townships	IGBC Green Townships Rating System is designed to address the issues of sprawl, automobile dependency, social and environmental disconnect	At least 25% of the total built-up area (in sq m) within the township should be earmarked for residential use
4	HQE2R	HQE2R was designed to provide an integrated approach with adapted methods and tools for use by local municipalities and their partners in their neighbourhood regeneration projects as in urban management projects	—
5	EarthCraft Communities	EarthCraft Communities (ECC) is a developer-certified, third-party verified program that recognizes responsibly designed and constructed communities. It is a regionally-specific tool utilized by land developers and local government agencies to promote smart growth, sustainable land development practices and healthier communities	The proposed ECC development must qualify as an infill location with walkable pedestrian connections to an existing sidewalk network

(Continued)

Table 1. (*Continued*)

Row	System	Main Aim	Type Space Included
6	GSAS	The primary objective of GSAS is to create a sustainable built environment that minimizes ecological impact while addressing the specific regional needs and environment of Qatar	Newly developing and existing neighbourhoods. Any combination of buildings can be assessed; there are no minimum or maximum sizes
7	Neighbourhood Sustainability Framework	The neighbourhood built environment is designed, constructed and managed to generate neighbourhoods that are adaptive and resilient places that allow people to create rich and satisfying lives while respecting the limitations of the natural environment	Can be applied to both new and existing neighbourhoods, Planned new neighbourhoods and neighbourhoods about to be built
8	GCAP	The GCAP provides a set of recommendations that are aimed at maintaining Melaka's competitiveness as a popular tourist and investment destination, keeping environmental challenges to a minimum, and establishing the state as a role model for liveability in the region	Just Maleka state (1663 km^2)
9	BCA	The BCA Green Mark Scheme was launched to promote environmental awareness in the construction and real-estate sectors. It is a benchmarking scheme which aims to achieve a sustainable built environment by incorporating best practices in environmental design and construction, and the adoption of green district technologies	>20 ha

10	Pearl Community rating system	The main aim of the Pearl system is to create more sustainable communities, cities and global enterprises and to balance the four pillars of Estidama: environmental, economic, cultural and social	At least 1,000 permanent residential population
11	EcoDistricts	To foster a new model and era of urban regeneration, EcoDistricts have created the EcoDistricts Protocol: a framework for achieving people-centred, economically vibrant, planet-loving neighbourhood- and district-scale sustainability	Existing neighbourhood, Brownfield site, Business district, Institutional campuses, Industrial lands, Mixed-used district
12	One Planet (Communities)	One Planet (Communities) Living is an initiative of Bioregional and its partners to make truly sustainable living a reality. One Planet (Communities) Living uses ecological footprinting and carbon footprinting as its headline indicators	It determines by each specific project
13	SPeAR	Arup has developed SPeAR® as a tool to assist in the consideration of sustainability issues, particularly with regard to decision-making and communication with stakeholders	SPeAR® is designed to cover all forms of projects, including design and delivery of new infrastructure, masterplans and individual buildings
14	Envirodevelopment	EnviroDevelopment is a scientifically-based branding system designed to make it easier for purchasers to recognize and, thereby, select more environmentally sustainable homes and lifestyles	From single detached houses all the way to whole suburbs, with a CityCat terminal in between

(Continued)

Table 1.   (Continued)

Row	System	Main Aim	Type Space Included
15	STAR	The STAR Community Rating System (STAR) is the certification program to recognize sustainable communities. Local leaders use STAR to assess their sustainability, set targets for moving forward, and measure progress along the way	—
16	Green Star (Communities)	The Green Star (Communities) evaluates the sustainability attributes of the planning, design, and construction of large scale development projects, at a precinct, neighbourhood, and/or community scale	A project's plan for development must include the development of at least four (4) buildings, of any size and mixture of Class 1–9 structures (except Class 4), as classified under the Building Code of Australia (BCA)
17	BREEAM (Communities)	BREEAM (Communities) aims to ensure that its standards provide social and economic benefits while mitigating the impacts of the built environment. In doing so, BREEAM (Communities) enables developments to be recognized according to their sustainability benefits and stimulates demand for sustainable developments	More useful for moderate or large mixed-use development; however, in some cases it can be used for single development too

18	CASBEE (For urban development)	CASBBE is aimed for comprehensive assessment of environmental performance of a construction project planned and conducted under the unified intention of development for a relatively large group of land sections such as a whole block or a district consisting of blocks	— A single block consisting of multiple building sites and public space adjacent to the block such as a road. — Multiple collective blocks and public space existing integrally with them such as roads
19	DGNB	The DGNB assesses buildings and urban districts which demonstrate an outstanding commitment to meeting sustainability objectives This takes the following fields into account: Environmental Quality, Economic Quality, Sociocultural and Functional Quality, Technical Quality, and Process Quality. Site Quality is integrated here as a criterion for assessment	1. The minimum size of an urban district is 2 ha of gross development area (GDA). 2. The district consists of a number of buildings and at least two development sites and has public or publicly accessible spaces and related infrastructure. 3. The residential area is no less than 10% no more than 90%.
20	LEED (ND)	LEED (ND) for Neighbourhood Development (LEED (ND)) was engineered to inspire and help create better, more sustainable, well-connected neighbourhoods. It looks beyond the scale of buildings to consider entire communities	Designed for neighbourhood-scale projects that are near completion, or were completed within the last three years

All these approaches use indicators as tools for generating usable and relevant information from the increasingly expanding volume of data that they acquire from a wide array of sources (Mitchell and Gordon, 1996). A plethora of research exists that focuses on the principles and indicators of sustainable urban and neighbourhood development (Sharifi and Murayama, 2015).

Green neighbourhood assessment systems are compared based on the sustainability framework which is reflected in their indicators and main topics and also on their special features (Yang *et al.*, 2017). This chapter will conduct a comparison between 20 main green neighbourhood systems based on their particular features. Feature-based comparison is helpful for those who want to apply their communities for inclusion in each system. This chapter specially indicates the information these communities may need before applying and as a snapshot for comparison. In the following, this comparison and the methodology used for are described.

2.1. *Feature-based comparison*

Many communities are beginning to recognize the impacts of their neighbourhoods on the environment and to provide significant changes for mitigating their environmental impact (Schüle *et al.*, 2017, Seiferling *et al.*, 2017). This shift in attitude has been driven in large part by a growing market demand for environmentally sound and energy-efficient products and services, initiated primarily from the non-profit sector and federal, state, and municipal building projects (GhaffarianHoseini *et al.*, 2013). A central issue in striving towards reduced environmental impact is the need for an applicable and mean-ingful yardstick for measuring environmental and energy performance (Smith *et al.*, 2006).

Some systems and mechanisms have been developed worldwide to help decision makers and communities assess their actual impacts on the environment as summarized in Table 1. These systems are different in sizes, reputation, years of publication, topics used and fees. Therefore, these types of diversities have been illustrated in this section.

The features which are under comparison in this section are: the region or country which they are in, the number of certificates/projects

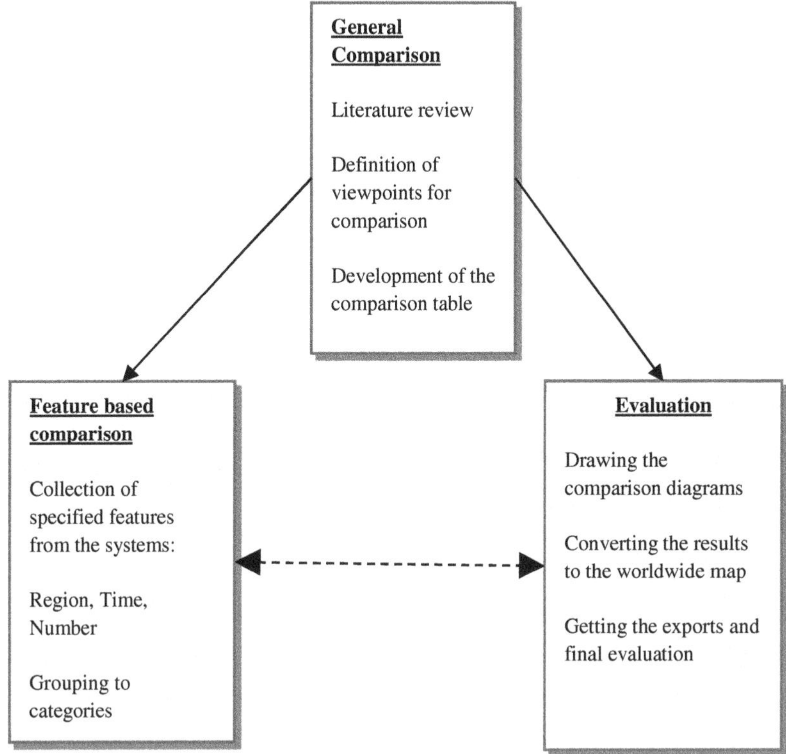

Fig. 1.   The procedure of the novel analysis of the international green neighbourhood assessment systems.

under each system and the first year of appearance of each system. This information is obtained from at least 60 resources, and it may be useful for those communities who have interests in assessing their environmental impacts and may obtain the certificates. Figure 1 summarizes the comparison process that has been undertaken in this chapter.

3. Results and Discussion

3.1. *Feature-based comparison*

This research has looked at the other side of the assessment system: "How framework is used (Zuo and Zhao 2014)? Instead of what framework is used?" This section conducted a review of the regions where each system appeared, the time of appearance, and the number of certificates/projects under each assessment system.

3.1.1. *Green neighbourhood systems distribution around the world*

The distribution of 20 green neighbourhood assessment systems around the world is shown in Fig. 2. It can be stated that the United States with four assessment systems followed by the United Kingdom and Australia with three assessment systems each have the most varied frameworks in the world. In addition, Germany, Japan, Singapore, India, United Arab of Emirates, New Zealand, Qatar, and Malaysia are provided with one framework each. In addition, European Union introduced two megaprojects including Ecocity and HQE2R; both are decision-making tools for the further upcoming systems. Most of the systems around the world are concentrated in North America, Europe and Oceania and some new systems have also developed in Eastern Asia and Middle East regions.

3.1.2. *Life time of the green neighbourhood systems*

Life time of the assessment systems is recognized as an indicator of emergence of this new approach around any area of the world. The result of

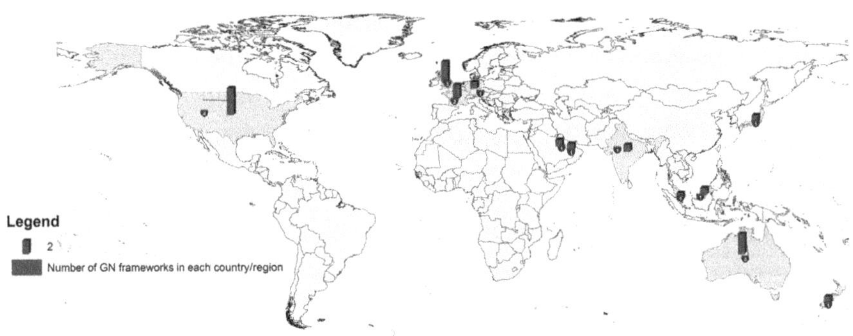

Fig. 2.   The number of green neighbourhood assessment systems around the world.

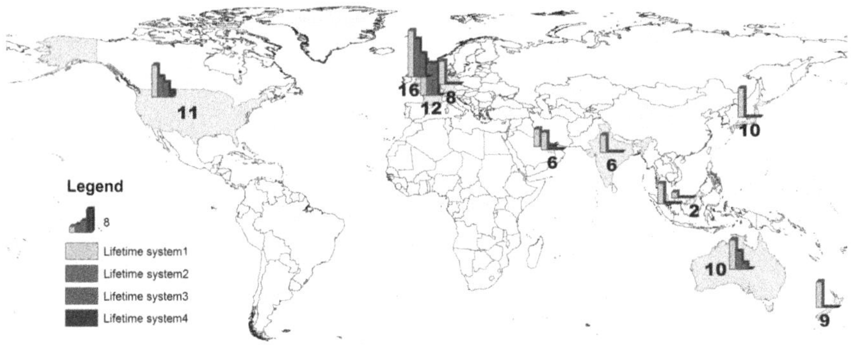

Fig. 3. Life time of the green neighbourhood assessment systems around the world.

this part is summarized in Fig. 3 and Table 2. The first systems on green neighbourhood were established in United Kingdom in 2000 and 2003. These systems were revised in the later years, but first they focused on a novel approach to communicate and deliver sustainability. As the first system, SPeAR was designed as a sustainability appraisal tool and as a framework to support project development and communicate outcomes according to the emerging sustainable development aspects; however, One Planet (Communities) which is the second emerging system mostly focused on implementing some green neighbourhood practices around United Kingdom.

After that and in 2004, HQE2R was invented in European Union as a methodological framework for sustainable regeneration projects reflecting the needs for these types of decision-making frameworks in urban development in the Europe. Ecocity was the other system that was launched in 2005 by the European Union resulting from the new trend to create sustainable cities in this continent. Emerging in 2005, Earthcraft (communities) is the first system in the world mainly focusing on neighbourhood's assessment according to their sustainability levels. However, it was a regionally specific tool utilized by land developers and local government agencies to promote smart growth, sustainable land development practices and healthier communities.

CASBEE (For Urban Development), DGNB, BREEAM (Communities) and LEED (ND) were introduced between 2006 and 2009 as a result

Table 2.   Life time of the green neighbourhood assessment systems around the world.

Year of First Version Released	Name of Systems	Country	Living Time of the System (Year)
2000	SPeAR	United Kingdom	16
2003	One Planet (Communities)	United Kingdom	13
2004	HQE2R	European Union	12
2005	Earthcraft Communities, Ecocity	United States, European Union	11
2006	Envirodevelopment, CASBEE (For Urban Development)	Australia, Japan	10
2007	Neighbourhood Sustainability Framework	New Zealand	9
2008	DGNB, BREEAM (Communities)	Germany, United Kingdom	8
2009	LEED (ND), BCA	United States, Singapore	7
2010	IGBC, QSAS, Pearl, Green Star (Communities)	India, Qatar, United Arab Emirates, Australia	6
2011	Ecodistrict	United States	5
2012	STAR, Green Infrastructure Project	United States	2
2014	GCAP	Malaysia	2

of demand for sustainable city development and its appraisal system and this new approach reached Australia and New Zealand in 2006 and 2007 with the introduction of Envirodevelopment and Neighbourhood Sustainability Framework, respectively. The Eastern part of Asia and the Middle East designed their own regional specific tools between 2008 and 2014 according to their particular requirements in sustainable urban development.

3.1.3. *Number of certificates/projects under each assessment system*

The information required for this part of the chapter cannot be easily obtained, as some systems did not distribute any statistics on their manuals and websites. Twelve assessment systems were compared, as shown in Fig. 4.

Among the systems, LEED (ND) in United States with about 184 neighbourhood certificates/projects is the most well-known system around the world. Envirodevelopment in Australia and SPeAR in United Kingdom each with 100 certificates/projects are the second high-demanded systems in the world and STAR Community and DGNB with 50 and 39 certificates/projects, respectively, receive the third highest ranking. Other systems with less than 30 certificates/projects are in the next positions.

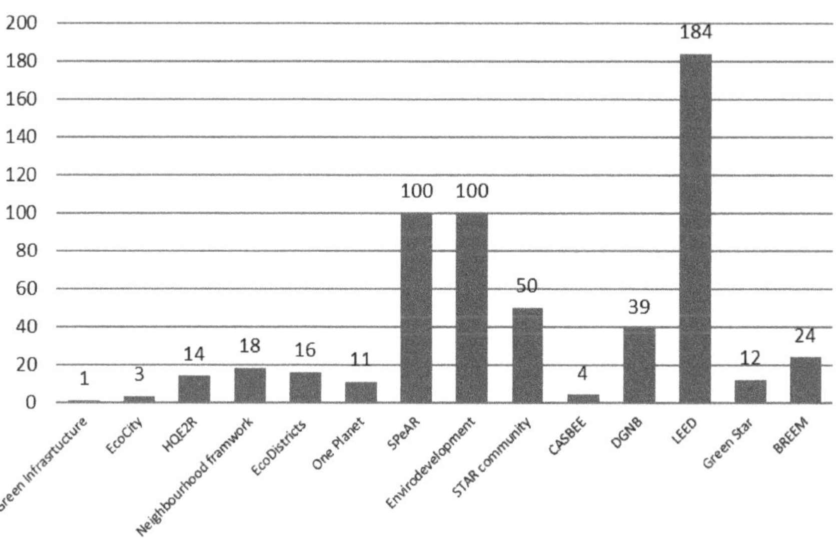

Fig. 4.   The number of certificates/projects under different green neighbourhood assessment systems around the world.

4. Conclusion

This chapter compared 20 international green neighbourhood assessment systems. The following featured were observed:

(1) About the regional distribution of the systems around the world, it can be said that the initiative of green neighbourhood system was first started in United Kingdom and European Union in 2000 and 2004, respectively, and then this trend reached United States in 2005. Australia, New Zealand introduced their own systems in 2006 and 2007, respectively, and then South-East Asia and Middle East started to commence their own regional systems between 2009 and 2014.

(2) United States with four main green neighbourhood systems owns the largest number of assessment systems in the world, followed by United Kingdom, Australia with three systems each, European Union with two systems and then the other countries each with one system.

(3) On the number of certificates/projects, LEED (ND) with 184 certificates is the main well-known system in the world followed by Envirodevelopment in Australia and SPeAR in United Kingdom each with 100 certificates/projects.

Future directions need to address research questions on whether a green neighbourhood certified as a specific level of greening means that it should get the same level of greening in the other systems as well. It means that they can investigate the actual results that affect a pilot green neighbourhood from different assessment systems and as a result, it might be surveyed whether they measure the same thing or there is a big difference in their perception of sustainability.

Acknowledgement

The authors wish to acknowledge the financial support from the Australian Research Council (ARC) Discovery Project under grant number DP150101015. We also wish to thank the following people for their gracious support in providing the necessary data and information, without which this work would not have been possible, including Mr. Nobufusa

Yoshizawa (CASBEE (For Urban Development), Japan), Mr. Stephan Anders (Product Manager Quartier from DGNB, Germany), Mr. Ben H Gill (from One Planet (Communities), United Kingdom), Ms. Melinda Walters (from Green Star (Communities), Australia), Ms. Ingrid Irby and Ms. Marci B. Reed (from Earthcraft Communities, United States), Mr. Glenn Crawford (from USGBS, United States), Mr. Taylor Hood (from Urban Development Institute of Australia) and EcoDistricts team from the United States.

References

Berardi, U. (2011). "Beyond Sustainability Assessment Systems: Upgrading Topics by Enlarging the Scale of Assessment", *International Journal of Sustainable Building Technology and Urban Development*, 2(4), 276–282.

Berardi, U. (2013). "Sustainability Assessment of Urban Communities through Rating Systems", *Environment, Development and Sustainability*, 15, 1573–1591.

Bond, A., Morrison-Saunders, A. and Pope, J. (2012). "Sustainability Assessment: The State of the Art", *Impact Assessment and Project Appraisal*, 30(1), 53–62.

Denis-Jacob, J. (2011). "Green Neighbourhoods: The Making of a Sustainable City". Available at: http://www.geographyjobs.ca/articles/green_neighbourhoods_the_making_of_a_sustainable_city.html (Accessed on March 29, 2011).

GhaffarianHoseini, A. H., Dahlan, N. D., Berardi, A., GhaffarianHoseini, A., Makaremi, N. and GhaffarianHoseini, M. (2013). "Sustainable Energy Performances of Green Buildings: A Review of Current Theories, Implementations and Challenges", *Renewable and Sustainable Energy Reviews*, 25, 1–17.

Hamedani, A. Z. and Huber, F. (2012). "A Comparative Study of "DGNB", "LEED" and "BREEAM" Certificate System in Urban Sustainability", in *7th International Conference on Urban Regeneration and Sustainability*.

Jamei, E., Rajagopalan, P., Seyedmahmoudian, M. and Jamei, Y. (2016). "Review on the Impact of Urban Geometry and Pedestrian Level Greening on Outdoor Thermal Comfort", *Renewable and Sustainable Energy Reviews*, 54, 1002–1017.

Karimipour, H., Mojtahedi, M. and Azari Dehkordi, F. (2015). "Introduction to a Quantitative Method for Assessment of Visual Impacts of Tehran Towers", *Soil Science and Environmental Management*, 6(6), 132–139.

Ksenija (2016). "Environmental Problems of Modern Cities", *Climatology & Meteorology*. Available at: https://owlcation.com/stem/Environmental-problems-of-modern-cities (Accessed on December 19, 2016).

Lee, M. G. and Kim, H. B. (2016). "A Cost–benefit Analysis for the Institutionalization of the Green Certification Scheme: The Case of the Development of the Sinjung District in Seoul, Korea", *International Journal of Urban Sciences*, 20(1), 88–106.

Liu, G., Nolte, I. Potapova, A. Michel, S. and Ruckert, K. (2010). "Comparison of Worldwide Certification Systems for Sustainable Building", in *9th International Conference on Sustainable Energy Technologies*, Shanghai, China, Baltic Sea Region.

Liu, Y., Meng, Q. Zhang, J. Zhang, L. Jancso, T. and Vatseva, R. (2015). "An effective Building Neighborhood Green Index Model for Measuring Urban Green Space", *International Journal of Digital Earth*, 9(4), 387–409.

Mitchell, G. (1996). "Problems and Fundamentals of Sustainable Development Indicators", *Sustainable Development*, 4(1), 1–11.

Ness, B., Urbel-Piirsalu, E. Anderberg, S. and Olsson, L. (2007). "Categorising Tools for Sustainability Assessment", *Ecological Economics*, 60(3), 498–508.

Reith, A. and Orova, M. (2015). "Do Green Neighbourhood Ratings Cover Sustainability?" *Ecological Indicators*, 48, 660–672.

Schüle, S. A., Gabriel, K. M. A. and Bolte, G. (2017). "Relationship between Neighbourhood Socioeconomic Position and Neighbourhood Public Green Space Availability: An Environmental Inequality Analysis in a Large German City Applying Generalized Linear Models", *International Journal of Hygiene and Environmental Health*, 220(4), 711–718.

Seiferling, I., Naik, N. Ratti, C. and Proulx, R. (2017). "Green Streets — Quantifying and Mapping Urban Trees with Street-level Imagery and Computer Vision", *Landscape and Urban Planning*, 165, 93–101.

Sharifi, A. and Murayama, A. (2013). "A Critical Review of Seven Selected Neighborhood Sustainability Assessment Tools", *Environmental Impact Assessment Review*, 38, 73–87.

Sharifi, A. and Murayama, A. (2014). "Neighborhood Sustainability Assessment in Action: Cross-evaluation of Three Assessment Systems and their Cases from the US, the UK, and Japan", *Building and Environment*, 72, 243–258.

Sharifi, A. and Murayama, A. (2015). "Viability of Using Global Standards for Neighbourhood Sustainability Assessment: Insights from a Comparative Case Study", *Journal of Environmental Planning and Management*, 58(1), 1–23.

Smith, T. M., Fischlein, M. Suh, S. and Huelman, P. (2006). Green Building Rating System: A Comparison of the LEED and Green Globes Systems in the US. United States, University of Minnesota.

Turcu, C. (2012). "Re-thinking Sustainability Indicators: Local Perspectives of Urban Sustainability", *Journal of Environmental Planning and Management*, 56(5), 695–719.

Yang, J., Sun, J., Ge, Q. and Li, X. (2017). "Assessing the Impacts of Urbanization-associated Green Space on Urban Land Surface Temperature: A Case Study of Dalian, China", *Urban Forestry & Urban Greening*, 22, 1–10.

Zuo, J. and Zhao, Z. (2014). "Green Building Research — Current Status And Future Agenda: A Review", *Renewable and Sustainable Energy Reviews*, 30, 271–281.

Chapter 27

A Study of Information Technology Adoption for Real-Estate Management: A System Dynamic Model

Fahim Ullah* and Samad M. E. Sepasgozar

Faculty of Built Environment, The University of New South Wales, Sydney, NSW 2052, Australia

**f.ullah@student.unsw.edu.au*

Abstract

Most of the information technologies in different sectors including construction and property management businesses have failed. Previous studies intended to identify a variety of factors influencing the technology adoption (TA) process, while the dynamics that govern the adoption of information technologies in real-estate business over time were ignored. This chapter aims to present a system dynamic (SD) model for TA in the real-estate domain using critical factors and clusters of real-estate websites including system quality, information quality, service quality

---

*Corresponding author.

and perceived ease of use. The model comprises key groups' perceived usefulness, user satisfaction and behavioural intention to use (BIU) a website. In addition, three quality components including system, information and service quality are included in the model. A total of 50 variables were identified and will be used for examining the model in different contexts. The model was developed based on an intensive literature review in two main domains: real-estate management and information systems or websites. The resulted model of this ongoing study will be examined in different contexts in future.

Keywords: Technology Adoption; Real Estate Websites; Systems Model.

1. Introduction

The dawn of technological era is reshaping the global industries. Information technology (IT) systems use more than 10% of global electricity (Mills, 2013). Globally, in the health industry, the use of IT is as high as 57% (Lee *et al.*, 2016). In academia, the use of technology is as high as 67% as per 2014 statistics (Dahlstrom *et al.*, 2014). This technological advancement penetrated the construction industry with the boom of IT in the early 2000s. Being referred to as the "Tipping Point" by Brandon *et al.* (2005), IT has extended the boundaries of the construction industry and has reshaped its outlook from traditional rigid 2D drawing-based industry to more modern virtual reality (VR) and 3D video-based industry. However, the technology adoption (TA) in construction is still localized to specific fields such as safety and project scheduling, and the holistic effects on many fields including real-estate websites and its customer perception is yet to be explored. Hewage *et al.* (2008) highlighted the use of IT for only 12% of construction-specialized training. Further, the authors pointed to the use of GPS and modern surveying IT tools to be as much as 33.3% of Alberta's construction industry. According to Dave *et al.* (2016), the corresponding Global Revenue Forecast of Internet-based businesses will increase from AUD 1.9 trillion in 2013 to AUD 7.1 trillion in 2020. This high percentage of contribution to the global GDP requires in-depth analysis of TA capabilities of global industries including construction and real estate.

The Australian construction industry contributes to 8% of GDP out of which the share of real-estate sector is 20% (AI-Group, 2015). Like its construction counter parts, real-estate management is evolving and adopting the modern technologies. The use of 3D images and GIS for real estate is getting special attention these days. In this context, it is imperative to work on its other parts along with making it a holistic and competitive technologically advanced industry. One such attention-seeking aspect is the state of TA in real-estate websites. The real-estate websites and their usage especially from customer perspective have not been explored till date. These websites need to be updated and upgraded to cope with the growing technological boom and adoption. Although a few technology adoption models (TAMs) do exist for websites in general, when it comes to real-estate websites, the state of practice and theory are not necessarily aligned. Not only is the holistic real-estate TAM missing but also the use of conceptual models such as advanced visualization and simulation-based dynamic modelling tools like systems dynamics (SD) has not been reported till date. Thus, this study attempts to collect the website usage key factors from customer' perspective and presents a novel conceptual systems model for TA in real estate and its websites. The model bases its concepts on the core factors and presents their dynamic interactions leading to better use or re-use of a real-estate website while capturing the general customer' requirements and needs.

2. Literature Review

With the advancements in construction technologies and era of Smart Cities (SCs) disruptive and advanced technologies are making their way into construction and real-estate industries (Ullah *et al.*, 2016; Ekman *et al.*, 2016; Ayub *et al.*, 2016). These technologies include the use of GIS, navigation tools, real-time locating systems, and virtual and augmented realities (Guo *et al.*, 2017). Construction real estate is gradually adopting the technologies such as 3D videos and 360 imageries but the state of its websites requires considerable improvements to be termed as the state of the art. In terms of TA, in real-estate websites, a few mentionable studies include Huang and Wang (2005) that presents a conceptual framework for early warnings and forecasting of real-estate development to inform the users about potential effects of regulatory policies on real-estate development. Ekman *et al.* (2016) evaluated the potential of smart commercial real-estate (CRE) using the case of a Swedish CRE firm that has developed and deployed a technology-based self-service system to help tenants reduce energy consumption. The results suggest that tenants need to enhance their competence in three areas: management of IT, within-organization and business-to-business networking skills for better usage of technology. Christensen *et al.* (2016) studied the role of real-estate profession in the preparedness, response, reconstruction, and recovery stages of a crisis by deploying the geospatial technology developed using geographic information, geographic information systems, gaming technology, and the creation of a user interface. The findings suggest considerable improvements in the decision-making process for those tasked with ensuring resilience in urban areas in case of crisis management of flooding, riots/protests, and terrorist events. These studies are useful in terms of real-estate management in general but specifically for real-estate websites management and usage, and the associated sophisticated modelling tools such as SD, the literature is scarce.

SD, an amalgam of graphical and mathematical simulations, has gained popularity in simulation-based studies. SD uses computer software such as Vensim®, AnyLogic® and Dynamo® for simulating and validating its models and achieving corresponding graphical results. It uses the "what if" scenarios for testing certain aspects of a complex model to understand its dynamic behaviour over time. It is applied to dynamic problems in complex economic, managerial, or social systems involving mutual interaction, interdependencies, causal influence and feedbacks (Ullah *et al.*, 2017). In general, SD approach begins with problem identification and corresponding definition. It proceeds through the mapping and modelling of the systems using simulations. Conceptually, SD is a mathematical system of stochastic and complex variables that are usually embedded in a nonlinear equation. In terms of functionality, it recognizes the structure of various dynamic systems, obtains feedback and simulates and iterates the associated loops to understand relationships of various system components. Following these iterations, graphical results are obtained of SD simulations (de Salles *et al.*, 2016).

The use of SD is almost non-existent in studies of real estate. Though increasingly adopted in the construction sector in recent times, real-estate research is yet to explore the benefits of SD. In construction, it has been used for project management and contractual arrangements such as PPP, and innovative methods like Six Sigma (Loosemore and Cheung, 2015; Ullah *et al.*, 2017). Bureš (2017) used SD for evaluating the social and economic behaviour of infrastructure projects using multiple complex systems models to derive a pattern of behaviour. For real estates, one of the very few reported studies is that of Ferreira *et al.* (2017) but it also does not use SD in holistic essence, rather some concepts of cognitive modelling are used.

When it comes to real-estate customers, the demands are always changing thus calling for the use of sophisticated tools and technologies, such as Virtual and Augmented Reality along with 3D models and location-based services (Sherman, 2017; Bhadsavie *et al.*, 2017). Similarly, the literature reports certain key factors of real-estate website management, such as immersion, customer behaviour, 3D visualization, perceived ease of use, location services, trust, reliability, information correctness, perceived risks, subjective norms and others. These critical factors add to user satisfaction and trust for using the real-estate services (Richardson and Zumpano, 2012; Arndt *et al.*, 2017; Yang *et al.*, 2017). In terms of TAM for website management, different studies have reported multiple critical factors for efficient performance and user satisfaction. These critical factors include but are not limited to customer experience, website design, loading speed, serviceability, feedback, customer support, customization, easy return and others (Ainsworth and Ballantine, 2017, Agrebi and Boncori, 2017).

The concepts of real estate combined with TAM and website management can be clustered together into three key groups that enhance the real-estate website usage from customer perspective: perceived usefulness (PU), user satisfaction (US) and behavioural intention to use (BIU). These groups have been used in various studies for online platform development keeping the customer perception as the pivotal point. Phua *et al.* (2012) used extended TAM to explore the behavioural intention of Home Economics teachers for using the Internet as a teaching tool. The finding highlighted four key factors as Internet attitude, PU, perceived ease of use and perceived enjoyment. Similarly, Rauniar *et al.* (2014) used TAM for studying the social media usage behaviour with Facebook as the target site. The study highlighted the perceived ease of use, site usage capability, playfulness, trustworthiness and PU as critical factors. Lee and Lehto (2013) used TAM for identifying determinants affecting BIU YouTube. The study focused on PU and the perceived ease of use mainly along with US, content richness, vividness and self-efficacy to highlight a 43.8% variance in BIU, provided the factors mentioned are catered for in terms of customer satisfaction.

Keeping the aforementioned relevant researches in focus, this study is aimed at proposing a conceptual SD model for TA in real-estate websites for enhancing its usage keeping the customer satisfaction as the central focus. The four key groups are used for model conceptualization along with additional factors that directly or indirectly affect these groups. Overall, the model is aimed to provide conceptual guidelines of incorporating technology in real-estate websites and a holistic SD-based validation framework.

3. Research Method

This study focuses on the development of a conceptual SD framework for identifying factors which may affect the process of TA in real-estate management. This section describes how the conceptual framework is developed and justifies the model implementation.

3.1. *Factors collection*

For collecting the factors of TA in real estates and websites in general, the online libraries and search engines of Google Scholar, Taylor & Francis, Emerald Insight, ASCE and ScienceDirect were used. The keywords used for searching included "TA", "TAMs", "Real estate technology", "Real estate website", "Technology models for website", and "Online real estate management". The search criteria were further restricted to papers published after 2012 to get the more recent publications in the field and get a know-how of the up-to-date happenings in the field of real estate management. Thus, a total of 50 factors were identified of the published literature that were further divided into groups, clusters and contributing factors. This allocation was performed using the published literature-based allocations and in-house discussion of the research group members involving one PhD Lecturer, one PhD Student, one Master's (research student) and three Master's (course work students) all working on real-estate TA with experience ranging from 1 to 10 years in field of construction. As a result, the grouping, clusters and factors as shown in Table 1 were obtained.

In addition to Table 1, the papers were also scrutinized for the models and associated techniques used by the authors to highlight the state of existing TAM in real estate management. Table 2 presents some of the useful studies in this context. The table highlights the topics, focus areas (TAM, real estates or websites), tools and methods used.

3.2. *Using SD for TAM in real estate*

As evident from the literature, SD has not been used so far in real estate's TAM. This not only makes it a novel approach to use but also highlights a critical research gap in the real-estate TAM context. Although alternative tools like simple graphical illustrations, structural equation modelling, fuzzy logics, and analytical hierarchical process can also be used in this context, their data collection, simulation restrictions and availability, and rigidity to potential changes makes SD a better candidate than its counterparts. SD offers many advantages making it a candidate for usage in real-estate TAM. It provides the loops and clusters enabling realistic simulations and conceptualization of factors. It can be loaded with multiple mathematical operations that will not only highlight the positivity or negativity of the loops but also make the resultant graphs more sensible, logical and understandable (de Salles *et al.*, 2016). The graphical results and sensitivity of SD simulations are a key advantage of SD that cannot be achieved by simple graphical tools used so far in the real-estate TAM. Further, a key advantage of the SD models is offering both the temporal and single point-based simulations and results that are not achievable with other contestants (Ullah *et al.*, 2017). Since this study is currently ongoing, in future the conceptual frameworks will be validated using real life case studies and data from experts, SD is the one of the best choices for air current study keeping its holistic simulations,

Table 1.   TAM groups, clusters and factors.

Groups	Clusters	Factors	References
Perceived usefulness	Systems Quality	Page location, Loading speed, Loading info structure, Website evaluation, Website design	Agrebi and Boncori (2017), Ainsworth and Ballantine (2017), Rauniar *et al.* (2014)
	Info Quality	Content richness, Familiar technology, Information novelty, 3D models, Accurate information, Updated information	Christensen *et al.* (2016), Ekman *et al.* (2016), Lee and Lehto (2013)
		Map features, Credibility, Content accuracy, Reliability, Feedbacks, Presence, Website serviceability, Website self efficacy, Self-organizing function, Immersion, Customer experience, Professionalism, Customer behaviour, Customer support, Real estate valuation, Effective searching, Mortgage calculator	Bhadsavie *et al.* (2017), Arndt *et al.* (2017), Ekman *et al.* (2016), Ferreira *et al.* (2017), Guo *et al.* (2017), Rauniar *et al.* (2014)
User satisfaction	Service Quality	Hyperlinks, Customization, Response time, Consistent graphics	Agrebi and Boncori (2017), Ainsworth and Ballantine (2017), Lee and Lehto (2013), Yang *et al.* (2017)
	Perceived Ease of Use	Easy return, Navigation tools, Finding information, Learning website	Bhadsavie *et al.* (2017), Arndt *et al.* (2017)
Behavioral intent to use		Purchase intention, 360 cameras, Sun pathways, Graphical statistics	Agrebi and Boncori (2017), Christensen *et al.* (2016), Guo *et al.* (2017), Richardson and Zumpano (2012)
Use/Reuse			Ainsworth and Ballantine (2017), Lee and Lehto (2013), Rauniar *et al.* (2014), Yang *et al.* (2017)

Table 2. Tools used for websites, real estates and TAM.

Focus Area	Topic	Tools Used	References
Websites	Website rationality	Website relationship, Proneness, Electronic customer relationship management	Agrebi and Boncori (2017)
	Customer cognitive response	Empirical conceptual models, Online experiments, Graphical models	Ainsworth and Ballantine (2017)
	Democratizing immersive storytelling	Immerj, Games and 3D Displays	Bhadsavie *et al.* (2017)
Real estates	Real estate agent target marketing	Online audio and visual tours	Arndt *et al.* (2017)
	Multi-purpose 3-D real estate	GPS, GIS, Games	Christensen *et al.* (2016)
	Developing smart commercial real estate	Technology-based self-service (TBSS), Theoretical service-dominant logic lexicon	Ekman *et al.* (2016)
	Real estate brokerage service evaluation	Cognitive maps, Decision expert approach	Ferreira *et al.* (2017)
	Internet usage for home search	Two-stage Heckman procedure	Richardson and Zumpano (2012)
TAM	User acceptance of YouTube for procedural learning	Conceptual graphical frameworks, Hypothesis testing	Lee and Lehto (2013)
	Social media and Facebook usage	Web-based questionnaire survey	Rauniar *et al.* (2014)

capabilities and advantages in view. All the figures in the conceptual models are constructed using Vensim® software, a handy tool for SD implementation that was previously used by different studies (Ullah *et al.*, 2017; Neuwirth, 2017).

3.3. *Framework development*

PU is the starting point for a user to start using a website. Customers will not use a website unless they think it is useful. This usefulness, if achieved, results in enhanced US that eventually leads to BIU of a user for using the product, website or service. These three groups starting from PU through US to BIU can be represented by a cyclic process with the central focus of increasing website use/reuse as shown in Fig. 1. As per Fig. 1, PU adds to US that in turn enhances the BIU. The elevated BIU eventually ensures the use of a website. The red delayed line in the figure represents a decision of the user: if they think the website is useful, they will use the cyclic process but if they somehow start using the website first without any knowledge about its usefulness, chances are that they would not reuse it again as they might find it less useful or useless for their target search.

The three key groups of website use/reuse comprise multiple clusters and factors in accordance with the core concepts of SD. A cluster is commonly referred to as a group of factors adding value to something whereas a factor is any contributing or affecting entity. Table 1 highlights these clusters and factors that are graphically represented in Fig. 2. It shows two types of relations and constituent variables. The factors connected by

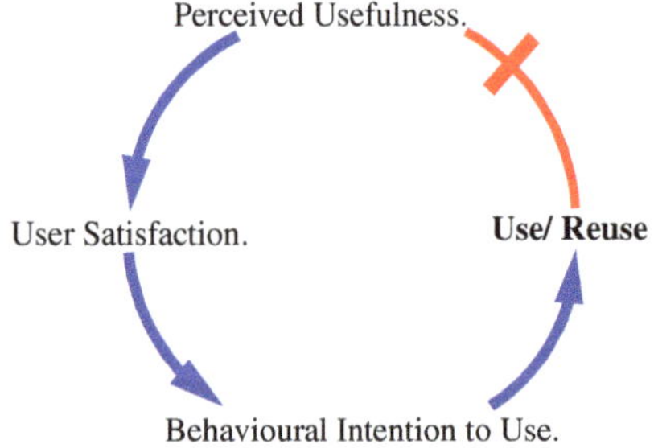

Fig. 1.   Website use/reuse influencing cycle.

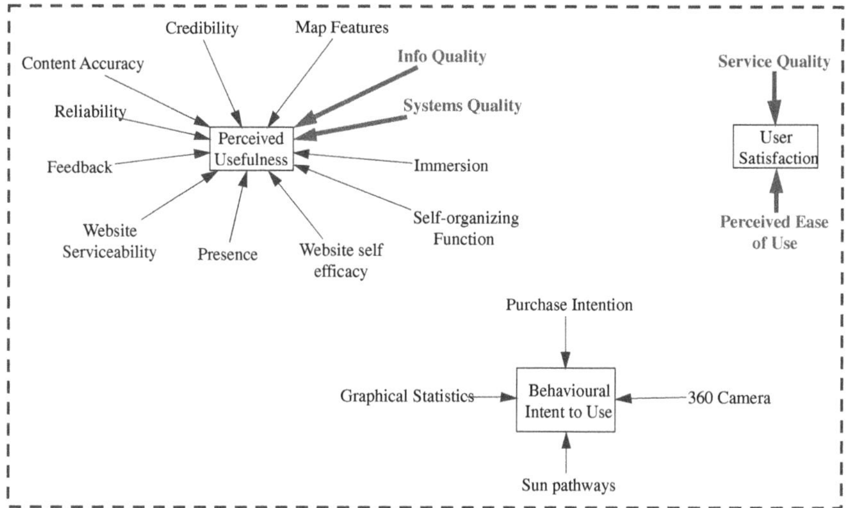

Fig. 2.   Three key groups constitution.

normal thin arrows are the factors, whereas the ones connected by thick dark arrows are the clusters. Thus, as shown in Table 1, PU consists of 10 factors and two clusters, US consists of two clusters and PU, and BIU consists of PU, US and four factors related to futuristic recommendation. In this way the cyclic flow shown in Fig. 1 is maintained.

As shown in Table 1, there are a total of four clusters in the model: two each for PU and US. These clusters are systems quality, information quality, service quality and perceived ease of use. Systems quality is the cluster associated with the website design and its speed. It incorporates the loading speed and information structure, the position at any time on a page and the go back option as evident from Table 1. Thus, in total, it houses five factors as shown in Fig. 3. Information quality cluster is related to the content and information displayed on the website. It includes the richness, novelty, accuracy and updating of information. Moreover, the 3D models and images including VR and AR technologies are also included in this cluster. Overall, this cluster houses six key factors as shown in Fig. 3. Service quality and perceived ease of use house four factors each. The former is related to quality of services offered such as response time, hyperlinks, quality graphics and the customization options,

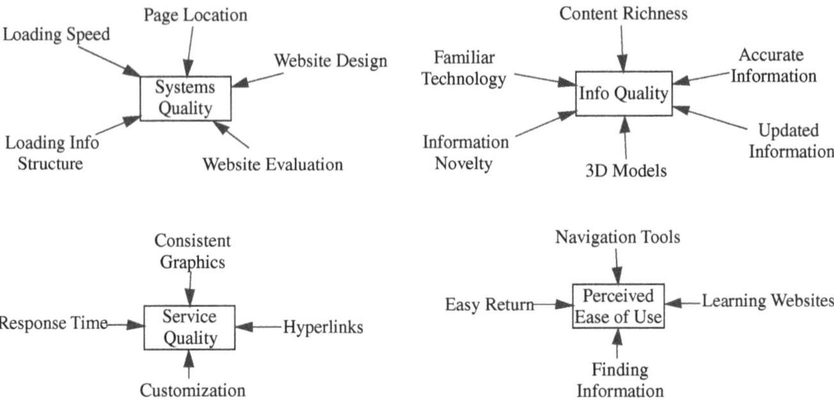

Fig. 3.   Clusters and their factors.

enabling the user to set a homepage preference especially in case of websites offering real-estate services in multiple countries. It also includes the language and regional support. The latter is aimed at user comfort thus providing options like navigation tools, go back and easy return options, saving recent searchers and information finding support, as presented in Fig. 3.

4. Results

The clusters combined with the factors and groups act as critical success criteria for ensuring the use/reuse of a real-estate website. Starting from the very basic level of website presence to its systems quality, subsequent usefulness, US and finally BIU, these factors, clusters and groups act as systems that must be dynamically managed due to their interdependencies, mix and robustness. Figure 4 presents the conceptual model for TA in real estate using holistic SD concepts. This model is developed by combining Figs. 1–4 keeping the relations of Fig. 1 as the model driver: PU adds to US which adds to BIU that in turn increases the use/reuse of a real-estate website.

The model has different colour codes and shapes referring to various groups, clusters and factors. Starting with the shapes, the circles represent clusters. Similarly, the green, blue and pink coloured rectangles represent

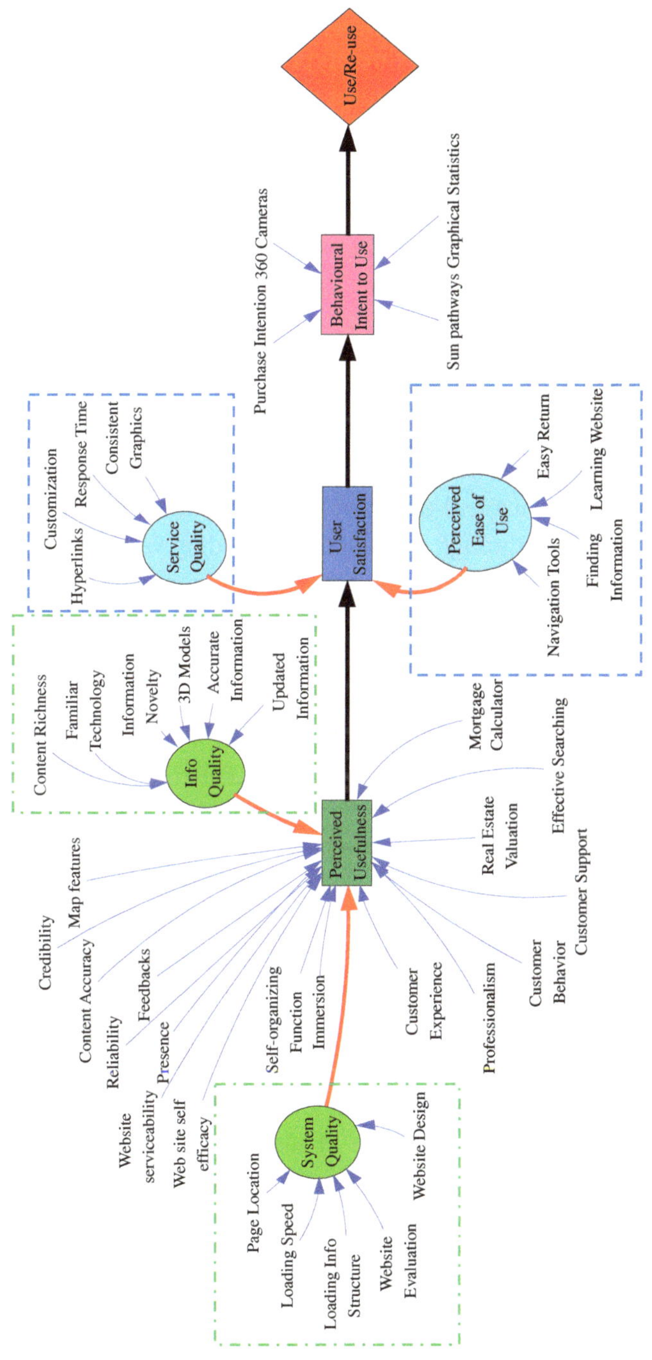

Fig. 4. Conceptual SD model for real-estate TA.

three key groups whereas the diamond shape at the end of the model corresponds to a decision. There are three types of arrows in the model: the normal blue arrows like previous figures represent the relation between a factor and its cluster or group, the thick dark red arrows represent the relation between a group and its cluster whereas the thick dark black arrows refer to the three key groups eventually leading to use/reuse decision by a customer for a real-estate website. Further, the model follows different colour codes for different groups and its clusters. The PU group is coloured green so its constituent clusters are represented by green circles and their constituent factors are enclosed in a green dashed rectangle. Similarly, blue colour is assigned to the US group, pink to BIU and red diamond for the use/reuse decision.

The models work in such a way that a factor should be improved, focused or catered for in a real-estate website. This factor will in turn add to the group directly or through its cluster, thus adding value to the group. The group will subsequently add up to the header group and finally these steps will ensure the increased use/reuse of the website. For example, consider the factor "loading speed". If the loading speed is optimized and increased, it will enhance the value of the "system quality". This system quality will increase the PU of the website that will make the user more satisfied by adding to US. This more satisfied user will develop or alter his behaviour from a casual or random user to a more dedicated, loyal and regular user when it comes to real-estate services usage. Thus, the use/reuse of the website is increased and more customers are attracted to the website. Similarly, the factor "hyperlinks" if added will increase the "service quality" that in turn will enhance the US thus convincing the user to develop usage behaviour and become a loyal user of the website. Finally, some factors like "360 cameras" and "sun pathways" are futuristic recommendations or upgrades to be added to real-estate website. These factors will directly add to positive behavioural alteration for website usage due to the novelty and comprehensiveness of ideas and technologies offered by a specific website. Thus, all the presented factors, clusters and groups if addressed properly, add to the positive behaviour of a customer for using a particular real-estate service.

This model is part of an ongoing study and is aimed at attracting more customers to real-estate websites for enhanced real-estate business. The concept presented considers the use of latest technologies and the ever-missing TAM for real estates. It uses the SD philosophy to incorporate the key factors of TA and the latest technologies such as 3D models, AR and VR and assigns them to technological groups of PU, US and BIU for enhancing the use of real-estate websites. The model will iterate user values and inputs from both customers and real-estate agents for validation and iteration to yield graphical results in future. The obtained results will not only give an insight into the current state of real-estate practice and its technologies but also provide a platform for in-depth analysis of individual factors. Thus, this conceptual model is expected to yield detailed frameworks for TA in real-estate websites and its individual factor exploration using modified methodologies and questionnaire surveys.

5. Conclusion

With the boom of technological era, different industries including construction are facing the effects of disruptive technologies and adopting it through various models and methods. Real-estate sector, in this context needs to be explored due to it being one of the highest financial markets (20% contribution to Australian construction industry) and having higher potential of technological adoption due to its more user and IT orientated nature. So far, the aspect of TA in real estate has not been explored.

Key groups and factors including PU, US and BIU for website use/reuse have been modelled in a conceptual SD model to show their interactions and flow of information for customer attraction and retention. The model has various clusters and constituent factors. All the factors reported are derived from published literature with a few suggestions for futuristic improvements. The model is equally beneficial for website managers and real-estate agents looking to excel in their businesses. Website managers can present new ideas to their clients such as illumination levels in the rooms and timely acquisition of the required expertise for real-time uploading and showcasing of such illumination levels on the websites. The real estate agents can benefit from the model in the form of holistic TAM and focus on the incorporation of latest technologies in the businesses. For example, use of 360 cameras for images, videos and

VR can help the real estate agents present their properties in an organized way to the new customers. This will help in getting new clients as well as retaining the existing customers.

The current study is the first step towards technological incorporation and adoption in real-estate websites. Further, the use of SD for TAM is another addition in the current study especially in the context of real-estate websites. The presented conceptual model is part of an ongoing research and is an evolving synthesis of real-estate TA and costumer perception incorporation. In future, the model will be validated using real-life data from both customers and real-estate agents and the path will be paved towards holistic TAM for real-estate websites. Further advancement in this context may involve the use of latest 3D technologies such as laser scanning and thermal sensors for incorporation of real-estate technologies.

References

Agrebi, M. and Boncori, A. L. (2017). "What Makes a Website Relational? The Experts' Viewpoint", *European Management Journal*, 35, 617–631.

Ai-Group (2015). "Australia's Construction Industry: Profile and Outlook".

Ainsworth, J. and Ballantine, P. W. (2017). "Consumers' Cognitive Response to Website Change", *Journal of Retailing and Consumer Services*, 37, 56–66.

Arndt, A., Harrison, D. M., Lane, M. A., Seiler, M. J. and Seiler, V. L. (2017). "Real Estate Agent Target Marketing: Are Buyers Drawn Towards Particular Real Estate Agents?" *Journal of Housing Research*, 26, 39–52.

Ayub, B., Ullah, F., Rasheed, F. and Sepasgozar, S. M. (2016). "Risks in EPC Hydropower Projects: A Case of Pakistan", in *8th International Civil Engineering Congress (ICEC) "Ensuring Technological Advancement through Innovation Based Knowledge Corridor"*, December 23–24, Karachi, Pakistan.

Bhadsavie, S. S., Yap, X. H. S., Segler, J., Jaisimha, R., Raman, N., Feng, Y., Biggs, S. J., Peoples, M., Brenner, R. B. and Mccann, B. C. (2017). "Immerj: A Novel System for Democratizing Immersive Storytelling", *In Virtual Reality (VR), 2017 IEEE*, IEEE, 367–368.

Brandon, P., Li, H. and Shen, Q. (2005). "Construction IT and the 'tipping point'", *Automation in Construction,* 14, 281–286.

Bureš, V. (2017). "A Method for Simplification of Complex Group Causal Loop Diagrams Based on Endogenisation, Encapsulation and Order-Oriented Reduction", *Systems*, 5(3), 1–22.

Christensen, P. H., Mcilhatton, D. and Blair, N. (2016). "Multi-Purpose 3-D Real Estate: Understanding the Role of 3-D Technology for Enhancing Resilience", *Journal of Real Estate Literature*, 24, 453–472.

Dahlstrom, E., Jacqueline, B. and Dziuban, C. (2014). ECAR Study of Undergraduate Students and Information Technology. Research report. Louisville, CO: ECAR, October 2014, 1–50. Retrieved from https://ctlt. sharepoint.illinoisstate.edu/Public/Handouts/Single%20Day%20 Workshops/Gee_YourClassroomTechPolicy/ECAR_Students%20and%20 IT%20Study.pdf.

Dave, B., Kubler, S., Främling, K. and Koskela, L. (2016). "Opportunities for Enhanced Lean Construction Management Using Internet of Things Standards", *Automation in Construction*, 61, 86–97.

De Salles, D. C., Neto, A. C. G. and Marujo, L. G. (2016). "Using Fuzzy Logic to Implement Decision Policies in System Dynamics Models", *Expert Systems With Applications*, 55, 172–183.

Ekman, P., Raggio, R. D. and Thompson, S. (2016). "Developing Smart Commercial Real Estate: Technology-based Self-service (TBSS) in Commercial Real Estate Facilities", *In Smart Cities Conference (ISC2), 2016 IEEE International*, IEEE, 1–6.

Ferreira, F. A., Spahr, R. W., Sunderman, M. A., Banaitis, A. and Ferreira, J. J. (2017). "A Learning-oriented Decision-making Process for Real Estate Brokerage Service Evaluation", *Service Business*, 11, 453–474.

Guo, H., Yu, Y. and Skitmore, M. (2017). "Visualization Technology-based Construction Safety Management: A Review", *Automation in Construction*, 73, 135–144.

Hewage, K. N., Ruwanpura, J. Y. and Jergeas, G. F. (2008). "IT Usage in Alberta's Building Construction Projects: Current Status and Challenges", *Automation in Construction*, 17, 940–947.

Huang, F. and Wang, F. (2005). "A System for Early-warning and Forecasting of Real Estate Development", *Automation in Construction*, 14, 333–342.

Lee, D. Y. and Lehto, M. R. (2013). "User Acceptance of YouTube for Procedural Learning: An Extension of the Technology Acceptance Model", *Computers & Education*, 61, 193–208.

Lee, J. L., Choudhry, N. K., Wu, A. W., Matlin, O. S., Brennan, T. A. and Shrank, W. H. (2016). "Patient Use of Email, Facebook, and Physician Websites to

Communicate With Physicians: A National Online Survey of Retail Pharmacy Users", *Journal of General Internal Medicine*, 31, 45–51.

Loosemore, M. and Cheung, E. (2015). "Implementing Systems Thinking to Manage Risk in Public Private Partnership Projects", *International Journal of Project Management*, 33, 1325–1334.

Mills, M. P. (2013). "The Cloud Begins with Coal", *Digital Power Group*. Available at: http://www. tech-pundit. com/wp-content/uploads/2013/07/Cloud_Begins_With_Coal. pdf.

Neuwirth, C. (2017). "System Dynamics Simulations for Data-intensive Applications", *Environmental Modelling & Software*, 96, 140–145.

Phua, P. L., Wong, S. L. and Abu, R. (2012). "Factors Influencing the Behavioural Intention to Use The Internet as a Teaching-Learning Tool in Home Economics", *Procedia-Social and Behavioral Sciences*, 59, 180–187.

Rauniar, R., Rawski, G., Yang, J. and Johnson, B. (2014). "Technology Acceptance Model (TAM) and Social Media Usage: An Empirical Study on Facebook", *Journal of Enterprise Information Management*, 27, 6–30.

Richardson, H. and Zumpano, L. (2012). "Further Assessment of the Efficiency Effects of Internet Use in Home Search", *Journal of Real Estate Research*, 34, 515–548.

Sherman, B. *Virtual Reality within the UITS Advanced Visualization Lab*. 16 Sep 2016/3 Feb 2017. Digital Arts & Humanities Workshop Series. Scholars' Commons, Wells Library, Indiana University, Bloomington. Retrieved from: http://hdl.handle.net/2022/21295.

Ullah, F., Thaheem, M. J. and Sepasgozar, S. M. E. (2016). "Sustainable Smart Cities: Evaluation of Australian Practice", *CONVR 2016 Proceedings of the 16th International Conference on Construction Applications of Virtual Reality*, December 11–13, Hong Kong.

Ullah, F., Thaheem, M. J., Siddiqui, S. Q. and Khurshid, M. B. (2017). "Influence of Six Sigma on Project Success in Construction Industry of Pakistan", *The TQM Journal*, 29, 276–309.

Yang, A. J.-F., Huang, Y.-C. and Chen, Y. J. (2017). "The Importance of Customer Participation for High-Contact Services: Evidence from a Real Estate Agency", *Total Quality Management & Business Excellence*, 1–17.

Chapter 28

Critical Review of Design for Maintainability Research in Buildings

Pramesh Krishnankutty*, Bon-Gang Hwang
and Ming Shan

*Department of Building, National University of Singapore,
Singapore 117566, Singapore*
**bdgkp@nus.edu.sg*

Lei Zhu

*Department of Construction and Real Estate,
Southeast University, Nanjing 210096, China*

Abstract

As maintenance of buildings and facilities is one of the key tasks in improving the sustainability of buildings, the design for maintainability (DfM) has recently become one of the hot issues due to the complexity associated with maintaining buildings and facilities. While the body of knowledge on DfM has started increasing, there is a lack of the systematic

---

*Corresponding author.

and comprehensive review on DfM-related issues. Thus, this chapter aims to conduct a comprehensive review, from the managerial and technical perspectives, in the relevant research areas. The findings from the managerial perspective were mainly: (1) maintainability implementation status and significant barriers and (2) critical management factors affecting DfM. Furthermore, the findings from the technical perspective involved (1) identification of maintainability considerations and criteria, (2) critical building components, (3) maintainability scoring systems, and (4) innovations and technical advancement to achieve DfM. This study also analyzed the state of the art and trends in DfM research and proposed future research directions. The findings from this study offer preliminary results of the critical review and will serve as a platform for both researchers and practitioners to retrieve the latest developments and trends in DfM.

Keywords: Design for Maintainability; Maintenance; Reviews; Future Research.

1. Introduction

Maintainability is defined by Smith (1981) as the probability that a failed item will be successfully restored to operational effectiveness by following the prescribed procedures within a given time frame. A similar definition was given by Feldman (1976) by further considering the reasonable effort and cost. These definitions were mainly from the reactive perspective of addressing the reparability of an item. In contrast to the reactive approach, Meier and Russell (2000) defined maintainability as the measure of the ease and ability to restore a defective item to its functional design state. In view of this, maintainability could be treated as a continuum rather than a discrete entity that describes the extent to which the gamut of maintenance could be seen as being achievable (Ikpo, 2009).

The estate and property management industry is facing increasing difficulties in maintaining the buildings, such as the increasing maintenance costs and tight maintenance budget (Lai, 2010), the increasing numbers of defects (Chong and Low, 2006), and increasing concerns about safety issues in maintenance (Moon *et al.*, 2015). Recently, researchers started realizing the importance of the maintainability. This is to achieve high-cost savings in operating and maintaining buildings facilities (De Silva *et al.*, 2004; Jaafar and Othman, 2016). During the past decade, the body of knowledge on the DfM has increased. However, there is a lack of systematic and comprehensive review of the diversified DfM-related research. Through a systematic review of the selected papers from well-known academic journals in the construction management and facility management, this study aims to (1) identify the major research areas of DfM, (2) analyze the state of the art and trends in DfM research and (3) propose future research directions. The findings from this study can serve as a platform for both researchers and practitioners to retrieve the latest developments and trends in DfM.

2. Literature Review

The need to improve the knowledge of maintainability during the design stage is pressing mainly because of the great effect of faulty design on the building maintenance which was identified after plenty of research (Chong and Low, 2006; Hassanain *et al.*, 2013; Lam, 2000; Mohammad *et al.*, 2014; Seeley, 1987). In addition, Lam (2000) identified that the causes of building defects under design, construction and maintenance were accountable for 40%, 30%, and 30%, respectively, in Hong Kong. Therefore, maintainability should be addressed right from the early planning and design stage at which all phases and features of a building production are combined (De Silva *et al.*, 2016). However, previous studies revealed that the main causes of the operation and maintenance issues were the designers (Al-Hammad *et al.*, 1997; Arditi and Hassanain, *et al.*, 2015), either because of their faulty design or because of their inadequate awareness of the maintainability problems during the service life (BCA, 200; Ishak *et al.*, 2007).

From another perspective, the early involvement of a facility manager will greatly improve a design in terms of maintainability. However, this is hard to achieve because, during the design stage, a facility management team may not have been set up (Meng, 2013). Alternatively, bringing the knowledge of a facility management team to the design stage is a helpful and feasible solution to this problem (Liu and Issa, 2014). As a result, design for maintainability (DfM) is drawing attention from both industry and academia over the years. DfM was also described as a decision-making tool to forecast the future maintainability of a building at design stages, which is of paramount importance to reduce the maintenance expenditure (De Silva *et al.*, 2016). In light of the above, maintainability should be addressed right from the design stage of a project (De Silva *et al.*, 2016; Ikpo, 2009). Because of the high flexibility in the design stage, early consideration of maintainability during the design and review stage could provide the potential for minimizing the occurrence of the defects (Hassanain *et al.*, 2015), operation costs, and thereby ensuring efficient building's performance (Jaafar and Othman, 2016).

3. Research Methodology

To gain insights into the current status and future trend of research on a particular topic, a methodical analysis of papers published in academic journals is vital and necessary (Darko and Chan, 2016). To achieve the research objectives explained above, this study adopted a systematic review approach that has been widely used in construction management review studies (Ke *et al.*, 2009; Yi and Chan, 2014; Zhang *et al.*, 2016). Based on the reviews, the DfM-related research trend and the key future research areas can be identified. The review approach comprises the following three steps: (1) the selection of academic journals; (2) the selection of relevant papers; and (3) analyzing the main findings and providing the future research directions. To acquire a more elaborated understanding of the research on DfM, a three-stage literature review was carried out, targeting the relevant studies published in the past two decades (i.e. from 1997 to 2016). At Stage 1, a comprehensive keyword search on the full context was performed in the search engine of Scopus. The keyword search included "DfM and building", "maintainability", "design", "building maintenance activities" and "maintenance works". The search was further limited to subject areas such as engineering, computer science, business management and accounting, materials science, environmental science, energy, economics, econometrics, and finance with the document type of "article". Finally, a total of 89 journals with 159 papers were identified at this stage.

At Stage 2, to further determine the target journals, this study not only considered the number of relevant papers published by journals but also referred to the top construction journals in the ranking list of Chau (1997), as well as the widely recognized journal list from the *Journal*

Citation Reports (JCR) and *SCImago Journal & Country Rank* (SJR). Moreover, the research on DfM is not limited to construction journals but also includes journals in the areas of facility management, property management, energy savings, etc. Finally, a total of 39 target journals were identified for this study.

4. Overview of DfM Publications

4.1. *Annual publication of DfM-related papers*

The number of DfM-related papers published between 1997 and 1999 only ranges from 1 to 3. During this period, DfM-relevant research may not have drawn the attention of the academic and industry fields. With the booming of the construction industry between 2000 and 2009, the number of DfM papers published ranged from 0 to 6. After 2010, the number of DfM papers increased up to 11 in 2015. The results demonstrated that DfM has attracted an increasing attention from researchers in the past few years. It may be because, several years later after buildings are constructed, people start to realize the huge amount of maintenance costs and the importance of maintenance considerations during the design and review stage (De Silva *et al.*, 2016; Minne and Crittenden, 2015). The findings showed that the research on DfM generally has an increasing trend. Certainly, the output of DfM-relevant research is not consistent, and there are fluctuations in the number of publications over the years. Nonetheless, DfM is getting more and more attention from researchers.

4.2. *Key journals in the DfM publications*

The number of publications from the top 18 journals considered for this study accounts for 81.16% of all identified DfM-relevant papers. The journals reviewed in this study are from *Journal of Performance of Constructed Facilities* (7) published the maximum number of DfM-related papers, followed by the *Jurnal Teknologi* (5), *Facilities* (5), *Architectural Science Review* (4), *Construction Management and Economics* (4), *Structural Survey* (4), *Journal of Construction Engineering and Management* (4), *ASHRAE Journal* (3), *Journal of*

Quality in Maintenance Engineering (2), *Journal of Architectural Engineering* (2), *Journal of Building Appraisal* (2), *International Journal of Precision Engineering and Manufacturing* (2), *AIJ Journal of Technology and Design (Japan)* (2), *Building Engineer, Building and Environment* (2), *Automation in Construction* (2), *Journal of Financial Management of Property and Construction* (2) and *Engineered Systems* (2).

5. Discussions of Research Topics in DfM

This study classified the research topics of DfM into two perspectives: management perspective and technical perspective. Within each perspective, this study summarized: (1) major topics, (2) gaps in the current body of knowledge/research and (3) future research.

5.1. *Management approaches to improve DfM*

The research focuses on the identification of improvement opportunities for DfM on the project delivery and development from the perspective of management and the areas were further categorized into findings from the management perspective and future research directions of DfM from the management perspective.

5.1.1. *Findings from the management perspective*

(1) Maintainability implementation status and significant barriers

The designers' level of maintenance-related knowledge has been increasing. Several investigation results showed that the maintenance-related knowledge level of the design firms and designers, to some extent, was good (Arditi and Nawakorawit, 1999; Kanniyapan *et al.*, 2015; Liu and Issa, 2016; Wu *et al.*, 2006). However, the extent of the designers' communication with facility managers or maintenance consultants was not sufficient (Arditi and Nawakorawit, 1999). In addition, the facility manager's involvement in the design stage still has great resistance in practice (Liu and Issa, 2016; Meng, 2013). Furthermore, there is still a lack of better practices and technologies that can transfer information from the design and construction phase to the operation and maintenance phase (Liu and Issa, 2016).

The two significant barriers for implementing DfM were the lack of information on maintainability knowledge and criteria, and the lack of failure/defect data and maintenance data, which were emphasized by both academic researchers and industrial practitioners (de Silva *et al.*, 2012; Kanniyapan *et al.*, 2015; Wu, 2010). For instance, Wu *et al.* (2006) identified that some standards, such as UK MoD Defence Standard (def-stan-00-41, 1993) and British Standards (BS 5760-14, 1992), were the major guides for maintainability design for the majority of respondents in the UK. Along with the advancement of technologies, these standards need to be updated in terms of maintainability. De Silva *et al.* (2012) also pointed out that poor specifications were one of the causes for the maintenance defects. According to Wu *et al.* (2006), although most companies in the UK kept maintenance records, few companies used historical data for the purpose of improving the system maintainability. Furthermore, most companies that developed maintenance-related policies were mainly based on their own experience, which could not be shared with other companies or nationally (Wu *et al.*, 2006). There is a pressing need to establish the policies and legal framework regarding building maintenance in many countries (Ikpo, 2009).

(2) Critical management factors affecting achieving DfM

Identifying the critical project management factors or practices that greatly affect achieving DfM could ensure the practical implementation of DfM practices by practitioners and the industry (Hwang *et al.*, 2016). The most frequently identified management factors or practices included: (1) using the integrated procurement methods, such as design and build (D and B), private finance initiative (PFI) and private–public partnership (PPP); (2) adopting life-cycle cost (LCC) criterion for design or tendering; (3) developing scoring system for maintainability; (4) developing maintainability guidelines; (5) providing training programs on maintainability; and (6) developing a shared maintainability knowledge database to enhance the communication among relevant parties and share the maintainability knowledge (Arditi and Nawakorawit, 1999; de Silva *et al.*, 2004). The integrated procurement methods give greater attention to the LCC including initial capital costs and costs for the operation, maintenance, and repair of a building within a contract, which is beneficial for achieving DfM (Wood, 2012). Whether the building maintenance is considered at the design stage is likely to depend on whether the owner will be the subsequent user that will benefit from the LCC consideration (Arditi and Nawakorawit, 1999). Moreover, among the factors, developing the knowledge on maintainability and developing the benchmarking or scoring systems for building maintainability were identified as the foremost factors that should be addressed (Chew and Das, 2008; De Silva *et al.*, 2004). A generic and holistic framework or rules will be not only beneficial for developing the body of knowledge on DfM but also beneficial for efficiently establishing the shared maintainability knowledge database because of the ease of interoperability in the whole building industry and among countries.

In addition, a shared knowledge database of maintainability is a possible solution to help different disciplines during the life cycle of a construction project to understand each other better and solve the problems at earlier project phases (Liu and Issa, 2016). Such knowledge database could include information of good maintainability practices, maintenance problems and defect, existing facility management requirements (Liu and Issa, 2016). Furthermore, without the establishment of DfM practices at the industry or national level, a formal method to address the maintainability during the project delivery process from the owner's perspective will benefit owners with a starting point for implementing maintainability. Considering all the factors, the Construction Industry Institute (CII) developed a model implementation process for implementing best practices of maintainability (Meier and Russell, 2000). The seven good maintainability approaches developed by CII (2002) could be references to develop the DfM good practices for the construction industry. Similarly, Das *et al.* (2009) developed a standard and reliable method for the acquisition of tacit knowledge from facility management (FM) and converted the knowledge into organizational records. Such a tool can help industry practitioners to start accumulating and implementing their maintainability knowledge without facing present constraints such as the current dearth of information and poor feedback leading to recurrent maintenance defects.

5.1.2. *Future research directions of DfM from the management perspective*

In light of the above, great efforts on the adjustment of management practices or approaches are needed to improve the implementation of DfM. Some essential questions are awaiting further investigation:

(1) How can the maintainability-related knowledge and practices be systematically acquired and properly documented?
(2) How can the maintainability-related standards and specifications be adjusted or updated?
(3) How can a generic and holistic framework or rules for developing benchmarking systems for maintainability be established?
(4) How can a shared knowledge database of maintainability be structured and developed?
(5) How can a motivation mechanism for implementing DfM be developed?

5.2. *Technical approaches to improve DfM*

The majority of the research focuses on the development of scoring and benchmarking systems for maintainability which could provide objective measures of maintainable design. The aim of this practice is to not only motivate the use of the knowledge on maintainability but also enable the decision makers to choose a better maintainable design option and minimize the increasing maintenance costs (de Silva *et al.*, 2004). The developed scoring systems are diverse not only because the building maintainability depends on a large number of building components, but also because the maintainability performance relies on multiple objective and subjective parameters and criteria (Das *et al.*, 2010). Therefore, the discussion is classified into the following three aspects: (1) primary maintainability considerations and criteria, (2) critical building areas/components, and (3) scoring calculation principles, representing the rule, object, and method to achieve the benchmarking of maintainability, respectively.

5.2.1. *Findings from the technical perspective*

(1) Identification of maintainability considerations and criteria

Currently, there is no consensus on the primary maintainability considerations and criteria. The top four most frequently mentioned and highly emphasized maintainability considerations and criteria were the following: accessibility; flexibility or adaptability of design; technological availability or feasibility; and material and component reliability and durability. First, accessibility is the ability to access or reach the parts or elements of a building for inspection and maintenance (BCA, 2004; Ganisen *et al.*, 2015). The less complex building shapes and features were identified as an indicator for accessibility (Ganisen *et al.*, 2015). Second, flexible and adaptable designs to accommodate current and future needs are desirable for a high maintainability (Ikpo, 2009). Therefore, another aspect of flexibility is that the design of elements that need continuous maintenance should avoid permanent fixation (Al-Hammad *et al.*, 1997). Third, the technological feasibility or availability means the feasible maintenance technology, the available equipment and materials on the market when a maintenance is needed, which is one of the prominent features of maintainability analysis (Al-Hammad *et al.*, 1997; Das *et al.*, 2010; Ikpo, 2009). Lastly, reliability and durability are the capability of a component or material to perform its necessary functions and without the replacement with a new material throughout its lifespan (Kanniyapan *et al.*, 2015). To quantify the extent of a design or building satisfying the maintainability considerations or criteria, the most frequently used method is the questionnaire survey using a Likert scale or an interview with industry experts. The overwhelming use of such method leads to the analysis that results are mostly subjective, which do not truly reflect the real world. For instance, whether the location of equipment is accessible or not, in most cases, can be quantified by its dimension. Therefore, integrating the subjective and objective evaluation of maintainability is necessary for future research.

(2) Critical building components

The critical building areas and components under constant maintenance were mainly generated from studies on building defect analysis, the effect of faulty design on building maintenance, and the labour-intensive maintenance activities (Al-Hammad *et al.*, 1997; Ali *et al.*, 2013; Chew, 2005a; Chew and de Silva, 2003). The top six most mentioned building areas and components were the following: façade; roof; wet area; water supply and drainage system (including sanitary); electrical system and heat; and ventilation and air conditioning (HVAC) system. These findings not only contributed to developing the maintainability scoring systems but also contributed to establishing a shared database of maintenance defects. Moreover, these findings are useful information for developing the maintainability design review checklists or guidelines. Das and Chew (2011) suggested that guidelines for achieving a higher maintainability through defect mitigation can be prepared in the form of a checklist. In Singapore, the BCA has published the "Building Maintenance (Strata Management) Regulations 2005" (BCA, 2005) and the "Design for Maintainability Checklist Version 1.3" (BCA, 2016). The maintainability checklists and guidelines could provide designers with correct good maintainability practices to follow during the design, ultimately reducing the occurrence of the identified defects. Moreover, such checklists and guidelines can help the designers, contractors, and suppliers strengthen their knowledge on maintainability and contribute to improving maintainability by providing information like the accessibility and durability of building elements or materials through maintenance manuals, as-built drawings, and material information leaflets (de Silva *et al.*, 2004; Ikpo, 2009).

(3) Maintainability scoring principles

Considering the characteristics of buildings, the maintainability score of a building is the aggregation of the maintainability estimates of each building area and component. Moreover, the maintainability estimate of each building area and component measures the satisfaction of all relevant maintainability considerations or criteria (Ganisen *et al.*, 2015). Generally, there are two perspectives to measure the maintainability of a building: cost-effective perspective and occupant's satisfactory perspective.

The majority of relevant research suggested that the maintainability should be calculated from the cost-effective perspective (de Silva *et al.*, 2016; Ganisen *et al.*, 2015; Garg, 1984; Minne and Crittenden, 2015; Thanaraju and Ali, 2015). Life-cycle costs (LCC) and life-cycle maintenance costs (LCMC) are the two most popularly used measurement criteria from the cost-effective perspective for assessing the maintainability. To estimate the maintenance costs, the historical data from the outsourced operation and maintenance contracts and maintenance budgeting experiences would be useful (Lai, 2010; Lai *et al.*, 2006). Currently, there is no consensus on the maintainability scoring principles. Cost-effective, energy efficient and occupant satisfactory are practically all important and possible calculation principles. However, the weight of each principle could be different considering the type of a project.

(4) Innovations and technical advancement to achieve DfM

Besides the maintainability scoring systems, 23% of research focuses on developing innovations and technologies to achieve DfM. Better practices, innovations and advanced technologies can improve the maintainability of buildings, efficiently transfer information among the development phases including design, construction, operation and maintenance, and to better implement the DfM concept. Among these innovations and advancement, Building Information Modelling (BIM)-related technologies and innovations were believed to be promising for supporting DfM and maintenance actions (Bouchlaghem *et al.*, 2005; Korman *et al.*, 2003; Liu and Issa, 2014). Although facility manager's early involvement in the design phase encounters numerous problems, it is possible to bring the knowledge of the facility management team through BIM software (Liu and Issa, 2014). For instance, Bouchlaghem *et al.* (2005) proposed three-dimensional models which can be used to check the integrity of services coordination, accessibility and maintainability. With the technology advancement, BIM software can partially fulfil these targets through conflict checking nowadays. Furthermore, BIM software could be a starting point for developing a BIM-facility management knowledge-sharing database which can be used as a reference for the future project design and construction (Liu and Issa, 2014).

5.2.2. *Future research directions of DfM from the technical perspective*

Efforts by various researchers on the technical approaches to improve DfM not only contribute to the body of knowledge of DfM but also show that the researchers believe in the promising future of DfM. The following concerns are likely to be addressed in the future research:

(1) How can the maintainability considerations and criteria be quantified by integrating the subjective and objective evaluation?
(2) How can the current research findings be combined to create maintainability checklists or guidelines in terms of different building areas which can be useful references for practitioners such as owners and designers?
(3) How can the score calculation principles be properly selected? Choosing one principle or integrating all principles, and assigning the importance weight according to different types of buildings should be determined in the future research.
(4) How can BIM-based technologies and software be developed to automatically detect maintainability problems and automatically evaluate the maintainability of a design?

6. Conclusions and Recommendations

Although concerns on the maintainability of buildings are increasing, there is currently no systematic review of DfM research. In light of this, this study conducted a systematic review to identify the major research areas and trends of DfM. The findings from this study first revealed that the annual publications of DfM were generally increasing, indicating the increased attention to this area. Moreover, the analysis of the selected papers showed that the *Journal of Performance of Constructed Facilities* published the maximum number of DfM-related papers.

This chapter reviewed DfM from two perspectives, namely management and technical perspectives. The major findings from the management perspectives are (1) maintainability implementation status and significant barriers, (2) critical management factors affecting achieving DfM, (3) maintainability scoring systems and (4) innovations and technical advancement to achieve DfM. Additionally, the major findings from the technical perspectives are (1) identification of maintainability considerations and criteria, (2) critical building components, (3) maintainability scoring systems and (4) innovations and technical advancement to achieve

DfM. This study also identified the future research directions for DfM research. In the managerial perspective, the future research areas would be (1) documenting maintainability-related knowledge and practices, (2) updating and improving the existing maintainability-related standards and specifications, (3) developing benchmarking systems and (4) developing motivation mechanism for implementing DfM. In the technical perspective, the future research areas would be (1) quantification of maintainability considerations and criteria, (2) creating maintainability checklists or guidelines for designers and contractors, (3) developing a scoring methodology for DfM and (4) developing BIM-based solutions for evaluating DfM.

Despite the limitations of the number of DfM publications reviewed, the findings from this chapter are still valuable. This study contributes to the body of knowledge by identifying future research areas of DfM from the managerial and the technical perspectives. The findings of this study could serve as a platform for both researchers and practitioners to retrieve and appreciate the latest developments and trends in DfM.

References

Al-Hammad, A., Assaf, S. and Al-Shihah, M. (1997). "Effect of Faulty Design on Building Maintenance", *Journal of Quality in Maintenance Engineering*, 3(1), 29–39.

Arditi, D. and Nawakorawit, M. (1999). "Designing Buildings for Maintenance: Designers' Perspective", *Journal of Architectural Engineering*, 5(4), 107–116.

BCA (Building and Construction Authority). (2004). "Maintainability of Buildings". Available at: https://www.bca.gov.sg/ResearchInnovation/others/Maintainability%20_2004-011_.pdf.

BCA (Building and Construction Authority). (2005). "Building Maintenance and Strata Management 2004". Available at: https://www.bca.gov.sg/publications/BMSM/BM_strata_mgt_reg.html (Accessed on October 25, 2016).

BCA (Building and Construction Authority). (2016). Design for Maintainability Checklist, Building and Construction Authority (BCA), Singapore, 49.

Bouchlaghem, D., Shang, H., Whyte, J. and Ganah, A. (2005). "Visualisation in Architecture, Engineering and Construction (AEC)", *Automation in Construction*, 14(3), 287–295.

Chau, K. W. (1997). "The Ranking of Construction Management Journals", *Construction Management and Economics*, 15(4), 387–398.

Chew, M. Y. L. (2005). "Defect Analysis in Wet Areas of Buildings", *Construction and Building Materials*, 19(3), 165–173.

Chew, M. Y. L. and Das, S. (2008). "Building Grading Systems: A Review of the State-of-the-art", *Architectural Science Review*, 51(1), 3–13.

Chew, M. Y. L. and De Silva, N. (2003). "Maintainability Problems of Wet Areas in High-rise Residential Buildings", *Building Research and Information*, 31(1), 60–69.

Chong, W. K. and Low, S. P. (2006). "Latent Building Defects: Causes and Design Strategies to Prevent Them", *Journal of Performance of Constructed Facilities*, 20(3), 213–221.

CII (2002). "Design for Maintainability", CII, United States.

Darko, A. and Chan, A. P. C. (2016). "Critical Analysis of Green Building Research Trend in Construction Journals", *Habitat International*, 57, 53–63.

Das, S. and Chew, M. Y. L. (2011). "Generic Method of Grading Building Defects Using FMECA to Improve Maintainability Decisions", *Journal of Performance of Constructed Facilities*, 25(6), 522–533.

Das, S., Chew, M. Y. L. and Poh, K. L. (2010). "Multi-criteria Decision Analysis in Building Maintainability Using Analytical Hierarchy Process", *Construction Management and Economics*, 28(10), 1043–1056.

Das, S., Poh, K. L. and Chew, M. Y. L. (2009). "Standardizing FM Knowledge Acquisition When Information is Inadequate", *Facilities*, 27(7), 315–330.

de Silva, N., Dulaimi, M. F., Ling, F. Y. Y. and Ofori, G. (2004). "Improving the Maintainability of Buildings in Singapore", *Building and Environment*, 39(10), 1243 1251.

de Silva, N., Ranasinghe, M. and de Silva, C. R. (2012). "Risk Factors Affecting Building Maintenance Under Tropical Conditions", *Journal of Financial Management of Property and Construction*, 17(3), 235–252.

de Silva, N., Ranasinghe, M. and de Silva, C. R. (2016). "Risk Analysis in Maintainability of High-rise Buildings under Tropical Conditions Using Ensemble Neural Network", *Facilities*, 34(1–2), 2–27.

Feldman, E. B. (1976). *Building Design for Maintainability*, New York: McGraw-Hill Companies.

Ganisen, S., Mohammad, I. S., Nesan, L. J., Mohammed, A. H. and Kanniyapan, G. (2015). "The Identification of Design for Maintainability Imperatives to Achieve Cost Effective Building Maintenance: A Delphi Study", *Jurnal Teknologi*, 77(30), 75–88.

Garg, N. P. (1984). "Building in Maintainability at the Design Stage", *Journal of the Institution of Engineers. India. Interdisciplinary and General Engineering*, 64, 106–110.

Hassanain, M. A., Assaf, S., Al-Ofi, K. and Al-Abdullah, A. (2013). "Factors Affecting Maintenance Cost of Hospital Facilities in Saudi Arabia", *Property Management*, 31(4), 297–310.

Hassanain, M. A., Fatayer, F. and Al-Hammad, A. M. (2015). "Design Phase Maintenance Checklist for Water Supply and Drainage Systems", *Journal of Performance of Constructed Facilities*, 29(3), 10.

Hwang, B.-G., Zhu, L. and Ming, J. T. T. (2016). "Factors Affecting Productivity in Green Building Construction Projects: The Case of Singapore", *Journal of Management in Engineering*, Articles In Press, 04016052,10.1061/(ASCE)ME.1943-5479.0000499.

Ikpo, I. J. (2009). "Maintainability Indices for Public Building Design", *Journal of Building Appraisal*, 4(4), 321–327.

Ishak, N. H., Chohan, A. H. and Ramly, A. (2007). "Implications of Design Deficiency on Building Maintenance at Post-Occupational Stage", *Journal of Building Appraisal*, 3(2), 115–124.

Jaafar, M. and Othman, N. L. (2016). "Maintainability and Design Aspect of Public Hospital", *Jurnal Teknologi*, 78(5), 107–113.

Kanniyapan, G., Mohammad, I. S., Nesan, L. J., Mohammed, A. H. and Ganisen, S. (2015). "Façade Material Selection Criteria for Optimising Building Maintainability", *Jurnal Teknologi*, 75(10), 17–25.

Ke, Y., Wang, S., Chan, A. P. C. and Cheung, E. (2009). "Research Trend of Public–Private Partnership in Construction Journals", *Journal of Construction Engineering and Management*, 135(10), 1076–1086.

Korman, T. M., Fischer, M. A. and Tatum, C. B. (2003). "Knowledge and Reasoning for MEP Coordination", *Journal of Construction Engineering and Management*, 129(6), 627–634.

Lai, J. H. K. (2010). "Operation and Maintenance Budgeting for Commercial Buildings in Hong Kong", *Construction Management and Economics*, 28(4), 415–427.

Lai, J. H. K., Yik, F. W. H. and Jones, P. (2006). "Critical Contractual Issues of Outsourced Operation and Maintenance Service for Commercial Buildings", *International Journal of Service Industry Management*, 17(4), 320–343.

Lam, K. (2000). "Quality Assurance in Management of Building Services Maintenance", Building Services Engineering Department, The Hong Kong Polytechnic University, in Southeast Asia Facility Management, May/June 2000 issue.

Liu, R. and Issa, R. R. A. (2014). "Design for Maintenance Accessibility using BIM Tools", *Facilities*, 32(3), 153–159.

Liu, R. and Issa, R. R. A. (2016). "Survey: Common Knowledge in BIM for Facility Maintenance", *Journal of Performance of Constructed Facilities*, 30(3), 10.

Meier, J. R. and Russell, J. S. (2000). "Model Process for Implementing Maintainability", *Journal of Construction Engineering and Management*, 126(6), 440–450.

Meng, X. (2013). "Involvement of Facilities Management Specialists in Building Design: United Kingdom Experience", *Journal of Performance of Constructed Facilities*, 27(5), 500–507.

Minne, E. and Crittenden, J. C. (2015). "Impact of Maintenance on Life Cycle Impact and Cost Assessment for Residential Flooring Options", *International Journal of Life Cycle Assessment*, 20(1), 36–45.

Mohammad, I. S., Zainol, N. N., Abdullah, S., Woon, N. B. and Ramli, N. A. (2014). "Critical Factors That Lead to Green Building Operations and Maintenance Problems in Malaysia", *Theoretical and Empirical Researches in Urban Management*, 9(2), 68–86.

Moon, S. M., Shin, C. Y., Huh, J., Oh, K. W. and Hong, D. (2015). "Window Cleaning System with Water Circulation For Building Façade Maintenance Robot and its Efficiency Analysis", *International Journal of Precision Engineering and Manufacturing — Green Technology*, 2(1), 65–72.

Seeley, I. H. (1987). *Building Maintenance*, London: Macmillan Education.

SJR (2016). "SCImago Journal & Country Rank". Available at: http://www.scimagojr.com/ (Accessed on November 14, 2016).

Smith, D. J. (1981). *Reliability and Maintainability in Perspective: Technical, Management And Commercial Aspects*, London: Macmillan.

Thanaraju, P. and Ali, H. M. (2015). "Important Activities in Activity-Based Life Cycle Cost in Building Facilities Maintenance: A Case Study", *Jurnal Teknologi*, 74(2), 79–85.

Wood, B. (2012). "Maintenance Integrated Design and Manufacture of Buildings: Toward a Sustainable Model", *Journal of Architectural Engineering*, 18(2), 192–197.

Wu, S. (2010). "Reviews and Case Studies: Research Opportunities in Maintenance of Office Building Services Systems", *Journal of Quality in Maintenance Engineering*, 16(1), 23–33.

Wu, S., Clements-Croome, D., Fairey, V., Albany, B., Sidhu, J., Desmond, D. and Neale, K. (2006). "Reliability in the Whole Life Cycle of Building Systems", *Engineering, Construction and Architectural Management*, 13(2), 136–153.

Yi, W. and Chan, A. P. C. (2014). "Critical Review of Labor Productivity Research in Construction Journals", *Journal of Management in Engineering*, 30(2), 214–225.

Zhang, S., Chan, A. P. C., Feng, Y., Duan, H. and Ke, Y. (2016). "Critical Review on PPP Research — A Search from the Chinese and International Journals", *International Journal of Project Management*, 34(4), 597–612.

Chapter 29

Are Underground Residential Buildings Feasible? A Preliminary Investigation for Innovations

Bon-Gang Hwang, Pramesh Krishnankutty*
and Ming Shan

*Department of Building, National University of Singapore,
Singapore 117566, Singapore
bdgkp@nus.edu.sg

Kristie Sze Ni Wong

*EC Harris Singapore Pte. Ltd., 1 Magazine Road,
Singapore 0595674, Singapore*

Abstract

As the space constraint in metropolises has continuously been an issue in many countries, a new idea of constructing underground facilities including residential buildings has recently emerged and attracted

---

*Corresponding author.

considerable attention from authorities. Under the circumstance, this study aims to investigate the possible advantages and disadvantages of underground residential buildings and to identify the critical risks in projects. To achieve the goals, a comprehensive literature review and pilot interviews were conducted, followed by an empirical questionnaire survey administered to 30 Singapore-based construction companies. The analysis results showed that "space saving" was the most significant advantage of underground residential buildings and "limited access to natural light" was the most severe disadvantage. Additionally, this study disclosed and discussed the top five critical risks of underground residential building projects, including "labour restrictions", "cost overruns", "local contractors' competence in underground construction", "material restrictions" and "economic fluctuations". This study has contributed to the body of knowledge by examining the advantages, disadvantages and critical risks in underground residential building projects innovatively. The findings from this study can also be informative to relevant project stakeholders and policymakers by enhancing their understanding of underground space utilization, which will facilitate their decision-making procedures accordingly.

***Keywords*:** Underground Space; Underground Residential Buildings; Risks; Decision-making.

1. Introduction

Over the recent years, a steady flow of population has been moving from rural areas to metropolises worldwide. According to the United Nations (2015), there used to be 126 cities around the world possessing one million populations or above in the year of 1970, while in 2014, this number increased to 417 sharply, suggesting a significant growth in the global metropolitan population over the past four decades. The influx of population can provide metropolises with ample human resources to support their social and economic development (Kaliampakos *et al.*, 2016; Li *et al.*, 2016a), however, it also makes the existing packed metropolises even more crowded and urges the authorities to explore new strategies to create space for those new metropolitan dwellers (Admiraal and Cornaro, 2016a, 2016b; Broere, 2016). Fortunately, the authorities found a good solution, namely developing underground spaces of the metropolises, and have devoted considerable efforts in this regard (Bartel and Janssen, 2016). Particularly, over the past two decades, numerous underground space structures such as underground shopping malls, railways, power plants, sewerage systems and depots have been constructed worldwide, which has indeed expanded urbanite's living space and relieved the land use pressure on those densely packed metropolises (Bobylev, 2009).

Singapore is a modern, prosperous metropolis, and a densely populated city-state that merely has a land area of 719 km^2 (Department of Statistics, 2016). To create more room for development, Singapore's traditional strategy is to reclaim land from the sea. However, this strategy is reaching its limit owing to several constraints such as the geographical boundaries, the increasing cost of sand supply (Zhou and Zhao, 2016). Thus, the local government has shifted its attention to the underground space of this dense urban metropolis. So far, attempts have been made by the state developer Jurong Town Corporation (JTC) that constructed a massive underground storage hub named Jurong Rock Caverns which can hold about eight million barrels of oil and free up approximately 60-hectare of land (JTC, 2017).

As Singapore's second-largest land user, the building sector has consumed 17% of the land in this country, just falling behind the front-runner, namely the defence sector by two percentage points (Ministry of National Development, 2013). Moreover, the National Population and Talent Division (2013) has predicted that the building sector would probably seize more land in the future as the population of this metropolis would continue to grow in the coming two decades. To accommodate the rising population, one of the solutions that is currently being considered by the local authority is to develop underground residential buildings. Despite the authorities' ambitious vision of underground residential buildings, very few trials on underground residential building projects are carried out worldwide. Hence, it is necessary to carry out a tentative research to learn more about underground residential buildings. This study, therefore, aims to explore the advantages and disadvantages of the underground residential buildings. Additionally, this study also attempts to investigate the critical risks that may occur during the construction of underground residential buildings.

2. Literature Review

Given the limited amount of research, the advantages, disadvantages and critical risks of underground residential building projects can hardly be extracted from the existing literature directly. However, research on underground industrial infrastructure and commercial building projects (e.g. underground railways, lifeline systems, and commercial centres) is abundant (e.g. Choi *et al.*, 2004; Ita, 2004; Lee *et al.*, 2016), which can provide valuable references to this study. Thus, a comprehensive literature review of underground industrial infrastructure and commercial building projects was conducted aiming to identify the potential advantages, disadvantages and risks of underground residential building projects. The literature review consists of two steps. In Step 1, two concise codes, namely TITLE "underground AND space" and TITLE "underground OR risk" were searched in Scopus. In Step 2, a careful visual examination was conducted with the attained papers to examine their relevance to the research topic of this study. At this stage, only the literatures tagged as journal articles and reviews were kept and reviewed. Eventually, a total of 29 valid papers were finalized, as listed in Tables 1 and 2.

2.1. *Potential advantages of underground residential buildings*

The comprehensive literature review revealed that underground residential buildings might have various advantages such as space and energy saving, high level of fire resistance, less influenced by natural disasters, lower operating costs, improved indoor thermal comfort, and more resistant to external noises. First of all, the development of underground residential buildings can effectively constrain the ever-increasing urban sprawl and meanwhile save space for the natural and heritage landscapes (Alkaff *et al.*, 2016; Bobylev, 2009). In addition, underground residential buildings are much more resistant to external noise than surface residential buildings as the soil is a solid matter which can prevent the transmission of noise much more effectively than liquid and gas (Alkaff *et al.*, 2016; Roberts *et al.*, 2016). Similarly, due to soil's fireproof nature, underground residential buildings face a low possibility of suffering from a major fire

Table 1.   Potential advantages and disadvantages of underground residential buildings.

Advantages and Disadvantages			References
Potential advantages	A1	Space saving	Alkaff *et al.* (2016), Bobylev (2009), Broere (2016), Carmody and Sterling (1983), Liu *et al.* (2015), Ronka *et al.* (1998)
	A2	Energy saving	Alkaff *et al.* (2016), Carmody and Sterling (1983), Chow *et al.* (2002), Goel *et al.* (2012), Liu *et al.* (2015), and Sugai *et al.* (2012)
	A3	High level of fire resistance	Alkaff *et al.* (2016), Carmody and Sterling (1983), Chow *et al.* (2002), Liu *et al.* (2015)
	A4	Less influenced by natural disasters	Alkaff *et al.* (2016), Goel *et al.* (2012), Liu *et al.* (2015)
	A5	Lower operating costs	Goel *et al.* (2012)
	A6	Improved indoor thermal comfort	Ronka *et al.* (1998), Roberts *et al.* (2016), Sugai *et al.* (2012), Liu *et al.* (2015), Alkaff *et al.* (2016)
	A7	More resistant to external noises	Alkaff *et al.* (2016), Broere (2016), Liu *et al.* (2015), Roberts *et al.* (2016), Ronka *et al.* (1998)
	A8	Increased level of privacy	Goel *et al.* (2012)

(*Continued*)

Table 1. *(Continued)*

Advantages and Disadvantages				References
Potential disadvantages	D1	Limited access to natural light		Carmody and Sterling (1983), Durmisevic (1999), Edwards and Torcellini (2002), Nakhaei *et al.* (2016), Roberts *et al.* (2016), Ronka *et al.* (1998)
	D2	Psychological resistance from people		Alkaff *et al.* (2016), Durmisevic and Sariyildiz (2001), Edelenbos *et al.* (1998), Goel *et al.* (2012), Lee *et al.* (2016), Liu *et al.* (2015), Nakhaei *et al.* (2016)
	D3	Environmental issues		Attanayake and Waterman (2006), Bobylev (2009), Goel *et al.* (2012)
	D4	High construction cost		Chow *et al.* (2002), Edelenbos *et al.* (1998), Goel *et al.* (2012), Kaliampakos *et al.* (2016), National Research Council and US National Committee on Tunneling Technology (1984), Sugai *et al.* (2012)
	D5	Climate isolation		Carmody and Sterling (1983), Goel *et al.* (2012), Roberts *et al.* (2016)
	D6	Safety concerns		Goel *et al.* (2012), Lee *et al.* (2016), Ogata *et al.* (1990), Qian and Lin (2016), Seo and Choi (2008), Sterling *et al.* (1992), Watanabe *et al.* (1992)
	D7	Restrictions due to existing structures		Goel *et al.* (2012)
	D8	Lack of a visible facade design		Goel *et al.* (2012), Roberts *et al.* (2016)

Table 2.   Potential risks in the underground residential building projects.

Code	Risks	References
R1	Labor restrictions	Smith, (1992), Choi *et al.* (2004)
R2	Material restrictions	Smith (1992)
R3	Economic fluctuations	Choi *et al.* (2004), Ghosh and Jintanapakanont (2004), Ita (2004)
R4	Inflation	Choi *et al.* (2004), Ghosh and Jintanapakanont (2004), Ita (2004)
R5	Pollution (e.g. underwater table pollution, smoke)	Ghosh and Jintanapakanont (2004)
R6	Act of God	Smith, (1992), Choi *et al.* (2004), Ghosh and Jintanapakanont (2004)
R7	Receptivity of underground housing	Alkaff *et al.* (2016), Durmisevic (1999), Edelenbos *et al.* (1998)
R8	Security problems at project site	Qian and Lin (2016), Seo and Choi (2008)
R9	Low productivity of workers	Smith, (1992), Choi *et al.* (2004), Ghosh and Jintanapakanont (2004)
R10	Resource cost fluctuations	Ghosh and Jintanapakanont (2004)
R11	Outdated skills and technology	Smith, (1992), Choi *et al.* (2004), Ita (2004), Sarkar and Dutta (2012)
R12	Decreasing market demand	Alkaff *et al.* (2016), Liu *et al.* (2015)
R13	Demands and variations to the project	Ghosh and Jintanapakanont (2004)

(Continued)

Table 2.   (*Continued*)

Code	Risks	References
R14	Cost overruns	Smith, (1992), Choi *et al.* (2004)
R15	Poor relationship between client and contractors/ sub-contractors	Choi *et al.* (2004), Ghosh and Jintanapakanont (2004)
R16	Local contractor's competency in underground construction	Choi *et al.* (2004), Ghosh and Jintanapakanont (2004), Ita (2004)
R17	Ground settlement	Smith, (1992), Choi *et al.* (2004), Ghosh and Jintanapakanont (2004), Ita (2004), Sarkar and Dutta (2012)
R18	Unforeseen site condition	Smith, (1992), Choi *et al.* (2004), Ghosh and Jintanapakanont (2004), Ita (2004), Sarkar and Dutta (2012)
R19	Adverse weather	Smith, (1992), Choi *et al.* (2004)
R20	High incident & accident rate	Smith, (1992), Ghosh and Jintanapakanont (2004), Seo and Choi (2008), Qian and Lin (2016)

disaster, although fire is still a significant hazard to the underground structures (Alkaff *et al.*, 2016; Liu *et al.*, 2015). Last but not least, compared to those residential buildings above the ground, underground residential buildings offer a safer living environment in the face of natural disasters, as they are constructed under the ground and thus are less likely to be damaged by high winds, hail storms, lightning strikes, tornadoes and other natural disasters (Alkaff *et al.*, 2016; Liu *et al.*, 2015). Table 1 summarizes and presents the diverse advantages briefly reviewed above.

2.2. *Possible disadvantages of underground residential buildings*

When comparing underground residential buildings with surface residential buildings, some disadvantages of the former should be realized. Based on the literature review, these disadvantages relate, first, to the comparatively high cost of construction. In addition, authorities are concerned with the psychological resistance from people, environmental issues, and various safety considerations (Alkaff *et al.*, 2016; Broere, 2016; Durmisevic, 1999; Lee *et al.*, 2016; Roberts *et al.*, 2016; Ronka *et al.*, 1998; Seo and Choi, 2008). Various disadvantages identified from the literature have also been summarized and presented in Table 1.

Normally, underground residential buildings cost far more to construct than equivalent surface residential buildings (Edelenbos *et al.*, 1998), as they involve numerous costly works such as excavation, support and permanent lining work (Goel *et al.*, 2012). In addition, the development of underground residential buildings may also lead to issues concerning the surrounding environment (Bobylev, 2009; Roberts *et al.*, 2016). Another significant disadvantage is people's psychological recognition of underground residential buildings. Normally, people hold negative associations of living underground, as they consider underground a place which is dark, damp, confined, easily disorientated, poorly ventilated and more related to death and burial (Lee *et al.*, 2016; Liu *et al.*, 2015; Nakhaei *et al.*, 2016). People are also concerned with the potential safety issues of living underground, which may comprise structure settlement, flood, shortage of oxygen, poisoning and the entrapment by the possible collapse of the underground structure (Edelenbos *et al.*, 1998;

Goel *et al.*, 2012; Watanabe *et al.*, 1992). Moreover, undoubtedly, living underground might also restrict people's access to the natural light and ventilation (Durmisevic, 1999; Nakhaei *et al.*, 2016; Roberts *et al.*, 2016; Ronka *et al.*, 1998).

2.3. *Potential risks in underground residential building projects*

Underground building projects take place in natural underground surroundings whose characteristics and behaviours are not always foreseeable. Thus, underground building projects confront a higher level of risk compared to the surface building projects (Klee, 2015). So far, considerable research efforts have been made to investigate the risks in underground building projects. For instance, Smith (1992) identified 19 risks in underground projects and grouped them into three categories including resource and project prerequisite risks, performance-related risks and outside influence-type risks. Choi *et al.* (2004) summarized 31 risks in subway construction projects and categorized them into four groups, namely construction-related risks, design-related risks, financial-related risks, and the act of God. Ita (2004) emphasized that the risks associated with cost, schedule, and quality were critical to underground construction projects. Concentrating on the construction stage of underground metro projects, Sarkar and Dutta (2012) systematically examined the risks related to the construction activities such as survey, excavation, piling, installation of structures and waterproofing. Qian and Lin (2016) examined the safety risk in Chinese underground building projects systematically and proposed risk mitigation strategies from six aspects including policy, law, administration, economy, education and technology. Based on the literature review above, a total of 20 risks compatible to the underground residential building projects were extracted and listed in Table 2.

3. Method and Data Presentation

3.1. *Data collection and presentation*

As a systematic method of collecting data, the technique of questionnaire survey has been widely used to gather professional views in construction engineering and management research (Fellows and Liu, 2015). Thus, this study developed a structured questionnaire to collect professionals' perceptions and judgements on underground residential buildings' advantages, disadvantages and risks. The first section of the questionnaire sought respondents' background information including their employer types, their designations, as well as the number of underground projects they have participated in. The second section solicited respondents' endorsements for the underground residential buildings' advantages and disadvantages, using a five-point rating scale (i.e. 1 = strongly disagree, 2 = disagree, 3 = neutral, 4 = agree, and 5 = strongly agree). The third section requested respondents to assess the risks in underground residential building projects regarding the likelihood of occurrence (LO) and magnitude of impact (MI) based on two five-point rating scales shown in Table 3.

As an institution-based questionnaire survey, 70 institutions that already had experiences in undertaking housing or underground projects in Singapore were first identified through the Building Construction Authority of (BCA) Singapore portal, and then the developed questionnaire was sent to them via emails. Finally, a total of 30 responses were received, representing a response rate of 43% for this survey. The sample size was relatively small because there were a limited number of companies with the underground project experience. However, statistical analysis could still be performed as the central limit theorem holds true with a sample size larger than 30 (Ott and Longnecker, 2001). The profiles of the institutions and the respondents are shown in Table 4. The respondents are

Table 3.   Rating scales for LO and MI.

	Likelihood of Occurrence			Magnitude of Impact
LO	**Linguistic Terms**	**Likelihood of References (Percentage)**	**MI**	**Linguistic Terms**
1	Least likely	<20	1	Least impactful
2	Less likely	20–40	2	Less impactful
3	Likely	40–60	3	Impactful
4	More likely	60–80	4	More impactful
5	Most likely	>80	5	Most impactful

Table 4.   Profile of the respondents.

Profile	Frequency	Percentage
Institution		
Governmental agency	2	7
Developer	2	7
Construction company	3	10
Consultancy	6	20
Architectural firm	13	43
Project management firm	3	10
Facility management firm	1	3
Respondent's designation		
Project manager	5	17
Architect	13	43
Quantity surveyor	3	10
Engineer	8	27
Facility manager	1	3
The number of the projects involved		
<5	8	27
5–10	13	43
>10	9	30

of different professional backgrounds such as project manager, architect, quantity surveyor, engineer and facility manager. The great diversity and heterogeneity of the respondent panel were able to help ensure the quality and reliability of the data collected (Harty, 2008).

3.2. *Data analysis methods*

3.2.1. *Intergroup comparison*

As the respondents are from different fields such as development, construction, architecture, consultancy, project management, and facility management firms, it is necessary to conduct an intergroup comparison to check whether significant differences exist among respondents from different institutions. Two different statistical methods, namely analysis of variance (ANOVA) and Kruskal-Wallis test, were considered to perform the intergroup comparison. ANOVA is a widely used parametric test that checks the differences among the means from three or above groups (e.g. Goodrum and Haas, 2002); while the Kruskal-Wallis test is a rank-based non-parametric statistical method that checks the potential differences among different groups, which requires no specific distribution on the tested data (Tixier *et al.*, 2014).

3.2.2. *Risk criticality indices*

The developed questionnaire requested respondents to assess the likelihood of occurrence (LO) and the magnitude of impact (MI) of each risk in underground residential building projects based on two five-point Likert rating scales indicated in Table 3. To depict each risk more comprehensively, this study used the risk criticality (RC) index to gauge each risk. As the assessments of LO and MI were both carried out with a five-point rating system, the RC is thus on a full scale of 25.

4. Results and Discussions

4.1. *Significant advantages of underground residential buildings*

Table 5 shows respondents' assessments of the potential advantages of underground residential buildings as well as the relevant statistical test results. Kruskal-Wallis test results in Table 5 reveal that there are no significant differences among the respondents of different professional backgrounds, implying that the assessments from the respondent panel can be treated as a whole for analysis. Assessment results in Table 5 reveal that the mean values of A1 "space saving", A6 "improved indoor thermal comfort", A7 "more resistant to external noises", and A8 "increased level of privacy" are greater than 3, indicating there are significant advantages of underground residential buildings.

Table 5.   Advantages of underground residential buildings.

Code	Advantage	Rank	Mean	SEM#	Kruskal-Wallis Test (*p*-value)
A1	Space saving	1	4.27	0.309	0.335
A6	Improved indoor thermal comfort	2	3.60	0.433	0.329
A7	More resistant to external noise	3	3.50	0.466	0.483
A8	Increased level of privacy	4	3.43	0.418	0.432
A4	Less influenced by natural disasters	5	2.90	0.516	0.077
A2	Energy saving	6	2.87	0.480	0.348
A3	High level of fire resistance	7	2.83	0.539	0.139
A5	Lower operating costs	8	2.33	0.429	0.673

Note: # SEM represents for standard error of measurement.

A1 "space saving" was assessed as the most significant advantage with a mean value of 4.27. This is an unsurprising result as the main purpose of developing underground residential buildings is to save urban spaces (Alkaff *et al.*, 2016; Bobylev, 2009; Liu *et al.*, 2015; Ronka *et al.*, 1998). A6 "improved indoor thermal comfort" was assessed as the second most significant advantage with a mean assessment of 3.60. A7 "more resistant to external noise" was ranked third with an assessment of 3.50. Surface buildings are often plagued by the noise outside of the buildings (Hongisto *et al.*, 2015). A8 "increased level of privacy" was ranked fourth. People living in surface housing buildings (especially in those high rise ones) often complain about their limited privacy (Yuen and Yeh, 2011), while underground residential buildings can settle this issue perfectly as it has blocked all the views outside and provided maximum privacy to its residents.

4.2. *Major disadvantages of underground residential buildings*

Table 6 presents respondents' assessments to the disadvantages of underground residential buildings and the relevant statistical test results.

Table 6. Disadvantages of underground residential buildings.

Code	Disadvantage	Rank	Mean	SEM#	Kruskal-Wallis Test (*p*-value)
D1	Limited access to natural light	1	4.47	0.419	0.541
D4	High construction cost	2	4.23	0.384	0.623
D5	Climate isolation	3	4.07	0.422	0.082
D2	Psychological resistance from the residents	4	3.97	0.491	0.806
D3	Environmental issues	5	3.77	0.465	0.427
D6	Safety concerns	6	3.60	0.464	0.403
D8	Lack of a visible facade design	7	3.33	0.602	0.815
D7	Restrictions due to existing structures	8	3.10	0.503	0.107

Note: # SEM represents for standard error of measurement.

As indicated in Table 6, the mean assessment results of all eight disadvantages are greater than 3, indicating all the disadvantages identified were severe obstacles to promoting underground residential buildings. According to Kruskal-Wallis test results, consistent views of the disadvantages were found among respondents of different professional backgrounds. However, more attention needs to be directed to D1 "limited access to natural light". This disadvantage can be regarded the most severe one as its assessment is statistically higher than all the rest disadvantages except for D4 "high construction cost".

D1 "limited access to natural light" was assessed as the most severe disadvantage of underground residential buildings with an assessment of 4.47. This result shows that respondents are deeply concerned about losing direct access to the natural light if choosing to live underground. D4 "high construction cost" was assessed as the second most severe disadvantage with an assessment of 4.23. The high construction cost has been a common obstacle to the promotion of any underground space development (Broere, 2016; Durmisevic, 1999; Ronka *et al.*, 1998). D5 "climate isolation" was ranked as the third most severe disadvantage. Receiving a high assessment of 3.97, D2 "psychological resistance from the residents" was assessed as the fourth most severe disadvantage. This result is in line with Alkaff *et al.* (2016) who have pointed out that the majority of people cannot accept living underground psychologically. D3 "environmental issues" were the fifth most severe disadvantage of underground residential buildings. Inevitably, the construction of underground residential buildings will lower the regional groundwater table (Goel *et al.*, 2012; Ronka *et al.*, 1998), which may damage the regional hydrologic system and result in settlement of surrounding structures. D6 "safety considerations" were ranked sixth with an assessment of 3.6.

4.3. *Likelihood, impact, and criticality of risks in underground residential building projects*

The test results in Table 7 show that the relevant data are not normally distributed. The Kruskal-Wallis test results in Table 7 show that there are no significant differences in the perceptions of LO and MI among the respondents of different professional backgrounds, suggesting consistent views among the respondent panel.

According to the LO values, the top five risks that are most likely to occur in underground residential building projects are R1 "labour restrictions," R14 "cost overruns," R3 "economic fluctuation," R16 "local contractor's competency in underground construction," and R2 "material restrictions." Based on the MI values, the top five risks that are most impactful on underground residential building projects are R16 "local contractor's competency in underground construction," R7 "receptivity of underground housing," R14 "cost overruns," R1 "labour restrictions," R2 "material restrictions," and R20 "high incident & accident rate." The RC values reveal that the top five critical risks in underground residential building projects are R1 "labour restrictions," R14 "cost overruns," R16 "local contractor's competency in underground construction," R2 "material restrictions," and R3 "economic fluctuation".

R1 "labour restrictions" were assessed as the most critical risk in underground residential building projects in Singapore with the highest RC value of 14.83. Being a small country with limited human resources, Singapore has to leverage foreign workforces to ensure its economic

Table 7. Assessments of risks in underground residential building projects.

Code	Risks	Risk Criticality		Likelihood of Occurrence				Magnitude of Impact			
		Rank	Mean	Rank	Mean	SEM#	Kruskal-Wallis Test	Rank	Mean	SEM#	Kruskal-Wallis Test
R1	Labor restrictions	1	14.83	1	3.67	0.474	0.727	4	3.80	0.543	0.698
R14	Cost overruns	2	14.57	2	3.50	0.572	0.084	2	3.83	0.526	0.329
R16	Local contractor's competence in underground construction	3	13.57	4	3.40	0.398	0.649	1	3.93	0.374	0.153
R2	Material restrictions	4	13.30	5	3.37	0.518	0.689	5	3.73	0.422	0.906
R3	Economic fluctuations	5	13.13	3	3.43	0.434	0.974	8	3.60	0.478	0.746
R7	Receptivity of underground housing	6	12.87	9	3.23	0.583	0.676	2	3.83	0.485	0.981
R18	Soil conditions	7	12.20	6	3.33	0.377	0.357	9	3.57	0.434	0.718

(Continued)

Table 7.   (*Continued*)

Code	Risks	Risk Criticality		Likelihood of Occurrence				Magnitude of Impact			
		Rank	Mean	Rank	Mean	SEM#	Kruskal-Wallis Test	Rank	Mean	SEM#	Kruskal-Wallis Test
R5	Pollution (e.g. underwater table pollution, smoke)	8	11.87	8	3.23	0.45	0.219	11	3.47	0.508	0.610
R20	High incident & accident rate	9	11.67	14	3.00	0.484	0.496	5	3.73	0.497	0.902
R10	Resource cost fluctuations	10	11.63	11	3.13	0.435	0.922	10	3.50	0.535	0.505
R4	Inflation	11	11.57	7	3.30	0.409	0.945	16	3.30	0.409	0.965
R13	Demands and variations to the project	12	11.53	10	3.17	0.499	0.478	11	3.47	0.494	0.422
R12	Decreasing market demand	13	10.97	13	3.03	0.491	0.043	17	3.27	0.537	0.684
R17	Ground settlement	14	10.93	11	3.13	0.366	0.639	13	3.40	0.433	0.593

R6	Force majeure (due to natural disasters)	15	10.57	20	2.63	0.604	0.226	7	3.70	0.623	0.950
R19	Unforeseeable weather	16	10.40	15	2.93	0.469	0.311	13	3.40	0.492	0.724
R9	Low productivity of workers	17	9.80	16	2.80	0.461	0.206	18	3.23	0.571	0.131
R11	Outdated skills and technology	17	9.80	17	2.77	0.494	0.501	15	3.37	0.447	0.341
R15	Poor relationship between client and contractors/ sub-contractors	19	8.93	18	2.73	0.388	0.923	19	3.13	0.451	0.313
R8	Security problems at project site	20	8.47	19	2.70	0.486	0.914	20	2.80	0.49	0.658

Note: # SEM represents standard error of measurement.

growth (Low, 2002). Receiving the second highest RC value of 14.57, R14 "cost overruns" were assessed as the second most critical risk in underground residential building projects, which suggests that the respondents have significant concerns with the cost issues of these projects.

R16 "local contractor's competence in underground construction" was ranked third in the RC evaluations, and it received the highest assessment in MI, indicating that it was the most impactful risk to underground residential building projects in Singapore. R2 "material restrictions" were assessed as the fourth most critical risk in underground residential building projects with an RC value of 13.30. Material restrictions hereby refer to two issues, one is the limited supply of the materials, and the other is the possible late delivery. R3 "economic fluctuations" were assessed as the fifth most critical risk in the underground residential building projects. This risk received the third highest evaluation in LO, which means that it is fairly likely to occur in Singapore. Singapore is a typical export-oriented economy which is very easily affected by the global economy (Monetary Authority of Singapore, 2017). The primary impact of economic fluctuations on the construction industry was to affect the developers' decision-making (Kabir *et al.*, 2014), for which some projects may be cancelled and even abandoned. Definitely, as a new type of project, underground residential building projects will be the first to be affected under this circumstance.

5. Conclusions and Recommendations

Over the past two decades, the underground built environment has achieved rapid development with a widespread global recognition of its distinct advantage in relieving land use pressure on densely packed metropolises. Recently, to address the settlement need of the increasing urban population, the authorities have started to consider the possibility of developing underground residential buildings. By administering an empirical questionnaire survey with the experienced professionals from Singapore's underground construction community, this study carried out a preliminary investigation of underground residential buildings, looking into its advantages, disadvantages and critical risks. Based on the survey results, the top four advantages of underground residential buildings were revealed as "space saving", "improved indoor thermal comfort", "more resistant to external noises", and "increased level of privacy". Particularly, "space saving" was found to be the most significant advantage. Also, the survey results disclosed six major disadvantages of underground residential buildings, which were "limited access to natural light", "high construction cost", "climate isolation", "psychological resistance from residents", "environmental issues" and "safety concerns". Furthermore, this study assessed the risks of underground residential building projects in terms of their likelihood of occurrence and magnitude of impact. The results reported that the top five critical risks were "labour restrictions", "cost overruns", "local contractors' competence in underground construction", "material restrictions" and "economic fluctuations".

From the findings of this study, a sceptical attitude towards underground residential buildings from the respondents can be sensed. This is because, among the eight identified advantages of underground residential buildings, only four of them were acknowledged by the respondents; while for the eight disadvantages, all of them were admitted as the factors hindering the implementation. Furthermore, among the 20 risks assessed, 18 of them (90%) were regarded as critical ones to underground residential building projects, suggesting that underground residential building projects would probably face significant risks during construction. These results of advantages, disadvantages, and critical risks remind the authorities that there are plenty of uncertainties in developing underground residential buildings, and thus an in-depth feasibility study must be carried out before the idea is materialized.

Despite the limitations of the opinion-based data and the relatively small sample size, the findings from this study are still valuable as they are the first investigations into the advantages, disadvantages and critical risks of underground residential building projects. Also, this study can provide the authorities with a thorough picture of the underground residential buildings which will facilitate their decision-making related to such type of projects. For future research, a series of mitigation measures, in both technical and managerial aspects that can tackle the risks residing in underground residential buildings can be proposed.

References

Admiraal, H. and Cornaro, A. (2016a). "Engaging Decision Makers for an Urban Underground Future", *Tunneling and Underground Space Technology*, 55, 221–223.

Admiraal, H. and Cornaro, A. (2016b). "Why Underground Space Should be Included in Urban Planning Policy — And How This Will Enhance an Urban Underground Future", *Tunnelling and Underground Space Technology*, 55, 214–220.

Alkaff, S. A., Sim, S. C. and Ervina Efzan, M. N. (2016). "A review of Underground Building Towards Thermal Energy Efficiency and Sustainable Development", *Renewable and Sustainable Energy Reviews*, 60, 692–713.

Attanayake, P. M. and Waterman, M. K. (2006). "Identifying Environmental Impacts of Underground Construction", *Hydrogeology Journal*, 14, 1160–1170.

Bartel, S. and Janssen, G. (2016). "Underground Spatial Planning — Perspectives and Current Research in Germany", *Tunnelling and Underground Space Technology*, 55, 112–117.

Bobylev, N. (2009). "Mainstreaming Sustainable Development into a City's Master Plan: A Case of Urban Underground Space Use", *Land Use Policy*, 26, 1128–1137.

Broere, W. (2016). "Urban Underground Space: Solving the Problems of Today's Cities", *Tunnelling and Underground Space Technology*, 55, 245–248.

Carmody, J. and Sterling, R. (1983). "Underground Building Design: Commercial and Institutional Structures", Underground Space Center, Van Nostrand Reinhold Co, New York United States.

Choi, H. H., Cho, H. N. and Seo, J. W. (2004). "Risk Assessment Methodology for Underground Construction Projects", *Journal of Construction Engineering and Management*, 130, 258–272.

Chow, F., Paul, T., Vahaaho, I., Sellberg, B. and Lemos, L. (2002). "Hidden Aspects of Urban Planning: Utilisation of Underground Space", *Proceedings of 2nd International Conference on Soil structure interaction in urban civil engineering; planning & engineering for the cities of tomorrow*; Swiss Federal Institute of Technology ETH Zurich, Zurich; 2002; 327–334.

Debrah, Y. A. and Ofori, G. (2001). "Subcontracting, Foreign Workers and Job Safety in the Singapore Construction Industry", *Asia Pacific Business Review*, 8, 145–166.

Department of Statistics (2016). Population Trends 2016, in: Department of Statistics, S. (Ed.), Singapore, pp. 3–6.

Durmisevic, S. (1999). "The Future of the Underground Space", *Cities*, 16, 233–245.

Durmisevic, S. and Sariyildiz, S. (2001). "A Systematic Quality Assessment of Underground Spaces — Public Transport Stations", *Cities*, 18, 13–23.

Edelenbos, J., Monnikhof, R., Haasnoot, J., Van Der Hoeven, F., Horvat, E. and Van Der Krogt, R. (1998). "Strategic Study on the Utilization of Underground Space in the Netherlands", *Tunnelling and Underground Space Technology*, 13, 159–165.

Edwards, L. and Torcellini, P. A. (2002). "A Literature Review of the Effects of Natural Light on Building Occupants", *National Renewable Energy Laboratory Golden*, CO.

Fellows, R.F. and Liu, A.M. (2015). *Research Methods for Construction*, Hoboken: John Wiley & Sons.

Ghosh, S. and Jintanapakanont, J. (2004). "Identifying and Assessing the Critical Risk Factors in an Underground Rail Project in Thailand: A Factor Analysis Approach", *International Journal of Project Management*, 22, 633–643.

Goel, R. K., Singh, B. and Zhao, J. (2012). *Underground Infrastructures: Planning, Design, and Construction*, Oxford: Butterworth-Heinemann.

Goodrum, P. and Haas, C. (2002). "Partial Factor Productivity and Equipment Technology Change at Activity Level in U.S. Construction Industry", *Journal of Construction Engineering and Management*, 128, 463–472.

Harty, C. (2008). "Implementing Innovation in Construction: Contexts, Relative Boundedness and Actor Network Theory", *Construction Management and Economics*, 26, 1029–1041.

Hongisto, V., Mäkilä, M. and Suokas, M. (2015). "Satisfaction with Sound Insulation in Residential Dwellings — The Effect of Wall Construction", *Building and Environment*, 85, 309–320.

Ita (2004). "Underground or aboveground? Making the choice for urban mass transit systems: A report by the International Tunnelling Association (ITA). Prepared by Working Group Number 13 (WG13). 'Direct and indirect advantages of underground structures'", *Tunnelling and Underground Space Technology*, 19, 3–28.

JTC, S. (2017). Jurong Rock Caverns, available from: https://www.jtc.gov.sg/industrial-land-and-space/Pages/jurong-rock-caverns.aspx (cited 30 Nov 2018).

Kabir, G., Sadiq, R. and Tesfamariam, S. (2014). "A Review of Multi-Criteria Decision-Making Methods for Infrastructure Management", *Structure and Infrastructure Engineering*, 10, 1176–1210.

Kaliampakos, D., Benardos, A. and Mavrikos, A. (2016). "A Review on the Economics of Underground Space Utilization", *Tunnelling and Underground Space Technology*, 55, 236–244.

Lee, E. H., Christopoulos, G. I., Lu, M., Heo, M. Q. and Soh, C. K. (2016). "Social Aspects of Working in Underground Spaces", *Tunnelling and Underground Space Technology*, 55, 135–145.

Liu, J., Zhao, X. and Yan, P. (2016). "Risk Paths in International Construction Projects: Case Study from Chinese Contractors", *Journal of Construction Engineering and Management*, 142, 05016002.

Liu, X., Wang, Z., Liu, J. and Yang, Z. (2015). "Concept Design and Development Model of Underground Villas", *Journal of Engineering Science and Technology Review*, 8, 46–52.

Low, L. (2002). "The Political Economy of Migrant Worker Policy in Singapore", *Asia Pacific Business Review*, 8, 95–118.

Ministry of National Development (2013). *A High Quality Living Environment for all Singaporeans*, Singapore: Ministry of National Development.

Monetary Authority of Singapore (2017). *The Singapore Economy*, Singapore: Monetary Authority of Singapore.

Nakhaei, J., Arefi, S. L., Bitarafan, M. and Kildiene, S. (2016). "Evaluation of Light Supply in the Public Underground Safe Spaces by Using of Copras-Swara Methods", *International Journal of Strategic Property Management*, 20, 198–206.

National Population and Talent Division (2013). "A Sustainable Population for a Dynamic Singapore-Population White Paper", Singapore.

Ogata, Y., Isei, T. and Kuriyagawa, M. (1990). "Safety Measures for Underground Space Utilization", *Tunnelling and Underground Space Technology*, 5, 245–256.

Ott, R. L. and Longnecker, M. (2001). *An Introduction to Statistical Methods and Data Analysis*, Duxbury, MA: Pacific Grove.

Ott, R. L. and Longnecker, M.T. (2008). *An Introduction to Statistical Methods and Data Analysis*, Boston: Cengage Learning.

Qian, Q. and Lin, P. (2016). "Safety Risk Management of Underground Engineering in China: Progress, Challenges and Strategies", *Journal of Rock Mechanics and Geotechnical Engineering*, 8, 423–442.

Roberts, A. C., Christopoulos, G. I., Car, J., Soh, C. K. and Lu, M. (2016). "Psycho-biological Factors Associated with Underground Spaces: What Can the New Era of Cognitive Neuroscience Offer to Their Study?" *Tunnelling and Underground Space Technology*, 55, 118–134.

Ronka, K., Ritola, J. and Rauhala, K. (1998). "Underground Space in Land-use Planning", *Tunnelling and Underground Space Technology*, 13, 39–49.

Sarkar, D. and Dutta, G. (2012). "A Framework for Project Risk Management for the Underground Corridor Construction of Metro Rail", *International Journal of Construction Project Management*, 4, 21–38.

Seo, J. W. and Choi, H. H. (2008). "Risk-based Safety Impact Assessment Methodology for Underground Construction Projects in Korea", *Journal of Construction Engineering and Management*, 134, 72–81.

Smith, R. J. (1992). "Risk Management for Underground Projects: Cost-Saving Techniques and Practices for Owners", *Tunnelling and Underground Space Technology incorporating Trenchless*, 7, 109–117.

Sterling, R., Carmody, J. and Rockenstein, W. H. (1992). "Case-Study of Life Safety Standards for a Large Mined Underground Space Facility in Minneapolis, Minnesota", *Tunnelling and Underground Space Technology*, 7, 119–125.

Sugai, Y., Sasaki, K., Yoshimura, K., Yukitake, T. and Muta, S. (2012). "Pilot Study on the Construction of Several Temperature-Controlled Multi-Purpose Rooms in a Disused Tunnel", *Tunnelling and Underground Space Technology*, 32, 180–189.

Tixier, A., Hallowell, M., Albert, A., van Boven, L. and Kleiner, B. (2014). "Psychological Antecedents of Risk-Taking Behavior in Construction", *Journal of Construction Engineering and Management*, 140, 04014052.

United Nations (2015). World Urbanization Prospects The 2014 Revision. Department of Economic and Social Affairs.

Watanabe, I., Ueno, S., Koga, M., Muramoto, K., Abe, T. and Goto, T. (1992). "Safety And Disaster Prevention Measures For Underground Space — An Analysis Of Disaster Cases", *Tunnelling and Underground Space Technology*, 7, 317–324.

Yuen, B. and Yeh, A. G. O. (2011). *High-Rise Living in Asian Cities*, Netherlands: Springer.

Zhou, Y. X. and Zhao, J. (2016). "Assessment and Planning of Underground Space use in Singapore", *Tunnelling and Underground Space Technology*, 55, 249–256.

Chapter 30

Economic-Efficiency Analysis of Rawalpindi Bypass Project: A Case Study

Yasir Mehmood and Hafiz Zahoor*

*Department of Construction Engineering and Management,
National University of Science and Technology,
Risalpur Campus 24080, Pakistan
hafiz.zahoor@mce.nust.edu.pk

Fahim Ullah

*Faculty of Built Environment,
The University of New South Wales, Sydney, Australia*

Abstract

Safe, efficient and user-friendly transportation of people and goods is the premier concern for highway agencies due to a growing tendency in urbanization, especially in developing countries such as Pakistan. This study presents economic-efficiency analysis of a proposed project, i.e. Rawalpindi Bypass, and an existing roadway structure in Pakistan. Considering the highway agency as well as road users, this study

---

* Corresponding author.

determines the transportation demand estimation (TDE), project cost and various savings such as travel time (TT), safety and vehicle operating cost (VOC). As the transportation decision-making (TDM) process usually involves the evaluation of an alternative decision with respect to a base-case DO-NOTHING Scenario, this study considers the existing roadway structure with no improvements as a DO-NOTHING Scenario, whereas the proposed bypass with two lanes in each direction is considered as an alternative. Economic-efficiency analysis is then carried out, for decision-making based on the monetary as well as non-monetary dimensions, using net present value (NPV) and benefit cost ratio (BCR) techniques. The study concludes that the proposed bypass is a viable alternative as compared to the existing roadway. The study's methodology and findings are expected to help the highway agencies and concerned stakeholders to conduct efficient TDM.

***Keywords*:** Economic-efficiency Analysis; Rawalpindi Bypass; Vehicle Operating Cost (VOC); Multi-criteria Decision-making (MCDM); Transportation Decision-making (TDM); Travel Time (TT); Safety Saving.

1. Introduction

Transportation legislation seeks multi-dimensional planning, taking into account various modes and mediums of transportation of goods and services. Additionally, today, making transportation laws requires a lot of deliberations, including the direct, indirect, tangible and intangible infrastructure impacts to the local community, environment, economy and most importantly the users (Sinha and Labi, 2007; Ullah *et al.*, 2016a). Similarly, safe, efficient and user-friendly transportation of people and goods have been a premier point of concern for highway agencies in the developed as well as developing countries. However, rapid development and urbanization impede the development of national and regional space corridors, as a rapid increase in urbanization creates an undesirable state of congestion and pollution, besides being a threat to the environment of a corridor (Sinha and Labi, 2007; Ullah *et al.*, 2016b; Ullah and Thaheem, 2017).

Transportation being one of the biggest sectors is also affected by the multi-staged and multi-dimensional propagation of urbanization (Hasnain *et al.*, 2017). The current state of traffic congestion in most of the urban cities of the world has created a podium for new challenges in the form of corporate jurisdiction and urban service delivery. Likewise, the urban challenges currently faced by the big cities of Pakistan, such as the twin cities of Rawalpindi–Islamabad, are equally indispensable and multi-dimensional (Ayub *et al.*, 2016a; Ullah *et al.*, 2017a). As a result, affluent and privileged classes of people living in the capital city of Islamabad are unable to access their attraction zones in time with comfort and convenience. In addition, lack of impartiality in the institutions' functionality acts as an impetus to the lapses in our structural hierarchy (Utama *et al.*, 2017). Similarly, poor mercantile growth, lack of venture capital and bankruptcy of civilian institutions are resulting in urban poverty and slumped employment. Such a situation has injected a heavy burden on the capital city road infrastructure, and at the same time, severely affected the credibility of road agencies to provide a safe, comfortable and convenient mode of transportation to the public (Ayub *et al.*, 2016b; Hasnain *et al.*, 2017; Ullah *et al.*, 2017b).

On the one hand, Islamabad is a planned city with extensive road networks resembling the grid structure, while Rawalpindi is a jargoned agglomeration of organic development representing a spider net structure with both the road standard and general traffic environment inferior to that of Islamabad. A practicable connection between these two cities remained as a big problem in the past due to the underestimation of travel demand (Zahoor and Choudhry, 2012). The traffic safety is also of significant concern. The traffic crashes per registered number of vehicles flung as high as 10 times any other developing city. The poor road design, bad drivers, higher road fares and absence of the understanding of other people rights negatively bolster the point of concern. A much needed programme is thus required to be formulated. The present public transport is mostly dominated by the private sector, which lessens the burden off the shoulders of the community but results in an inefficient and unequal system (Ullah *et al.*, 2017; Winston, 1991). A substantial effort has been directed towards practical and low-cost traffic management projects. A narrowing and rehabilitation of two-sided carriageway should be a continuous process together with strong juxtaposition from a diatribe transportation infrastructure policy at network level (Hasnain *et al.*, 2017). The planned extension of the framework in Islamabad is expected to adjust this increasing urban population unlike Rawalpindi. The long-term scenarios highlight the due essence of connectivity of the twin cities as a solution towards the detainment of specific characteristics to produce common labour markets. Of all such interior traffic engineering and management problems in the two cities, the main heavy vehicle traffic on the Grand Trunk Road is causing serious problems to the inside traffic conditions in the twin cities.

On the other hand, the heavy passenger and freight traffic resulting from the construction of Metro project and Pakistan Motorway Network have injected a huge burden on the mediocre road structure of both the cities. The heavy vehicles and freight traffic aiming to traverse between Taxila and Rawat have to pass through the congested and inefficient roads of the capital city together with passing from the central business district of Rawalpindi downtown. For the time being, the heavy vehicle traffic is bifurcated to reach Rawat via Kashmir Highway and Islamabad Expressway. However, a long-term proper and adequate connectivity infrastructure project needs to be constructed in between the two cities for proper carriage of passengers and freight traffic, and at same time, to lessen the burden on the street and road networks of twin cities.

Hence, to meet the travel demand and reduce the traffic congestion in the twin cities, Rawalpindi Development Authority and Capital Development Authority decided to conduct a detailed feasibility study of the bypass project. This study, therefore, presents a systematic decision-making model for the Rawalpindi Bypass transportation project using economic-efficiency analysis. In addition, it explains the significant engineering factors that can help the planners to comment about the overall appropriateness, efficiency, effectiveness and feasibility of the proposed project. It also presents the evaluation of all key costs and benefits related to agencies and/or users, and a mechanism of how these respective costs and benefits could be worked out for achieving a defensible, appropriate and rationale solution to the underscored transportation problem.

2. Research Methodology

This research highlights the complex nature of transportation planning and presents a systematic decision-making model that can aid the engineers and planners. The study embraces transportation demand estimation (TDE) of the proposed bypass, i.e. Rawalpindi bypass project, and all the segments of the existing roadway structure. It determines the project cost and various savings such as travel time (TT), safety and vehicle operating cost (VOC) savings, using the economic-efficiency analysis. As the transportation decision-making (TDM) process usually involves the evaluation of effectiveness and efficiency of an alternative decision with respect to a base-case DO-NOTHING scenario, the existing roadway structure with no improvements is taken as DO-NOTHING scenario, whereas the construction of Rawalpindi bypass with two lanes in each direction is taken as an alternative scenario.

The model presented in this research analyzes the most sensitive issues simultaneously, such as VOC savings, TT savings, safety savings, and air quality impacts. Hence, this research demonstrates an in-depth evaluation of an important transportation project using economic-efficiency analysis and most modern methods of multi-criteria decision-making analysis. The analysis would help to comment about the overall appropriateness, efficiency, effectiveness and feasibility of the project.

2.1. *Economic-efficiency analysis*

The overall framework for carrying out an economic-efficiency analysis on the basis of benefit and cost considerations is illustrated in Fig. 1.

2.2. *Demand estimation*

TDM is the key aspect of transportation evaluation process used for predicting the need for transportation in terms of passenger, freight or vehicle volumes. It also includes user benefits and costs, cash flow patterns, economic efficiency, effectiveness and equity influenced by volume of traffic using the facility. One of the important steps is the calculation of real-time affected volumes of road users on all the road segments of Islamabad and Rawalpindi that are expected to contribute to the utility usage of Rawalpindi Bypass. For the transportation system evaluation, two scenarios are considered, base-case scenario (without any intervention) and alternative scenario (with intervention).

2.3. *Base-case scenario (do nothing)*

National Highway N-5 is the main urban arterial road passing through the centre of Rawalpindi city. The major freight movement through heavy vehicles takes place through this road that causes severe

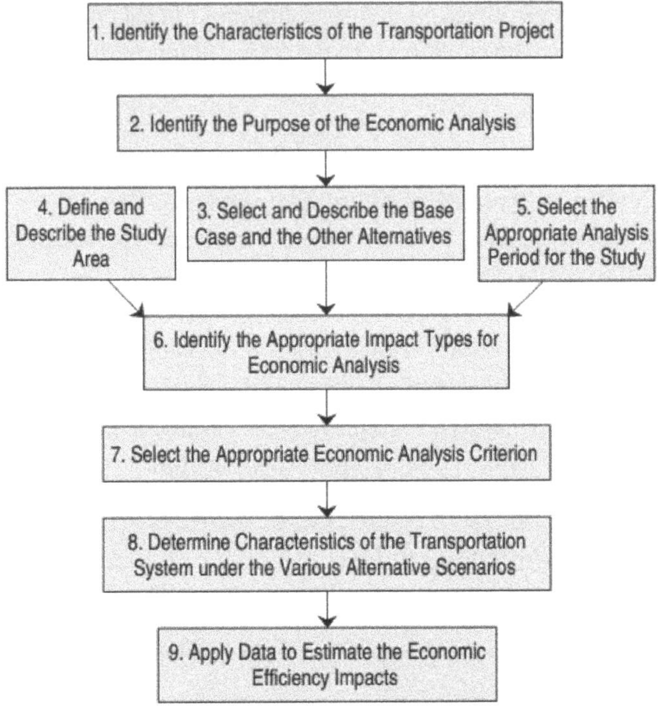

Fig. 1.   Framework for economic-efficiency analysis.

Source: Sinha and Labi (2007).

congestion in the area. Commercial vehicles, mostly the trucks, are diverted to IJP road which passes through Faizabad joining the Islamabad Express way and travel to Rawat. The base-case scenario is related to the study of different impacts of traffic that can be easily monetized like TT impacts, safety impacts and VOCs and benefits without any construction. In this scenario, we have assumed only one route for Rawalpindi commercial and non-commercial traffic. It covers a distance of 34 miles, and starts from Rawat, travels on Islamabad highway, IJP road and National Highway N-5. Traffic on its different segments is shown in Fig. 2.

2.4. *Alternative scenario (proposed Rawalpindi city bypass)*

Alternative scenario is related to the study of traffic impacts after the construction of Rawalpindi Bypass (32.58 miles), as shown in Fig. 3.

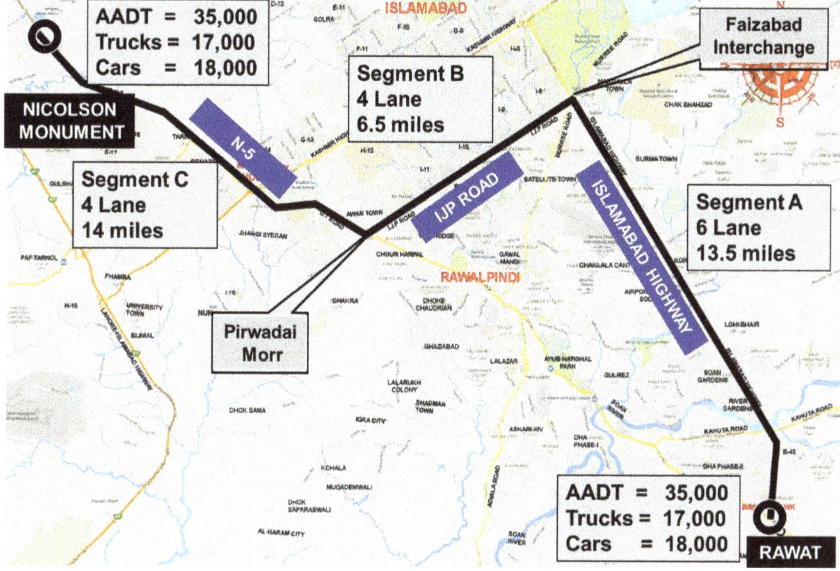

Fig. 2.   Base-case scenario (do nothing option).

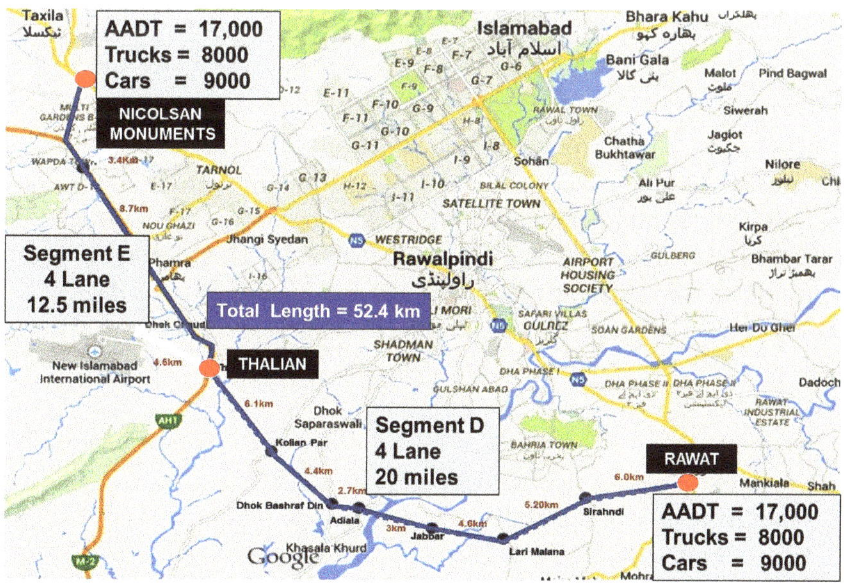

Fig. 3.   Alternative scenario (proposed Rawalpindi city bypass).

The proposed bypass will provide an alternate route for heavy vehicles to travel outside the city, resulting in a reduction of congestion. For the purpose of calculations, the data is gathered on the assumption of annual average daily traffic (AADT).

3. Analysis

3.1. *Calculating TT savings*

TT benefits achieved due to introduction of intervention were calculated using the following formula:

$$\text{TT Benefit} = (\text{No. of vehicles} \times \text{Occupancy}) \times (\text{Difference of time (min)}/60) \times \text{Rate}$$

Car = Rs. 1,535,121.00

Truck = Rs. 790,206.67

Total per day = Rs. 2,325,327.67

Total TT saving per annum = Rs. 848.75 M = Rs. 0.848 billion

The impact of TT savings is evaluated after the intervention, as shown in Tables 1–3.

3.2. *Safety savings*

It includes the evaluation of direct or indirect safety component that reduces the rate or severity of crashes. An in-depth study about the safety (tangible and intangible) components has been performed relative to the proposed bypass. Finally, the monetized benefit of safety savings is calculated and is incorporated in the final multi-criteria decision-making tool. The steps involved in determining safety impacts are as follows:

- Determine the vehicle mile travelled (VMT) in 100 million for each section = (AADT × Section Length × 365)/108.
- Find the fatal and non-fatal crash rate per 100 million VMT for each type of highway (Tables 4–7).

Table 1. Assumption for evaluation of TT savings.

S/No	Islamabad Highway	IJP, N-5 Road	Proposed Rawalpindi Bypass
Number of lanes (on each side)	3	2	2
AADT	35,000 (Trucks 17,000 & Cars 18,000)	35,000 (Trucks 17,000 & Cars 18,000)	17,000 (Trucks 8,000 & Cars 9,000)
Signals	1 Traffic signal/mile	1 Traffic signal/mile	Signal free
Vehicle occupancy	Trucks: 1, Cars: 1.2	Trucks: 1, Cars: 1.2	Trucks: 1, Cars: 1.2
FFS	50 mph	50 mph	60 mph
AWDT	$1.0991 \times$ AADT	$1.0991 \times$ AADT	$1.0991 \times$ AADT
TT costs	Cars: 295/h	Cars: 295 /h	Cars: 295 /h
(2015 constant)	Truck: 205/h	Truck: 205/h	Truck: 205/h

Table 2.   TT impacts of existing do nothing scenario.

Seg.	Length (miles)	No. of Lanes	AADT (veh/day)	Capacity (vphpl)	Assumed FF Speed (mph)	Demand (vph), V	2-Directional Capacity (vph), C	V/C	Avg Speed (mph)	TT (min) Dist/ Speed
A	13.5	6	35,000	2,000	50	35,000/24 = 1,459	12,000	0.12	33	(13.5/33) *60 = 24.5
B	6.5	4	35,000	2,000	50	1,459	8,000	0.18	33	11.8
C	14	4	35,000	2,000	50	1,459	8,000	0.18	33	25.45
						Total time				61.75

Table 3.   TT impacts of proposed bypass.

Seg.	Length (miles)	No. of Lanes	AADT (veh/ day)	Capacity (vphpl)	Assumed FF Speed (mph)	AWDT (veh/ day), T	2-Directional Capacity (vph), C	AWDT/C	Avg Speed (mph)	TT (min) Dist/ Speed
D	20	4	17,000	2,200	60	18,685	8,800	2.12	59.37	20.21
E	12.5	4	17,000	2,200	60	18,685	8,800	2.12	59.37	12.63
					Total time					32.84

Table 4.   Fatal and non-fatal crashes on existing route without bypass.

Seg.	Length (mile)	AADT	100 Million VMT	No. of Fatal Crashes Per 100 Million VMT	No. of Non-fatal Crashes Per 100 VMT
A	13.5	35,000	$(35{,}000 \times 13.5 \times 365/10^8)$ = 1.724	(1.724×1.30) = 2.241	(1.724×124.69) = 214.96
B	6.5	35,000	0.830	1.08	103.5
C	14	35,000	1.788	2.32	223

Table 5.   Fatal and non-fatal crashes on existing route in presence of bypass.

Seg.	Length (mile)	AADT	100 Million VMT	No. of Fatal Crashes Per 100 Million VMT	No. of Non-fatal Crashes Per 100 VMT
A	13.5	18,000	0.887	1.153	110.6
B	6.5	18,000	0.427	0.555	53.24
C	14	18,000	0.924	1.196	114.72

Table 6.   Fatal and non-fatal crashes on bypass.

Seg.	Length (mile)	AADT	100 Million VMT	No. of fatal Crashes Per 100 Million VMT	No. of Non-fatal Crashes Per 100 VMT
D	20	17,000	1.241	0.931	85.13
E	12.5	17,000	0.776	0.582	53.23

Table 7.   User safety benefits.

Crashes	Principal Arterial (without Bypass)	Principal Arterial (with Bypass)	Proposed Rawalpindi Bypass (Expressway)	Crash Saving	Crash Cost Saving (Millions Rs.)
Fatal	5.641	2.904	1.513	1.23	(1.23×10^7) = 12.33
Non-Fatal	543.64	278.56	139.258	125.1	(125.1×2.5^6) = 312.71
Total user safety savings					325

- Calculate the total fatal and non-fatal crashes for each section by simply multiplying the unit crash rate with the calculated VMT.
- Sum the crashes (both fatal and non-fatal separately) of all the constituent sections before intervention as well as after the intervention.
- Find the safety savings = crashes before intervention − crashes after intervention (Sinha and Labi, 2007).
- Assumptions for the evaluation of user safety benefits are as follows:

Unit crash cost in year 2015 (2015 Constant Rs.)
Per capita income of Pakistan = \$1250 (2015)
Fatal crash = 70 times GDP per capita
$\qquad$ = 70 × US\$ 1250 = US\$ 87,500 = Rs. 10,000,000
$\qquad$ (approx.)
Non-fatal crash = 18 times GDP per capita
$\qquad$ = 18 × US\$ 1250 = US\$ 22,500 = Rs. 2,500,000
$\qquad$ (approx.)

Hence, total user safety savings: 325 million [2015 constant Pakistan Rupees (PKR)].

3.3. *VOC savings*

It includes the direct expenses comprising the costs of vehicle ownership and vehicle operation. VOC savings of a transportation improvement or intervention simply refer to the reduction in VOCs compared to a base-case alternative (Sinha and Labi, 2007). For Rawalpindi bypass project, an overall monetized VOC saving of various alternative decisions, relative to DO Nothing alternative, is calculated as follows:

- Find vehicle mile travel per day (VMT/day) with and without intervention.
- Find unit VOC (dollars/vehicle-mile).
- Find the speed of the segment.
- If speed <50 mph, use formula = $(C+D)/S$.
- If speed >50 mph, use formula = $a_0 - a_1 S - a_2 S^2$ (further details can be read in Sinha and Labi (2007, Table 7.3, p. 163)).

- Find daily VOC by multiplying VMT/day with unit VOC.
- Calculate user VOC benefits of intervention and annual savings of VOC by adding all values obtained from daily VOC calculations and multiplying it with 365 (Tables 8–10).

> VOC user saving = [Existing route w/o Intervention] – [Existing route with intervention + Expressway/Bypass] = [131.77] – [68.03 + 59.59]
> VOC user saving = US$ 4.15 M (2005 Constant $)
> Dollar rate (2005): 1 Dollar = 59.7 PKR
> VOC saving (2005 PKR) = 4.15 × 59.7 = 247.75 M
> VOC saving (2015 PKR) = 247.75 × 233/100 = 577.27 M
> Total VOC saving: Rs. 577.27 Million (2015 Constant PKR)

4. Results

4.1. *Economic-efficiency analysis*

Economic-efficiency analysis of Rawalpindi Bypass is carried out based on net present value (NPV). The detailed cost and benefits (present worth) of the bypass are calculated by converting it to the Present/Base year, i.e. year 2015. Benefits of the project include TT benefits, user safety benefits, VOC benefits, and salvage value for pavements, bridges, interchanges, bypass and culverts.

4.2. *TT savings*

TT benefits for the 20-year period, i.e. 2015–2035 are shown in Table 11 and Fig. 4.

4.3. *User safety, VOC benefits, salvage value and costs of the project*

User safety benefits, VOC savings, salvage value and project cost for 20-year period (2015–2035) are shown in Tables 12–15 and Figs. 5–6.

Table 8.   VOC savings on existing route without bypass.

Seg.	Length (mile)	FF Speed (mph)	AADT	VMT		VOC [Using Hepburn's Model] (cents/ veh-mile)		Daily Cost ($)		Annual Cost (2005 Constant $)
				Car	Truck	Car	Truck	Car	Truck	
A	13.5	33	35,000	$(18,000 \times 13.5) =$ 243,000	$(17,000 \times 13.5) =$ 229,500	26.17	34.75	63,593	79,751	$361.015 \times 365 =$ 131,770,657
B	6.5	33	35,000	117,000	110,500	26.17	34.75	30,618	38,398	
C	14	33	35,000	252,000	238,000	26.17	34.75	65,948	82,705	
					Total					131.77 M

Table 9. VOC savings on existing route in presence of bypass.

Seg.	Length (mile)	FF Speed (mph)	AADT	VMT		VOC [Using Hepburn's Model] (cents/veh-mile)		Daily Cost ($)		Annual Cost (2005 Constant $)
				Car	Truck	Car	Truck	Car	Truck	
A	13.5	33	18,000	$(9{,}000 \times 13.5) =$ 121,500	$(9{,}000 \times 13.5) =$ 121,500	26.17	34.75	31,797	42,222	$186{,}388 \times 365 =$ 68,031,693
B	6.5	33	18,000	58,500	58,500	26.17	34.75	15,309	20,328	
C	14	33	18,000	126,000	126,000	26.17	34.75	32,947	43,785	
					Total					68.03 M

Table 10. VOC savings on proposed bypass.

Seg.	Length (mile)	FF Speed (mph)	AADT	VMT		VOC [Using Hepburn's Model] (cents/veh-mile)		Daily cost ($)		Annual Cost (2005 Constant $)
				Car	Truck	Car	Truck	Car	Truck	
D	20	60	17,000	(9,000 × 20) = 180,000	(8,000 × 20) = 160,000	25.85	33.71	46,530	53,936	163,257 × 365 = 59,588,896
E	12.5	60	17,000	112,500	10,000	25.85	33.71	29,081	33,710	
				Total						59.59 M

Table 11.   TT benefits.

Year	TT Savings (2015 Constant PKR)	SPPWF at 5% Factor	PWC at 5%
2015–2035	24627130460.7	1 – 0.3769	14933117752.7
		Rs. (Billions)	14.9331177527

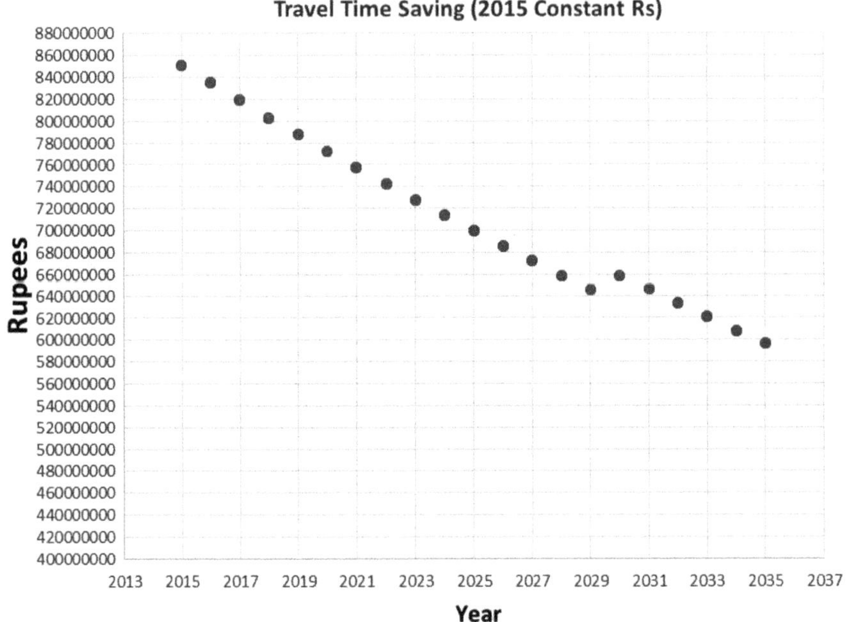

Fig. 4.   TT savings.

Table 12.   User safety benefits.

Year	Safety Saving Cost	SPPWF at 5% Factor	PWC at 5%
2015–2035	9,246,960,750	1 – 0.3769	5,629,205,478
		Rs. (Billion)	5.629205478

Table 13.   VOC savings.

Year	Safety Saving Cost	SPPWF at 5% Factor	PWC at 5%
2015–2035	13911896934.8	1 – 0.3769	8462479717
		Rs. (Billion)	8.462479717

Table 14.   Salvage values of pavement inventories.

Type	Service Life	Remaining Life	Construction Cost (Rs. Million)	Salvage Value (Rs. Million)
Pavement	35	15	7,169	3,072
Bridges	50	30	1,994	1,196
Interchange + overpass + curves	50	30	12,152	7,291
Total salvage value				11,560

Table 15.   Costs of the project.

Type	Incursion Year	Value
Construction cost	1 (start of project)	31,570
Operation and maintenance cost	Each year (annually)	28.6
Periodic maintenance cost	5,10,15 (every 5th year)	45.795
Overlay cost	10 (end of 10th year)	8,550

4.4. *Cash flow diagrams*

The annual benefits and costs related to the project under study are expressed as a cash flow diagram in Fig. 7.

4.5. *Net present value*

4.5.1. *Present worth of benefits and costs*

Present worth of benefits and present worth of cost are summarized in Tables 16 and 17. The NPV of proposed bypass is calculated by

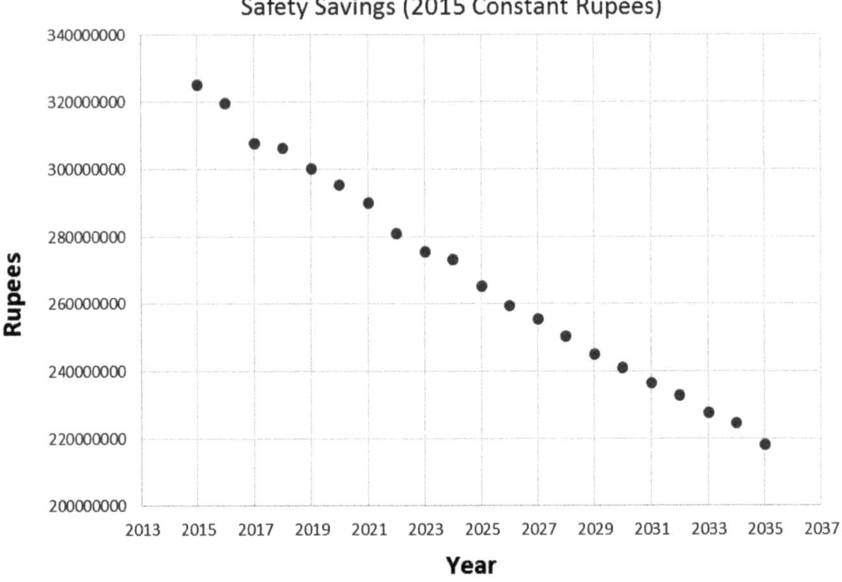

Fig. 5.   Safety savings.

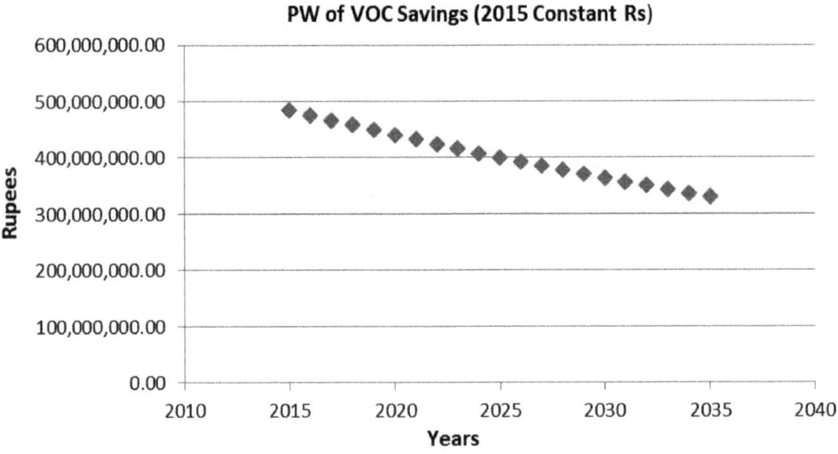

Fig. 6.   Present worth of VOC savings.

subtracting the present worth of cost (Rs. 37,233) from the present worth of benefit (Rs. 40,559). It shows that the construction of bypass is a viable option and economically cost effective. It has more benefits as compared to the cost associated with it.

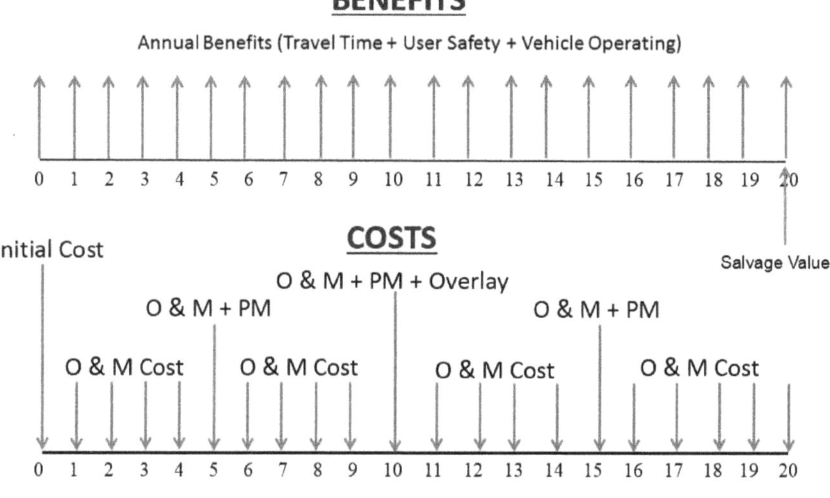

Fig. 7.   Cash flow diagram.

Table 16.   Present worth of benefits.

Type	Salvage Value (Rs. Million)	Factor	Present Worth Value (Rs. Million)
TT	14,908	1	14,908
User safety	5,629	1	5,629
VOC	8,462	1	8,462
Pavement	3,072	1	3,072
Bridges	1,196	1	1,196
Interchanges + overpasses + culverts	7,291	1	7,291
Total present worth of benefits			40,559

4.5.2. *Benefit cost ratio*

The benefit cost ratio (BCR) is a ratio of the equivalent uniform annual value (or NPV) of all benefits to that of all costs incurred over the analysis period.

$$\text{BCR} = \text{PW}_{\text{Benefits}} / \text{PW}_{\text{Costs}} = 40{,}560/37{,}233 = 1.09$$

Table 17.   Present worth of costs.

Type	Salvage Value (Rs. Million)	Value	Factor	Present Worth Value (Rs. Million)
Construction cost	1 (start of project)	31,570	1	31,570
Operation and maintenance cost	Each year (annually)	28.6	12.46 (USPWF)	357
Periodic maintenance cost	5, 10, 15 (every 5th year)	45.8	0.7835 (SPPWF 5%,5)	58
			0.6139 (SPPWF 5%,5)	
			0.4810 (SPPWF 5%,5)	
Overlay cost	10 (end of 10th year)	8,550	0.6139 (SPPWF 5%,5)	5249
	Total present worth of costs			37,233

As BCR is more than 1, it clearly shows that adopting the alternative, i.e. construction of Rawalpindi bypass, is a feasible option.

5. Discussion

The new millennium is characterized by continued growth in commercial and personal travel demand. The transportation agencies and service providers are continuously striving to keep their assets in acceptable conditions to offer desirable level of service in the most cost-effective manner and within available resources. This research produces an engineering and scientific comparison of various costs and benefits associated with the road agency and users about the construction of alternatives. Consistent with such efforts is the need for best-practices evaluation and monitoring of the expected impacts of alternative investment decisions, policies and other stimuli on the operations of existing or planned transportation systems and their environments. Such impacts involve economics (including quantified benefits and costs), economic development (job increase), environmental or ecological impacts, community impacts and technical impacts.

The study has carried out the economic-efficiency analysis and estimated the TDE of different segments of a national highway including project cost, operation and maintenance cost, TT savings, safety savings, VOC savings, using multi-criteria TDM process. Moreover, multi-criteria decision-making technique was used as it can simultaneously analyze multiple options with different dimensions including monetary and non-monetary benefits. However, the study has a limitation in that the traditional analysis techniques have not been included in its scope.

Besides sharing the twin city's heavy traffic load, the analysis revealed that the proposed bypass project will provide an enhanced speed of 60 mph as compared to 50 mph on the existing roads (do nothing scenario). Moreover, it will provide a signal-free corridor as compared to one traffic signal per mile. TT will also reduce from 61.75 min to 32.84 min. The bypass will not only optimize on VOC savings but will significantly reduce the number of fatal and non-fatal crashes per 100 Million VMT (see Sections 3.1–3.3). Overall, the benefits of the bypass project are more than its construction, and operation and maintenance cost of thus achieving a B/C ratio of more than 1 (*B/C* = 1.09). Hence, the feasibility of bypass project is confirmed in terms of all the stakeholders, including road user and agency.

6. Conclusion

Economic-efficiency analysis is a key criterion for evaluation of most engineering problems. It is a decision-making tool that assesses the efficiency of investments from a monetary standpoint and incorporates the monetized costs and benefits associated with alternative decisions and actions. It is useful for evaluating systems on the basis of one or more performance measures that are monetizeable. A comprehensive decision-making model is presented in this study to compare the costs and benefits of a transportation project, i.e. Rawalpindi bypass. An existing roadway structure with no improvements was considered as a DO-NOTHING scenario, whereas the construction of a bypass was considered as an alternative scenario. The economic-efficiency analysis was carried out comprised of TDE, project cost, and measurement of various savings such as TT, safety and VOC savings. The study concludes that the proposed bypass project is an economically feasible project. The systematic decision-making model presented in this study can aid the engineers and planners to efficiently plan their highway projects. Besides, its comprehensive methodology can be replicated to economically analyze any transportation project.

References

Ayub, B., Ullah, F., Rasheed, F. and Sepasgozar, S. M. (2016a). "Risks in EPC Hydropower Projects: A Case of Pakistan", in *Proceedings of the 8th International Civil Engineering Congress (ICEC-2016) "Ensuring Technological Advancement through Innovation Based Knowledge Corridor"*, December 23–24, Karachi, Pakistan.

Ayub, B., Thaheem, M. J. and Ud Din, Z. (2016b). "Dynamic Management of Cost Contingency: Impact of KPIs and Risk Perception", *Procedia Engineering*, 145, 82–87.

Hasnain, M., Thaheem, M. J. and Ullah, F. (2017). "Best Value contractor selection in road construction Projects: ANP-based Decision Support System", *International Journal of Civil Engineering*, 16(6), 695–714.

Sinha, K. C. and Labi, S. (2007). *Transportation Decision Making: Principles of Project Evaluation And Programming*, Chapter 8, Economic Efficiency Impacts, Hoboken: John Wiley and Sons, pp. 197–228.

Ullah, F., Thaheem, M. J. and Sepasgozar, S. M. E. (2016a). "Sustainable Smart Cities: Evaluation of Australian practice", in *Proceedings of the 16th International Conference on Construction Applications of Virtual Reality CONVR 2016*, December 11–13, Hong Kong.

Ullah, F., Ayub, B., Siddiqui, S. Q. and Thaheem, M. J. (2016b). "A Review Of Public–private Partnership: Critical Factors of Concession Period", *Journal of Financial Management of Property and Construction*, 21(3), 269–300.

Ullah, F. and Thaheem, M. J. (2017). "Concession Period of Public Private Partnership Projects: Industry —Academia Gap Analysis", *International Journal of Construction Management*, 18(5), 418–429.

Ullah, F., Thaheem, M. J., Siddiqui, S. Q. and Khurshid, M. B. (2017a). "Influence of Six Sigma on Project Success in Construction Industry of Pakistan", *The TQM Journal*, 29, 276–309.

Ullah, F., Thaheem, M. J. and Umar, M. (2017b). "Chapter Six Public-private Partnerships in Pakistan: A Nascent Evolution", *Public–Private Partnerships in Transitional Nations: Policy, Governance and Praxis,* N Mouraviev, N Kakabadse, ed., Vol. 1, 127–150.

Utama, W. P., Chan, A. P., Gao, R. and Zahoor, H. (2017). "Making International Expansion Decision for Construction Enterprises with Multiple Criteria: A Literature Review Approach", *International Journal of Construction Management*, 1–11.

Winston, C. (1991). "Efficient Transportation Infrastructure Policy", *The Journal of Economic Perspectives*, 5(1), 113–127.

Zahoor, H. and Choudhry, R. M. (2012). "The Most Neglected Construction Safety Practices in Rawalpindi/Islamabad", in *Proceedings of the CIB W099 International Conference on Modeling and Building Health and Safety*, September 10–11, Singapore, pp. 312–322, http://goo.gl/bpwghH.

Chapter 31

BIM-based Operation and Maintenance Management System for Smart Industrial Parks — A Case Study in China

Xiaoyan Jiang*, Kang Hu and Xinhua Zheng

*Department of Civil Engineering,
Hefei University of Technology,
Hefei, Anhui 23009, China
jiangxiaoyan@hfut.edu.cn

Yong Liu

*Department of Engineering Management,
Zhejiang Sci-Tech University, Hangzhou,
Zhejiang 310018, China*

Bo Xia and Martin Skitmore

*School of Civil Engineering and Built Environment,
Queensland University of Technology (QUT),
Gardens Point, Brisbane, QLD 4001, Australia*

---

*Corresponding author.

Abstract

As the subsystem of a Smart City (SC), Smart Industrial Parks (SIP) not only improve industry-clustering ability and enhance economic strength but also facilitate social sustainable development. Although SIP construction has gained popularity globally, there are still various problems and flaws in their development and implementation. This chapter presents a case study of the Shanghai Qiantan SIP and proposes a framework and implementation plan for an operation and maintenance (O&M) management system based on building information modelling (BIM). This comprises six layers (i.e. portal layer, application layer, platform layer, data layer, model layer and facility layer) to attain SIP requirements and functions. The steps in O&M system development comprise demand research, plan implementation and BIM model building. The case study helps to develop a strategic view of SIP technology, particularly on how to expand the case database for the reference of future research.

Keywords: Smart City; Smart Industrial Parks; Shanghai Qiantan; Operation and Maintenance; BIM.

1. Introduction

Since IBM advanced the concept of wisdom Earth in late 2008 (Harrison and Donnelly, 2011), that is, a more instrumented, interconnected and intelligent world, Smart City (SC) has been integrating information and communication technology (ICT) and Internet of things (IoT) technology in a secure fashion to manage the assets and resources of cities, economies and people from the perspective of urban development. These assets include infrastructures, transportation systems, schools, libraries, hospitals, buildings and various industrial facilities (Paroutis *et al.*, 2014; Cocchia, 2014). In China, the concept of an SC, which aims to enhance the life quality of citizens, has been gaining increasing attention in the agendas of policymakers, but there are not yet any comprehensive SCs even at the pilot level. On the contrary, many standalone projects have been thriving based on a few core applications, such as smart transportation, smart government management platforms, smart communities and especially Smart Industrial Parks (SIP) (Kim and Wang, 2014). The relationship between SIP and the SC is shown in Fig. 1 (adapted from Angelidou, 2015).

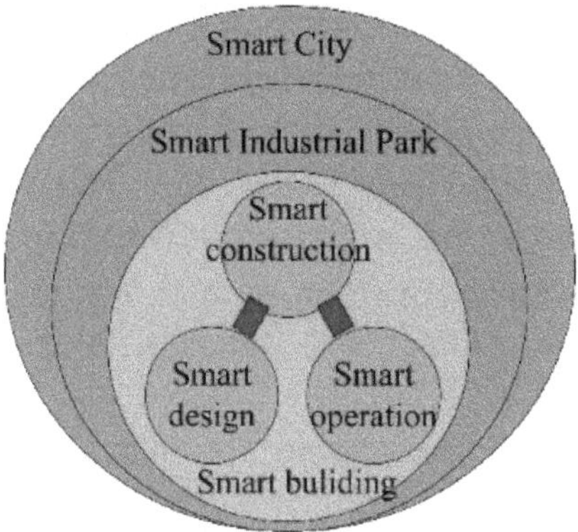

Fig. 1.   SC and SIP.

SIP is the miniature and subsystem of the SC and an important platform for a city that not only improves its industry clustering ability and enhances economic strength but also facilitates social sustainable development. As an advanced industrial park paradigm, the SIP is a combination of new information and communication technology and the sustainable development idea that can provide rapid information collection, high-speed information transmission, strong centralized computing, intelligence transaction processing and ubiquitous service ability. The integration of building information modelling (BIM), big data, cloud calculating, IoT and industrial park provides integrated resources and shares the information of the whole industrial park, which will attract more enterprises and provide a more liveable environment (Neirotti, *et al.*, 2014; Zanella *et al.*, 2014). Many applications have shown that SIP construction centring on information services promotes traditional industrial park agglomeration and improves social sustainability. It is becoming a popular developing trend across the world. The 2017 China Smart Park Innovation Forum Annual Meeting reported that over 15,000 industrial parks were established in China by 2015 and they have accounted for 30% of China's GDP. A quarter of the national industrial parks have launched SIP to date.

However, there are still many problems and flaws in the process of implementation and operation of SIP. While great emphasis has been placed on the applicability of information and communication technologies

(ICT) to SIP projects, there has been little analysis or evaluation of such planning processes of smart projects. In addition, the development of SIP has encountered various problems, such as repetitive construction, lack of intelligence and people-oriented services, low interaction with industry, and incomplete operation and maintenance (O&M) management. This is mainly because SIP employs a rather unclear definition and work scope (Kim and Wang, 2017; Gomez and Paradells, 2015), especially for its O&M management.

In view of the above background, it is necessary to introduce SIP, especially its information system for operations and maintenance management. Based on the Shanghai Qiantan SIP case in China, this chapter provides a framework for SIP in general and an O&M management system in particular based on BIM. The findings from the case study of Qiantan SIP help reflect the implementation status and prospect of SIP in China.

2. Design of the Qiantan Smart Industrial Park's O&M System

2.1. *Brief description of the Qiantan SIP*

The Shanghai Qiantan SIP is located in the western part of the Huangpu River and Xuhui District Riverside Zone, spanning an area over 2.83 km^2 and including a sports centre, schools and business districts with associated municipal infrastructure facilities. The SIP aims to be an international business district and new industrial support for Shanghai through the settlement of the business, cultural media, sports leisure and e-commerce industries. Qiantan SIP is expected to become a centre for the regional headquarters of multinational corporations and domestic enterprises. An Aerial View of Qiantan SIP is shown in Fig. 2.

2.2. *Demand analysis of Qiantan SIP system*

A demand analysis of the Qiantan SIP system needs to be completed before the general framework is designed. The SIP O&M management system is based on the improvement and extension of the traditional building O&M system. The traditional O&M system has many

Fig. 2.   Shanghai Qiantan SIP.

advantages, such as good applicability and mature technology, but it is difficult to provide community management and government management information and connect to cloud services and GIS data with the BIM model. According to the advantages and disadvantages of the traditional O&M system, the demand of Qiantan SIP is as follows:

(1) Achieving a comprehensive management function involving buildings, infrastructure, environment, community, space, facility, monitoring, energy consumption, parking, security, emergency, and O&M management. Due to information complexity and multi-management function, the solution needs to be designed according to the actual situation and management requirements.
(2) Adopting standard information coding, a unified data interface and network communication protocol.
(3) Supporting BIM model importation and transformation.
(4) Keeping O&M information updated dynamically.

2.3. *General framework of the Qiantan SIP O&M system*

In view of the above demands, the general framework (as shown in Fig. 3) of the Qiantan SIP O&M management system consists of six-layer components, including a portal layer, application layer, platform layer, data layer, model layer and facility layer.

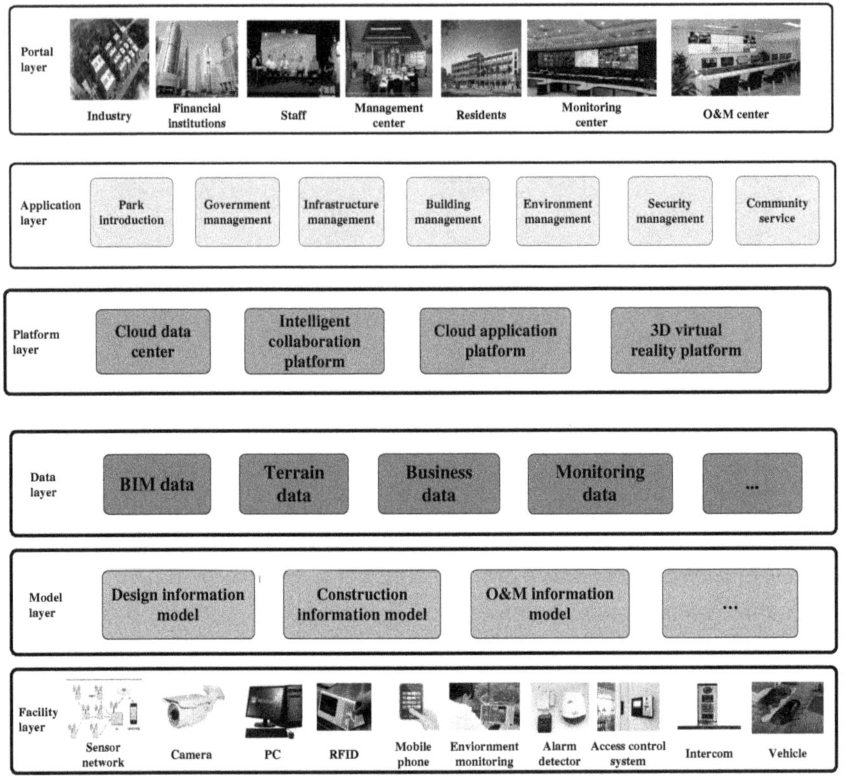

Fig. 3.   General framework of the Smart Qiantan O&M management system.

(1) **Portal layer:** The portal layer is the top layer of the framework. Its role is to set the permissions for the different types and different grades of configured users reading relevant information within the scope of authorization.

(2) **Facility layer:** The facility layer is a basic layer at the bottom of the framework, which acquires various data from different kinds of terminal devices.

(3) **Data layer:** The data layer is used to store and modify the data required for the application layer, including BIM data, terrain data, business data and monitoring data. According to the disparate sources and different types, the data from terminal devices enter into the data layer to be classified, among which the BIM data are entered into the model layer.

(4) **Model layer:** The model layer, which connects the data layer and application layer, is the core of the O&M management system. It includes powerful editing tools for importing, exporting and making large-scale or bulk changes to the model.

(5) **Application layer:** The application layer works directly for users and terminal equipment, and provides specific intelligent applications and services for different types of SIP users. It is responsible for data disposal and analysis to accomplish such functions as park introduction and the management functions of government management, infrastructure management, security management, environmental management and community management.

(6) **Platform layer:** The platform layer is a comprehensive information management service platform and illustrates the data in various ways by human–machine interactive interface, including a cloud data centre, intelligent collaboration platform, and cloud application and 3D virtual reality platform, by which users can realize such various management functions of the application layer as information transmission, real-time running data feedback and human–computer interaction.

The Qiantan SIP O&M system has many characteristics, such as clearness in design, modularity and ease of expansion.

3. Implementation Plan of the SIP O&M System-based BIM

3.1. *Development platform*

According to the current practices of China's smart projects, the implementation method of the O&M management system based on BIM can be summarized into four categories as follows:

(1) a simple O&M management system based on as-built BIM model;
(2) a secondary development O&M management system based on a mature platform such as Autodesk Design Review, Navisworks or Revit;

(3) using O&M management software (Archibus, etc.) to analyze BIM data;
(4) developing the O&M management system based on BIM in accordance with project characteristics.

When existing methods are incapable of meeting the specific requirements of project O&M management, a new O&M management system should be developed. However, it is necessary to convert BIM data into IFC format in order to make it compatible with the new development platform. The BIM model should be modified before inserting it into the new platform due to the loss of data during the translation process.

Considering actual demand, Qiantan SIP uses *Archibus* software and the *CE* engine as the system operation platform by combining with BIM technology, FM technology, Internet, GIS, cloud technology, Internet of Things, and other advanced technologies to build a smart O&M management system (Shou *et al.*, 2014; Jin *et al.*, 2014; Wenge *et al.*, 2014).

3.2. *Implementation plan*

During its design and construction stages, the O&M sector should firstly focus on requirement analysis, as shown in Section 2.2, then start data collection and processing to build the BIM model, finally develop the O&M system, and improve and evaluate system function. After the completion of this work, real-time O&M dynamic data that are connected to the finished system and the O&M management system start running. The implementation plan is shown in Fig. 4.

3.3. *Formation of O&M BIM*

The formation of BIM from the design model to the as-built model is a complex process. The initial model derived from the design drawings should not be connected to the O&M system. The BIM design will be converted to construction BIM via data transformation and adding project parameters. The serviced formation will be stored once the construction BIM has been deepened by *Tekla* and other software through integrating purchasing information and construction information, and finally, the as-built BIM will be formed.

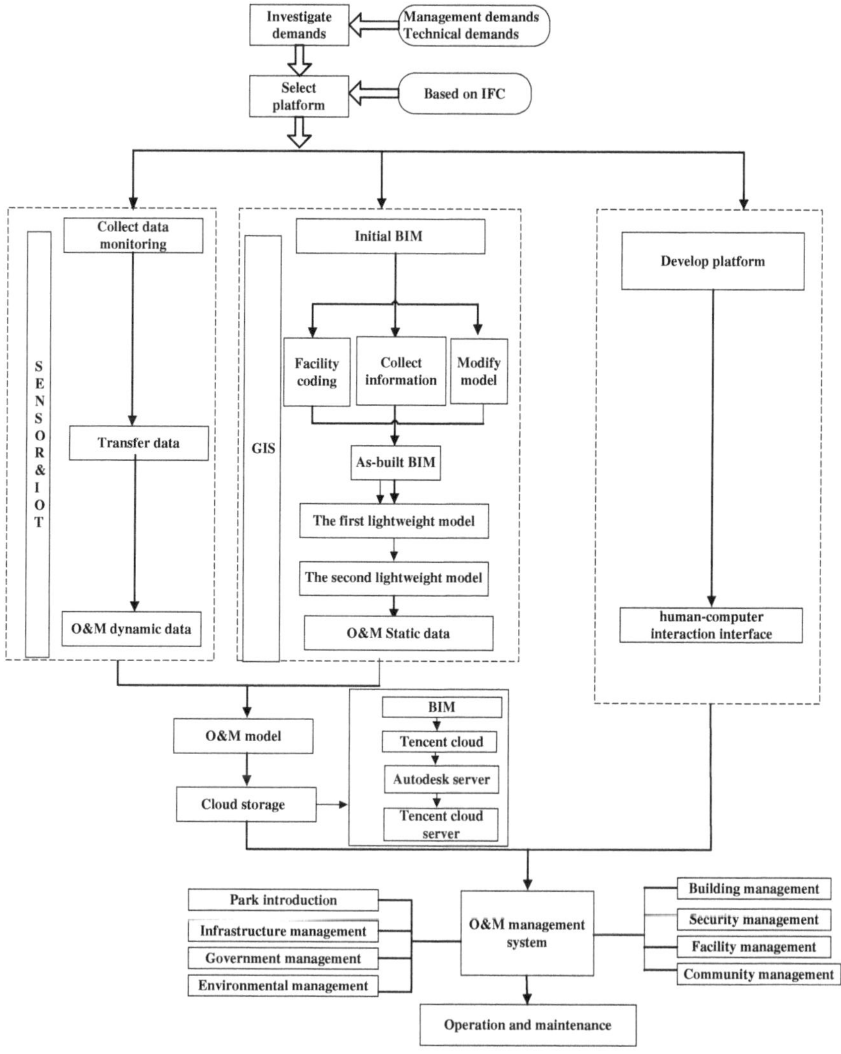

Fig. 4.   Implementation plan of the Smart Qiantan O&M management system.

However, separate parameter information from the component entities of BIM that needs to be integrated with them is very important and a lightweight BIM is in demand as the as-built BIM is too big to use in practice (Kang and Hong, 2014). Considering the redundancy, semantics, and the parameterization of BIM data under the limited resources of network bandwidth and web browsers, lightweight solution for real-time

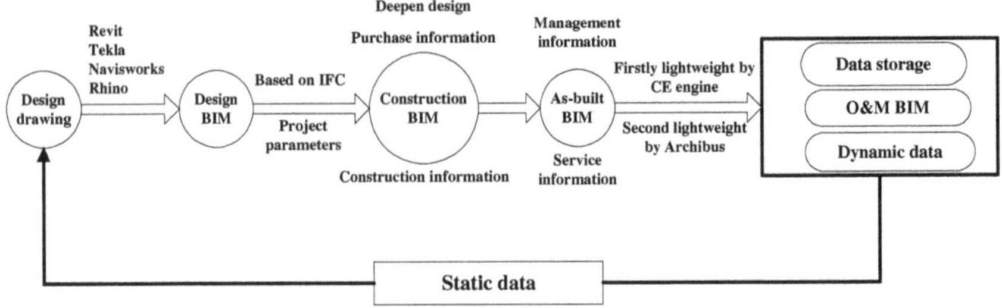

Fig. 5.   Formation of the O&M BIM.

visualization of large-scale BIM scenes is imperative (Liu *et al.*, 2016). Firstly, the information will be separated by *Revit*, then the first light-weighting will be conducted using the CE engine and the second light-weighting by *Archibus* software. The O&M BIM will then be shaped. The lightweight BIM will then be built up by connecting O&M BIM and dynamic data, and storing information to *Tencent* cloud to realize real-time updating as shown in Fig. 5.

4. Conclusions

SIP plays an important role in the SC and subsystem. This chapter introduces the relationship between the SC and SIP and proposes a framework and implementation plan for an SIP O&M management system. The framework comprises six layers (i.e. portal layer, application layer, platform layer, data layer, model layer and facility layer) to attain SIP requirements and functions. The steps in designing the O&M system comprise demand research, plan implementation and building the BIM model. The core task of the Qiantan SIP O&M management system is to separate parameter information from the component entities of BIM and integrate with them. The main reason for the separation is that an enormous amount of as-built BIM data is hard to put in use. By repeat testing, Smart Qiantan adopted the *CE* engine to meet the requirements of lightweight BIM and employed *Archibus* to obtain a good human–computer interaction. Building a BIM model is not a one-time effort but a complex multistep process including the design BIM, construction BIM, O&M BIM and model modification, and lightweight BIM.

The case of Shanghai Qiantan SIP O&M system provides a sample for similar smart projects in future. This chapter also identifies the current difficulty with BIM modelling, and more lightweight problems will be further explored in future studies.

References

Angelidou, M. (2015). "Smart Cities: A Conjuncture of Four Forces", *Cities*, 47, 95–106.

Cocchia, A. (2014). "Smart and Digital City: A Systematic Literature Review", *Smart City*, Springer International Publishing, pp. 13–43.

Gomez, C. and Paradells, J. (2015). "Urban Automation Networks: Current and Emerging Solutions for Sensed Data Collection and Actuation in Smart Cities", *Sensors*, 15(9), 22874–22898.

Harrison, C. and Donnelly, I. A. (2011). "A Theory of Smart Cities', in *Proceedings of the 55th Annual Meeting of the ISSS-2011*, Vol. 55, No. 1, Hull, UK.

Jin, J., Gubbi, J., Marusic, S., and Palaniswami, M. (2014). "An Information Framework for Creating a Smart City Through Internet of Things", *IEEE Internet of Things Journal*, 1(2), 112–121.

Kang, T. W. and Hong, C. H. (2014). "Development of Lightweight BIM Shape Format Structure to Represent Large Volume Geometry Objects Using GIS with Facility Management", in *ISARC. Proceedings of the International Symposium on Automation and Robotics in Construction*, Vol. 31, Vilnius Gediminas Technical University, Department of Construction Economics & Property, p. 1.

Kim, J. S. and Wang, X. (2017). "Smart City Strategies towards Sustainable Transformation of China's Industrial Parks", Sustainable Cities in Asia., 1st. London: Routledge. pp. 248–260.

Kim, J. S. and Wang, X. (2014). "Rethinking the Strategic Dimensions of Smart Cities in China's Industrial Park Developments: the Experience of Suzhou Industrial Park, Suzhou, China. In REAL CORP 2014–PLAN IT SMART! Clever Solutions for Smart Cities", in *Proceedings of 19th International Conference on Urban Planning, Regional Development and Information Society*, CORP–Competence Center of Urban and Regional Planning, pp. 487–496.

Liu, X., Xie, N., Tang, K. and Jia, J. (2016). "Lightweighting for Web3d Visualization of Large-Scale Bim Scenes in Real-Time", *Graphical Models*, 88, 40–56.

Neirotti, P., De Marco, A., Cagliano, A. C., Mangano, G. and Scorrano, F. (2014). "Current Trends in Smart City Initiatives: Some Stylised Facts", *Cities*, 38, 25–36.

Paroutis, S., Bennett, M. and Heracleous, L. (2014). "A Strategic View on Smart City Technology: The Case of IBM Smarter Cities During a Recession", *Technological Forecasting and Social Change*, 89, 262–272.

Shou, W., Wang, X., Wang, J., Hou, L. and Truijens, M. (2014). "Integration of BIM and Lean Concepts to Improve Maintenance Efficiency: A case study", *Computing in Civil and Building Engineering*, 373–380.

Wenge, R., Zhang, X., Dave, C., Chao, L. and Hao, S. (2014). "Smart City Architecture: A Technology Guide for Implementation and Design Challenges", *China Communications*, 11(3), 56–69.

Zanella, A., Bui, N., Castellani, A., Vangelista, L. and Zorzi, M. (2014). "Internet of Things for Smart Cities", *IEEE Internet of Things journal*, 1(1), 22–32.

Chapter 32

Study on Subcontract Decision-making of International Construction Enterprises Considering Spillover Effect

Jianbo Zhu, Qianqian Shi* and Zhaohan Sheng

*School of Management and Engineering,
Nanjing University, Nanjing 210093, China*
qqs1991@126.com

Peng Wu

*School of Design and the Built Environment,
Curtin University, Perth, Australia*

Abstract

In the process of internationalization of construction enterprises, the international enterprises need to consider whether part of the contract should be subcontracted to the local enterprises to gain a competitive advantage in the international contractor market; thus, forming a decentralized project subcontracting structure. Assuming that the local

*Corresponding author.

subcontracting market has a monopoly characteristic, the owner has a quality perception difference for the products provided by different enterprises, and the spillover effect will occur in the cooperation of subcontracting. The Hotelling model is used to study the impact of the subcontracting strategy on international enterprises and the whole market. The results show that local enterprises choose to subcontract to reduce the cost, and can always improve their profit level, intensify market competition, squeeze the market share of international contractors and reduce their profits; international enterprises in the local market need to balance the cost reduction caused by the subcontracting and the negative effect of the quality level brought by the counterparty through subcontracting, the differentiated competition equilibrium results appear in this model.

Keywords: Spillover Effect; Subcontract Decision-making; Hotelling Model; Internationalization.

1. Introduction

In recent years, China has made great achievements in infrastructure construction. The construction capacity of China in the fields of large-span bridge, underground rail transit and high-speed railway have represented the international advanced standards. The large number of infrastructure construction projects in China has played an important role in cultivating domestic construction contractors. After gaining the ability to compete in the international market, numerous domestic contractors have gone abroad to compete in the international market.

The localization of production and operation elements is an important mode of overseas investment for those contractors with the purpose of better adapting to the fast-changing political, economic and social environment in the international market. By outsourcing some of their projects or production processes to the enterprise partners in the host country, they can effectively realize the asset-light operation, reduce the operational risk and maintain the core competitive advantage. This chapter has been devoted to this topic. When contracting international projects, international contractors often choose the outsourcing strategy so as to reduce the risk and gain the cost advantage. However, outsourcing will necessarily lead to the spillover effect of technology and knowledge, thus weakening the competitive advantage of the contractor. As a result, many international contractors are stuck in a dilemma.

Although outsourcing can bring rich external resources to the contractors, it will also pose many risks to the enterprises. The contract-issuing party should share the knowledge to assist the contract-receiving party to solve relevant problems (Zhao *et al.*, 2004). The contract-receiving party is also faced with technology theft and spillover risk (Das *et al.*, 1998, 2001). With the weakening and leakage of core knowledge, the enterprise will also gradually lose its core competitiveness (Grimpe *et al.*, 2010). It is doubtful whether the resources of those enterprises gathering together can be effectively integrated (Weigelt *et al.*, 2009). There is a typical principal–agent relationship between contract-issuing party and contract-receiving party. Both of them have a large quantity of confidential information. From the angle of unilateral moral risk, Nakamura proposed that the relevant system should be improved to prevent the moral risk (Nakamura, 2009). Marceau suggested controlling the moral risk through governmental subsidy and incentive (Marceau, 2002). Despite reducing the cost, outsourcing will also involve the potential transaction cost; thus, weakening the cooperative motives of both the parties (Barthélemy *et al.*, 2010).

At the early stage, outsourcing can have a positive effect on the enterprise. The following spillover risk will negatively impact the cooperation. As noted by Dussauge, the participants in the cooperation could not control the input boundary of soft resources well (Dussauge *et al.*, 2000). Hsuan and Mahnke also pointed out that the input into the outsourcing research and development would obviously increase the explicit cost. The loss of implicit knowledge spillover may also occur (Hsuan *et al.*, 2011). Some scholars have made an in-depth research into the spillover effect in the enterprise cooperation from the perspective of mathematical modelling. D'Aspremont and Jacquemin built a classical AJ model. According to their research results, although a manufacturer could reduce its own cost by investing in the research and development, it would also help to reduce the cost of its competitors (D'Aspremont *et al.*, 1988). The effect of investment activities on the marginal cost of the project was also further analyzed. Based on the Cournot duopoly model, Kamien built a two-stage game model for spillover effect, namely KMZ model (Kamien *et al.*, 1992). The influence of the spillover effect on the input, output and profit of enterprise members was further analyzed.

Under the background of the China-proposed Belt and Road Initiative, there is a great difference in the political environment and construction technologies for the contracting of international projects between different countries in Asia, Africa, Europe and America. This project will introduce the game model to the competitive decisions about international project contracting. The price competition and outsourcing decision of those contractors dealing with the market competition will also be analyzed. For example, Hotelling model is a classical model for the research of differentiated product competition. Since it was proposed, numerous scholars have made some expansions in different forms, such as traffic expense function (Economide *et al.*, 1986; Neven *et al.*, 1986) and consumers' preference distribution (Tabuchi *et al.*, 1995; Harter *et al.*, 1997).

2. Basic Assumption

It is assumed that the service outsourcing decision of the contractors is based on Hotelling model. This model has the feature of being to the point and concise when it is used for the research into market positioning and pricing. It can help us focus on the research topic.

It is also assumed that the preference of the owner is linearly and evenly distributed on the Hotelling line with a length of 1. The reservation unity u of the owner should be large enough to ensure that the contractors will still have the interest in purchasing the services under different strategies. When the owner with a preference which is set as x purchases the service contract from the contractor, he will gain the utility of $u + \theta q_i - p_i - t(x - x_i)^2$. θq_i is the utility perception brought by the services with different qualities. $-t(x - x_i)^2$ is the walking cost. For the ease of forming the equilibrium, it is expressed in the form of quadratic function. p_i is the price of the service contract.

If it is assumed that there are two contractors competing with each other in the market, the pricing of the contractor lies in x_i of the Hotelling line. It can be assumed that enterprise 1 is on the left side of enterprise 2, which means $x_1 < x_2$. It is further assumed that the cost of the two contractors is both set as c.

It is also assumed that the two competitive contractors can cooperate with each other to subcontract the services of the upstream enterprise. Once the cooperation happens, the extensive exchange of knowledge and technologies will also happen between the contractors and upstream enterprise. It will lead to a certain degree of spillover effect, thus reducing the production lost. In other words, the production cost of enterprise i will be $\dfrac{c}{1+r_i}$. r_i is the coefficient of spillover effect.

3. Both Contractors Adopt the Centralized Structure

First of all, we assume that both of the contractors have adopted the centralized structure. In this case, all of the production services will take the form of self-provision rather than outsourcing. In order to research the influence of decentralized outsourcing structure and spillover effect on the contractors, we have made the comparative analysis with the benchmark of centralized production structure. The changing profit of the contractor using different strategies has also been compared.

The contractor has the autonomy for pricing the services. According to the price of different contractors, the owner can maximize his own residual value to make the decision. In other words, there is a point x^* on the Hotelling line. It can satisfy the condition that the residual value of the two contractors should be the same. The formula is $u + \theta q_1 - p_1 - t(x^* - x_1)^2 = u + \theta q_2 - p_2 - t(x^* - x_2)^2$. The owner with a preference on the left side of the point can purchase the service from the contractor 1. The owner with a preference on the right side of the point can purchase the service from the contractor 2. The formula $x^* = \dfrac{p_2 - p_1 - \theta \Delta q}{2t(x_2 - x_1)} + \dfrac{x_2 + x_1}{2}$ can be gained easily. For ease of calculation, the difference in quality between the two contractors is set as Δq. At this point, the profit of the contractors is as follows:

$$E_1 = (p_1 - c)x^*,$$
$$E_2 = (p_2 - c)(1 - x^*).$$

The indifference position of the residual value of the owner is

$$x^* = \frac{x_2 + x_1 + 2}{6} - \frac{\theta \Delta q}{6t(x_2 - x_1)}.$$

The prices of contractors 1 and 2 are set as

$$p_1 = \frac{t(x_2{}^2 - x_1{}^2) + 2t(x_2 - x_1) + 3c - \theta \Delta q}{3},$$

$$p_2 = \frac{-t(x_2{}^2 - x_1{}^2) + 4t(x_2 - x_1) + 3c + \theta \Delta q}{3}.$$

Given the price and market share above, the profits of contractors 1 and 2 are

$$E_1^1 = \frac{\left[t(x_2^2 - x_1^2) + 2t(x_2 - x_1) - \theta \Delta q \right]^2}{18t(x_2 - x_1)},$$

$$E_2^1 = \frac{\left[-t(x_2^2 - x_1^2) + 4t(x_2 - x_1) + \theta \Delta q \right]^2}{18t(x_2 - x_1)}.$$

4. One Contractor Adopts the Decentralized Structure

This part of the chapter is concerned with the case where one contractor uses the centralized structure and the other one uses the decentralized structure. In this case, the profit function of the contractor using the centralized structure remains unchanged. The contractor choosing the decentralized structure (or rather service outsourcing) will enjoy the spillover effect brought by the cooperation with the third party. The third party is the third enterprise providing professional subcontracting services. It can provide solutions for enterprise customers rather than directly participate in the market competition.

In the case of enterprise 1 choosing the service outsourcing, the indifference position of the residual value of the owner is similar to that with centralized structure. The formula is still $x^* = \dfrac{p_2 - p_1 - \theta \Delta q}{2t(x_2 - x_1)} + \dfrac{x_2 + x_1}{2}.$

But the profit of enterprise 1 will benefit from the spillover effect. To be more specific, the cost will be reduced to a certain extent. The extent of cost reduction is determined by the cooperative depth between the two enterprises. At this point, the profit of the enterprise with decentralized structure is expressed as the formula $E_1 = \left(p_1 - \dfrac{c}{1+r_1} \right) x^*$. The profit of the enterprise with centralized structure is still expressed as the formula $E_2 = (p_2 - c)(1 - x^*)$.

Proposition 1. *With the given market positioning and cost spillover effect, enterprise 1 chooses service outsourcing. Enterprise 2 chooses the centralized structure. The game equilibrium between contractors 1 and 2 is as follows*:

The prices of contractors 1 and 2 are

$$p_1 = \frac{t(x_2{}^2 - x_1{}^2) + 2t(x_2 - x_1) - \theta \Delta q + c + \dfrac{2c}{1+r_1}}{3},$$

$$p_2 = \frac{-t(x_2{}^2 - x_1{}^2) + 4t(x_2 - x_1) + \theta \Delta q + 2c + \dfrac{c}{1+r_1}}{3}.$$

The indifference position of the owner is

$$x^* = \frac{x_2 + x_1 + 2}{6} - \frac{\theta \Delta q - c + \dfrac{c}{1+r_1}}{6t(x_2 - x_1)}.$$

The profits of contractors 1 and 2 are

$$E_1^2 = \frac{\left[t(x_2{}^2 - x_1{}^2) + 2t(x_2 - x_1) - \theta \Delta q + \dfrac{cr_1}{1+r_1} \right]^2}{18t(x_2 - x_1)},$$

$$E_2^2 = \frac{\left[-t(x_2{}^2 - x_1{}^2) + 4t(x_2 - x_1) + \theta \Delta q - \dfrac{cr_1}{1+r_1} \right]^2}{18t(x_2 - x_1)}.$$

Proposition 2. *When contractor 1 chooses service outsourcing and contractor 2 chooses the centralized strategy, the profit of contractor 1 will increase and the profit of contractor 2 will decrease. The profit in the market will improve as a whole.*

Judging from the price changes of the two contractors, contractor 1 has gained some cost advantage through the spillover effect during the process of outsourcing. During the competition, contractor 1 can transform cost advantage into a lower price; thus, increasing the market share and occupying the market space of contractor 2. At this point, the two contractors have adopted the centralized structure. The price of enterprise 1 has been decreased by $\Delta p_{11} = \dfrac{2r_1 c}{1+r_1}$. The price of enterprise 2 has been reduced by $\Delta p_{21} = \dfrac{r_1 c}{1+r_1}$. The indifference position of the owner has moved to the right side: $\Delta x^*_1 = \dfrac{r_1 c}{6t(x_2 - x_1)(1+r_1)}$.

The profit of contractor 1 increases by

$$\Delta E_1 = \frac{\left[2t(x_2^2 - x_1^2) + 4t(x_2 - x_1) - 2\theta \Delta q + \dfrac{r_1 c}{1+r_1} \right] \dfrac{r_1 c}{1+r_1}}{18t(x_2 - x_1)}.$$

The profit of contractor 2 decreases by

$$\Delta E_2 = \frac{\left[-2t(x_2^2 - x_1^2) + 8t(x_2 - x_1) + 2\theta \Delta q - \dfrac{r_1 c}{1+r_1} \right] \dfrac{r_1 c}{1+r_1}}{18t(x_2 - x_1)}.$$

The increased value of the profit of contractor 1 is higher than the decreased value of the profit of contractor 2, which can be seen easily from $\left| \Delta E_1 \right| - \left| \Delta E_2 \right| > 0$. It means that the market has benefited from the outsourcing strategy of contractor 1. The profit in the market has increased as a whole. As long as the spillover effect exists, it must be profitable for contractor 1 who is sure to adopt the outsourcing strategy. After being left at a disadvantage in the competition, contractor 2 has to follow the strategy of lowering the price to keep the market share from being occupied by contractor 1.

In the other case where enterprise 2 adopts the outsourcing strategy, it is similar to the outcome of the first case. To be more specific, contractor 1 will be under competitive pressure from contractor 2 to improve the cost. Both of them will lower the price. Compared with the centralized structure, the market share, which is originally held by contractor 1 on the Hotelling line, will move to contractor 2. The profit change of the two contractors is

$$E_1^3 = \frac{\left[t(x_2^2 - x_1^2) + 2t(x_2 - x_1) - \theta \Delta q - c + \dfrac{c}{1 + r_2} \right]^2}{18t(x_2 - x_1)},$$

$$E_2^3 = \frac{\left[-t(x_2^2 - x_1^2) + 4t(x_2 - x_1) + \theta \Delta q + c - \dfrac{c}{1 + r_2} \right]^2}{18t(x_2 - x_1)}.$$

5. Both Contractors Adopt the Outsourcing Strategy

In this part, we will further consider the case where both contractors adopt the decentralized outsourcing strategy. In this case, the two contractors will cooperate with the same professional subcontracting enterprise. As a matter of fact, it is a common thing. For example, although there are many companies capable of bridge construction in the domestic systems of railway and road, the number of those excellent manufacturers capable of producing steel box girder or stayed cable is still limited. By means of professional subcontracting, the spillover effect brought by the vertical cooperation of the two contractors can be reflected in the decrease of the cost. For the sake of simplification, it is assumed that the coefficient of spillover effect is consistent with the vertical cooperation of the single enterprise.

In this case, because of the improved service of the contractor with a lower quality, the indifference position of the residual value of the owner is changed into $x^* = \dfrac{p_2 - p_1 - \theta \Delta q}{2t(x_2 - x_1)} + \dfrac{x_2 + x_1}{2}$. The two contractors will benefit from the cost spillover effect. The profit of enterprise 1 is $E_1 = \left(p_1 - \dfrac{c}{1 + r_1} \right) x^*$ and the profit of enterprise 2 is $E_2 = \left(p_2 - \dfrac{c}{1 + r_2} \right)(1 - x^*)$.

Proposition 3. *With the given coefficients for market positioning, cost and quality spillover effect, the two contractors choose the outsourcing strategy. The result of the game equilibrium between the two contractors is as follows:*

The prices of contractors 1 and 2 are, respectively,

$$p_1 = \frac{t(x_2^2 - x_1^2) + 2t(x_2 - x_1) - \theta \Delta q + \dfrac{2c}{1+r_1} + \dfrac{c}{1+r_2}}{3},$$

$$p_2 = \frac{-t(x_2^2 - x_1^2) + 4t(x_2 - x_1) + \theta \Delta q + \dfrac{c}{1+r_1} + \dfrac{2c}{1+r_2}}{3}.$$

The indifference position of the owner is

$$x^* = \frac{x_2 + x_1 + 2}{6} - \frac{\theta \Delta q + \dfrac{c}{1+r_1} - \dfrac{c}{1+r_2}}{6t(x_2 - x_1)}.$$

The profits of contractors 1 and 2 are, respectively,

$$E_1^4 = \frac{\left[t(x_2^2 - x_1^2) + 2t(x_2 - x_1) - \theta \Delta q + \dfrac{c}{1+r_2} - \dfrac{c}{1+r_1} \right]^2}{18t(x_2 - x_1)},$$

$$E_2^4 = \frac{\left[-t(x_2^2 - x_1^2) + 4t(x_2 - x_1) + \theta \Delta q + \dfrac{c}{1+r_1} - \dfrac{c}{1+r_2} \right]^2}{18t(x_2 - x_1)}.$$

Compared with the case where the two contractors adopt the centralized strategy, the competition for the market share between them on the Hotelling line will be more intensive. At this point, the price of enterprise 1 decreases by $\Delta p_{11} = \dfrac{2cr_1}{3(1+r_1)} + \dfrac{cr_2}{3(1+r_2)}$, and the price of enterprise

2 decreases by $\Delta p_{21} = \dfrac{cr_1}{3(1+r_1)} + \dfrac{2cr_2}{3(1+r_2)}$. The indifference position of

the owner's department will move to $\Delta x^*_1 = \dfrac{-\dfrac{c}{1+r_1} + \dfrac{c}{1+r_2}}{6t(x_2 - x_1)}$.

As shown by the comparison between Propositions 2 and 3, when the two contractors adopt the outsourcing strategy, contractor 2 will regain a portion of the market share occupied by contractor 1. Some owners also turn to contractor 2 for services. The price of the two contractors will further decrease to compete for the market share. Compared with the situation in Proposition 2, the price of contractor 1 decreases by $\dfrac{cr_2}{(1+r_2)}$, while the price of contractor 2 decreases by $\dfrac{2cr_2}{(1+r_2)}$. In other words, during the outsourcing process, the spillover effect of the cost will cause the two contractors to further lower the price with the purpose of maximizing the profit. Compared with the case where the two contractors adopt the centralized contracting structure, the two contractors choosing the outsourcing strategy will return to the same point; thus, intensifying the price competition. Hence, outsourcing will not necessarily bring more profits. At this point, the two contractors will be faced with a great challenge. In what follows, Nash equilibrium for the gaming between the two parties will be analyzed.

As shown in the payment matrix (Table 1), when contractor 2 chooses the centralized strategy, the profit of contractor 1 adopting the decentralized outsourcing structure is higher than when the centralized structure is used. That is, $E_1^1 < E_1^2$. When contractor 2 chooses the decentralized strategy, the profit of contractor 1 adopting the decentralized outsourcing structure is higher than when the centralized structure is adopted. That is, $E_1^3 < E_1^4$. In other words, it is better for contractor 1 to adopt the decentralized outsourcing strategy. The same conclusion can be drawn for contractor 2. Hence, (decentralized, decentralized) is the

Table 1. The payment matrix of the two contractors in the international contracting competition.

Parties in the Game		Contractor II	
		Centralized Structure	Decentralized Structure
Contractor I	Centralized structure	E_1^1, E_2^1	E_1^3, E_2^3
	Decentralized structure	E_1^2, E_2^2	E_1^4, E_2^4

dominant equilibrium for the prisoner's dilemma. It means that the two contractors will adopt outsourcing to maximize their profits.

6. The Quality Spillover Effect Under the Decentralized Structure

As shown above, it is a wise decision for the two contractors to adopt the decentralized outsourcing structure when only the cost reduction brought by the share of knowledge and professional division of labour in the cooperation are considered. Moreover, during the cooperation involving the same subcontracting enterprise, the enterprise with a higher quality will inevitably face the spillover of technology and knowledge, thus improving the service quality of the lower-quality contractor. If it is assumed that the horizontal spillover effect occurs, the lower-quality enterprise can improve its service quality by a degree of $\alpha \Delta q$ ($0 \leq \alpha \leq 1$). Will it influence the equilibrium structure of the competition?

In this case, because of the improvement on the service quality of the lower-quality contractor, the indifference position of the residual value of the owner changes into $x^* = \dfrac{p_2 - p_1 - \theta \Delta q (1-\alpha)}{2t(x_2 - x_1)} + \dfrac{x_2 + x_1}{2}$. The two contractors will benefit from the cost spillover effect of vertical cooperation. Then the profit of enterprise 1 is $E_1 = \left(p_1 - \dfrac{c}{1+r_1} \right) x^*$ and the profit of enterprise 2 is $E_2 = \left(p_2 - \dfrac{c}{1+r_2} \right)(1 - x^*)$.

Proposition 4. *With the given coefficients for the market positioning, cost and quality spillover effect, the two contractors will choose the service outsourcing strategy. The result of the game equilibrium of the two contractors is as follows:*

The prices of contractors 1 and 2 are, respectively,

$$p_1 = \frac{t(x_2^2 - x_1^2) + 2t(x_2 - x_1) - \theta \Delta q(1-\alpha) + \dfrac{2c}{1+r_1} + \dfrac{c}{1+r_2}}{3},$$

$$p_2 = \frac{-t(x_2^2 - x_1^2) + 4t(x_2 - x_1) + \theta \Delta q(1-\alpha) + \dfrac{c}{1+r_1} + \dfrac{2c}{1+r_2}}{3}.$$

The indifference position of the owner is

$$x^* = \frac{x_2 + x_1 + 2}{6} - \frac{\theta \Delta q(1-\alpha) + \dfrac{c}{1+r_1} - \dfrac{c}{1+r_2}}{6t(x_2 - x_1)}.$$

The profits of contractors 1 and 2 are, respectively,

$$E_1^5 = \frac{\left[t(x_2{}^2 - x_1{}^2) + 2t(x_2 - x_1) - \theta \Delta q(1-\alpha) + \dfrac{c}{1+r_2} - \dfrac{c}{1+r_1} \right]^2}{18t(x_2 - x_1)},$$

$$E_2^5 = \frac{\left[-t(x_2{}^2 - x_1{}^2) + 4t(x_2 - x_1) + \theta \Delta q(1-\alpha) + \dfrac{c}{1+r_1} - \dfrac{c}{1+r_2} \right]^2}{18t(x_2 - x_1)}.$$

The conclusion of Proposition 4 is compared. In this case, contractor 1 benefits from the quality spillover of contractor 2. With the quality of contractor 1 improved, the gap of quality between the two contractors has been narrowed, thus winning the favour of more owners. This is because the improvement on the quality can ensure a stable market share. At this point, contractor 1 adopts the strategy of raising the price. Although the market share of contractor 1 will decrease slightly, more profits can be gained. The price of contractor 1 has increased by $\dfrac{\alpha \Delta q}{3}$. Because of the weakening of the quality advantage, contractor 2 will lower the price to gain more market share, thus reaping more profits. The price of contractor 2 has increased by $\dfrac{\alpha \Delta q}{3}$.

Proposition 5. *With the given coefficients for market positioning, cost and quality spillover effect, the game between the two contractors has displayed a differentiated equilibrium.*

In the payment matrix, profit of contractor 1 and contractor 2 of strategy (decentralized, decentralized) changed into E_1^5 and E_2^5, when contractor 2 chooses the centralized and decentralized strategy, the profit of contractor 1 adopting the decentralized structure is higher than when the centralized structure is used. That is, $E_1^1 < E_1^2$ and $E_1^3 < E_1^5$. Hence, adopting a decentralized outsourcing structure is the strictly dominant strategy for contractor 1. This is because contractor 1 can obviously benefit from the cost spillover effect and thus gain the cost advantage during the process of cooperating with the upstream subcontractor. When contractor 2 also adopts the subcontracting strategy, contractor 1 can adopt the same strategy to reduce the cost and compete with contractor 2 in the price war. Moreover, contractor 1 can benefit from contractor 2's knowledge input into the upstream subcontractor. With the product quality of contractor 1 improved, the gap of quality between the two contractors has been narrowed. It means that contractor 1 can win a larger market share and thus reap more profits. Contractor 2 is left at a relative disadvantage. On one hand, contractor 2 hopes to reduce the cost by means of subcontracting. On the other hand, contractor 2 is concerned about whether the knowledge spillover brought by the subcontracting will improve the quality of the competitor. Therefore, contractor 2 must fully consider the positive effect of cost spillover and the negative effect of quality spillover before making the decision. With the coefficients of cost and quality considered, the conclusions will be drawn.

Conclusion: When the quality spillover effect exceeds a certain level $\alpha > \dfrac{r_2 c}{\theta \Delta q(1+r_2)}$, contractor 2 will adopt the centralized strategy. In this case, the strategy for the two contractors is (decentralized, centralized). When the quality spillover effect is beneath a certain level $\alpha < \dfrac{r_2 c}{\theta \Delta q(1+r_2)}$, contractor 2 will adopt the decentralized strategy. In other words, the strategy for the two contractors is (decentralized, decentralized).

In the situation of the $\alpha > \dfrac{r_2 c}{\theta \Delta q(1+r_2)}$, when the contractor cooperates with the upstream subcontractor, the assistance given to the subcontractor in many ways will lead to the knowledge spillover. As a result, it will not only reduce the cost of the contractor but also improve the manufacturing and service ability of the subcontractor. Moreover, the quality difference between the contractor and the competitor will be reduced. In other words, the competitive advantage of the contractor will be weakened. Because of $E_2^2 > E_2^5$, it is not a wise decision for contractor 2 to choose the decentralized structure. Therefore, contractor 2 adopts the centralized strategy. In order to maximize the profit, the contractor needs to finish all work without resorting to subcontracting. In the situation of the $\alpha < \dfrac{r_2 c}{\theta \Delta q(1+r_2)}$, despite the negative effect of the knowledge spillover, for contractor 2, the profit increase due to subcontracting is still significant, and contractor 2 will adopt the decentralized strategy.

7. Conclusions

In the international market of project contracting, subcontracting has become an important means of reducing the cost and diversifying the risks, such as political and technological. During the subcontracting process, international contractors are also faced with the risk of knowledge spillover. In extreme cases, it may even pose a threat to the core competitiveness of the enterprise. The subcontracting decision of international contractors, which is analyzed in this chapter, is consistent with the background of the enterprise relying on the domestic resources and the host country's resources. After considering the cost spillover effect and quality spillover effect brought by subcontracting as well as analyzing the optimal strategy for international contractors, we have drawn the main conclusions as follows.

Firstly, when only the cost spillover is considered, Nash equilibrium for both contractors is to adopt subcontracting strategy. In this case, the contractor with a greater cost spillover effect will reap more profits. We also find that the overall profit of the market will increase when the strategy of (centralized, decentralized) is used. It has been against our intuitive judgment. Secondly, when the cost spillover and quality spillover exist at the same time, the decentralized structure can help international contractors gain the competitive advantage with a lower cost. However, the quality spillover will follow. The outsourcing can also increase the market share of the competitor, which has a negative effect on the contractor. Hence, the international contractor must take all aspects of the profit into consideration before choosing the outsourcing strategy. When the quality spillover stays at an overly high level, higher priority should be given to using the internal resource of the enterprise than outsourcing.

Acknowledgement

This work is supported by Major Program of National Natural Science Foundation of China (71390520, 71390521), National Natural Science Foundation of China (71571098, 71732003, 71671088, 71671078, 71701090), Nanjing University Innovation and Creative Program for PhD candidate (2016010), and the program A&B for Outstanding PhD candidate of Nanjing University (201801A001, 201701B009, 201701B010).

References

Barthélemy, J. and Quélin, B. V. (2010). "Complexity of Outsourcing Contracts and Ex Post, Transaction Costs: An Empirical Investigation", *Journal of Management Studies*, 43(8), 1775–1797.

D'Aspremont, C. and Jacquemin, A. (1988). "Cooperative and Noncooperative R&D in Duopoly with Spillovers", *American Economic Review*, 78(5), 1133–1137.

Das, T. K. and Teng, B.-S. (1998). "Resource and Risk Management in the Strategic Alliance Making Process", *Journal of Management*, 24(1), 21–42.

Das, T. K. and Teng, B.-S. (2001). "Trust, Control, and Risk in Strategic Alliances: An Integrated Framework", *Organization Studies*, 22(2), 251–283.

Dussauge, P., Garrette, B. and Mitchell, W. (2000). "Learning from Competing Partners: Outcomes and Durations of Scale and Link Alliances in Europe, North America and Asia", *Strategic Management Journal*, 21(2), 99–126.

Economides, N. (1986). "Minimal and Maximal Product Differentiation in Hotelling's Duopoly", *Economics Letters*, 21(1), 67–71.

Grimpe, C. and Kaiser, U. (2010). "Balancing Internal and External Knowledge Acquisition: The Gains and Pains from R&D Outsourcing", *Journal of Management Studies*, 47(8), 1483–1509.

Harter, J. F. R. (1997). "Hotelling's Competition with Demand Location Uncertainty", *International Journal of Industrial Organization*, 15(3), 327–334.

Hsuan, J. and Mahnke, V. (2011). "Outsourcing R&D: A Review, Model, and Research Agenda", *R&D Management*, 41(1), 1–7.

Kamien, M. I., Muller, E. and Zang, I. (1992). "Research Joint Ventures and R&D Cartels", *American Economic Review*, 82(5), 1293–1306.

Marceau, J. (2002). "Divining Directions for Development: A Cooperative Industry–government–public Sector Research Approach to Establishing R&D Priorities", *R&D Management*, 32(3), 209–221.

Nakamura, T. (2003). "The Mechanism of Bluesilk®, An Assistant Tool for Industry-University Collaboration", *Journal of Information Processing & Management*, 46(7), 455–462.

Neven, D. J. (1985). "On Hotelling's Competition with Non-uniform Customer Distributions", *Economics Letters*, 21(2), 121–126.

Tabuchi, T. and Thisse, J. F. (1995). "Asymmetric equilibria in spatial competition", *International Journal of Industrial Organization*, 13(2), 213–227.

Weigelt, C. (2009). "The Impact of Outsourcing New Technologies on Integrative Capabilities and Performance", *Strategic Management Journal*, 30(6), 595–616.

Zhao, L., Yim-Teo, T. H. and Yeo, K. T. (2004). "Knowledge Management Issues in Outsourcing", in *Proceedings 2004 IEEE International Engineering Management Conference*, Vol. 2.

Chapter 33

Research on Technologically Managed Integrated Systems of Steel Box Girder Manufacturing in the Hong Kong–Zhuhai–Macao Bridge of China

Yumin Qiu* and Zhaohan Sheng[†]

*School of Management and Engineering,
Nanjing University, Nanjing 210093, China*
ThomasChiuNju@hotmail.com
[†]*zhsheng@nju.edu.cn*

Abstract

Currently, China's traditional steel manufacturing technology and equipment cannot meet the requirements of the project quality and manufacturing progress of the Hong Kong–Zhuhai–Macao Bridge (HZMB). Further challenges are faced by foreign steel manufacturing (i.e. high costs, long cycles and human sea tactics are not suitable for the engineering reality of the HZMB). As a result of these challenges, the steel box girder of the HZMB has achieved an optimal allocation of

---

*Corresponding author.

resources through technological innovation and management as a complete collaborative innovation system. This has improved the management ability to handle complex problems through the formation of a steel box girder, which has formed a typical case with regard to management systems. This chapter performs a complete case analysis and theoretical analysis regarding the HZMB steel box girder manufacturing technological management system from four aspects: background and countermeasures, technologically managed comprehensive system design, the strategic position and responsibilities of the owner, and the effect of the implementation of these aspects. This has provided China's major infrastructure construction and management system with a valuable case of system management and experience, enriching innovation with regard to management ideas and theories concerning the complexity of major engineering projects based on China's reality.

Keywords: The Hong Kong–Zhuhai–Macao Bridge; Steel Box Girder Manufacture; Technologically Managed Integrated System; System Management.

1. Introduction

The HZMB of China is a large-scale transportation hub project across clusters of seas. The bridge crosses the Lingdingyang Waters of the South China Sea in the Pearl River Delta and connects the Hong Kong Special Administrative Region, the City of Zhuhai in Guangdong Province, and the Macao Special Administrative Region. It includes elevated roads, sea bridges, artificial islands and submarine tunnels. The bridge tunnel boasts a total length of about 35.6 km and the length of the main bridge is 22.9 km. The total financial investment in the bridge project has been 100 billion yuan. The project began at the end of 2009 and is expected to be completed in 2016. It is currently the largest transportation infrastructure project in China, with the highest level of technical difficulty. It represents the highest levels of technology and management with regard to bridge construction in contemporary China. At the same time, it is also a milestone with regard to bridge construction in the contemporary world.

The superstructure of the HZMB is made of steel. The steel structure is an important part of the main component of the bridge project with a manufacturing scale of 425,000 tonnes, and the project quantity of which is equivalent to the steel structure of 12 Hong Kong–Ngong Shuen Chau Bridges. The HZMB is designed to last for as long as 120 years in line with the British standard (BS). The quality of the products meets or exceeds the Hong Kong–Ngong Shuen Chau Bridge of China and the Oakland Bay Bridge of the United States. Among other aspects, the orthotropic steel deck structure of the HZMB is the key to this anti-fatigue effect, while the steel deck U-rib fillet weld is significant, as the quality of it is difficult to control. Take the Ngong Shuen Chau Bridge as an example: they took three years to make 30,000 tonnes of steel. Under the same quality standard requirements, the problem facing the construction of the HZMB is how to complete the manufacturing of 425,000 tonnes of steel in three years, both at the technical and management levels. This is a typical problem regarding the system management of a major engineering project site, which calls for systematic analysis and solutions under multi-level and multi-scale system considerations.

Existing research has studied management systems from different angles. Scholars from abroad have mainly studied system integration management with regard to driving factors, advantages, disadvantages, measures and model construction (Domingues *et al.*, 2012; Flyvbjerg, 2006; Qiu *et al.*, 2017; Too and Weaver, 2014; Zeng *et al.*, 2007). Among Chinese scholars' research, Gang (2006) has put forward all-around control countermeasures covering the whole process according to the connotations and characteristics of current industrial construction projects. Gang's study focuses on system management, analyzing investment control measures in all stages of the project, including in the planning and decision-making stages, design stage, bidding stage, construction stage, and completion stage. Liu and Liang (2003) based on an analysis of the relationship between internal and external factors of a company and the efficiency of investment in research and development, point out that, in order to improve the success rate of the project, system management must be used so as to strengthen the management involved in the initial decision, the project process control and input factors, decision and risk control, and externality and internalization. Liu and Lu (2011) have, through system dynamics, provided a top-down management method from the strategic level to describe project responsibility systems, putting this into practice with the help of real-life examples from industry project management. Xie and Dong (2010) have put forward a study of engineering project management, which concerns system management activities, such as the effective planning, organization, control and coordination of the whole process of the construction project, according to objective economic law. Tsien Hsueshen put forward a theory of metasynthesis to deal with the open complex giant system. From qualitative and quantitative approaches, Sheng and colleagues (Sheng, 2009; Sheng and You, 2007; Sheng, You, and Li, 2008) used a comprehensive integration method theory to put forward an integrated management system, which has obtained good results in practice for large-scale engineering in China. Based on these studies, and through the construction and application of the technological management of the comprehensive manufacturing system of the steel box girder in the HZMB, this chapter makes a case study and theoretical analysis of a typical management system.

2. Background and Countermeasures

Considering the current situation of the domestic steel structure manufacturing industry, key processes, such as assembly and welding, are mainly based on manual or semi-mechanized operations. Although industrialized production has already been realized, the level of automation is still low, the quality of the steel structure produced is restricted by the level of welding and mechanical equipment with poor stability and the production efficiency is relatively low. This is to say, both the quality and the schedule of the steel plate unit of China's traditional manufacturing technology and equipment are unable to meet the engineering needs of the HZMB.

When it comes to situations abroad — for example, in Japan — steel bridge structure manufacturing mainly adopts an artificial and robotic mode of production. With its excellent workers and leading robotics technology, Japan has reached a high level of manufacturing quality. With regard to bridge engineering projects in Europe, the United States, South Korea, and Hong Kong (aside from a small number of orders falling into Japan), most manufacturing orders are handled by several

Chinese companies. The management method adopted here generally involves a long production period (once or twice longer than domestic projects) and high costs (50–100% higher than domestic projects). These processes are strengthened using human sea tactics under strict supervision (10 times more personnel than domestic projects) and with a higher frequency of quality control (five to 10 times more than domestic projects). Super-compulsive artificial supervision, such as in these examples, ensures a quality-guaranteed system.

Therefore, the current domestic steel plate unit of traditional manufacturing technology and equipment cannot meet the needs of the project quality and progress of the HZMB, while the foreign steel manufacturing system, with its high costs, lengthy construction periods, and human sea tactics, is not suitable for the project reality of the HZMB. The only solution is to combine technological and managerial innovation into a comprehensive and collaborative system in order to cope with the manufacturing challenges created by the HZMB.

Specifically, first of all, the key to large quantities of plate element manufacturing lies in the stability of quality. Due to the low degree of automation in the traditional manufacturing process, the direct participation level of workers is high, and so, regardless of the high skill level of the workers, the quality of the products will be unstable because of fluctuations in physiological, psychological, emotional and environmental factors, such as the manufacturing site. Therefore, to make the quality of large quantities of plate element stable, the minimum level of participation of the people in the manufacturing process must be realized. Specifically, the automation and robotic levels of the manufacturing process must be vigorously improved, which can be realized during the production of small-sized units, as long as the automated assembly line of the manufacturing mode is used.

Secondly, although the integral components of the steel box girder are of great mass, the assembling and welding of the box beams can still be directly processed in a linear, synchronous way, according to the manufacturing process of the bridge section. The assembling period can also be greatly shortened on the basis of effectively ensuring geometric precision. As for large segment assembly, as long as the construction is relatively closed, stable and in a standardized environment (i.e. all completed in the workshop), adverse impacts caused by factors such as weather on the progress and quality of the assembly can be avoided. Moreover, the standardized assembly of the units is also beneficial in achieving the target of zero damage, zero pollution, and zero accidents (health, safety and environmental management; HSE). Additionally, with regard to the mental tension of high-level workers on the steel beam assembly, as the automation manufacturing mode of board units has saved a great deal of human resources, high-level personnel can be effectively transferred to the steel beam assembly as much as possible in order to ease difficulties and achieve a balanced allocation of human resources in steel manufacturing, which is yet another aspect of management for dealing with the manufacturing challenges posed by the steel manufacturing of the HZMB.

To summarize, for the challenges and difficulties created by the steel manufacturing of the HZMB, based on system analysis, the system management theory can be summarized as: through "factorization", to elevate to "automization" and "intellization", to realize the "least personnel" aspect in plate element manufacturing and the "best personnel" aspect in the whole assembly (i.e., through technical management), to create a comprehensive innovation system, to optimize the resource allocation system, and to improve the ability to tackle the problem of complex management in steel manufacturing.

3. Technologically Managed Integrated System Design

In summary, in the steel manufacturing and engineering of the HZMB, a system of managerial innovation needs to be built in accordance with technological innovation under the guidance of the owners. The comprehensive ability also needs to be enhanced through a combination of innovation and guidance. Therefore, the top-level design of this comprehensive system needs to be completed in accordance with the following principles:

(1) The main system will consist of the HZMB authorities (owners) and multiple suppliers. Among these main bodies, the owners (i.e. the HZMB administration officials in leading and dominant positions) become the "main body in order" of this group.

(2) The supplier, as the main body of a technological innovation system, will, according to the actual situation of the enterprises and the requirements of the steel manufacturing involved (e.g. "high standards", "automization", and "intellization"), put into practice the technological and managerial innovation of the entire enterprise. This may entail adjustments in the development strategies of the company, innovation in the concept of enterprise management, enhancements in the level of technology, and optimization of the allocation of resources and other aspects, which, as a whole, can be defined as self-organization development within the enterprise. Therefore, in this process, owners should not only play an important role in guiding and nurturing the project, but should also not enforce additional mandatory administrative interventions. Instead, the owners should create a self-organizing environment for the company from the perspective of the "main sequence".

(3) For self-organizing enterprises, the evolution process of the project can only be realized under a fair and reasonable market contract. During this process, the role of the owners represents, to a considerable extent, the function of the government in guiding and cultivating the enterprises to carry out technological innovation and to enhance core competitiveness.

The aforementioned top-level design principles reflect a normative relationship between the owners of major projects (government agents) and suppliers (enterprises), based on "government-market" synergy.

4. The Strategic Positioning and Responsibility of the Government (the Owners)

Under general circumstances, the owners, as "Party A" in the supply contract are mainly responsible for the selection of the suppliers, the bidding processes and so on. The targeting of the steel manufacturing of the HZMB gives many new responsibilities to the owners.

First of all, the HZMB is a major project with the government as its main investor. It is a major infrastructure project designed to meet the needs of the public and has strong attributes of public good. In such a major project, the public entrusts the government with performing important management activities. Specific construction management works are then progressively commissioned by the government-to-government agencies at all levels, as well as professional departments and units responsible for the project so as to form a multi-level, principal–agent chain. In this multi-entrusting relationship, the government appears both as an agent and as an entruster; this relationship can therefore be called a "government style" principal–agency relationship (Wang, 2002; Zhang and Sheng, 2014). The project owners of the HZMB will, in many cases, perform the functions of the government in this government style principal–agency relationship. This is fully reflected in the technological and managerial innovation within the manufacturing enterprises of the steel beams of the HZMB.

Secondly, major projects are generally valuable resources of technological innovation, while important technological innovations in engineering tend to go beyond the requirements of the construction itself to radiate and develop into further enhancements in the profession and the industry and even with regard to national competitiveness. Therefore, in the process of important technological innovation in engineering, the owners of the projects usually represent the state and the government in the strategic organization and arrangement of the relevant technological innovation activities, such as the selection and optimization of the main body of technological innovation, the building of the technological innovation platform, the design of the technological innovation system, the cultivation of the main body of the technological innovation, and the supervision of the whole process of technological innovation.

Furthermore, according to the forecast of the HZMB steel demand and the production capacity of the relevant enterprises, steel manufacturing enterprise groups will form a strong coupling and distributed network. In order to ensure the owners are equipped with strong analysis, control, coordination, and execution abilities, organizational innovation is necessary. To meet this end, the HZMB Management Bureau set up a special cross-department in July 2011, the "HZMB Project Steel Structure Project Management Office", and has clearly stipulated that the office is responsible for the entire comprehensive process of the steel beam manufacturing. Through the supplement of the system and a clear interface between the authorities, the office has ensured the necessary authorities over the entire office function effectively in the implementation of the project management. This decision has changed the traditional mode of multi-department management in steel manufacturing, reduced work and management interfaces, helped to improve efficiency and execution of the project, and reflects the principles of the owners — when confronted with a complex management problem, design a variable, structured management organization to match the situation.

Finally, although it is a self-organizational behaviour of the company that the suppliers achieve integrated technological and managerial innovation, the owners must provide the necessary environment and resources during this process to cultivate a dynamic mechanism. To summarize, these provisions mainly include the following aspects:

(1) Strengthening market research and optimizing manufacturing enterprises

For the former, in-depth and extensive market research in terms of the steel plate unit manufacturing automation has been carried out. After a full and detailed argumentation, it has been agreed that the automation technology of the plate unit is not only feasible, but it is also possible that the overall steel manufacturing quality has reached an internationally advanced level and the overall positioning and target of steel manufacturing have therefore been made clear.

For the latter, the steel structure manufacturing engineering office has compared and screened domestic steel enterprises from many aspects and fields; through a rigorous bidding process, four manufacturing enterprises with good performances, high technology levels and rich management experience have been determined.

(2) Regulating market mechanisms and giving full play to enterprise enthusiasm

Steel enterprises undertaking manufacturing for the HZMB not only face production tasks depending on resources stocks, but in order to go beyond a general sense of productive behaviour and to realize the enterprise's and the industry's target of technological innovation, the enterprises must also maintain strong and lasting technological innovation enthusiasm and be equipped with grand visions and goals. One of the manufacturers of the HZMB, the China Shanhaiguan Railway Bridge Group, has made it clear that they are determined to achieve the "Shanqiao Reform" by undertaking the historical mission of the HZMB steel manufacturing business, making the enterprise leap "from the No. 1 in China to an international leader". Another company, the Wuchang Shipyard, wants to achieve a new integration of people, equipment, technology, and management through this project, enabling the Wuchang Shipyard to become a world-class steel manufacturing enterprise. Practice has proven that owners can guide enterprises to firmly establish the values and principles which are integral to the success of the overall situation, in which the owners should not only consider "punishing the evil" as a surveillance measure, but should also consider "promoting the good". The owners must fully mobilize and release positive factors in the enterprise to push the spirit of innovation to a new level.

Of course, steel manufacturing is essentially still a market behaviour, so the market value of enterprises should be respected from the beginning to the end. The stipulations of contracts signed by all parties should also be respected along with the basic rules of the market economy, the spirit of which should especially be embodied in the design of related contracts by the owners. Specifically, the owners have strengthened the following points in the relevant bidding contracts.

First, mandatory requirements of enterprises to add automation equipment necessary in steel manufacturing have been put forward; as long as a device does not meet the requirements, it cannot go through the bidding qualification examination. At the same time, the automation level of the independent score items has been established in the evaluation details, which accounts for a large proportion of the technology score in order to restrain and encourage the bidding enterprises to increase investment in resources as early as the bidding stage. This will also introduce automation technology and equipment and finally ease the schedule pressure and project risks after winning the bid as much as possible.

Second, a reasonable product price can effectively drive the enterprise's technological innovation behaviour. In order to enable enterprises to maintain enthusiasm and a sense of responsibility in technological innovation, the owners have improved product price standards in the contract. The governments of Hong Kong, Zhuhai and Macau have considered, from the general picture and practical levels, unifying the bidding project budget at about 30% higher than conventional steel, with the increase in the amount of expense mainly spent on the purchase of automation equipment and site construction. After further optimization, the final contract price reaches a level about 20% higher than conventional domestic projects, both to ensure the needs of the project and to realize the cost control target.

Third, in aspects such as steel beam manufacturing drawings and technical specifications, the owners have also actively carried out a lot of service work, providing enterprises with a strong guarantee as to the standardization and quality of production. For the bid-winning enterprises that have already carried out the implementation, the owners have arranged staff members of the steel box girder office and internationally famous steel structure quality management experts to visit the scene in order to conduct inspections and examinations in person with guidance and supervision from the equipment, facilities, organization, and management systems.

Fourth, in addition to the cultivation, support, and help of enterprises, the services of the owners are also represented in strict supervision. For example, enterprises are only allowed to start the project after meeting the starting standards with regard to equipment, personnel, materials, technology and management systems. Certain enterprises have shown subcontracting intentions after winning a bid, which may have a serious impact on the stability of the steel box girder quality. At this time, the bridge project owners have clearly used important political influence and authority with unambiguous attitudes, adhering to the bottom line that no subcontracting is allowed. After serious discussions, related businesses have stated that they will never carry out subcontracting and have ordered a set of automated production lines, meeting the basic conditions of starting the construction.

The aforementioned series of designing the principles and guiding the work of the owners has prompted enterprises to straighten out the relationship between long-term goals and immediate economic benefits at a strategic level, enabling the enterprises to reform technology, add equipment, remould production processes, and improve manufacturing automation from a strategic perspective.

In the environment of market economy, it is reasonable, with regard to the market demands for enterprises as the main body of the market with independent decision-making powers, to conduct a balanced consideration of costs and benefits, while the owners can only encourage the enthusiastic and innovative spirit of the enterprises through market leverages. At the same time, the owners must ensure this model and relationship are fully reflected in the contracts signed with the company because contracts, as an agreement of market relations, fully guarantee the market in optimizing the allocation of resources within the system and regulate clearly and reasonably the legal rights and responsibilities of both parties.

5. The Implementation Effect of Technologically Managed Comprehensive Systems

The technologically managed comprehensive system is based on the steel manufacturing of the HZMB and has been implemented for more than three years. Although the project is of a large scale, strict quality standards are enforced led by the owners and with the active participation of the enterprises involved. As a result of this, a brand new industrial pattern of "new plant, sophisticated equipment, advanced technology, first-class quality, scientific management, and system innovation" has been formed. Compared with the traditional manufacturing process (Figs. 1 and 2), the production efficiency of the project has been improved by more than 30% with stable welding quality and one-time inspection pass rates higher than 99.9%.

The first production line of plate unit manufacturing automation in China has been built for use, for example, in the processing of non-standard materials and U-rib grooves, plate unit assembly and welding, all of which has achieved automation and intelligence.

(1) In the steel manufacturing process, the "three non-product" concept of quality management has been achieved, which means that, during the phase of board unit and steel structure assembly, the technology of

Fig. 1. Comparison of a traditional plate unit production workshop with an automatic demonstration model production line.

Fig. 2. Comparison of a traditional outdoor assembly with a total assembly in the workshop.

"no motor lifting and welding, and non-destructive lifting and non-destructive support" has been utilized. This can effectively avoid damage of the parent material and improve work efficiency.

(2) Ultrasonic phased array inspection technology has been applied to the penetration rate and the internal quality inspection of the U-rib welding seam, which has realized the detection of full coverage and provided a guarantee for improving the fatigue resistance of the steel bridge.

(3) Through the advanced welding data management system and the LAN, to monitor the welding work online, the welding quality of each welding seam has been controlled and made traceable.

On the basis of the automation and intelligent production in the manufacturing of steel box girder plate units, and through product quality inspections of the first manufactured product and the ultrasonically phased array inspections of the roof U-rib welding, mass production conditions have been achieved. The steel beam assembly was completed in Guangdong Province, the first in China to achieve the workshop operations of large-scale steel box girder assembly from the traditional site and extensive transformation into the factory and fine management mode, which mainly includes the following.

(1) Total assembly in the workshop. The entire assembly has adopted the "long line method". In accordance with the bridge manufacturing, the assembly and welding of the steel box girder in small sections have been achieved in a linearly synchronous way. Based on effective precision in the assembly size, the assembly period has been greatly shortened; the assembly of a 6×110-m long continuous steel box girder, with a total weight of more than 10,000 tonnes, can be completed within 60 days. Large segment assembly is totally completed in the factory to avoid any negative impacts from bad weather and sunshine on the assembling schedule, quality, and linearity in large segments.

(2) Coating in the workshop. The coating of the steel box girder is realized in the workshop, which avoids the influence of outdoor operations on the quality of the welding and coating, improves the working environment, and enables the promotion and stability of the overall coating quality. The coating of the steel structure of the HZMB pursues zero damage, zero pollution and zero accidents, having reached an advanced level in the world with respect to health, safety and environmental management (HSE).

(3) In particular, as a result of the automation and intellectualization of plate unit manufacturing, high-level frontline technicians and workers have been greatly saved, which enables the transfer of these valuable human resources into the stage of steel box girder assembly and improves the quality of the total assembly.

At present, the overall level of manufacturing technology innovation for the steel box girder structure of the HZMB in China has reached a world-leading level.

6. Conclusion

The steel manufacturing engineering project of the HZMB has, through the design and implementation of the technologically managed comprehensive system, achieved success in project site management. This success has not only created stable, high-quality steel products at the material and engineering level for the bridge's construction, but has also offered valuable experience in technological and managerial innovation and the comprehensive pattern of engineering resources, leading to theoretical analysis in many aspects.

First, the key technology breakthroughs in major construction projects, when seen directly, seem to be only in the category of technological innovation. As the systematic integration of resources is the core connotation of the technology, technological innovation naturally requires the environment and mechanisms of resource integration. Therefore, technological innovation must be in cooperation with and requires the support of managerial innovation. It needs to be completed on the basis of a comprehensive system of technology management. To do this effectively, the owners must have a close understanding of technology management problems and build an effective comprehensive system on this basis.

Second, in this system, the government (the owner) is a social and public agent with a greater ability to integrate and allocate resources than other main bodies. At the same time, they should play a leading and guiding role at a higher level.

Third, projects involving important technological innovation generally have a breakthrough in the development process, which not only has direct significance for the construction itself but, because of the cross-border and radiation effect, can also improve scientific and technological progress in the related industry of science and technology. Therefore, with regard to this type of technological innovation, we must go beyond the engineering and enterprise level. On behalf of the government, owners need to establish innovative strategic goals at higher levels to combine technological innovation, economic and social benefits, and the long-term strategies of the state, striving to "make big things big", rather than "make big things small".

Concretely speaking, the role of the government in major engineering technological innovation activities mainly includes the following.

(1) The government (the owner) needs to determine the innovation orientation of the major engineering technology and strategy selection to design technological innovation with the "spillover effect" in mind and to enhance technical progress in the industry and achieve national strategic objectives with a consistent roadmap.

(2) The government (the owner) needs to provide an effective guarantee for resources and the environment with regard to innovation, especially when the demand of resources goes beyond the enterprise's capabilities. If market behaviour fails, the government (the owner) is incumbent in playing a leading role in the organization, including with regard to the determination of innovation objectives, the selection of innovative enterprises, and establishing an innovative system and standardizing innovative behaviour.

(3) The government (the owner) plays a leading role in the technological innovation of major projects but it cannot replace enterprises in carrying out technological innovation activities. Enterprises are always the main part of engineering technological innovation; they are not only the direct beneficiaries of technological innovation success, but they must also undertake the necessary risks with regard to innovation as well. Therefore, the "responsibility, rights, and benefits" of the innovation are very clear; the government (the owner) needs to protect relevant resources, support the environment, and share risks. When the conditions of the enterprise innovation are not fully mature, the government, through supporting policies, needs to strengthen the innovation capacity of the enterprises. To summarize, in this process, the government designs the basic framework of the integrated system through the work of the owners, while the market allocates resources through the enterprise within the framework.

Major engineering technology innovation activities need a more comprehensive system of technology management in order to provide greater support. The so-called "technologically managed comprehensive system" systematically analyzes and designs the essential innovation management involved in technology innovation, such as innovation platforms, innovation organization, innovation systems, and innovation mechanisms in order to form a complete, effective, and operatable management system. In this system, the leading role of the government, alongside the enterprises and the supporting effects of other units, can be given full play. The mechanisms, processes, and procedures involved can also be operated clearly and effectively, enabling the entire technological activity to constantly approach the goal of innovation. The HZMB manufacturing process benefits from the guidance of managerial innovation, having designed an effective technologically managed integrated system. It has also integrated the problem of technical innovation with the problem of managerial innovation, which, after mutual fusion and transformation, has transformed into an overall problem. In this system, the owners and the enterprises both retain their own interests and aspirations, as well as their shared vision and common goals, especially with regard to the way in which the system mechanism effectively enables a partnership between subjects. Through the market rules and order and standard contracts, the behaviour and responsibilities of each party can also be standardized, not only with regard to the economical, but also through cultural communication and institutional norms. Within the comprehensive system, the organization and self-organization, macro- and micro-consultation, coordination and synergy of the owners and enterprises are put into full play. It is especially important for enterprises, in a full understanding of the market as the foundation of any resource allocation system, to accept the government's (the owner) role in strategic leading.

The innovation in the automation and intelligent technology of the HZMB steel manufacturing project not only guarantees the high standards, stability, and progress of the manufacturing of the steel box girder of the HZMB, but also gives a strong impetus to the technological progress in the Chinese steel structure manufacturing industry. The experience and the prompted theoretical thinking achieved in the technologically managed comprehensive system in the bridge steel beam project is continuously deepening and being perfected. The steel beam manufacturing of the HZMB has provided a valuable and successful case with regard to management systems for China's major infrastructure construction projects. This is especially true with regard to technology management innovation problems, which have greatly enriched the creative thoughts and theories of the management of this project with regard to the complexity of major engineering situations based on the real situation in China.

References

Domingues, P., Sampaio, P. and Arezes, P. M. (2012). "New Organisational Issues and Macroergonomics: Integrating Management Systems", *International Journal of Human Factors and Ergonomics*, 1(4), 351–375.

Flyvbjerg, B. (2006). "From Nobel Prize to Project Management: Getting Risks Right", *Project Management Journal*, 37(3), 5–15.

Gang, L. (2006). "Study on the Entire Process Controlling of Industry Items of Construction Investment", *East China Economic Management*, 20(5), 104–105 (in Chinese).

Liu, J. and Liang, L. (2003). "R&D Investment System Management", *R&D Management*, 15(1), 56–62 (in Chinese).

Liu, Y. and Lu, M. (2011). "A System Dynamics Method of Project Management Theory", *Research on Science and Technology Management*, 31(8), 183–186 (in Chinese).

Qiu, Y., Chen, H., Sheng, Z. and Cheng, S. (2017). How to Govern the Conflicting Institutional Logics in Mega-Infrastructure Project: Evidence From the Hong Kong–Zhuhai–Macao Bridge in China, Working paper.

Sheng, Z. (2009). "A Case Study on the Integrated Management Mode of Large-Scale Complex Project: The Theoretical Development on the Management of Sutong Bridge", *Construction Economics*, (5), 20–22 (in Chinese).

Sheng, Z. and You, Q. (2007). "Integrated Management Theory: Exploration of Engineering Management of Sutong Bridge", *Complex Systems and Complexity Science*, 4(2), 1–9 (in Chinese).

Sheng, Z., You, Q. and Li, Q. (2008). "Integrated Management Theory: Methodology and Method of Large Complex Project Management", *Science and Technology Progress and Policy*, 25(10), 193–197 (in Chinese).

Too, E. G. and Weaver, P. (2014). "The Management of Project Management: A Conceptual Framework for Project Governance", *International Journal of Project Management*. Elsevier Ltd., 32(8), 1382–1394. Available at: http://dx.doi.org/10.1016/j.ijproman.2013.07.006.

Wang, J. (2002). "The Construct Modelling of 'Government-Style' Principal-Agent Theory", *Management World*, (1), 139–140 (in Chinese).

Xie, Y. and Dong, X. (2010). "Research Trend of Engineering Project Management in China", *Operation and Management*, (12), 27–28 (in Chinese).

Zeng, S., Shi, J. J. and Lou, G. (2007). "A Synergetic Model for Implementing an Integrated Management System: An Empirical Study in China", *Journal of Cleaner Production*, 15(18), 1760–1767.

Zhang, J. and Sheng, Z. (2014). "The 'Government-Style' Principal–Agent Relationship of Mega-Project Decision-Making: A Case Study Based on the Hong Kong–Zhuhai–Macao Bridge Project", *Scientific Decision-Making*, (12), 23–34 (in Chinese).

Chapter 34

A Critical Review of System Dynamics Modelling in Construction Management Research

Mingqiang Liu*, Yun Le and Yi Hu

*School of Economics and Management,
Research Institute of Complex Engineering and
Management, Tongji University, Shanghai, China*
**liu_mq163@163.com*

Bo Xia

*School of Civil Engineering and Built Environment,
Queensland University of Technology, 2 George Street,
Brisbane City, QLD 4000, Australia*

Abstract

As an effective modelling method for large-scale systems of high complexities, the system dynamics (SD), mainly based on feedback structures, has been increasingly used in the construction management research. However, a systematic review of this area is unavailable until date.

---

*Corresponding author.

This chapter presents a systematic analysis of 100 papers from 40 peer-reviewed journals. Useful data have been found in terms of years of publication, contribution by individuals and research institutions, citations of the selected papers and relevant research topics. These results are beneficial for future research of SD in the construction management field.

Keywords: System Dynamics Modelling; Construction Management Review.

1. Introduction

System dynamics (SD) is a modelling method used to explore and understand a complex system in a holistic manner by ascertaining its feedback structure and the resultant behaviour (Sterman, 2000). This method was first proposed by Jay Forrester in the 1960s at the Massachusetts Institute of Technology to deal with large-scale systems of high complexities (Yuan and Wang, 2014). It acknowledges the behaviour model and characters of a complex system in terms of its internal feedback. This method can solve macro-level problems and avoid the micro-level fragmented details. Thus, this method is suitable for the modern corporations and social organizations with significant dynamic complexity (Ko and Chung, 2014).

Nowadays, SD has been widely accepted as a discipline rather than a mathematical modelling method, which combines the system theory and the computer simulation technique. It can also incorporate technical, organizational, human and environmental factors in dynamic system processes while simulating the behaviour of major outputs of a system over time. Enabled by the rapid development of SD-related software, SD has been increasingly applied in many areas, such as transportation (Shepherd, 2014), project management (Lyneis and Ford, 2007), organizational accident (Yang *et al.*, 2012), complex word learning (Sterman, 2001) among others.

The last two decades have witnessed a rapid growth of research studies applying SD modelling in construction management (CM) research, with the primary aim to explore feedback and interaction of factors in the construction-related complex system. The SD can provide a strategic viewpoint that emphasizes on the behavioural aspects of projects and their relation with managerial strategies (Rodrigues and

Bowers, 1996). In addition, construction processes evolve over time as a result of feedback responses to maintain required project performance, which involve nonlinear relationships, and accumulations of project progress and resources (Lyneis and Ford, 2007). SD can streamline the complex relationships between construction organizations and processes; thus, it leads to the increasing use of the SD model in CM research (Lyneis and Ford, 2007), especially for the complicated construction systems in megaproject delivery.

Despite the significance of the SD modelling for addressing construction and project management-related problems and the care needed for the correct use of this technique, a critical review has not yet been undertaken to provide a systemic understanding of the SD use in the CM field. Therefore, this chapter aims to fill this gap by conducting a comprehensive review of SD modelling research in construction project management areas.

2. SD Modelling in Construction Management Research

Traditional modelling methods in CM research (such as Work Breakdown Structure, Gantt Charts, PERT/CPM networks, Project Crashing Analysis, Trade-off Analysis, etc.) are not entirely adequate to address challenges from complex projects owing to their limited function. Numerous researchers suggest the use of SD in contemporary CM research, such as project activity planning and identification of causes of rework in construction projects. The project organization can be also viewed as a complex system; thus, SD can make the sub-systems interrelate to pursue and achieve the project goals (Love *et al.*, 2002).

SD has a strong ability of simulating a complex construction system with different types of resources and analyzing the process of construction method selection under changing project conditions and objectives (Ozcan-Deniz and Zhu, 2016). SD modelling is applied in construction project management research because construction issues involve a great amount of complex and dynamic interdependent components and involve multiple feedback processes, nonlinear relationships, and hard (quantitative) and soft (qualitative) data (Ogunlana *et al.*, 2003).

SD can analyze the interrelationships and feedback existing within any complex system (Sterman, 2000; Thomas *et al.*, 2016). It can deal with issues within systems with high dynamic complexity, derived from interactions among interrelated components that evolve over time, such as stocks and flows, time delays, nonlinearities and feedback loop structures (Terouhid and Ries, 2016). Both internal and external environments of construction projects are dynamic or relatively unstable. Changes inevitably occur during construction projects, which have significant and often unpredictable effects on its organization and management. With the help of SD modelling, construction project managers can react appropriately to changes and understand how it can influence the behaviour of the project system (Love *et al.*, 2002).

3. Research Methodology

To ensure the comprehensiveness of the research, the Scopus database is selected as the literature search source because it includes all the leading journals in the CM field.

In order to achieve the research objectives, this study adopted a research process with reference to Xiong *et al.* (2015). First, two key items, i.e. "SD" and "construction" were used to search in Scopus database. The full search code is as follows:

(TITLE-ABS-KEY ("system dynamics") AND TITLE-ABS-KEY (construction) AND LANGUAGE (english)) AND DOCTYPE (ar OR re) AND SUBJAREA (mult OR ceng OR chem OR comp OR eart OR ener OR engi OR envi OR mate OR math OR phys OR mult OR arts OR busi OR deci OR econ OR psyc OR soci) (time:2016-11-21). 348 papers were identified at this stage.

Second, another review on the papers was implemented to ensure that all SD papers are related to CM issues. As a result, 172 papers were identified. In addition, an in-depth content analysis was conducted to examine the research methods, topics and studied nature of projects so as to filter the selected papers (Zhang *et al.*, 2016). After that, 100 relevant papers were identified for overview.

Finally, the selected 100 papers were closely examined to reveal the trends of SD-based modelling and topic coverage of SD-based research to obtain a holistic picture of SD modelling application in CM research.

4. Overview of SD Research

4.1. *Trends of SD-based model in CM*

4.1.1. *Publications by years*

Table 1 shows the number of SD-based papers published in each year between 1997 and 2016. Majority of the published SD-based papers

Table 1.   Number of articles in journals.

	Journals	**Number**
1	*Journal of Construction Engineering and Management* (JCEM)	17
2	*Construction Management and Economics* (CME)	8
3	*International Journal of Project Management* (IJPM)	6
4	*Resources, Conservation and Recycling* (RCR)	6
5	*Construction Innovation* (CI)	5
6	*Engineering, Construction and Architectural Management* (ECAM)	5
7	*Journal of Management in Engineering* (JME)	5
8	*Automation in Construction* (AIC)	4
9	*Accident Analysis and Prevention* (AAP)	3
10	*Canadian Journal of Civil Engineering* (CJCE)	3
11	*Journal of Computing in Civil Engineering* (JCCE)	3
12	*Waste Management* (WM)	3
13	*Journal of Cleaner Production* (JCP)	2
14	*KSCE Journal of Civil Engineering* (KSCE JCE)	2
15	*Mathematical and Computer Modelling* (MCM)	2
16	*System Dynamics Review* (SDR)	2
17	*Building and Environment* (BE)	1
18	*Cities*	1
19	*Computer-Aided Civil and Infrastructure Engineering*	1
20	*Electronic Journal of Information Technology in Construction*	1
21	*Energy and Buildings*	1
22	*Energy Policy*	1

(Continued)

Table 1.   *(Continued)*

	Journals	**Number**
23	*European Journal of Operational Research*	1
24	*IEEE Transactions on Engineering Management* (IEEE TEM)	1
25	*International Journal of Civil Engineering*	1
26	*International Journal of Disaster Risk Reduction*	1
27	*International Journal of Production Research*	1
28	*International Journal of Productivity and Performance Management*	1
29	*Journal of Engineering, Design and Technology*	1
30	*Journal of Industrial Engineering and Management*	1
31	*Journal of Infrastructure Systems*	1
32	*Journal of Modelling in Management*	1
33	*Journal of Operations Management*	1
34	*Renewable and Sustainable Energy Reviews*	1
35	*Safety Science*	1
36	*Sustainability (Switzerland)*	1
37	*Sustainable Cities and Society*	1
38	*Total Quality Management and Business Excellence*	1
39	*Waste Management and Research*	1
40	*Water Resources Management*	1

appear in the leading CM journals (see Wing, 1997), such as *Journal of Construction Engineering and Management* (JCEM), *Construction Management and Economics* (CME) and *International Journal of Project Management* (IJPM). As shown in Fig. 1, the number of SD-based journal papers shows an increasing trend since 2016, indicating the increasing application of SD modelling in the CM field.

4.1.2. *Contributions of countries and institutions to SD research*

To obtain the contribution scores of authors and the countries/ regions, the selected articles were quantitatively analyzed following

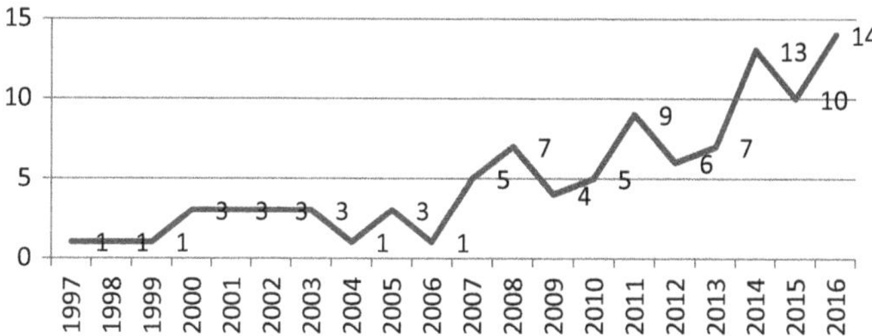

Fig. 1.   The number of SD-based paper in CM in journal papers.

Howard *et al.*'s (1987) method (see Eq. (1)), which has been widely used in the CM field (Hu *et al.*, 2015; Zheng *et al.*, 2016).

$$\text{Score} = \frac{1.5^{n-i}}{\sum_{i=1}^{n} 1.5^{n-i}}, \tag{1}$$

where n means the number of the authors and i is the order of the specific author.

The citation of a journal paper to some extent reflects the paper quality and the influence in the research field. Owing to the coverage limitations of the Web of Science and Scopus, the Google Scholar database, which can provide indirect citation, was used as the source of citation for the selected papers. The powerful search function of Google Scholar is a simple yet thorough channel to make a comparative citation analysis of all the identified papers (Hu *et al.*, 2015).

For each of the identified papers, the score of the authors can be calculated and then the scores of authors' institutions and the countries can also be obtained. Table 2 shows the analysis results.

As shown in Table 2, the USA, with the largest number of researchers and organizations publishing SD papers in the CM field, has the highest score, followed by Mainland China, Hong Kong, Taiwan, Korea, Australia, UK and Singapore.

Table 3 shows the top 10 institutions in this area. The Hong Kong Polytechnic University receives the highest score (8.58) with 14 researchers involved. The second and third highest scores go to the Seoul National University in Korea and the Asian Institute of Technology in Taiwan. The Massachusetts Institute of Technology, University of Illinois at

Table 2.   Research Origins of SD about CM.

Countries	Scores	University/ Organization	The Number of Researcher
USA	20.88	22	47
Mainland China	11.79	14	26
Hong Kong	10.86	5	20
Taiwan	10.66	8	17
Korea	10.29	9	22
Australia	8.57	10	14
UK	7.38	14	22
Iran	5.76	5	9
Canada	5.09	3	8
Singapore	2.07	1	5
Egypt	1.42	1	3
New Zealand	1	1	3
Italy	0.79	1	2
South Africa	0.79	2	2
Palestine	0.6	1	1
Norway	0.6	1	1
Pakistan	0.42	2	2
Kenya	0.42	1	1
Netherlands	0.4	1	1
Germany	0.21	1	1

Table 3.   Top 10 Research institutions publishing SD articles in construction management.

Ranking	University/Institution	Country/Region	Researcher	Scores
1	The Hong Kong Polytechnic University	HK	14	8.58
2	Seoul National University	Korea	13	7.12
3	Asian Institute of Technology	Taiwan	6	5.27

(*Continued*)

Table 3.   (*Continued*)

Ranking	University/Institution	Country/Region	Researcher	Scores
4	Massachusetts Institute of Technology	USA	5	3.44
5	University of Alberta	Canada	5	3.41
6	University of Illinois at Urbana-Champaign	USA	3	2.77
7	University of Michigan	USA	6	2.43
8	Iran University of Science and Technology	Iran	5	2.41
9	Shenzhen University	China	6	2.41
10	Southwest Jiaotong University	China	1	2.3

Urbana-Champaign and University of Michigan in USA are also the leading institutions in publishing SD-related papers. All the selected papers involve a total of 536 researchers.

Citation has been increasingly used as a key indicator to measure the quality of papers (Hu *et al.*, 2015; Zheng *et al.*, 2016). Thus, the numbers of citations of the identified papers were analyzed using the Google Scholar database. The top 10 mostly cited articles are listed in Table 4. Love *et al.* (2002) is the most cited paper with 229 citations. With other five highly cited papers, the same author used SD as the major tool to investigate the rework problem in construction projects. Pena-Mora has two highly cited papers on fast-track issues. In addition, 9 of the 10 papers were published before 2010, and only 1 paper by Yuan *et al.* (2011) after 2010, which may attract more citations.

4.2. *Topic coverage of SD-based research*

SD modelling has been widely used in the CM field. Rodrigues and Williams (1998) categorized applications of SD modelling in three areas: risk analysis, cost estimation and schedule monitoring and diagnosis. Moonseo and Peña-Mora (2003) stated that SD modelling could also be applied to other topics in construction management research, such as

Table 4. Top 10 journal articles ranked by the citation.

Ranking	Authors (year)	Document Title	Journal	Citations
1	Love *et al.* (2002)	Using SD to better understand change and rework in construction project management systems	IJPM	229
2	Love *et al.* (1999)	Determining the causal structure of rework influences in construction	CME	195
3	Pena-Mora and Li (2001)	Dynamic planning and control methodology for design/build fast-track construction projects	JCEM	145
4	Ford (2001)	Waiting for the boom: A simulation study of power plant construction in California	Energy Policy	144
5	Motawa *et al.* (2007)	An integrated system for change management in construction	AIC	118
6	Love *et al.* (2000)	Modelling the dynamics of design error induced rework in construction	CME	103
7	Lee *et al.* (2005)	Quality and change management model for large-scale concurrent design and construction projects	JCEM	94
8	Pena-Mora and Park (2001)	Dynamic planning for fast-tracking building construction projects	JCEM	88
9	Love *et al.* (2008)	Forensic project management: An exploratory examination of the causal behaviour of design-induced rework	IEEE TEM	87
10	Yuan *et al.* (2011)	A model for cost–benefit analysis of construction and demolition waste management throughout the waste chain	RCR	86

project scope management, resource acquisition and allocation, project performance and policies management.

As an increasing number of publications applied SD modelling in CM research, categorizing the topic coverage of the identified papers is necessary (Zheng *et al.*, 2016). Keywords in papers can express the core content and reflect the topic coverage. According to Zheng *et al.* (2016), paper topic categorization can be determined based on the frequency, change over time and tendency of the keywords. In addition, the scientometric technique can be advantageous to analyze the keyword co-occurrence network of the identified papers.

4.2.1. *Keyword co-occurrence network*

According to Zheng *et al.* (2016), the process of keyword occurrence analysis comprises: (1) date acquisition and pre-processing, (2) analysis and modelling, and (3) communication, visualization and layout. The Citespace III, the third generation of the Citespace software, is used as the analysis tool to deal with the data (Chen, 2006). As only the data from Web of Science (WoS) can be directly keyed into the software, data from other databases (Scopus or others) were keyed into the software manually. The data collection is not sufficient and pre-processing is necessary. Thus, the keywords with minor differences in the spellings were edited. For example, "project management" and "project-management" have almost the same meaning, which can be normalized. Keywords, such as "SD", "model" and "construction (industry)" can be removed so as to identify the topic coverage.

It is significant to set the terms of time slicing, term source, term type, nodding types, pruning and visualization before the keyword co-occurrence network is visualized. According to Chinowsky and Taylor (2012), the time span was set for 20 years from 1997 to 2016 and was divided into four five-year time slices. Term sources were from the "Title", "Abstract", "Descriptors", "Identifiers" and term type was "Burst terms". Owing to the big data in every time slicing, data fitting is necessary to depict the knowledge map more clearly, so only top 100 most occurred items were selected from the each slice. The pruning way was selected combining with the minimum spanning tree and pruning sliced networks. At last, other options were considered to choose the default.

When all options were set, CiteSpace III was used to visualize the keyword co-occurrence network. Finally, 227 nodes and 314 edges were identified and the density of the network was 0.0093, which indicates a sparse network. The node size means the frequency of the keyword and the edge means the two different words which together reflect a main theme. In addition, the 10 most frequent keywords are (a) performance, (b) project management, (c) rework, (d) waste management, (e) feedback, (f) design, (g) change management, (h) demolition waste, (i) culture, and (j) policy. The high frequency keywords also included US, China and Hong Kong, which indicated the popularity of SD application in these countries and regions.

5. Conclusions

The SD modelling method has been widely used in the CM research over the past two decades, but no critical review is available up to date. To fill the research gap, this study identified the 100 relevant papers published in 40 journals between 1997 and 2016. The analysis results revealed that most of these SD-based papers appeared in the leading CM journals, such as JCEM,CME, IJPM, JME, ECAM and AIC. The top three countries/ regions publishing the most SD-related papers in CM are the USA, China and HK SAR; the top three organizations are The Hong Kong Polytechnic University, Seoul National University and Asian Institute of Technology in Taiwan. With the help of CiteSpace software, the top 10 keywords of the identified SD papers are (a) performance, (b) project management, (c) rework, (d) waste management, (e) feedback, (f) design, (g) change management, (h) demolition waste, (i) culture, and (j) policy. The research findings present a holistic knowledge map of SD application in the CM field and pave the way for further research.

Acknowledgement

The work described in this chapter has been funded by the National Natural Science Foundation of China (Grant Nos. 71390523 & 71501142) and the Australian Research Council (ARC) Discovery Project (Grant No. DP170101208).

References

Chen, C. (2006). "CiteSpace II: Detecting and Visualizing Emerging Trends and Transient Patterns in Scientific Literature", *Journal of the China Society for Scientific & Technical Information*, 57(3), 359–377.

Chinowsky, P. and Taylor, J. E. (2012). "Networks in Engineering: An Emerging Approach to Project Organization Studies", *Engineering Project Organization Journal*, 2(1–2), 15–26.

Howard, G. S., Cole, D. A. and Maxwell, S. E. (1987). "Research Productivity in Psychology Based on Publication in the Journals of the American Psychological Association", *American Psychologist*, 42(42), 975–986.

Hu, Y., Chan, A. P. C., Le, Y. and Jin, R. Z. (2015). "From Construction Megaproject Management to Complex Project Management: Bibliographic Analysis", *Journal of Management in Engineering*, 31(4), 04014052.

Ko, C. H. and Chung, N. F. (2014). "Lean Design Process", *Journal of Construction Engineering & Management*, 140(6), 04014011.

Love, P. E. D., Holt, G. D., Shen, L. Y., Li, H. and Irani, Z. (2002). "Using Systems Dynamics to Better Understand Change and Rework in Construction Project Management Systems", *International Journal of Project Management*, 20(6), 425–436.

Lyneis, J. M. and Ford, D. N. (2007). "System Dynamics Applied to Project Management: A Survey, Assessment, and Directions for Future Research", *System Dynamics Review*, 23(2–3), 157–189.

Moonseo, P. and Peña-Mora, F. (2003). "Dynamic Change Management for Construction: Introducing the Change Cycle into Model-based Project Management", *System Dynamics Review*, 19(3), 213–242.

Ogunlana, S. O., Li, H. and Sukhera, F. A. (2003). "System Dynamics Approach to Exploring Performance Enhancement in a Construction Organization", *Journal of Construction Engineering & Management* 129(5), 528–536.

Ozcan-Deniz, G. and Zhu, Y. (2016). "A System Dynamics Model for Construction Method Selection with Sustainability Considerations", *Journal of Cleaner Production*, 121, 33–44.

Rodrigues, A. and Bowers, J. (1996). "System Dynamics in Project Management: A Comparative Analysis with Traditional Methods", *System Dynamics Review*, 12(2), 121–139.

Rodrigues, A. G. and Williams, T. M. (1998). "System Dynamics in Project Management: Assessing the Impacts of Client Behaviour on Project Performance", *Journal of the Operational Research Society*, 49(1), 2–15.

Shepherd, S. P. (2014). "A Review of System Dynamics Models Applied in Transportation", *Transportmetrica B*, 2(2): 83–105.

Sterman, J. D. (2000). *Business Dynamics: Systems Thinking and Modeling for a Complex World*, New York: McGraw Hill.

Sterman, J. D. (2001). "System Dynamics Modeling: Tools for Learning in a Complex World", *California Management Review*, 43(4), 8–25.

Terouhid, S. A. and Ries, R. (2016). "People Capability: A Strategic Capability for Enhancing Organizational Excellence of Construction Firms", *Journal of Modelling in Management*, 11(3): 811–841.

Thomas, A., Menassa, C. C. and Kamat, V. R. (2016). "System Dynamics Framework to Study the Effect of Material Performance on a Building's Lifecycle Energy Requirements", *Journal of Computing in Civil Engineering*, 30(6): 04016034.

Wing, C. K. (1997). "The Ranking of Construction Management Journals", *Construction Management & Economics*, 15(4), 387–398.

Xiong, B., Skitmore, M. and Xia, B. (2015). "A Critical Review of Structural Equation Modeling Applications in Construction Research", *Automation in Construction*, 49, 59–70.

Yang, M. G., Love, P. E. D., Brown, H. and Spickett, J. (2012). "Organizational Accidents: A Systemic Model of Production versus Protection", *Journal of Management Studies*, 49(1), 52–76.

Yuan, H. and Wang, J. (2014). "A System Dynamics Model for Determining the Waste Disposal Charging Fee in Construction", *European Journal of Operational Research*, 237(3), 988–996.

Yuan, H. P., Shen, L. Y., Hao, J. J. L. and Lu, W. S. (2011). "A Model for Cost–Benefit Analysis of Construction and Demolition Waste Management Throughout the Waste Chain", *Resources Conservation & Recycling*, 55(6), 604–612.

Zhang, S., Chan, A. P. C., Feng, Y., Duan, H. and Ke, Y. (2016). "Critical review on PPP Research — A search from the Chinese and International Journals", *International Journal of Project Management*, 34(4), 597–612.

Zheng, X., Y. Le, A. P. C. Chan, Y. Hu and Y. Li (2016). "Review of the Application of Social Network Analysis (SNA) in Construction Project Management Research", *International Journal of Project Management*, 34(7), 1214–1225.

Chapter 35

Human Error Identification and Analysis for Shield Tunnel Construction

Jue Li and Hongwei Wang*

*School of Automation, Huazhong University of
Science and Technology, Wuhan 430074, China*
**hwwang@hust.edu.cn*

Abstract

Human error is the main root cause of shield tunnel construction (STC) accidents. Human error mitigation and reduction is an important aspect for STC safety management. Human error identification (HEI) is an effective technique in addressing this issue. Since the informatization and automation of construction have transformed workers' manual-based behaviours into cognitive-based behaviours, this suggests a need for managing human error from a cognitive perspective. This chapter adopts and develops an HEI and analysis approach for safety improvements on the STC site. A cognitive-based HEI framework was developed and implemented for shield machine operation task by integrating the various data resources from the shield machine operation handbook, interviews with the subject matter experts and STC accident reports. Firstly, the shield machine operation task was decomposed into over 100 task steps through the use of hierarchical task analysis (HTA) method. Secondly, the operator activities were identified to describe the task step, and five behaviour modes that summarize the main types of the operator's task behaviours were obtained. Finally, when combined with the information, decision and action (IDA) cognitive model and current HEI researches, each operator activity was associated with the specific cognitive phases, and the corresponding operator failure modes were proposed. In this research, 45 shield operation accident cases are reviewed under the proposed HEI framework. The major human error issues in the shield machine operation task were explored through a retrospective analysis of these accident cases. This research provides insights into effective cognitive strategies to support the mitigation and reduction of human errors. This can further provide decision support on the improvement of construction safety performance in other construction projects.

Keywords: Human Error; Shield Tunnel Construction; Human Error Identification; Accident Analysis.

1. Introduction

Humans are the core component of construction site, and human behaviour has a dominant influence on the construction process. According to the statistics, human error is the main cause of accidents in the construction industry (Haslam *et al.*, 2005). Human error identification (HEI), analysis and control are essential for construction safety management. In recent years, with the development of informatization and automation in construction, the construction industry starts to have the characteristics of the manufacturing industry: worker's construction tasks have been shifting from physical activities to operation and monitoring activities. This change transforms workers' manual-based behaviours into cognitive-based behaviours. As a result, it becomes more and more necessary to identify and analyze workers' human errors from a cognitive perspective.

There are three major research directions of construction human error/ unsafety behaviours according to Fang *et al.* (2016): safety climate/ culture research, behaviour-based safety research, and cognitive-based safety research. Safety climate/culture research and behaviour-based safety research account for a large portion of the current construction safety research. However, these studies do not expound the reason that safety interventions are effective for construction safety, meanwhile, these studies cannot reveal the mechanism of workers' error. Cognitive-based safety research can solve the problems in safety climate/culture and behaviour-based safety research by explaining the cause of human errors. However, the external behavioural responses of workers and the associated cognitive reaction mechanisms are still not completely understood. Furthermore, lacking a cognitive-based framework for analyzing risk factors would result in the inaccurate assessment of human errors and poor safety management.

This study aims to fill the research gaps by investigating the types of human errors in construction industry, explaining the psychological aspects of human errors, and capturing the context associated with human errors. HEI is an effective technique to address these issues. Based on cognitive models for human error, HEI has been used to identify the types of cognitive failure that are likely to occur in the interaction between humans and systems (Baber and Stanton, 1996). Currently, a number of HEI techniques have been widely applied in many industries, including the nuclear power industry, the chemical industry, the oil and gas industry, and the healthcare industry. However, they are not universal for each domain because the development process must match different industrial characteristics. Construction industry is quite different from other industries in that it has unique construction task characteristics and worker behaviours, complicated construction processes, poor construction environment and so on. Therefore, the identification process, the analysis methods and the specific details of HEI must adapt to the construction industry.

With the shield tunnel construction (STC) as research background, this chapter proposes a framework of HEI for STC, qualitatively analyzes the accident cases related to the shield machine operation, and discusses the implications for construction safety improvement.

2. Background

STC is widely used in current metro construction projects in China. Shield tunnelling is a complex construction method by which workers operate the shield machine technical system to excavate and form the tunnel structure by using shield shell and segments. During the tunnelling process, the shield machine operation is the most crucial task for STC. The shield machine operator must performs multiple tasks (e.g. shield machine attitude control, cutter state monitor, earth discharge control, segment assembling and grouting reinforcement) to ensure the shield machine technical system maintains at a safety level. Thus, STC safety is heavily dependent upon the capability of the operator, and the majority of STC-related accidents were characterized by operator's human errors.

While applying HEI for STC, it is vital to learn more about the shield machine operation task, operator behaviour and the STC context. The main characteristics of these aspects are as follows:

(1) There is a wide variety of subtasks/task steps in the shield machine operation task. Subtasks and task steps are related to the specific duties of operators, construction methods, technical systems, and working environment, which are direct causal factors of human errors. As a result, the range of activities of shield machine operation is much more diverse than other industries mentioned above.

(2) The behaviour of operators has various patterns. In the shield machine task, shield machine operators not only rely on skill-based operation activities, but also follow detailed step-by-step procedures for some routine task steps. In some abnormal construction situations, the operators may handle the abnormal situations by using their knowledge. Therefore, the cognitive characteristics of operators usually are varied during the construction process, and the relevant human errors may have different producing mechanisms.

3. Methodology

This section provides a description of the research methods used to develop an HEI and an analysis framework, which is illustrated in Fig. 1. The research integrates various domain data, HEI techniques and cognitive model to develop a structural HEI and analysis methodology and to propose some effective measures that would help to mitigate and reduce the human error in the shield machine operation task. The details of the methods and steps followed in this study are presented in the following Sections.

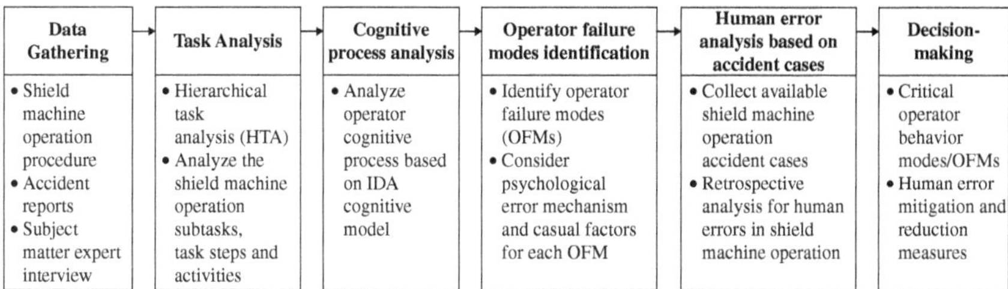

Fig. 1.   The framework of an HEI and analysis for STC.

3.1. *Shield machine operation task analysis*

Task analysis is aimed at identifying subtasks and task steps associated with the operator activities. Operator activity refers to operator's interaction with the objective reality on the STC site, it includes physical process and mental process. For example, "monitor" is a typical activity involved in the "shield machine guiding" subtask. The shield machine operation subtasks, task steps and operator activities of shield machine operation task are derived from the hierarchical task analysis (HTA). HTA is used to break down and study the individual elements of a task (Stanton, 2006). In this study, the shield machine operation task was decomposed into 4 first-level subtasks, 23 second-level subtasks and over 100 task steps. Meanwhile, 16 types of operator activities were identified. Figure 2 illustrates the process of task analysis and the partial results.

As mentioned before, the number of shield machine operation subtasks/task steps presents a challenge for identifying the human errors for a wide range of the identified objects. After the HTA for the shield machine operation task, we found that the task steps share some common features in the combination mode of the activities associated with them. Through a deliberate analysis process, five stable combination modes of the activities are identified, which are called *behaviour modes* in this study. These behaviour modes represent the types of behaviours generally carried out by the operator. The detailed descriptions of the behaviour modes are shown in Table 1. Considering the behaviour mode as the analysis object can increase the effectiveness and ease of HEI and analysis.

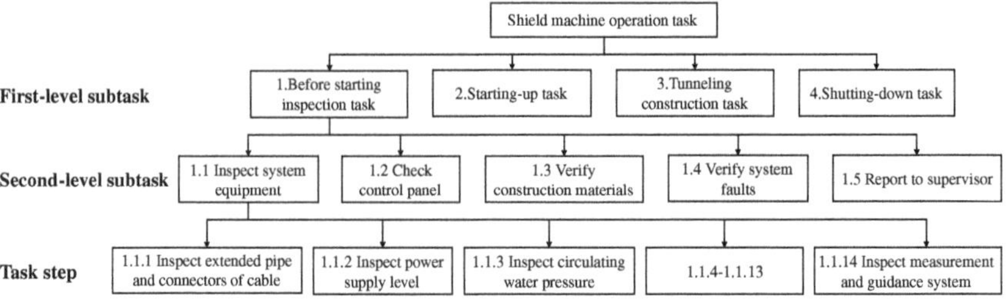

Fig. 2.   HTA of shield machine operation task.

Table 1.   Combination of activities and characteristics of cognitive process for each behaviour mode.

Behaviour Modes	The Combination of Activities	The Characteristics of Cognitive Process		
		I	D	A
Routine inspection (RI)	Observe, Scan, Evaluate, Identify, Adhere, Record	IG	RB	OB
System state monitoring (SSM)	Monitor, Evaluate, Identify, Record	V	SB	OB
System state adjustment (SSA)	Execute, Regulate, Maintain, Verify, Evaluate, Identify, Decide, Adhere	IG	SB/RB	OP
Diagnose and reasoning (DR)	Monitor, Observe, Diagnose, Adhere, Decide, Plan	IG	DB	OB
Passive information reception (PIR)	Detect, Evaluate, Identify	UIG	SB/RB	OB/CO

3.2. *Operator's cognitive process analysis*

When combined with the IDA cognitive model, each operator activity can be associated with the specific IDA phases (Ekanem *et al.*, 2016), such as information pre-processing, situation assessment, decision-making and action execution. Thus, the behaviour mode characterized with the combination of operator activities can be related to cognitive phases and detailed cognitive functions. When analyzing the cognitive process of each behaviour mode, we considered the cognitive functions of activities that the behaviour mode contains and corresponding specific task steps performance together as a whole. As a result, several main characteristics of operator's cognitive process were captured. In the information pre-processing process, there are three types of information gathering and processing: intentionally gathering (IG), unintentionally gathering (UIG) and Vigilance (V). In the decision-making process, there are three levels of operator's decision-making: skill-based (SB), rule-based (RB) and knowledge-based (KB) (Rasmussen, 1983). There are three basic actions in the action executing process: observation (OB), operation (OP) and communication (CO). Different behaviour modes are associated with different characteristics of cognitive process, the details are shown in Table 1.

3.3. *Operator failure mode identification*

A classification of operator failure modes is proposed to describe the possible types of operator failures in the three cognitive phases (information pre-processing, decision-making and action executing). According to the Phoenix method (Ekanem *et al.*, 2016), the operator failure modes (OFMs) can be defined as the generic functional modes of failure of the operator in his interactions with the shield machine technical system, and they represent the manifestation of the operator failure mechanisms and corresponding causes of failure. Moreover, the OFMs can be related to the operator activities based on the IDA cognitive model. The detailed introductions of the failure modes and their relationships with each activities can be reviewed in CREAM (Hollnagel, 1998) and Phoenix (Ekanem *et al.*, 2016).

In order to identify the specific OFMs for each behaviour mode in the shield machine operation task, the meanings of the identified OFM must match the specific types of operator activity and the cognitive process characteristics of the behaviour mode. In this way, the specific OFMs of each behaviour mode can be identified, as shown in Table 2. Furthermore, the producing mechanisms and the causes of OFMs can be explained based on engineering psychology and cognitive psychology research. It would improve the efficacy of the human error control measures on STC sites.

Table 2.   Relationship between behaviour modes and operator failure modes.

Operator Cognitive Process		Behaviour Modes				
IDA Phases	**Operator Failure Modes**	**RI**	**SSM**	**SSA**	**DR**	**PIR**
Information pre-processing	I1 Key alarm not responded to (intentional and unintentional)					✓
	I2 Data not obtained (intentional and unintentional)		✓	✓	✓	✓
	I3 Data discounted	✓	✓	✓	✓	✓
	I4 Decision to stop gathering data				✓	
	I5 Wrong data source attended to	✓	✓	✓		✓
	I6 Data not checked with appropriate frequency	✓	✓			
	I7 Information miscommunicated					✓
	I8 Reading error	✓	✓	✓	✓	✓
Decision-making	D1 System state misevaluated	✓	✓	✓		✓
	D2 Inappropriate strategy chosen	✓	✓	✓	✓	✓
	D3 Failure to adapt procedures to the situation			✓		
	D4 Procedure step omitted (intentional and unintentional)	✓		✓		
	D5 Decision to delay action				✓	✓
	D6 System state misdiagnosed				✓	
	D7 Faulty reasoning				✓	
	D8 Inadequate plan				✓	
Action executing	A1 Incorrect timing of action			✓		
	A2 Incorrect operation of component/object			✓		
	A3 Action on wrong component/object			✓		

3.4. *Human error analysis based on accident cases*

Publicly available shield machine operation related accident reports are retrieved from Internet reports, accident records, and academic literature.

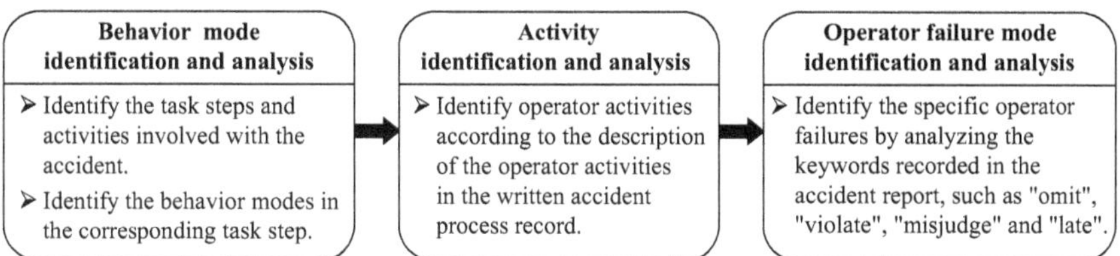

Fig. 3.   Retrospective accident analysis procedure.

There are over 140 accident reports gathered form these data resources. In order to obtain the cases which are suitable for HEI and analysis, the targeted accident cases must be relevant to at least one human error, besides that, the human error and the accident process should be analyzed in detail. As a result, 45 available accident cases are filtrated.

All 45 cases are reviewed under the HEI framework. The accident retrospective analysis procedure is demonstrated in Fig. 3. Through the retrospective analysis, the actual human error problems in STC can be reflected. The producing laws of human error can be explored from the retrospective analysis results with the use of statistical analysis, cluster analysis, association analysis and other data analysis methods.

4. Results and Implications

4.1. *Behaviour modes in shield machine operation task*

According to the retrospective analysis procedure in Section 3.4, the behaviour modes that are involved with the 45 available accident cases should be identified from the accident written records. Table 3 shows the number of accidents in which a behaviour mode was identified. Diagnose and reasoning behaviour mode occurs most frequently (39/45). The second and the third most frequently occurring behaviour modes are System State Monitoring (35/45) and System State Adjustment (26/45), respectively. Passive Information Reception (9/45) and Routine Inspection (6/45) modes occur much less frequently than the other three behaviour modes.

Table 3.   Number of accidents (out of 45 accident cases) in which a behaviour mode was identified.

Behaviour Modes	Number of Cases Identified	Percentage (%)
Routine Inspection (RI)	6	13
System State Monitoring (SSM)	35	78
System State Adjustment (SSA)	26	58
Diagnose and Reasoning (DR)	39	87
Passive Information Reception (PIR)	9	20

4.2. *Operator failure modes in shield machine operation task*

In order to understand the overall distribution of the OFMs in the shield machine operation task, the OFMs occurring in each behaviour mode need to be analyzed. Table 4 shows the number of OFMs identified in corresponding behaviour modes. As for the *Routine Inspection* behaviour mode, "data not checked with appropriate frequency", "reading error" and "procedure step omitted" are the main OFMs. In the *System State Monitoring* behaviour mode, "system state misevaluated", "data discounted" and "data not obtained" are the top three OFMs that frequently occur.

4.3. *Implications for human error mitigation and reduction*

The research identified several major human error issues in the shield machine operation task from the three most frequently identified behaviour modes. The safety management for the shield machine operation task deserves special emphasis on the situations in which the operator diagnoses the shield machine technical system states, continually monitors the system states, or controls the shield machine. According to the result, we can ascertain several crucial shield machine operation subtasks and task steps associated with the situations mentioned before. The subtasks/task steps that involve frequent monitoring, evaluation, regulation and diagnostic activities should be concerned about their most frequently occurring OFMs, such as "system state misevaluated", "data not obtained" and "data discounted".

Table 4.   Number of OFMs identified in corresponding behaviour modes.

Operator Failure Modes	Number of OFMs Identified				
	RI	SSM	SSA	DR	PIR
I1 Key alarm not responded to (intentional and unintentional)	—	—	—	—	4 (**2nd**)
I2 Data not obtained (intentional and unintentional)	—	24 (**3rd**)*	5	30 (**2nd**)	4 (**2nd**)
I3 Data discounted	2	26 (**2nd**)*	7	13	5 (**1st**)
I4 Decision to stop gathering data	—	—		14	—
I5 Wrong data source attended to	1	2	1	—	1
I6 Data not checked with appropriate frequency	3 (**1st**)*	12	—	—	—
I7 Information miscommunicated	—	—	—	—	0
I8 Reading error	3 (**1st**)	4	0	0	2
D1 System state misevaluated	0	32 (**1st**)	18 (**1st**)	—	3
D2 Inappropriate strategy chosen	0	0	16 (**3rd**)	22 (**3rd**)	0
D3 Failure to adapt procedures to the situation	—	—	10	—	—
D4 Procedure step omitted (intentional and unintentional)	3 (**1st**)	—	4	—	—
D5 Decision to delay action	—	—	—	5	1
D6 System state misdiagnosed	—	—	—	34 (**1st**)	—
D7 Faulty reasoning	—	—	—	16	—
D8 Inadequate plan	—	—	—	9	—
A1 Incorrect timing of action	—	—	3	—	—
A2 Incorrect operation of component/object	—	—	17 (**2nd**)	—	—
A3 Action on wrong component/object	—	—	7	—	—

Note: *1st, 2nd and 3rd refer to the sorted order of OFMs of each behaviour mode based on the number of identified OFMs.

Combining the cognitive mechanisms and the causes of those OFMs, we are able to propose several effective cognitive-based measures for human error mitigation and reduction of the shield machine operation task.

(1) **Diagnose and reasoning behaviour mode:** The "Data not obtained" is a primary cause of "System state misdiagnosed" and "Inappropriate strategy chosen" considering the interrelationship among their cognitive functions. The cognitive mechanism of "data not obtained" in the Diagnose and Reasoning behaviour mode is called attention narrowing, which has a negative effect on individual situation awareness performance (Li, 2013). While facing a complicated abnormal situation, the operator is prone to pay too much attention to the abnormal symptom (e.g. system parameters, alarms, components), it might also reduce his chance to get some important clues that are out of the plan.

Measures and implication: A mandatory monitor procedure will need to be developed to prevent missing important clues; the assistant operator and other worker will need to closely monitor the other system symptom to guarantee global system safety.

(2) **System state monitoring behaviour mode:** Cognitive load is the main cause of "System state misevaluated", "Data discounted" and "Data not obtained" in the System State Monitoring behaviour mode. The shield machine operator's cognitive load is associated with the perception, processing, and communication of information (Chang and Mosleh, 2007). During a complicated or abnormal construction situation, a great deal of construction information would cause a high cognitive load, and then a mental stress state. Operator tends to intentionally abandon his attempt to gather more information or make irrational decisions under the undesirable mental state.

Measures and implication: Targeted emergency response procedure needs to be developed to help operators handle complicated abnormal situation and relieve their stress. Furthermore, several useful cognitive strategies can be proposed to help operators maintain a state of mindfulness and introspection, such as rehearsing tasks for future execution, using technological solutions to monitor how operator's mental state changes during construction, and building self-introspection procedure into operation tasks.

There is a great potential for using cognitive-based strategies to manage human error on the STC site. It is necessary to develop and come up with more systematic analysis results to build practical human error mitigation and reduction measures through further study.

5. Conclusion

This study proposed a cognitive-based HEI framework to detect, identify and analyze the human errors in the construction industry. The HEI framework was successfully applied to the STC context with a special focus on shield machine operation accidents in China. A deeper understanding of the cognitive mechanism of human error in STC project was obtained from the retrospective analysis of the accident cases. In this way, the current study has made a step forwards into proposing effective cognitive strategies to support the mitigation and reduction of human errors.

Future research should look into how to mine the accident analysis results to gain more knowledge about the cognitive laws of human errors, for example, the interrelationships among OFMs. The validation of the HEI framework should be conducted to demonstrate its reliability of application in the construction industry. In addition, a quantitative risk analysis approach should be used to evaluate the risks of human error for critical steps in the construction task. A Bayesian-network-based risk analysis method could be a feasible solution to address the issue (Khakzad *et al.*, 2011).

Acknowledgements

This work was supported by National Natural Science Foundation of China Grant (Project No. 71390524).

References

Baber, C. and Stanton, N. A. (1996). "Human Error Identification Techniques Applied to Public Technology: Predictions Compared With Observed Use", *Applied Ergonomics*, 27(2), 119–131.

Chang, Y. H. J. and Mosleh, A. (2007). "Cognitive Modeling and Dynamic Probabilistic Simulation of Operating Crew Response to Complex System Accidents. Part 2: IDAC Performance Influencing Factors Model", *Reliability Engineering & System Safety*, 92(8), 1014–1040.

Ekanem, N. J., Mosleh, A. and Shen, S. H. (2016). "Phoenix — A Model-Based Human Reliability Analysis Methodology: Qualitative Analysis Procedure", *Reliability Engineering & System Safety*, 145, 301–315.

Fang, D., Zhao, C. and Zhang, M. (2016). "A Cognitive Model of Construction Workers' Unsafe Behaviors", *Journal of Construction Engineering and Management*, 142(9), 04016039.

Haslam, R. A., Hide, S. A., Gibb, A. G., Gyi, D. E., Pavitt, T., Atkinson, S. and Duff, A. R. (2005). "Contributing Factors in Construction Accidents", *Applied Ergonomics*, 36(4), 401–415.

Hollnagel, E. (1998). *Cognitive Reliability and Error Analysis Method (CREAM)*, New York, YN: Elsevier.

Khakzad, N., Khan, F. and Amyotte, P. (2011). "Safety Analysis in Process Facilities: Comparison of Fault Tree and Bayesian Network Approaches", *Reliability Engineering & System Safety*, 96(8), 925–932.

Li, Y. (2013). Modeling and Simulation of Operator Knowledge-Based Behavior. PhD Dissertation, College Park, MD: University of Maryland.

Rasmussen, J. (1983). "Skills, Rules, and Knowledge; Signals, Signs, and Symbols, and Other Distinctions in Human Performance Models", *IEEE transactions on Systems, Man, and Cybernetics*, (3), 257–266.

Stanton, N. A. (2006). "Hierarchical Task Analysis: Developments, Applications, and Extensions", *Applied Ergonomics*, 37(1), 55–79.

Chapter 36

Research on Multi-Level Principal–Agent Relationships and Government Behaviour Analysis of Decision-making in Megaprojects

Huimin Liu* and Lanjun Wang[†]

*School of Management and Engineering,
Nanjing University, Nanjing 210093, China*
liuhm@edu.cn
[†]*2653774056@qq.com*

Abstract

The planning and decision-making in the phase of initial a mega construction project is led by government in general, which is related to the national economic development, scientific and technological development and the guarantee of people's livelihood. With a gradual increase in the number and scale of mega construction in China, scientificity and democracy in the decision-making process are both improved remarkably. However, there exist agent crisis and dissimilation of public power, including opportunism behaviours. In this chapter, more than 30 field investigations have been conducted on the long-span bridges across seas, rivers and gorges in China in accordance with the complexity and

---

*Corresponding author.

uncertainty of mega construction decision-making. A conceptual model of hierarchical principal–agent in mega construction decision-making has been established, which mainly focuses on the two core subjects, that is, superior and subordinate governments in the model. It analyzes adverse selection and moral hazard behaviours under the principal–agent relationship of the subordinate government in order to motivate agent behaviours and avoid issues related to the traditional management.

Keywords: Megaprojects; Principal–agent Theory; Upper and Lower Government.

1. Introduction

Mega construction usually aims to satisfy social public needs. It has a relatively obvious attribute of public good. With the continuous development of market economy and the need to provide people's livelihood, more and more mega construction projects in the fields like water engineering, military affairs, aerospace and transportation will be established in China in the future (Lu, 2009).

There are also some examples of failure in mega construction decision-making at home and abroad, such as the Aswan Dam project in Egypt and the Three Gorges project in China. With the continuous development for several years, scientificity and democracy in the decision-making process are both improved remarkably, but there are still some problems (Huang and Xu, 2012). For example, the central government and local governments have different expectations and objectives. The former one mainly considers maintaining its own political legitimacy, promoting development and balance of national economy and actively providing social welfare, while the latter ones mainly consider position promotion of agents and the economic development in their periods from the perspective of regional economic development. During the competition process, because of the lack of information transparency and imperfect supervision mechanism, some local governments have a tendency to misrepresent economic data, demand information and social service quality. In this situation, it is likely that the agent crisis including opportunism behaviours which are under the control of local bureaucrats may appear. Sometimes, many "jelly-built projects" and even public power dissimilation cases may emerge endlessly.

Therefore, it is possible that risks appear in the process like information transmission, when the project involves communications and cooperation between superior and subordinate governments. Once this kind of risk increases, it is inevitable to make immeasurable damage on economic development and social security. Thus, it is necessary and important to introduce the principal–agent theory into research on the decision-making management process of mega construction.

At present, for research on decision-making management of mega construction in China, the involved decision-making contents are mainly the influence factors of decision making, decision-making mechanism and decision-making methods. In the aspect of the influence of engineering decision-making, scholars have done research from multiple angles. Qi Yanxia tried to study the ethical restriction by which engineering decision-making is influenced (Qi, 2010); Ye Pei and other scholars have studied the decision-making mechanism of the uncertain environment on engineering project's decision-making (Ye and Wang, 2014). In the aspect of decision-making mechanism and methods, Zhang Jinwen and other scholars based on the practice of Hong Kong–Zhuahai–Macau Bridge (HZMB) conducted a qualitative analysis of government behaviours during decision-making of mega construction by combining principal–agent theories (Zhang and Sheng, 2014); Ding Xiang and other scholars put forward an analysis method of mega construction decision-making on the basis of calculation experiment (Ding *et al.*, 2015); Lu Guangyan, Hu Ping, He Jie and other scholars studied decision-making mechanism and group decision-making methods of mega construction projects from the perspective of systems engineering theories and methods (Lu, 2009; Hu, 2010; He, 2011); however, the current studies about decision-making management of mega construction are all just at the level of concept analysis and theoretical model. Most of them adapt the method of one single case, not yet drawing conclusions from the basic experience of mega construction decision-making from the perspective of multiple cases. There is a lack in research on descriptions and analyses of superior and subordinate governments' behaviours that consider China's situations and mega construction decision-making situations. Therefore, this chapter will conclude multi-level relationships of mega construction decision-making under the unique situation of China based on

30 long-span bridge project cases in China. It will also analyze "adverse selection" and "moral hazard" issues between the central government and local governments in order to provide policy suggestions for decision-making subjects of mega construction through the descriptions and analyses of superior and subordinate governments' principal–agent behaviours.

2. Conceptual Model of Government Principal–Agent Relationships of Mega Construction

In general, substantial boundary of engineering projects is limited. It only exists in a certain or some regions under government's jurisdiction. During the actual process of project construction and management, it is inevitable that the superior government authorizes the subordinate government of the project location to handle and solve relevant issues in many situations. In this way, principal–agent relationships between superior and subordinate governments (including administrative departments in superior and subordinate governments) have been formed (Zhang, 2005). Therefore, after the agent of local government takes office, he/she starts the principal–agent relationship with the central government in all aspects. This is the common government principal–agent relationship (Wang, 2002). The principal–agent relationship relating to megaconstruction is a special part of the common one.

Megaconstruction has many characteristics, like great influence, long-term period, and huge investment (Lu *et al.*, 2008). Compared with the common government principal–agent relationship, the government principal–agent of mega construction not only has features of "government" and "hierarchy", but also has features of variety and dynamics (Yan and Zhao, 2005). So, it is much more complicated. In the process of mega construction decision-making and construction, there are at least four principal–agent relationships: (1) principal–agent relationships between the public and local governments, (2) principal–agent relationships between government agencies at all levels, (3) principal–agent relationships between governments and professional institutions, and (4) principal–agent relationships between governments and project managers, as shown in Fig. 1.

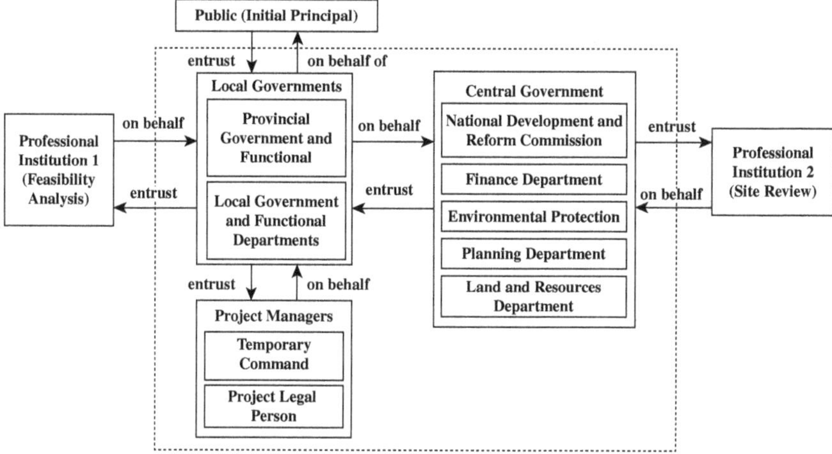

Fig. 1. Principal–agent relationships of decision-making in megaprojects.

There are at least six decision-making subjects involved. Except "the weak principal" — the public, there are at least five direct stakeholders. In a slightly more complicated principal–agent relationship, it even involves many governments, multiple investors, multi-level project leaders, project supervision agencies and other subjects. In addition, the change in the external environment, continuous improvement in the cognitive ability of decision-making subjects, constantly changing decision-making objects and continuous progress in the decision-making technology determine the dynamics of mega construction decision-making subjects.

3. Superior and Subordinate Governments Principal–Agent Relationships of Mega Construction Decision-making

Mega construction decision-making is mainly led by the government, but there are also the typical "government" principal–agent relationships. So, this paper, through the investigation and analysis of many cases, mainly focuses on the second principal–agent relationship, that is, research on the multi-level government principal–agent (Grossman, 1983; Rogerson, 1985).

3.1. *Long-span bridge cases*

In the process of case selection, the following principles need to be followed: (1) the case should be in accordance with the subject of research; (2) it should be representative; (3) it should have sufficient data and information. Based on the principles above, it can be found that the bridge construction projects in China, in particular mega infrastructure construction, have achieved remarkable achievements around the world. The management system is relatively mature and the data are comparatively complete. Therefore, in the process of case selection, long-span bridge projects are taken as the research objects in this chapter. The projects are mainly located at the eastern coastal areas from the south to the north and Yangtze River Basin from the east to the west. The selected cases are shown in the appendix.

Many methods are used in this chapter to collect data, such as interviews, records of original files, field investigations, search of online information, search of books and literature. It has realized the combination of public materials and field research. The collected data are abundant and complete.

Before the field investigation, data were collected from library materials and website information for the first part of the work. Information about basic situations and main features of the project were sorted out. Information about bridge construction was divided into several modules. It means that if you want to learn about a bridge project, you can start from four parts, namely decision-making management, construction management, worksite management and science, technology and innovation management in order to determine report framework of every bridge case base.

After the collection and sorting out of public materials, it is time to start field investigation and go to the proprietors' companies in 14 provinces for research. In the process of investigation, leaders who had participated in the decision-making process of bridge projects were interviewed. All the interviewees were once involved in the preliminary planning and decision-making of mega construction.

3.2. *Analysis of principal–agent relationships between superior and subordinate governments*

Based on the research results mentioned above, it is found that due to the fragmentation of China's administration management, the government subjects involved in the preliminary decision-making of mega construction have shown a feature of diversity, including the related government administrative departments, competent departments of the areas to which projects belong, and local and provincial governments, where projects are located.

(1) From the horizontal perspective, government administrative departments assume the examination and approval responsibilities in different functional areas. For instance, the National Development and Reform Commission is responsible for organizing related departments to examine and approve investments in public projects. It has the right of giving final approval for the establishment of mega construction. But the finance department, planning department, environmental protection department, land and resources department and competent department of the areas to which projects belong, respectively, undertake the examination and approval of investment scheme, construction planning scheme, environmental impact assessment, land use and management, technologies and standards of related industries with partial right of approval and decision-making.

(2) From the long-term perspective, governments have a clear hierarchy as decision-making subjects in most mega infrastructure projects:
   (a) *The first level*: The National Development and Reform Commission has the right to final decision on the establishment of construction;
   (b) *The second level*: Provincial governments are responsible for the overall planning and scientific decision-making for the establishment of construction, and approving the implementation of project feasibility studies;
   (c) *The third level*: Municipal governments where projects are located carry out specific project feasibility studies.

In addition, due to the public nature of mega construction, governments cannot be ignored in the process of project construction and management. Through the investigation and study on construction and management modes of over 30 bridges across rivers, sea or gorges built in the three decades after China's reform and opening up, it can be seen that

Table 1. Management mode properties of leading organizations in long-span bridge projects.

Related Powers	Government-led Self-management Mode (Temporary Command)	Project Legal Person System Mode
Administrative power	Central and local governments	Central and local governments
Routine power	Competent authorities	Competent authorities
Financial power	Local governments	Local governments or local government funding vehicles
Executive power	Temporary command formed by governments	Project companies with legal person status

the administrative power, routine power and financial power and other important powers are still in the hands of governments whether in the government-led self-management mode in the initial reform and opening-up period, or in the project legal person system mode which is widely adopted today (Table 1).

4. Analysis of Superior and Subordinate Governments' Behaviours in the Decision-making Process of Mega Construction

In general, in the principal–agent relationship between superior and subordinate governments during the decision-making process of mega construction, the central government is the principal and the local government is the agent. The former reviews and makes decisions on the project applications submitted by the latter one. The latter chooses whether to act honestly and reliably as it is supposed to be. However, in view of the diversity and multi-level nature of our governments in the decision making process of mega construction, this chapter will choose governments and functional departments at the central level that have the right to project approval as principles while the provincial and municipal governments and related functional departments as agents in order to simplify the model. It should be noted that the project owners, whether they are the project legal person entities or temporary command, are mostly composed of government staff and therefore regarded as the agents like local governments in this chapter.

Therefore, before and after the formation of the contract, behaviours of the superior and subordinate governments are in a three-stage competition, which is shown as the extensive-form competition tree in Fig. 2. In the extension above, the two competition parties are, respectively, the central government and the local government. The first stage is a selection phase for the central government, during which it chooses whether to entrust or not (and motivate contract, described as below). If it does not choose to entrust, it would not receive service from agents. $R(0)$ denotes the profits of the central government in the absence of an agent, and the local government has no revenue under this circumstance. If the central government chooses to entrust, they will enter the second stage where the local government chooses whether to accept the contract or not. If the local government refuses it, the result will be the same as that in the first stage. But if it accepts, then the competition enters the third stage where the local government chooses whether to strive or not.

There are two scenarios in the third stage that need to be discussed, that is, where the local government works hard and where it does not.

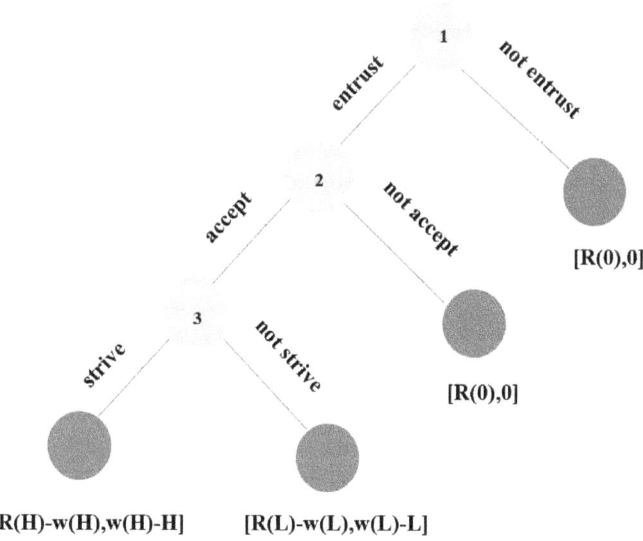

Fig. 2.   Competition behaviours of two-level governments.

During the decision-making process of mega construction, local governments' efforts are mainly reflected in the demonstration and supervision work. In the first scenario, the central government can get a higher revenue $R(H)$. Accordingly, the local government can also receive higher wages (paid or shared by the central government) $w(H)$, but it also needs to shoulder more costs, such as resource utilization, new technology development and supervision, which are recorded as high cost H. Therefore, revenues of the central government and the local government are, respectively, $R(H) - w(H)$ and $w(H) - H$. However, in the second scenario, the central government gets lower revenue $R(L)$. The local government receives lower wages $w(L)$ correspondingly and shoulders lower cost L. At this point, revenues of the two sides are $R(L) - w(L)$ and $w(L) - L$, respectively.

The central government and the local government have inconsistent goals and information asymmetry. The central government focuses on balancing the needs of construction at the national level and solving the long-standing imbalance in development between regions as well as between urban and rural areas. While the local government puts its emphasis on how to maximize regional economic growth by investing in more mega construction. In addition, given that the power and responsibility of government departments are essentially exercised by specific government officials, principals and agents involved in this chapter not only include government departments in the virtual sense but also individuals in the physical sense. Under this circumstance, the agents will act to maximize their own interests and the so-called "agent crisis" will emerge at the same time. Further analysis of government behaviours from two aspects — *ex ante* adverse selection and *ex post facto* moral hazard — is as follows:

(1) The adverse selection behaviours in the process of mega construction decision-making are mainly reflected in the following situations: the local government uses its advantage of information to conceal facts or deceive the central government, such as exaggerating its executive ability and economic ability and increasing local needs of public services excessively; and even "bad officers" would expel "good officers". Over the long run, the local government will tend to make false report when planning mega constructions in order to make profits.

Local governments' behaviours may be induced by several factors, such as the one-sided and short-term government performance management, the closed process and the arbitrary manner.

(2) The moral hazard behaviours in the process of mega construction decision-making are as follows: after the project is established, the local government conceals or misrepresents themselves to the superior government for the sake of their own interests. To be specific, it starts from the contractor selection through bidding, implementation of the project to the acceptance check and project maintenance and management, which may result in achievement exaggeration and concealing of problems.

5. Conclusions

Firstly, a number of mega infrastructure projects in China are taken as examples in this paper to summarize the process and derive the characteristics of decision-making in mega construction under the unique circumstances in China. The model of relationships between superior and subordinate governments in the decision-making process of mega construction and their principal–agent behaviours are studied as well. Additionally, "adverse selection" and "moral hazard" existing in relationships between the central government and the local government are analyzed.

Based on the above, it is possible to build principal–agent models, introduce reasonable parameters, construct objective functions and constraints of superior and subordinate governments and solve them in the future. Meanwhile, the optimal incentive mechanism for the central government can be designed to motivate the local government, that is, how the central government guides the local government to act in the best interest of its principle under the natural circumstances of inconsistent objective functions and information asymmetry.

Acknowledgements

This research was supported by The Natural Science Foundation of China (Grant No. 71671088 and Grant No. 71390521).

Appendix: Overview of Long Bridges

Type	Name	Period	Investment (RMB: billion)	Length (km)	Region
Cable-stayed Bridge	Langqi MinJiang Bridge	2010.9–2014.1	2.256	2.675	Eastern China
	Nanpu Bridge	1988.12–1991.12	0.820	8.346	Eastern China
	Tongling Yangtze River Highway Bridge	1991.12–1995.12	0.450	2.592	Central China
	Yangtze River. Tunnel Bridge	1996.2–2009.10	123	25.6	Eastern China
	Nanjing Third Yangtze River Bridge	2003.1–2005.10	3.09	15.6	Eastern China
	The Donghai Bridge	2002.6–2005.5	7.11	32.5	Eastern China
	Hangzhou Bay Sea Cross Bridge	2003.11–2007.6	11.8	36	Eastern China
	Su Tong Yangtze River Highway Bridge	2003.7–2008-6	8.4	8.2	Eastern China
	Wuhan Tianxingzhou Yangtze River Bridge	2004.9–2008.9	11.06	4.657	Central China
	Jintang Bridge	2006.4–2009.11	7.7	21.02	Eastern China
	Hubei East Hubei Yangtze River Bridge	2006.8–2010.9	3	5.762	Central China
	Jiashao Bridge	2008.12–2013.7	6.35	10.137	Eastern China
	XiamenZhang Across-sea Bridge	2008.11–2013.5	4.986	11.7	Eastern China
	Jiujiang Yangtze River Highway Bridge	2009.10–2013.10	4.478	25	Central China

(Continued)

(Continued)

Type	Name	Period	Investment (RMB: billion)	Length (km)	Region
Suspension Bridge	HuMen Bridge	1992.10–1999.4	3.02	15.76	Eastern China
	Haicang Bridge	1996.12–1999.12	2.87	5.926	Eastern China
	Runyang Yangtze River Bridge	2000.10–2005.10	5.3	35.66	Eastern China
	Jiangyin Sanjiang Bridge	1994.11–1999.9	3.625	3.071	Eastern China
	Wuhan Yangluo ChangJiang River Bridge	2003.11–2007.12	2.0	10	Eastern China
	Ma'anshan Yangtze River Bridge	2008.12–2013.12	7.078	36.274	Central China
	Baling River Bridge	2005.4–2009.12	1.48	1.564	Western China
	Taizhou Yangtze River Bridge	2007.12–2012.11	9.37	62.088	Eastern China
	Aizhai large Suspension Bridge	2007.10–2012.3	1.5	1.073	Western China
	Fourth Nanjing Yangtze Bridge	2008.12–2012.12	6.8	28.996	Eastern China
	Longjiang Suspension Bridge	2011.5–2016.5	1.465	2.470	Western China
Archbridge	Lupu Bridge	2000.10–2003.6	2.5	8.722	Eastern region
	Caiyuanba Yangtze River Bridge	2004.11–2007.10	0.95	1.651	Western China
	Chaotianmen Yangtze Grand Bridge	2006.3–2009.4	1.32	4.881	Western China

References

Ding, X., Sheng, Z. H. and Li, Z. (2015). "Decision Analysis for mega-projects based on Computation Experiments", *Journal of Systems & Management*, 4, 545–551.

Grossman, S. J. and Hart, O. D. (1983). "An Analysis of the Principal–Agent Problem", *Econometrica*, 51(1), 7–45.

He, J. (2011). Group Decision-making Mechanism and Methods under Large-scale Construction Projects, Nanjing University.

Huang, J. and Xu, X (2012). "Analysis of Government Decision-making Capability: The Decision-Making Horizon of State Key Construction Projects — Taking the Decision-making of the Three Gorges Project as an Example", *Journal of Jiangsu Administration Institute*, 01, 89–97.

Hu, P. (2010). "Application of System Engineering Methodology in Decision-making Mechanism of Major Engineering Projects", *Water Conservancy Economy*, 05, 57–60, 79.

Lu, G. (2009). "Failure of Major Engineering Decision Making and Construction of Major Engineering Decision-making Mechanism", *Forum on Science and Technology in China*, 04, 30–35.

Lu, G. Fu, C., Wu, J. and Liu, Y. (2008). "Research on Decision Making Process and Decision Characteristics of Major Projects — Taking the Three Gorges Project as an Example", *Forum on Science and Technology in China*, 08, 20–24.

Qi, Y. (2010). Research on Ethics Statute of Engineering Decision, Dalian University of Technology.

Rogerson, W. (1985). "The First-order Approach to Principal–agent Principal–agent Problems", *Econometria*, 53, 1357–1367.

Wang, J. (2002). "Theoretical Analysis of the State Budgetary Principal–agent Principal–agent Relationship", *Public Finance Research*, 01, 54–56.

Ye, P. and Wang, J. (2014). "Research on the Influence Mechanism of Uncertainty Environment on the Decision of Construction Engineering Project", *Digest of Management Science*, 36, 74–76, 79.

Yan, L. and Zhao, L. (2005). "Analysis of Two-level and Multi-level Principal — Agent Chain in Government Investment Projects", *Finance Issues Research*, 12, 41–47.

Zhang, J. and Sheng, Z. H. (2014). "Research Principal–Agent Relationship 'government type' Major Engineering Decisions Based on our Hong Kong–Zhuahai–Macau Bridge Engineering Practice", *Scientific Decision*, 12, 23–34.

Zhang. J. Y. (2005). "Principal-agent relationship, levels of government and the burden on farmers", *Reform of Economic System [J]*, 03, 72–75.

Chapter 37

Review of Building Maintenance Practices and Sustainability at the Government Level

Peng Wu*, Xin Hu and Xiangyu Wang

*The Australasian Joint Research Centre for
Building Information Modelling, Curtin University,
Kent Street, Bentley, WA6102, Australia*
**Peng.wu@curtin.edu.au*

Abstract

Building maintenance is considered as an invaluable process in preserving buildings' continual use and retaining their value and quality. The fast development of building maintenance industry has resulted in the participation of governments in its practices. Learning from their practices can benefit building maintenance stakeholders significantly. However, there is still a lack of systematic review of the current building maintenance practices of governments for the purpose of organizational learning. The study aims to address this research gap by delivering a review of building maintenance practices in developed economics. The research benefits building maintenance stakeholders

---

*Corresponding author.

by delivering them knowledge regarding governments' maintenance practices in the industry.

Keywords: Building maintenance; Sustainability; Policy; Evaluation.

1. Introduction

Building maintenance is an invaluable process in preserving buildings' continual use and retaining their value and quality (Vijverberg, 2002). It is estimated that more than 50% of the total output of the construction industry is from building repair and maintenance in countries or regions, such as the United Kingdom (UK), and Hong Kong (Woodsworth, 2001; Kherun, 2002; Chiang, 2015). In order to maintain buildings, governments need to invest a number of financial resources and labour force each year.

It is argued that a useful performance assessment model for building maintenance should be developed to improve productivity. However, building maintenance can be influenced by a variety of factors and its assessment is complicated and hard. Researchers also stated that building maintenance is an underresearched area, and it is essential to explore the crucial factors that impact the implementation of building maintenance to benefit the future improvement of building maintenance activities (Wood, 2005).

This study aims to provide a review of the literature regarding the government-level practices of implementing building maintenance from developed economics, including the United States (US), Hong Kong, UK and Australia. The research findings of this study can benefit building maintenance stakeholders by delivering them knowledge about best practices in the industry, which will facilitate the sector's sustainable development.

2. Literature Review

Building maintenance means improving all parts of a facility, its services and surroundings to an acceptable standard to sustain its utility and value (Allen, 1993). This can be divided into three categories: corrective, preventive and condition-based maintenance (Horner *et al.*, 1997). The aims of building maintenance are diverse, such as preserving buildings for continual use, improving investment values and ensuring a good appearance (Allen, 1993). Cleaning, inspecting, repairing and replacing systems and components are the major activities involved (Arditi and Nawakorawit, 1999). There are a variety of factors influencing building maintenance practices, such as building features, tenant characteristics, and government policies (El-Haram, 2002). To evaluate building maintenance performance, a project success index including time, cost, quality, functionality, safety and environmental friendliness was suggested (Lam *et al.*, 2010).

Issues are also reported in building maintenance practices, such as building design inefficiencies, poor performance of the contracted company and poor construction quality (Arditi and Nawakorawit, 1999). To improve the performance of building maintenance activities, a variety of building maintenance strategies are suggested. For instance, Lind and Muyingo (2012) suggested a new structure for building maintenance focusing on long-term strategies, goals and short-run adjustments. Jones and Sharp (2007) suggested a performance-based process model which helps managers assess their actions before maintenance implementation and monitor the influence of their actions.

Over the past few years, sustainability has been an emerging focus for government authorities to integrate. As indicated by Wu *et al.* (2016, 2017), government-related buildings occupy one of the largest market shares of green building design and construction. However, it is unclear how sustainability can be integrated in the maintenance stage of the building's life cycle. This chapter also aims to investigate the important sustainability factors in the building's maintenance stage.

3. Research Method

To achieve the research aim, the research method of literature review is adopted in this study. Literature review analyses and summarizes published literature and suggests action plans for future studies and practices. A comprehensive literature review can not only facilitate the development of theories but also provide a platform for knowledge dissemination (Webster and Watson, 2002).

There are two important activities involved in the implementation of literature review, including identifying relevant literature and analysing literature. First, to identify relevant literature in this study, government websites were searched to retrieve related building maintenance documents. This study focuses on the countries/regions in US, Hong Kong and UK and also the local government of Queensland in Australia. These countries/regions were chosen for analysis due to their advanced building maintenance practices. In addition, it covers best practices in both the federal and local government levels. To analyse the collected data, content analysis was employed. Content analysis is a method to describe and quantify phenomena by analysing written, verbal or visual documents (Elo and Kyngäs, 2008). It has been widely used in the literature review studies (Hu *et al.*, 2016) and provides a robust method to achieve the research aim.

4. Results

4.1. *The United States*

Performance-based approach programs for managing government-related buildings have been widely adopted in the US. One of the most important standards in regulating the maintenance of buildings is the standard classification for serviceability of an office facility (ASTM E1670-95a) published by the US government. Four distinct factors are provided in the standard:

- **Strategy and program for maintenance:** This factor aims to assess the availability and adequacy of a strategic maintenance program for managing government-related buildings.
- **Competences of in-house staff:** This factor aims to assess whether in-house staff are provided to assist the building maintenance and the competency level of these in-house staff.
- **Occupant satisfaction:** This factor aims to assess the occupant satisfaction level and identify strategies to improve the occupant satisfaction level.
- **Unit costs and consumption:** This factor aims to use existing information to assess the unit costs and consumption of maintenance activities.

The ASTM E1670-95a is practical to implement because it classifies the service level before establishing relevant factors to assess maintenance performance. For example, in the category of strategy and program for maintenance, five service levels are first defined, including Level 1 (indicating that there are no major consequences if the level of maintenance is not satisfactory), Level 3 (indicating that minimum building maintenance is required and there are no major consequences of failed maintenance), Level 5 (indicating that acceptable building maintenance is required and there are minor consequences if the building is not well maintained), Level 7 (indicating that a higher than average level of building maintenance is needed most of the time) and Level 9 (indicating that a higher than average level of maintenance is needed at all times). Based on such classifications, the standard provides different performance levels for assessing the strategy and program aspects of the maintenance, including the availability of such program, budget, human resources, availability of replacement items, and the availability of contractors and subcontractors for conducting the maintenance.

Similarly, in assessing the competency level of in-house staff, the standard first classifies the service requirement into five levels from Level 1 (indicating that there are required skilled maintenance operation for in-house staff) to Level 9 (indicating that well-trained in-house staff are available for conducting cost-effective and reliable maintenance). Accordingly, the detailed assessment criteria of the competence level includes training provided for these in-house staff, whether these staff have obtained trade qualifications, whether staff are able to read and operate the electrical system, HVAC equipment and piping systems.

It can be seen that a very effective performance-based building maintenance program has been established in the US to help government authorities to assess the maintenance of government-related office buildings. However, one important observation from the results is that sustainability has now been integrated into the current evaluation. Sustainability areas that are commonly adopted in the green building design and construction including the use of regionally manufactured materials, the use of existing building components, the achievement of energy efficiency and the use of renewable energies are yet to be implemented in the maintenance stage. This points out new research directions of how to integrate sustainability into the maintenance programme, how to assess the sustainability performance of building maintenance and how to use the assessment to support decision-making (Zhong and Wu, 2015; Wu *et al.*, 2014a).

4.2. *Hong Kong*

In Hong Kong, a one-stop maintenance service is provided by its building maintenance scheme. The scheme aims to implement the inspection of building maintenance, such as damage and hazard elimination, maintenance cost reduction and breakdown minimization.

The criteria include five dimensions of tangibility, reliability, responsiveness, assurance and empathy. The tangible criterion is associated with the necessary resources (e.g. facilities, equipment and manpower) allocated by contractors to implement the scheme. The reliability criterion refers to the contractors' capability to achieve client requirements within appropriate time. In addition, the responsiveness criterion refers to the contractors' willingness to provide suitable services to tenants and customers. Moreover, the assurance criterion refers to the contractors' abilities to address enquiries and inspire stakeholders. At last, the empathy criterion is related to monitoring and understanding customers' feelings. A detailed assessment of these criteria reveals that none of the assessment criteria is related to sustainability and further improvement is required to integrate sustainability into the maintenance scheme.

4.3. *The United Kingdom*

Building maintenance management in the UK is regulated by different councils. The maintenance evaluation relies on the owner's consideration. However, it is centrally guided by the British Standard 8210:2012 Guide to facilities maintenance management. The British Standard Institution recommends the use of tools, such as value management, the EFQM Excellence Model, balanced scorecards and benchmarking, to provide guides to measure performance and indicate potential improvement opportunities. The commonly adopted methods of measurement include the following:

- **Compliance:** Services will be delivered in accordance with relevant statutory legislation and appropriate guidelines including British

Standards, Manufacturers Recommended Instructions, IEE Recommendations, Gas Safety Regulations, Construction (Design Management) Regulations including Duties of the Client, Health & Safety at Work Act and other relevant statutory legislation.

- **Value for money:** Random sample checks will be performed on 20% of all maintenance works carried out by contactors. In addition, all works over £2,000 in value will also be the subject of checks. The check will include a desktop review of invoices to check that the amounts charged are consistent with the tendered rates and the hours charged are reasonable for the type of work. A site visit will also be arranged to inspect the work.

- **Quality of service:** The client will monitor contractors'/consultants' performance annually or as required against quality indicators including timeliness of response, health and safety, safeguarding and quality of works on site.

- **Customer satisfaction:** The client will appoint a dedicated liaison officer for each building, who will undertake regular liaison with the building consistent with the scale and complexity of works being undertaken. They will undertake site visits, attend meetings as required and will seek feedback from the school and keep records of customer satisfaction on a job-by-job basis.

4.4. *Queensland government in Australia*

The frameworks related to the maintenance of government buildings published by the Queensland Department of Housing and Public Works (QDHPW) include the following: (1) Maintenance Management Framework and (2) Building Asset Performance Assessment Framework and the asset management frameworks. Table 1 summarized the key factors considered in the building maintenance frameworks and polices from QDHPW.

4.5. *Sustainability in building maintenance*

Based on the above current maintenance schemes adopted in various countries, it can be concluded that sustainability and environmental protection are not often considered in the maintenance stage of buildings. This can be caused by the relatively low environmental impact when

Table 1.   Key factors considered in the building maintenance frameworks and polices from QDHPW.

Tier 1 Factors	Tier 2 Factors	Tier 3 Factors
Asset performance	• Building condition assessment • Asbestos surveys • Building asset register • Building reviews • Risk management audits • Data collection for life-cycle planning • Energy management audits • Engineering investigations • Environmental audits • Water management audits	asbestos audits; asbestos inspections building codes audits; fire safety audits, town planning code audits; health and amenity audits; functionality audits; utilization audits; postoccupancy evaluations geotechnical investigations, structural integrity investigations, electrical investigations
Environmental sustainability	• Heating, ventilation and air conditioning systems audit • Building sealing audit • Glazing and shading audit • Energy Efficient Control systems audit • Waste, energy and water monitoring equipment audit	
Lifecycle maintenance costs	• Agency management/ administration costs • Condition assessment costs • Planning maintenance costs • Unplanned maintenance costs	Computerized maintenance systems Preventative; statutory; condition-based

compared with the design and construction stage. The construction stage of buildings, in terms of the extraction of raw materials, the manufacture of raw materials, material transportation and on-site construction, has a significant environmental impact. The cement sector alone accounts for almost 5% of global carbon emissions (Wu *et al.*, 2014a). The manufacturing of steel and aluminum, which are commonly adopted in the construction industry, is also energy-intensive. The integration of sustainability into the design and construction stage is therefore the focus of the current studies. However, it should be noted that many new schemes have emerged to assess the performance of existing buildings, including maintenance, over the past few years. One of the most notable schemes is the Building Research Establishment Environmental Assessment Method (BREEAM) In-Use. The scheme includes 10 KPIs which help assess the sustainability performance at the operational and maintenance stages. The 10 KPIs can be categorized into the following groups:

- **Carbon emissions:** Due to the rising recognition of global climate change, carbon emissions are commonly used as indicators for assessing the sustainability performance of products, processes and activities. The indicators related to carbon emissions include the building CO_2 (i.e. the assessment of carbon emission levels per unit of floor area), occupant CO_2 (i.e. the assessment of carbon emission levels per full time equivalent), and energy CO_2 (i.e. the assessment of carbon emission levels per unit of energy consumption). These indicators can help reveal the effectiveness of operation and maintenance programs in maintaining certain level of performance towards sustainability.
- **Energy consumption:** One of the most commonly adopted indicators to assess environmental sustainability is energy consumption and it is not surprising that the indicator is used in BREEAM In-Use for assessing sustainability. The indicator is represented by the energy consumption per unit of floor area.
- **Water consumption:** Water usage and water reduction have been the targeted areas of building design and construction, as demonstrated from the green building rating systems, such as the Leadership in Energy and Environmental Design (LEED) (Wu *et al.*, 2014b). The water consumption is assessed by consumption per unit of floor area.
- **Waste:** The waste is assessed at two levels: recycled waste and waste sent to landfill.

In addition to the BREEAM In-Use, LEED for existing building operation and maintenance is one of the most successful assessment frameworks for building operation and maintenance. Similar to the LEED rating system for green building, green operation and maintenance include the following assessment categories:

- **Sustainable sites:** This category is related to the assessment of the availability of building management plan, the provision of alternative transportation to reduce environmental impact, as well as the reduction of heat island effect and light pollution.
- **Water efficiency:** Compared to the assessment of building design and construction, which commonly focus on water usage and water reduction strategies, some new factors are now added to this category. The new factors include a system for water performance measurement and the management of cooling tower.
- **Energy and atmosphere:** This category is related to the continuous monitoring of the building's performance on energy efficiency and the use of renewable energy to achieve emission reduction.
- **Materials and resources:** This category is mostly related to sustainable purchasing and solid waste management, such as the availability of relevant policy and purchasing durable goods.
- **Indoor environmental quality:** One of the most notable additions to this category is green cleaning, which is related to the availability of a green cleaning program including assessment, performance monitoring and green procurement.

5. Discussions

The suite of documents available on the QDHPW clearly articulates the building maintenance and asset management frameworks and requirements. Agencies from the other developed economies rely heavily on outsourcing the work. For instance, the US government relies on the standards published by the American Society for Testing and Materials (ASTM) and Government Services Agency (GSA) for establishing building maintenance standards. However, these standards were not clearly communicated with the public through their website or other communication methods.

The study suggests that building maintenance management is transforming from a prescriptive-based specification towards a performance-based assessment. The US and Hong Kong have developed a detailed set of metrics to evaluate building maintenance work. The way of a prescriptive standard is the traditional building maintenance, which is based on preventive maintenance of the mechanical systems. This requires that facility managers play major roles in managing related activities to achieve aims such as sustainability. However, it does not show performance levels by which the performance of the building is monitored. In addition, it also does not take owners' considerations into account.

Research findings also indicated that it is beneficial to separate the evaluation criteria for different levels. For example, the US experience suggests five levels of building maintenance performance. The owners can choose the correct levels in order to identify the appropriate evaluation criteria. This is useful to maintain a high level of maintenance efficiency.

This review also finds that supportive facilities (including both management and technology aspects) are needed for the purpose of facilitating building maintenance. How to successfully capture related information in information technology is a research trend at the time of this study. This will promote maintenance teams' organizational learning and understand building maintenance history and factors that can impact maintenance activities.

A significant finding of this review is that although sustainability has not been successfully into the maintenance schemes of government authorities, it has been widely adopted when the building is targeted to achieve a green operation and maintenance certification through certification programs such as BREEAM In-Use and LEED for existing building operations and maintenance. As government-related buildings constitute one of the largest markets for green building design and construction, it can be expected in the future that it will also be a major market for green building operation and maintenance. As such, understanding the sustainability indicators used for green building operation and maintenance is imperative. This study provides useful guidance on this issue and can guide future decision-making, e.g. selecting appropriate materials (Zhong *et al.*, 2016).

6. Conclusions

The study provides a review of government literature on implementing building maintenance in developed economics, including the US, Hong Kong, UK, and Australia. It was found that maintenance management is transforming towards a performance-based assessment. Additionally, it is shown that separating the evaluation criteria for different levels of building maintenance is beneficial. The review also suggested that supportive facilities are needed to promote the effectiveness of building maintenance activities. The findings of this study depict the current practices of implementing building maintenance at the government level. It provides important implications for future studies such as the development of agency-specific building maintenance framework. It is recommended that policymakers consider this review findings and take actions towards identifying and capturing building maintenance data in the most appropriate data format for decision-making.

References

Allen, D. (1993). "What is Building Maintenance?" *Facilities*, 11, 7–12.

Arditi, D. and Nawakorawit, M. (1999). "Issues in Building Maintenance: Property Managers' Perspective", *Journal of Architectural Engineering*, 5, 117–132.

Chiang, Y. H., Li, J., Zhou, L., Wong, F. K. and Lam, P. T. (2015). "The Nexus Among Employment Opportunities, Life-Cycle Costs, and Carbon Emissions: A Case Study of Sustainable Building Maintenance in Hong Kong", *Journal of Cleaner Production*, 109, 326–335.

El-Haram, M. A. and Horner, M. W. (2002). "Factors Affecting Housing Maintenance Cost", *Journal of Quality in maintenance Engineering*, 8, 115–123.

Elo, S. and Kyngäs, H. (2008). "The Qualitative Content Analysis Process", *Journal of Advanced Nursing*, 62, 107–115.

Horner, R., El-Haram, M. and Munns, A. (1997). "Building Maintenance Strategy: A New Management Approach", *Journal of Quality in maintenance Engineering*, 3, 273–280.

Hu, X., Xia, B., Skitmore, M. and Chen, Q. (2016). "The Application of Case-Based Reasoning in Construction Management Research: An Overview", *Automation in Construction*, 72, 65–74.

Jones, K. and Sharp, M. (2007). "A New Performance-Based Process Model for Built Asset Maintenance", *Facilities*, 25, 525–535.

Kherun, N. A., Ming, S., Petley, G. and Barrett, P. (2002). "Improving the Business Process of Reactive Maintenance Projects", *Facilities*, 20, 251–261.

Lam, E. W., Chan, A. P. and Chan, D. W. (2010). "Benchmarking Success of Building Maintenance Projects", *Facilities*, 28, 290–305.

Lind, H. and Muyingo, H. (2012). "Building Maintenance Strategies: Planning Under Uncertainty", *Property Management*, 30, 14–28.

Vijverberg, G. (2002). "Renovation of Offices in the Netherlands: Reasons, Points for Attention and Obstacles", in *Proceedings of the CIB Working Commission 070, CABER*, Glasgow Caledonian University, September, pp. 728–737.

Webster, J. and Watson, R. T. (2002). "Analyzing the Past to Prepare for the Future: Writing a Literature Review", *MIS Quarterly*, xiii–xxiii.

Wood, B. R. (2005). "Towards Innovative Building Maintenance", *Structural Survey*, 23, 291–297.

Wordsworth, P. (2001). *Lee's Building Maintenance Management*, London: Blackwell Science.

Wu, P. and Low S.P. (2011). "Managing the Embodied Carbon of Precast Concrete Columns", *Journal of Materials in Civil Engineering*, 23(8), 1192–1199.

Wu, P., Mao, C., Wang, J., Song, Y. and Wang, X. (2016). "A Decade Review of the Credits Obtained by LEED V2. 2 Certified Green Building Projects", *Building and Environment*, 102, 167–178.

Wu, P., Song, Y., Shou, W., Chi, H., Chong, H. Y. and Sutrisna, M. (2017). "A Comprehensive Analysis of the Credits Obtained by LEED 2009 Certified Green Buildings", *Renewable and Sustainable Energy Reviews*, 68, 370–379.

Wu, P., Xia, B. and Zhao, X. (2014a). "The Importance of Use and End-of-Life Phases to the Life Cycle Greenhouse Gas (GHG) Emissions of Concrete — A Review", *Renewable and Sustainable Energy Reviews*, 37, 360–369.

Wu, P., Xia, B., Pienaar, J. and Zhao, X. (2014b). "The Past, Present and Future of Carbon Labelling for Construction Materials — A Rreview", *Building and Environment*, 77, 160–168.

Zhong, Y. and Wu, P. (2015). "Economic Sustainability, Environmental Sustainability and Constructability Indicators Related to Concrete-and Steel-projects", *Journal of Cleaner Production*, 108, 748–756.

Zhong, Y., Ling, F. Y. Y. and Wu, P. (2016). "Using Multiple Attribute Value Technique for the Selection of Structural Frame Material to Achieve Sustainability and Constructability", *Journal of Construction Engineering and Management*, 143, p.04016098.

Chapter 38

A Process Model of Information Sharing and Knowledge Integration for Project Performance

Qian Li* and Jiayuan Yan[†]

*School of Engineering Management,
Nanjing University, Nanjing, China*
**qianli@nju.edu.cn*
[†]17826029910@163.com

Heap-Yih Chong

*School of Built Environment, Curtin University,
Perth, Australia*
heap-yih.chong@curtin.edu.au

Abstract

Engineering project consists of full life-cycle construction and business activities. Many stakeholders will collaborate together throughout the process. Information sharing and knowledge integration are the key drivers for the stakeholders to cooperate in the activities. It is worth conducting an in-depth study on how to effectively control the information

---

[†] Corresponding author.

and knowledge between different subjects in the project, with the aim to realize knowledge innovation and improved project performance. Therefore, the chapter aims to develop a process model of information sharing and knowledge integration between different organizations based on analyses of the theoretical environment. It would then discuss the influence path of information sharing, knowledge sharing and integration on the project performance in order to provide advice on cross-organizational knowledge integration and innovation for project managers.

***Keywords*:** Information Sharing; Knowledge Integration and Creation; Project Performance.

1. Introduction

Uniqueness, uncertainty and other characteristics of the project make it necessary to carry out collaboration and innovation in the implementation phase. But the simple information sharing and communication between various parties cannot meet the needs of collaboration and innovation. Further, it needs to introduce concepts of knowledge management, knowledge integration and creation into project management, and conduct unified management on heterogeneous knowledge of the innovation subjects, that is, the application of integrated innovation management in engineering practice (Ruan, 2012).

Information sharing is a basic prerequisite for collaboration and innovation of multiple subjects. Within the organization, information sharing refers to sharing and transmitting information from different sources in order to enhance the communication and cooperation among the organization members; at the organizational level, the project team can realize information sharing through personal interactions between subjects or organizations after establishing effective information-sharing channels and forming team knowledge (Becerra, 2008). Information obtained in a shared environment will be transformed into valuable knowledge after information processing and finally comes into new knowledge through knowledge creation (Nonaka *et al.*, 1996). Knowledge creation consists of two aspects. On the one hand, it converts from tacit knowledge to explicit knowledge and finally becomes tacit knowledge. On the other hand, it transmits information from the individual level to the team level and later to the internal organization or even across organizations.

Most project activities require a series of knowledge skills from different organizations in the industry. Carrillo (2004) believed that effective knowledge management processes would provide solutions to improving project performance. Senaratne and Sexton (2008) thought that as project phases evolve, subjects need to keep collecting information and knowledge from members of the project team in the problem-solving process. At present, research mainly focuses on the information sharing and knowledge management within the project organization, while research on information sharing and knowledge integration across organizations is relatively little. In fact, during the process of solving complex issues, the participating subjects in the cross-organizational projects will realize the interaction and coordination of the entirety of distributed and heterogeneous knowledge sources through joint discussion, communication and learning, and finally realize the project performance.

This chapter aims to develop a process model of information sharing and knowledge integration among projects and organizations. It further analyzes their relationships with the project performance. In Section 2, the study objectives are introduced; Section 3 introduces the theoretical background of research; an integration model of project business activities, organization reconstruction and knowledge integration is constructed in Section 4; in Section 5, a process model of information sharing, knowledge management and knowledge integration effects and project performance is established; and Section 6 summarizes the whole passage.

2. Aims

The aims of this chapter are as follows:

(1) To define connotations of information sharing and knowledge management between project groups;
(2) To analyze the integration model of project business activities, organization reconstruction and knowledge integration;
(3) To develop a process model of information sharing and knowledge integration towards the project performance.

3. Background

3.1. *Information sharing*

Information sharing refers to sharing and transmitting information from different sources to enhance the communication and cooperation among organization members. In this chapter, information sharing of the engineering project is defined as the sharing of the "explicit" texts (documents) and other relevant project materials between the participants. Based on the theory of social exchange and trust, establishing an effective information-sharing channel in the project team can help participants obtain relevant information and further process it into useful knowledge. Relationships among data, information and knowledge will be analyzed. The data need to be transformed into information first and then from information to knowledge (Frey, 2001). Guodong Yu (2017) built a knowledge model consisting of four-layer 2D structures. The data source level contains a variety of information sources. The information level aims to realize the classification and management of information; the knowledge level aims to realize the description and integration of organizational knowledge; the application level aims to transform the existing knowledge into the key points of product innovation. Data are transformed into useful information, and information is transformed into knowledge through further analyses. Once individuals absorb the information and memories are formed in the mind, the information will become knowledge (Ruan, 2012).

In order to complete a project, it is necessary to collect a large amount of information and knowledge, and the participants' information, and tasks, technologies, resources and individuals should interact with each other (Tsoukas, 1996). Especially in the aspect of technology research and development, it is crucial to manage the information flow process effectivity in order to promote the production, creation and application of knowledge to facilitate innovation and further realize breakthroughs in technology innovation.

3.2. *Knowledge management and knowledge integration*

From the perspective of attributes, knowledge can be divided into two categories: explicit knowledge and tacit knowledge (Choo *et al.*, 2000).

Because of the broad coverage of knowledge management, it needs to not only conduct effective information management but also create an environment within the organization where knowledge sharing and learning are encouraged in order to promote the creation of knowledge. Connotations of knowledge management can be analyzed from three perspectives, namely knowledge acquisition, knowledge storage and knowledge learning. Knowledge acquisition refers to tacit knowledge, explicit knowledge or the combination of both, such as related technical knowledge, craft and process information (Kasvi *et al.*, 2003). Knowledge learning and transfer can be realized through the acquisition, sharing and integration of knowledge from experience with project members. Knowledge storage can be realized by project members through the use of files, models and knowledge bases.

Academia's definition of knowledge integration has not been unified yet. From the perspective of knowledge assimilation and application, Cohen and Levinthal (1990) thought that knowledge integration includes sifting valuable external knowledge out, internalizing it into individuals' abilities during the process of learning, and then playing its vital role in business innovation. Alavi and Tiwana (2015) believed that knowledge integration is the process of integrating individuals' knowledge into a knowledge system in a specific situation. By exchanging the existing information and knowledge, the various participants in the project will realize the interaction and coordination of the distributed and heterogeneous knowledge source integration and then generate new knowledge, namely knowledge integration.

4. Integration Model of Business–Organization–Knowledge Between Project Organizations

An engineering project is a full life-cycle construction activity, which involves coordination of several subjects. A variety of business activities are included in the four phases of the project, that is, the definition, planning, implementation and completion. It can be found that different knowledge is required in different business phases after decomposing the business in each project phase. It is critical for the overall goal realization to maximize the use of different knowledge and experiences in business activities (Arditi *et al.*, 2002). Griffith and Sidwell (1997) believed that the knowledge required for the project completion should be not only concentrated in the phase of design and construction but also used in the project of the full life-cycle. Based on the research results of Griffith and Sidwell (1997), the knowledge and stakeholders required in all business activities in each phase are sorted out in Table 1.

The tasks in each phase of the project need to be decomposed into specialized units and thus form the information and knowledge required to complete the unit. On one hand, the knowledge includes contract texts, technical specifications, and archival sources. On the other hand, the experience in historical projects, experiences of managers and experts, etc. are also included; in order to finish the unit task of project business, clients need to coordinate with the design organization, construction organization, consulting organization and the scientific institution for joint participation. They also need to decompose the project organizations and

Table 1.   Knowledge required in engineering project activities of all phases.

Project Phases	Required Information and Knowledge	Project Stakeholders
Definition	Client's corporate objectives; client's project requirements; project strategy; project priorities — time, cost, quality; project team selection; definition of relationships, responsibilities and authority; type of project; location, site conditions and environment; resources; legislation and regulation; climatic influences; project risk; form of contract; contract negotiation	Clients; project establishment organizations; governments
Planning	Design concepts; construction details; feasibility analyses; task requirements; key technologies; design drawings; construction drawings	Design organizations; scientific institutions; consulting organizations; contractors

(Continued)

Table 1. (*Continued*)

Project Phases	Required Information and Knowledge	Project Stakeholders
Implementation	Construction knowledge; skill base; construction methods; sequencing; resource deployment; standards and control of quality; organizational structure; management and supervisory style; industrial relations; planning and progressing methods; material procurement; use of plant and equipment; site layout and temporary facilities; site safety	Contractors; subcontractors; construction organizations; supervision organizations; design organizations; scientific institutions
Completion	Installation and commissioning; operational requirements; user requirements; life cycle provision; maintenance	Clients; construction organization; supervision organization; governments

recompose personnel in accordance with the professional information and knowledge requirements, which makes every business unit implemented by the professional team from the organizations. The knowledge integration platform will be built through the combination of different personnel from clients, design organizations, construction organizations and consulting organizations. On this platform, personnel from different organizations will realize knowledge creation by sharing their respective information and knowledge and through the innovation of association integration and those of semi-association integration in order to guarantee the realization of project objectives.

As a result, the process of knowledge integration is divided into four phases, namely decomposition of business units, understanding of information and knowledge requirements, reconstruction of organization units, and knowledge integration and innovation. Finally, it will promote knowledge circulation, oriented by the project quality, objective and processing, as shown in Fig. 1.

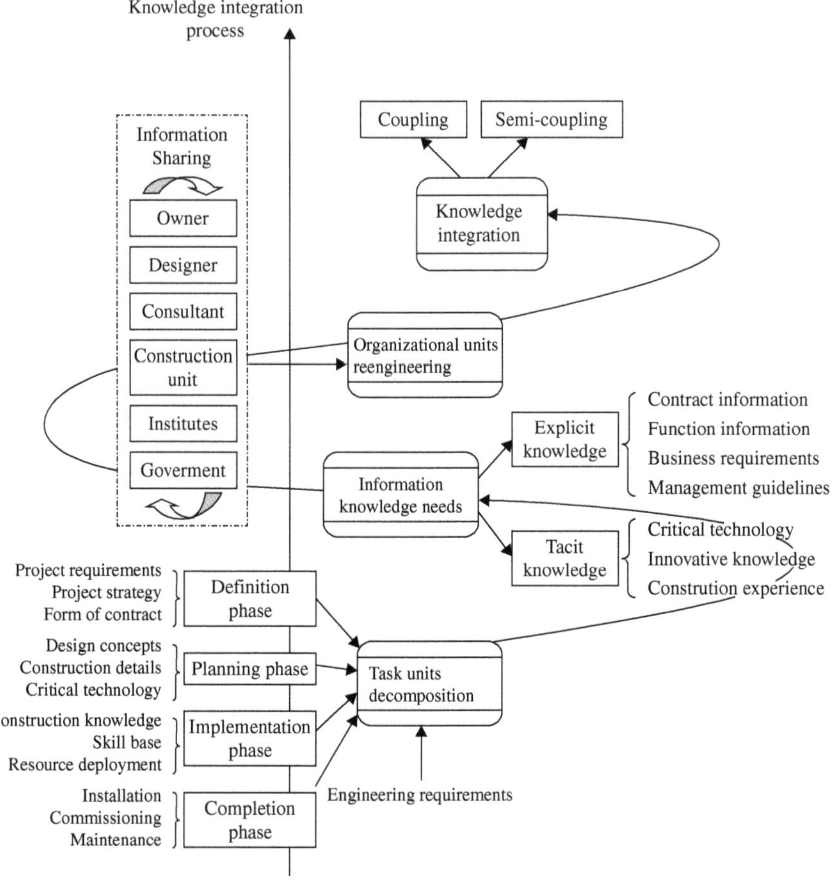

Fig. 1.   Process of knowledge integration.

5. Process Model of Information Sharing, Knowledge Integration and Project Performance

5.1. *Information sharing and knowledge management*

Information sharing and knowledge management are prerequisites for developing new technologies or new knowledge. In the different stages of the project, the stakeholders usually have various kinds of information and technical schemes. As the project progresses, the stakeholders usually communicate with each other face to face in the process of coordination and management among all participants, which is beneficial to the acquisition of tacit knowledge (Koskinen *et al.*, 2003). Hansen (2002) believed that information sharing can promote trust among different organizations and enable stakeholders to archive and store knowledge according to task requirements. In particular, the encoded or textual explicit knowledge is more conducive to the realization of knowledge storage. In addition, the information-sharing mechanism among the project organizations can supervise and urge the organizations to learn from each other, further improving the knowledge system within each organization, promoting the project knowledge innovation and increasing the performance goals.

Therefore, information sharing among the engineering project participants can promote different stakeholders' knowledge acquisition, knowledge storage and knowledge learning and improve their level of knowledge management.

5.2. *Knowledge management and knowledge integration*

Effective knowledge management can add value to knowledge. This process facilitates the knowledge integration and sharing among project members (Samir *et al.*, 2014). It is necessary for different stakeholders in the project to strengthen their ability of acquiring knowledge to promote the realization of knowledge integration and to improve their ability of solving project problems to realize the performance goal. Zahra (2002) pointed out that the organizations with the outstanding knowledge acquisition ability can promote the realization of knowledge integration and improve the realization of organization performance. Renzl (2008) pointed out that the correct archive and storage of knowledge can realize efficient knowledge transfer, which is to further integrate knowledge resources. Additionally, the organizations in the project start from the perspective of solving their own problems, make further analyses, tests and demonstration of the acquired knowledge and realize knowledge integration, knowledge creation and project objectives by the innovation of association integration and those of semi-association integration.

Therefore, the knowledge management of knowledge acquisition, knowledge storage and knowledge learning between the engineering project participants is helpful to the digestion, absorption, association and integration, and innovation of knowledge.

5.3. *Information sharing and project performance*

The atmosphere and environment of information sharing in engineering construction can promote the information exchange among members. Summarizing experiences and lessons can provide a foundation for the targeted performance achievement of project progress, cost and quality. Maurer (2010) and others believed that in project management, the establishment of cultural orientation of information sharing among cooperative organizations is helpful to facilitate communications between all participants, making them willing to contribute relevant knowledge in order to meet the requirements of project innovation.

Therefore, the information sharing among the project participants can facilitate the achievement of targeted performance.

5.4. *Knowledge integration and project performance*

The targeted performance will be achieved not only through the simple exchange of existing information but also through knowledge management and knowledge integration. Knowledge integration generates new knowledge in order to solve the problems that appear in the implementation process of the project. Levinthal and March (1993) believed that integrating tacit knowledge such as the past knowledge and experience and then combining it into implemented projects will improve the innovation performance to some extent. Besides, project construction carries out cross-organizational cooperation dependent on the external complementary resources. Whether the organizations are willing to share information and knowledge becomes the key to cooperative innovation. Katila and Ahuja (2002) believed that using the existing knowledge and exploring new knowledge makes organizations qualified to integrate resources and collaborate together. Then new products, new technologies and new craft will be developed. Technical difficulties in the construction process will be solved as well.

Therefore, information sharing is a basic prerequisite for collaborative innovation between project stakeholders. The influence path of information sharing on the project performance is mainly conveyed by knowledge management and knowledge integration.

5.5. *Trust and knowledge integration between organizations*

Engineering project is a complicated undertaking, which requires coordination between multiple subjects, coordination of multiple objectives, and even the technology innovation integrated through multiple methods. Therefore, in the project organization network, effective relationship management is needed if the knowledge management of all stakeholders is planned to transform into the knowledge integration among multiple subjects. In particular, trust between organizations needs to be built. Renzl (2008) pointed out that trust among team members facilitates knowledge archiving and storage, and realizing efficient knowledge transfer helps to integrate knowledge resources and to realize knowledge integration. Kucharska (2016) believed that sharing tacit knowledge is an informal voluntary behaviour of knowledge owners. Trust strongly accelerates the sharing of tacit knowledge and positively advances knowledge integration activities in the team.

Therefore, cross-organizational trust has a moderating effect on the relationships between knowledge management and knowledge integration and creation.

5.6. *Knowledge integration and complexity of the project*

Projects under different technical environments and organizational environments have different focuses on the unsolved issues. For the general projects, the implementation process of the project needs to develop a scientific plan and conduct strict site management if it aims to achieve the project objectives. But for the complicated projects, stakeholders need to coordinate with each other to solve the issues in the implementation process. To be more specific, they should use information and knowledge to explore, make analyses, integrate and innovate. In this process, tacit knowledge needs to be explicit; descriptive knowledge needs to be programmed; unit knowledge needs to be integrated. Thus, for the projects with greater complexity, frequent information and knowledge exchange between project members or organizations can enhance the organizing ability, which further facilitates project success. The project objectives can be achieved at a relatively low cost and a relatively fast speed (Sher and Lee, 2004). But for the projects with less complexity, it is enough to use the common programmed management to guarantee the project realization. In this situation, knowledge integration has a slight effect on the project performance.

Therefore, the complexity of projects has a moderating effect on the relationships between knowledge integration and project performance.

Based on the conclusions above, the process model of information sharing, knowledge integration and project performance is shown in Fig. 2.

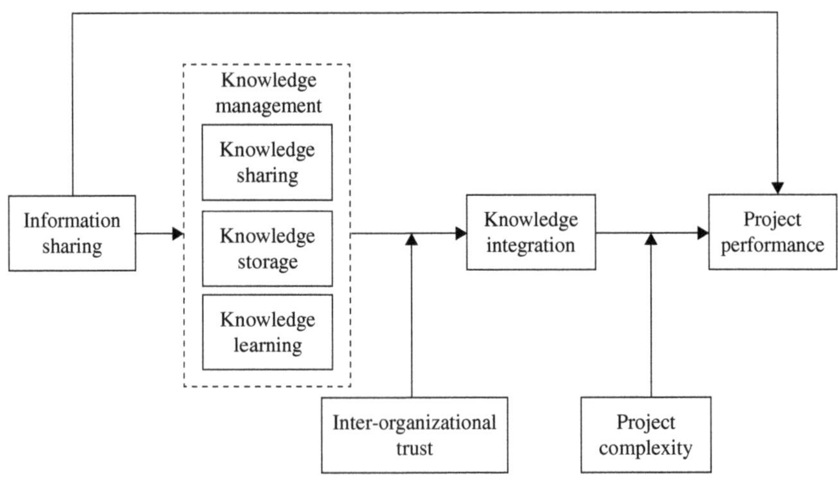

Fig. 2. Process model of information sharing, knowledge integration and project performance.

6. Conclusions

In this chapter, the authors take the engineering project as the background of multi-subject coordination and multi-disciplinary knowledge as the study object and discuss the relationships between information sharing and knowledge integration towards the project performance. After the literature review, information sharing can be defined as the sharing of "explicit" texts (documents) and other materials related to the project among all participants. By processing and transferring information, we have the valuable knowledge needed for the project completion. Additionally, it effectively controls the knowledge flow and diffusion in the project and conducts a unified management of the heterogeneous knowledge distributed among all innovative subjects, that is, knowledge management. In order to solve project difficulties, especially those at the technical level, the information sharing and knowledge management within the organization is not enough. Cooperation between organizations is required. The knowledge complementarity among different knowledge bases will be utilized in order to integrate the knowledge of all organizations, form new knowledge, and further promote technology innovation.

Knowledge integration is the key to successful completion of a cross-organizational project. In order to represent the process of knowledge integration more intuitively and clearly, a process model of "business unit decomposition → information and knowledge requirement → reconstruction of organizational unit → knowledge integration" is put forward in this paper. Each project phase is decomposed. Relevant information and knowledge are required for the completion of each business activity. But it is unlikely for a certain organization to complete all the activities. Knowledge demand facilitates the reconstruction of organizational unit. The members equipped with the knowledge about different industries share explicit and tacit knowledge, then integrate knowledge, and finally realize knowledge integration and innovation.

Based on the conclusions above, the relationships among information sharing, knowledge management, knowledge integration and project performance, and the transfer path of functions are further analyzed and a process model involving the three phases is put forward in this chapter. It can be found that information sharing is a basic prerequisite of collaborative innovation among project stakeholders. Information sharing can advance the knowledge acquisition, knowledge storage and knowledge learning of different stakeholders and improve the knowledge management. While knowledge management is conducive to the digestion, absorption, association and integration, and innovation of knowledge. Therefore, information sharing ultimately facilitates the achievement of targeted project performance. The influence path is delivered mainly through knowledge management and knowledge integration. In addition, trust among organizations and the complexity of projects play a moderating role in the model. In the future research, empirical research will be useful to validate this theoretical framework.

Acknowledgement

The authors would like to thank the anonymous referees very much for their value comments and suggestions. This work was supported by the National Natural Science Foundation of China (71571098), Major Program of the National Natural Science Foundation of China (71390520, 71390521), Fundamental Research Funds for the Central Universities (011814380023), Jiangsu Provincial Public Key Laboratory of Public Works open topics (GGSS2016-01), and the Program B for Outstanding PhD candidate of Nanjing University 406 (201701B009).

References

Alavi, M. and Tiwana, A. (2015). "Knowledge Integration in Virtual Teams: The Potential Role of KMS", *Journal of the American Society for Information Science & Technology*, 53(12), 1029–1037.

Arditi, D., Elhassan, A. and Toklu, Y. C. (2002). "Constructability Analysis in the Design Firm", *Journal of Construction, Engineering and Management*, 128(2), 117–126.

Becerra-Fernandez, I., Xia, W., Gudi, A. *et al.* (2008). "Task Characteristics, Knowledge Sharing and Integration, and Emergency Management Performance: Research Agenda and Challenges", in *Proceedings of the 5th International Conference on Information Systems for Crisis Response and Management (ISCRAM)*, pp. 88–92.

Carrillo, P. (2004). "Managing Knowledge: Lessons from the Oil and Gas Sector", *Construction Management and Economics*, 22(6), 631–642.

Choo, C. W. (1998). "The Knowing Organization: How Organization Use Information to Construct Meaning, Create Knowledge, and Make Decisions", *Homer Encyclopedia*, 37(5), 721–724.

Cohen, W. M. and Levinthal, D. A. (1990). "Absorptive Capacity: A New Perspective on Learning and Innovation", *Administrative Science Quarterly*, 128–152.

Frey, R. S. (2001). "Knowledge Management, Proposal Development, and Small Businesses", *Journal of Management Development*, 20(20), 38–54.

Griffith, A. and Sidwell, A. C. (1997). "Development of Constructability Concepts, Principles and Practices", *Engineering, Construction and Architectural Management*, 4(4), 295–310.

Gupta, S. and Polonsky, M. (2014). "Inter-firm Learning and Knowledge-Sharing in Multinational Networks: An Outsourced Organization's Perspective", *Journal of Business Research*, 67(4), 615–622.

Handy, C. (1994). *The Age of Paradox*, Boston: Harvard Business School Press.

Hansen, S. (2002). "Web Information Systems: The Changing Landscape of Management Models and Web Applications", in *Proceedings of the 14th International Conference on Software Engineering and Knowledge Engineering*. ACM, pp. 747–753.

Kasvi, J. J. J., Vartiainen, M. and Hailikari, M. (2003). "Managing Knowledge and Knowledge Competences in Projects and Project Organisations", *International Journal of Project Management*, 21(8), 571–582.

Katila, R. (2002). "New Product Search Overtime: Past Ideas in their Prime?", *Academy of Management Journal*, 45(5), 995–1010.

Koskinenm J. P., Vallinen, J. and Siren, J. (2003). Method and network system for charging a roaming network subscriber: US, WO 2003092317 A1.

Kucharska, W. *et al.* (2016). "Trust, Collaborative Culture and Tacit Knowledge Sharing in Project Management: A Relationship Model". United Kingdom: International Conference on Intellectual Capital and Knowledge Management and Organisational Learning.

Levinthal, D. A. and March, J. G. (1993). "The Myopia of Learning", *Strategic Management Journal*, 14(S2), 95–112.

Maurer, I. (2010). "How to Build Trust in Inter-Organizational Projects: the Impact of Project Staffing and Project Rewards on the Formation of Trust, Knowledge Acquisition and Product Innovation", *International Journal of Project Management*, 28(7), 629–637.

Nonaka, I., Takeuchi, H. and Umemoto, K. (1996). "Theory of Organization Knowledge Creation", *International Journal of Technology Management*, 100(2), 105–109.

Renzl, B. (2008). "Trust in Management and Knowledge Sharing: The Mediating Effects of Fear and Knowledge Documentation", *Omega*, 36(2), 206–220.

Ruan, X., Ochieng, E. G., Price, A. D. F. *et al.* (2012). "Knowledge Integration Process in Construction Projects: A Social Network Analysis Approach to Compare Competitive and Collaborative Working", *Construction Management & Economics*, 30(1), 5–19.

Senaratne, S. and Sexton, M. (2008). "Managing Construction Project Change: A Knowledge Management Perspective", *Construction Management and Economics*, 26(12), 1303–1311.

Sher, P. J. and Lee, V. C. (2004). "Information Technology as a Facilitator for Enhancing Dynamic Capabilities Through Knowledge Management", *Information & Management*, 41(8), 933–945.

Tsoukas, H. and Papoulias, D. B. (1996). "Understanding Social Reforms: A Conceptual Analysis", *Journal of the Operational Research Society*, 47(7), 853–863.

Yu, G., Yang, Y. and Xing, Q. (2017). "An Ontology-based Approach for Knowledge Integration in Product Collaborative Development", *Journal of Intelligent Systems*, 26(1), 35–46.

Index

biomimicry proponents, 348

biomimicry solution-based approach, 342

biomimicry thinking, 341

biopolymers, 197

bitumen, 232

bitumen prices, 239

block works, 296

BREEAM (Communities), 456

bridge, 327

bridge health monitoring system, 332

bridge model, 333

Building Code of Australia, 135

building component, 195

buildings components/facilities, 28

Building Construction Authority (BCA), 515

building costs, 227

building design, 83

building elements, 138

building environmental assessment tool, 169

building evaluation, 19

building facilities, 25, 429, 488

building for environmental and economic sustainability, 169

building industries, 264

building information, 249

building information modelling (BIM), 2, 8, 17–19, 24–25, 27, 39, 79, 204, 215, 248, 279, 321, 354, 429, 558–559, 566

building information modelling (BIM)-related technologies and innovations, 498

building information module, 30

building inspectors, 266

buildings location, 305

building maintainability, 494–495

building maintenance, 430, 494, 496, 650, 652–654, 660

building maintenance frameworks, 655

Building Maintenance Guidebook, 437

building maintenance management, 659

building maintenance practices, 649, 651

building maintenance stakeholders, 649

Building Material, 226–228

building material prices, 225, 229, 242

building operation and maintenance, 658

building operations, 247

building orientation, 80

building performance, 19, 28

building performance analysis, 93

building performance evaluation, 25

Building performance indicators, 32

building rating tool, 167

building sector, 337

building simulation, 304

building spaces, 32

building's performance, 18, 490

building sustainability, 167

building systems, 311

built environment, 80, 215, 342

bulldozer, 70

business-as-usual thinking, 119

C

360 cameras, 4, 7, 482

CAD system, 42, 204, 249

capital city road infrastructure, 533

Lightning Source UK Ltd.
Milton Keynes UK
UKHW021037210719
346520UK00001B/21/P